SUSTAINABLE ENERGY CONVERSION FOR ELECTRICITY AND COPRODUCTS

SUSTAINABLE ENERGY CONVERSION FOR ELECTRICITY AND COPRODUCTS

Principles, Technologies, and Equipment

ASHOK RAO, PH.D.
Advanced Power and Energy Program
University of California
Irvine, CA, USA

WILEY

Published by John Wiley & Sons, Inc., Hoboken, New Jersey
Published simultaneously in Canada

For general information on our other products and services or for technical support, please contact our Customer Care Department within the United States at (800) 762-2974, outside the United States at (317) 572-3993 or fax (317) 572-4002.

Wiley also publishes its books in a variety of electronic formats. Some content that appears in print may not be available in electronic formats. For more information about Wiley products, visit our web site at www.wiley.com.

Library of Congress Cataloging-in-Publication Data applied for.

Set in 10/12pt Times by SPi Global, Pondicherry, India

Printed in the United States of America

10 9 8 7 6 5 4 3 2 1

1 2015

CONTENTS

PREFACE

We need both electricity and chemicals for our modern way of life but environmental impacts have to be factored in to sustain this type of lifestyle. The book provides a unified, comprehensive, and a fundamental approach to the study of the multidisciplinary field of sustainable energy conversion to generate electricity and optionally coproduce synthetic fuels and chemicals. Modern power plants with these objectives differ in many significant respects to the traditional methods of generating electricity from fossil fuels such as coal, that is, by "simply" burning the fuel to generate steam and producing electricity via a Rankine cycle. Such "steam plants" were traditionally designed by mechanical engineers, while modern power plants, with more and more emphasis being placed on sustainability that impacts both the thermal performance as well as the environmental signature, are incorporating processes that have traditionally been handled by chemical engineers. Furthermore, the harsh environments that some of the equipment are exposed to consisting of very high operating temperatures, pressures, and corrosive atmospheres require advanced materials capable of exhibiting good mechanical properties at those conditions, as well as suitable chemical properties. Thus, this subject of sustainable power plant development and design is of current interest not only to mechanical, chemical, and industrial engineers and chemists but also to material scientists. Many of the topics covered in this book should also be useful to electrical engineer involved in the development of the modern power plant.

Some of the principles from each of these fields are essential in developing a more complete understanding of energy conversion systems for electricity generation. While an exhaustive discussion of all of the basic multidisciplinary principles required in a book of reasonable length is not possible to provide, this book does provide adequate depth in order to be largely self-contained. Each topic is covered in sufficient detail from a theoretical and practical applications standpoint essential for engineers. This book could serve as a textbook for a senior- or a graduate-level course, especially in chemical, mechanical, and industrial engineering and is assumed that the student has had undergraduate courses in thermodynamics, fluid mechanics, and heat transfer. Researchers and practicing industry professionals in energy conversion field will also find this quite useful as a reference book.

The book starts with an introduction to energy systems (Chapter 1) and describes the various forms of energy sources: natural gas, petroleum, coal, biomass, other renewables and nuclear. Their distribution is discussed in order to emphasize the uneven availability and finiteness of some of these resources. Impact on the environment is also included along with an introduction to the supply chain and life cycle analyses in order to emphasize the holistic approach required for sustainability. The next set of chapters discusses the underlying principles of physics and their application to engineering and is as follows:

- Chapter 2: Thermodynamics and its application to combustion and power cycles, and the first and second law analyses are discussed.
- Chapter 3: Fluid flow with an introduction to both incompressible flow and compressible flow followed by applications to flow through pipes and fittings, droplet separation, fluidization, and turbomachinery are presented.
- Chapter 4: The three modes of heat transfer, conduction, convection, and radiation, followed by application of these principles to heat exchange equipment design are discussed.
- Chapter 5: Mass transfer and chemical reaction engineering that includes fundamentals of diffusive and convection mass transfer and reaction kinetics are provided followed by application to design of both mass transfer equipment and reactors.

After covering the fundamentals of equipment design, the next set of chapters dwells in more specific subjects dealing with “energy plants” (i.e., plants in which the principal product is energy such as electrical or thermal) and is as follows:

- Chapter 6: Prime movers which are at the heart of a power plant that includes steam turbines, gas turbines, reciprocating internal combustion engines, and hydraulic turbines are discussed.
- Chapter 7: Systems engineering introduces the reader practical aspects of systems or process design. Topics covered in this chapter include at an introductory level, systems integration and application of exergy analysis and pinch technology, dynamic modeling and process control, development of process flow diagrams, cost estimation and economics, and application of life cycle assessment.
- Chapters 8 and 9: With an understanding of systems design and integration, the reader is then introduced to major power cycles, the Rankine cycle and the Brayton–Rankine combined cycle.

- Chapter 10: Coproduction of fuels and chemicals, which is gaining significant attention more recently due to the synergy and the ability to change the split between electricity generation and coproduct synthesis with intermittent renewables supplying a larger fraction of power to the grid, is next introduced. Synthesis of some key coproducts is described and specific examples of coproduction in both natural gas and coal or biomass based integrated gasification combined cycles are presented.
- Chapter 11: Advanced systems such as fuel cells along with hybrid cycles employing fuel cells and membrane separators and reactors are discussed.
- Chapter 12: An introduction to renewables such as wind, solar, and geothermal as well as nuclear energy is presented. Also included in this chapter is a discussion of the dependence of intermittent renewables such as solar and wind on fossil fueled plants for maintaining electrical power grid stability at least in the foreseeable future before large-scale energy storage devices are commercially available. This chapter is included to provide the reader a background on some of the other means of generating power sustainably, especially since their contribution to the energy mix will be increasing as time progresses.

In summary, the comprehensive and fundamental nature of the book that addresses both the practical issues and theoretical considerations will thus make it attractive to a broad range of practitioners and students alike, serving as a textbook for a senior- or graduate-level course related to the energy conversion disciplines of chemical, mechanical, and industrial engineering, as well as a reference or monograph for the professional engineer or researcher in the field including electrical engineers. Sustainable energy conversion is an extremely active field of research at this time. By covering multidisciplinary fundamentals in sufficient depth, this book is largely self-contained and suitable for different engineering disciplines, as well as chemists working in this field of sustainable energy conversion. The professional societies with interest in this field include the AIChE, ACS, ASME, SME, and IEEE.

University of California, Irvine, USA ASHOK RAO

ABOUT THE BOOK

There are only few comprehensive books that cover the various aspects of energy conversion for electricity generation. Such books, however, have focused on either overall systems or hardware with less emphasis given to the physical principles. Dr. Rao with his vast practical experience working in industry designing "real systems" and his theoretical understanding of the underlying principles brings to the table a unique and synergistic blend of these two essential knowledge bases. His involvement at the university gives the author the advantage of writing a book that is useful not only to the practicing professional but also to the student.

The contents of this book are based on Dr. Rao's experience gained both in industry which was for over more than 30 years and working in a university setting which was for about 10 years. The industry experience included training junior staff in developing and designing power systems, while the university setting involved teaching short courses and a senior-/graduate-level course in sustainable energy, as well as guiding graduate students in their research.

ABOUT THE AUTHOR

Dr. Ashok Rao is a well-acknowledged national and international leader in the field of energy conversion for generation of electricity and coproduction of chemicals and has made wide-ranging contributions in these fields over the past 40 years in industry as well as at the University of California, Irvine's Advanced Power and Energy Program, where he is currently its chief scientist for Power Systems. Prior to joining the university, Dr. Rao had worked in industry for more than 30 years, and due to this unique combination of industry experience and academia, he has been able to make significant contributions at the university as exemplified by his various publications in the energy conversion technologies area. His combination of scientific activity with practical solutions has resulted in high-quality publications that have always stimulated other scientists and engineers to study and develop his ideas. A variety of energy systems studied by Dr. Rao in his scientific activity range from advanced gas turbines to integrated gasification combined cycles to fuel cell–based power systems. His hands-on experience in working with today's young engineers and students points out the gaps in their knowledge bases and forms a good basis for making this book complete.

Prior to joining the university, he had worked for 25 years at Fluor Corporation, a world-class engineering company that employed more than 40,000 international employees. Due to his leadership role and expertise in energy technology, he was made a director in Process Engineering and his responsibilities included the development of a variety of energy conversion processes while minimizing the impact on the environment, for electric power generation using gas turbines, reciprocating internal combustion engines, combined cycles, and fuel cells as well as the production of hydrogen, synthesis gas[1] (or syngas for short), Fischer–Tropsch liquids, ammonia, alcohols, and dimethyl ether from coal, petroleum coke, biomass, liquid hydrocarbons, and natural gas. He was honored by Fluor in 1994 for his pioneering work in the advancement of energy systems including his work on the Humid Air Turbine cycle, a major internationally acknowledged achievement, by making him a technical fellow. He was later made a senior fellow at Fluor for continuing his significant contributions in the area of energy conversion. He was also honored by the California Engineering Council for his contributions in the area of energy conversion. Being recognized as a world-class leader in power cycles, he was invited to be the associate editor for the *ASME Journal of Engineering for Gas Turbines and Power* and a keynote speaker at the 2011 International Conference on Applied Energy, Perugia, Italy. He also has a number of patents to his credit in the field of energy conversion. He has authored a chapter titled "Gas fired combined cycle plants," for a book *Advanced Power Plant Materials, Design and Technology*, Woodhead Publishing, and completed a book as its principal editor titled *Combined Cycle Systems for Near-Zero Emission Power Generation*, also for Woodhead Publishing. More recently, he completed a chapter titled "Evaporative gas turbine (EvGT)/humid air turbine (HAT) cycles," for a book *Handbook of Clean Energy Systems*, for John-Wiley.

[1] This is the name borrowed from the petrochemical industry, the gas composition being similar to gas used to synthesize petrochemicals.

1

INTRODUCTION TO ENERGY SYSTEMS

Webster's dictionary defines energy as "capacity for action or performing work." Both potential energy stored in a body held in a force field (say the gravitational field at an elevation) and kinetic energy of a moving body are forms of energy from which useful work may be extracted. Energy associated with the chemical bonds that hold atoms and molecules together is potential energy. In the case of fuel combustion, this form of energy is transformed primarily into kinetic energy of the molecules as vibrational, rotational, and translational and referred to as thermal energy, by exothermic combustion reactions, and may be transferred as heat and a portion of it may be converted into work utilizing a heat engine. The first and the second laws of thermodynamics govern these energy transformation processes. The first law, a law of experience, recognizes energy conservation in nature while the second law, also a law of experience, recognizes that the overall quality of energy cannot be increased. In fact, for real processes, the overall quality of energy decreases due to the requirement for a reasonable potential difference to drive the energy conversion process at finite (nonzero) rates (e.g., heat or mass transfer with finite driving forces, temperature in the case of heat, and chemical potential in the case of mass) resulting in inefficiencies (or irreversibilities) such as friction.

Thus, in summary, any action occurring in nature is accompanied by a reduction in the overall quality of energy (a consequence of the second law) while energy is neither destroyed nor created (a consequence of the first law), but transformed into state(s) or form(s) with an overall lower quality or potential (again a consequence of the second

Sustainable Energy Conversion for Electricity and Coproducts: Principles, Technologies, and Equipment, First Edition. Ashok Rao.

law). Per Einstein's famous equation, $E = mc^2$ relating mass m and energy E (c is the speed of light), matter should also be included in the energy conservation principle stated earlier when involving nuclear energy.

Natural or human processes harness higher forms of energy, convert a fraction of it to useful form (the upper limit of this fraction being constrained by the second law), and ultimately all the energy is downgraded to lower forms of energy and rendered often useless. Thus, there is a continuous depletion of useful energy forms and the rate of depletion depends on how close the fraction of energy converted to useful forms approaches the limit set by the second law; in other words, the energy resource depletion rate depends on how efficient the conversion processes are. The pollutants generation rate is also increased in proportion to energy resource usage rate with a given set of technologies. Thus, it is imperative that higher efficiency processes, that is, processes that convert a higher fraction of energy supplied into useful forms, be pursued for the sustainability of our planet, that is, for the conservation of resources as well as reduction of the environmental impact.

1.1 ENERGY SOURCES AND DISTRIBUTION OF RESOURCES

Fossil fuels such as natural gas, petroleum, and coal are the primary sources of energy at the current time while oil shale resources (not to be confused with shale gas or oil) are being exploited only to a limited extent at the present time (primarily in Estonia) due to the relatively higher costs. Nuclear by fission of radioactive elements has been used to a lesser extent in most countries than fossil fuels except in France where the majority of electricity is produced in nuclear power plants. Renewables include biomass, geothermal, wind for turning turbines to generate electricity, solar used directly to supply heat or in photovoltaic cells, and rivers that provide hydroelectricity, which in certain regions such as the U.S. Pacific Northwest and in Norway and Sweden is quite significant. It should be appreciated that it is the solar energy that actually "drives" wind as well as the rivers by evaporating water and transporting it back as clouds to form rain at the higher elevations.

1.1.1 Fossil Fuels

According to the biogenic theory, fossil fuels are remains ("fossils") of life forms such as marine organisms and plant life that flourished on our planet millions of years ago. This energy is thus a stored form of solar energy that accumulated over millions of years. At the current and projected rates of usage however, it is unfortunate that fossil fuels will be used up in a fraction of time compared to the time it took to collect the energy from the sun. Ramifications include losing valuable resources for synthesizing chemicals as well as shocking the environment by pumping the very large quantities of CO_2 into the atmosphere at rates much higher than the rates at which the CO_2 can be fixed by life forms on our planet. Buildup of CO_2 in the atmosphere results in global warming, the sun's radiation falling on earth's surface that is reradiated as infrared heat is intercepted by the CO_2 (and other greenhouse gases such as CH_4) present in the earth's atmosphere resulting in energy accumulation, the increase in earth's temperature dependent on the concentration of such gases.

Among the various fossil fuels, treated natural gas as supplied which is mainly CH_4 is the cleanest. The naturally occurring sulfur compounds are removed (a small amount of odorant consisting of a sulfur bearing organic compound such as a mercaptan is however added for safety since a leak can be detected from the odor), while there is no ash present in this gaseous fuel. Only minor amounts of other elements or compounds are present. With a hydrogen to carbon (H/C) ratio being highest among all fossil fuels and the fact that natural gas can be utilized for generating power at a significantly higher overall plant efficiency (a measure of the fraction of the energy contained in the fuel converted into net electrical energy output by the plant), the greenhouse gas CO_2 emission can be lowest among the various fossil fuels.

The United States with its vast coal reserves can meet its energy demands for the next 300 years, but coal is the dirtiest fossil fuel since it contains sulfur and nitrogen that form their respective oxides during combustion and known pollutants, ash that can also cause environmental damage, and has low H/C ratio, which means high CO_2 emissions.

Natural gas is simplest in terms of composition and being a gas mixes readily in the combustor and burns relatively cleanly. Furthermore, pollution abatement measures are more easily implemented in natural gas-fueled plants. On the other hand, coal which has a very complex structure containing various elements in addition to carbon and hydrogen and being a solid is more difficult to burn cleanly. Coal combustion undergoes a complex process of drying/devolatilization followed by oxidation of the released gases and the char formed, while a significant fraction of the ash forms particles typically 10 μm or less in size called fly ash. The low visibility around some of the older coal-fired power plants that are not equipped with particulate removal systems from the flue gas or removal systems that are not very efficient is due to the fly ash.

Liquid fuels derived from petroleum are intermediate between natural gas and coal with respect to the environmental signature. From a combustion efficiency standpoint that also relates to minimizing formation and emission of unwanted pollutants, oil being a liquid needs to be atomized typically to less than 10 μm within the combustor to provide a large surface area so that it can vaporize and mix before combustion can occur. Combustion of oil can produce soot particles as well as other pollutants such as unburned hydrocarbons depending on how well the combustion process is designed. Ash particles may also be emitted depending on the type of oil being burnt.

The overall efficiency for a fuel-based power plant is expressed by the ratio of the net power produced by the plant and the "heating value" of the fuel (as a measure of the energy supplied by the fuel). There are two types of heating values used in industry, the higher heating value (HHV) and the lower heating value (LHV). The HHV is obtained by measuring the heat released by combustion of a unit amount of fuel with air at an initial defined standard temperature[1] and cooling the combustion

[1] 25°C per the American Petroleum Institute while 60°F per the Gas Processors Suppliers Association.

products to that initial temperature while condensing the H_2O vapor formed (and adding the released latent heat of condensation). It may also be calculated from the heat of formation data as explained in Chapter 2. Empirical correlations have been developed for fuels such as coal where the heat of formation data is not available. The Dulong–Petit formula is sometimes used in estimating the HHV of coal when experimental data is not available. It is also used for estimating the HHV of other fuels such as biomass again when experimental data is not available but with limited success. Other correlations have been developed and a summary of various such formulae are presented by García et al. (2014).

The LHV as defined in North America consists of not taking credit for the heat of condensation of the H_2O formed by combustion, that is, LHV = HHV – Latent heat of water vapor formed. In Europe, the International Energy Agency's definition of LHV for fuels such as coal is used instead, which is calculated from the HHV as follows: $LHV = HHV - 212 \times H - 24 \times H_2O - 0.8 \times O$, where H and O are the weight percentages of hydrogen and oxygen in the fuel on an "as-received" (i.e., in the condition received at the plant, which includes its moisture and ash contents) basis, H_2O is the weight percentage of moisture in the as received fuel, and both HHV and LHV are in kilojoule per kilogram of coal on an as received basis. This definition results in a value for the LHV that is typically lower than that as defined in North America, making the LHV-based efficiency reported by European authors higher. Thus, this difference in the definition of LHV should be taken into account while comparing the efficiency of energy conversion systems as reported by European and North American authors, in addition to whether the efficiency is on an LHV or HHV basis. Typically, the LHV efficiency is used for reporting the overall plant efficiency of gas-fired power systems, while in the United States, the HHV efficiency is used for solid fuel (e.g., coal and biomass) fired power systems.

Since the difference between HHV and LHV of a fuel depends on its hydrogen content, the ratio of LHV to HHV is lowest for H_2 at 0.846, highest for CO at 1.00, while for CH_4 it is 0.901. For a liquid fuel such as diesel, this ratio is about 0.93–0.94 while that for ethanol is 0.906. For a bituminous coal such as the Illinois No. 6 coal, this ratio is around 0.968; for a subbituminous coal from the Powder River Basin, Wyoming, this ratio is around 0.964; and for a lignite from North Dakota, this ratio is around 0.962. Since the difference between the HHV and the LHV is significant for most fuels, it is important to specify whether the efficiency reported is on an HHV or LHV basis.

The characteristics of these fossil fuel resources are discussed next in more detail including brief descriptions of the initial processing required before these resources can be utilized for energy recovery. This front-end processing requirement should be taken into account to get a complete picture of the environmental impacts of these resources, that is, from a "cradle to grave" standpoint, which in the case of transportation fuels is referred to as "well to wheel" or "mine to wheel."

1.1.1.1 Natural Gas Natural gas like petroleum is believed to be derived from deposits of remains of plants and animals (probably microorganisms) from millions of years ago according to the biogenic theory. Natural gas may be found along with petroleum or by itself as in many gas fields where little or no oil is found. According to another theory called the abiogenic theory however, natural gas was produced from nonliving matter citing the presence of CH_4 on some of the other planets and moons in our solar system.

Along with CH_4 that is by far the major combustible constituent of natural gas, other light hydrocarbons present in natural gas include C_2H_6, C_3H_8, and C_4H_{10}. Raw natural gas may contain CO_2 and sometimes N_2 and these gases have no heating value. CO_2 is typically removed from the natural gas while C_2H_6, C_3H_8, and C_4H_{10} are usually removed and marketed separately as special fuels or as feedstocks for the manufacture of petrochemicals. A number of other elements and compounds are also found in natural gas such as H_2, H_2S, and He. H_2S is also removed from the natural gas before it is pipelined for sale. H_2S is a toxic gas and its oxides formed during combustion of the fuel are pollutants as discussed later in this chapter. Table 1.1 (Rao et al., 1993) shows typical contract specifications for natural gas.

The composition of natural gas can vary significantly as shown by the data presented in Table 1.2 (Rao et al., 1993), however. Variability in composition of the natural gas can also occur during peak demand months in certain areas of the United States (the northeast). During such periods, natural gas may contain as much as 4% by volume O_2. The gas supply company may blend in propane or butane to extend the fuel supply during such peak demand and air is added as a diluent to hold the Wobbe Index (a measure of the relative amount of energy entering the combustor for a fixed pressure drop across a nozzle and discussed in more detail in Chapter 6) within limits.

Recent discoveries of "unconventional natural gas" have increased the supply of this fossil fuel, especially in the United States, while the potential for a smaller carbon footprint when compared to other fossil fuels has created a renewed interest in its use.

TABLE 1.1 Typical U.S. contract specifications for natural gas

Specification	Limits for custody transfer[a]
H_2S	0.25–1.0 grains/100 standard ft^3
Mercaptans (odorants)	1.0–10.0 grains/100 standard ft^3
Total sulfur	10–20 grains/100 standard ft^3
CO_2	2% by volume
O_2	0.2% by volume
N_2	3% by volume
Total inert gases	4% by volume
H_2	400 parts per million by volume (ppmv)
CO	None
Halogens	None
Unsaturated hydrocarbons	None
Water	7 lbs/10^6 standard ft^3 (118 kg/10^6 normal m^3)
Hydrocarbon dewpoint	45°F at 400 psig (7.2°C at 27.6 barg)
HHV	975 Btu/standard ft^3, minimum (38.4 MJ/normal m^3)

[a] 15.43 grains = g; 100 standard ft^3 (at 60°F, 1 atm) = 2.679 normal m^3 (at 0°C, 1 atm).

TABLE 1.2 Variation in composition of natural gas

	Volume (%)			
Component	Mean	Standard deviation	Minimum	Maximum
CH_4	93.0	5.5	73	99
C_2H_6	3.0	2.6	0	13
C_3H_8	1.0	1.4	0	8
C_4H_{10}	0.5	1.0	0	7
C_5H_{12}	0.1	0.3	0	3
C_6H_{14}	0.1	0.1	0	1
N_2	1.5	2.9	0	17
CO_2	0.5	0.5	0	2

For a 100-year time horizon with mean fugitive emissions of CH_4 as estimated by Howarth et al. (2011), a state-of-the-art coal-fired power plant that is described in Chapter 8 (assuming a bituminous coal) requires approximately 40% carbon capture and sequestration for similar greenhouse gas emissions per net MW as a state-of-the-art shale derived natural gas fired gas turbine-based combined cycle power plant that is described in Chapter 9. In such combined cycles, exhaust heat from the gas turbine (Brayton cycle) is utilized for generating steam that is then used in a steam turbine (Rankine cycle) to generate additional power. Compounding the challenge of CO_2 capture and sequestration from a coal-fired power plant is the associated decrease in the overall plant efficiency that further increases the amount of the required CO_2 capture amount for similar greenhouse gas emissions on a net MW basis for the two types of fuels. Existing coal-fired boiler plants may be converted to natural gas-fired combined cycles while utilizing a significant portion of the existing equipment.

Natural gas-fired combined cycles with their lower capital cost as compared to coal-based power can complement renewables such as solar and wind that are intermittent in nature. Other advantages include their suitability to small-scale applications such as distributed power generation where the power plant is located close to the load and does not depend on the conventional mega scale electric supply grid, and for combined heat and power (CHP) that is discussed in Chapter 10.

Unconventional natural gas consists of (i) "tight gas" that is contained in low permeability reservoirs, (ii) coal bed methane that is adsorbed in coal, and (iii) shale-associated gas that is contained in low permeability shale formations (whose permeability is even lower than that of tight gas reservoirs) and whose supply is rapidly rising. Conventional reservoirs of natural gas are contained in porous rock formations of sandstone and carbonates such as limestone and dolomite, and contain gas in interconnected pore spaces through which the gas can flow ultimately to the wellbore. In the case of coal bed methane, the gas is recovered from coal seams by usually releasing pressure. CO_2 injection to displace the adsorbed natural gas is also being considered. Recovery of tight gas as well as shale-associated gas typically requires stimulation of the reservoir to create additional permeability. Directional drilling in the horizontal direction exposes a much larger portion of the reservoir than conventional vertical wells making it cheaper on a well to well basis and is thus more conducive to developing tight and shale gas resources (see Fig. 1.1). Horizontal drilling consists of drilling down to say 600 m, then horizontally for approximately 1½ km by which the well runs along the formation, opening up more opportunities for gas to enter the wellbore. Permeability is increased by hydraulic fracturing after the well has been drilled and completed. It involves breaking apart (shale) rocks in the formation by pumping large quantities of liquid (~20,000 m^3 of water with 9.5% proppant particles such as sand and 0.5–1% chemicals additives) into the well under high pressure to break up the rocks in the reservoir. The proppant particles keep an induced hydraulic fracture open after the fracking liquid pressure is released. The gas is recovered as pressure is released. Some 750 chemical compounds are added to the water that include lubricants, biocides to prevent bacterial growth, scale inhibitors to prevent mineral precipitation, and corrosion inhibitors and clay stabilizers to prevent swelling of expandable clay minerals.

The U.S. Energy Information Administration's (2013a) estimates of technically recoverable shale gas in 2012 were the highest for China at 32×10^{12} m^3 followed by the United States at 20×10^{12} m^3, while 207×10^{12} m^3 on a worldwide basis, which is a significant fraction of the total worldwide natural gas reserve, about one-third of the total. To put these numbers in perspective, the annual consumption of the gas in the year 2012 was 3.4×10^{12} m^3 on a worldwide basis.

Shale gas typically has significantly more higher hydrocarbon (higher than CH_4) content than conventional natural gas that helps improve the overall economics since these higher hydrocarbons can be recovered and sold separately at a higher value.

In addition to the higher hydrocarbons, natural gas at the well head typically contains acid or "sour" gases such as CO_2 and H_2S as mentioned previously, moisture,

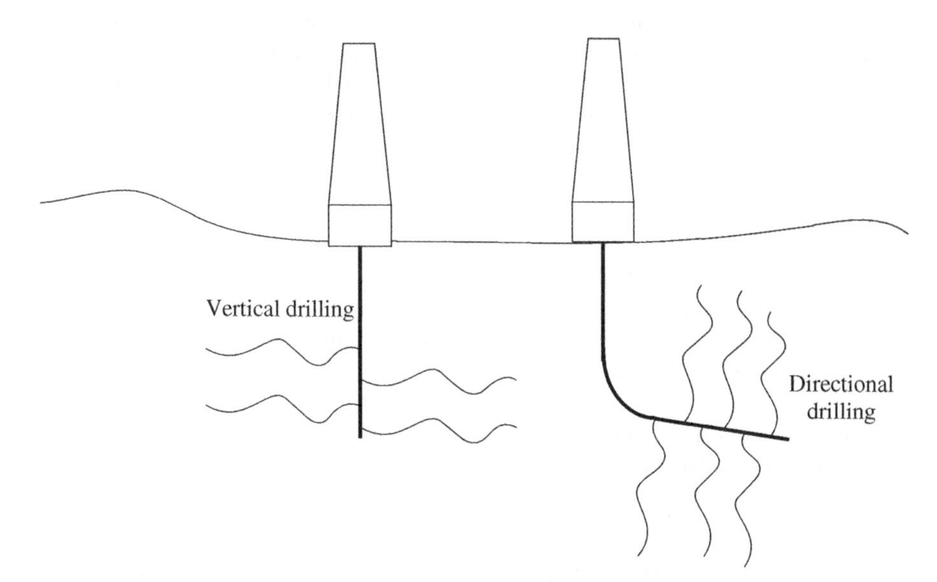

FIGURE 1.1 Vertical versus directional (essentially horizontal) drilling

and range of other unwanted components that must be removed. Treatment involves a combination of adsorptive, absorptive, and cryogenic steps. The sour gases form corrosive acids when combined with H_2O, for example, CO_2 forms carbonic acid. Additionally, CO_2 decreases the heating value of the natural gas. In fact, natural gas is not marketable when the concentration of CO_2 is in excess of 2–3%. H_2S in addition to being an extremely toxic gas is very corrosive in the presence of moisture. Sour gases may be separated from the natural gas by contacting with monoethanolamine (MEA) in a low pressure operation when requiring stringent outlet gas specifications. Diethanolamine (DEA) is used in medium to high pressure treating. DEA does not require solvent reclaiming as does MEA where unwanted stable compounds are formed from the natural gas contaminants. Hg may also be present in natural gas in which case the treatment process includes adsorption. It is preferred to use a regenerative process so that the generation of a hazardous stream consisting of the Hg laden adsorbent is avoided. These types of mass transfer processes are described in Chapter 5.

Another component that may be present in natural gas is He and finds industrial uses. It is cryogenically separated and recovered when its concentration is greater than 0.4% by volume in the natural gas. It may be further purified in a pressure swing adsorption (PSA) process. Both distillation and PSA are described in Chapter 5. The higher hydrocarbons: C_2H_6, C_3H_8, and C_4H_{10} represent added value over regular pipeline gas while the "condensate" consisting of C_5+ fraction can be used as gasoline. Also, condensate removal is often required to meet dew point specifications of pipeline gas, that is, to avoid condensation within the pipeline.

Example 1.1. Assuming ideal gas behavior, calculate the LHV and the HHV of natural gas in kilojoule per standard[2] cubic meter (kJ/Sm^3) corresponding to the "mean" composition shown in Table 1.2 utilizing LHV and HHV data provided for each of the components in Table 1.3. Assume that the C_4 through C_6 hydrocarbons are all straight chained (i.e., they are "normal" designated by an "*n*" hydrocarbons), the heating value as well as other thermophysical properties being dependent on the degree of branching of the carbon chain. Then compare the LHV of natural gas thus calculated with that calculated from the HHV.

Solution

Since ideal gas behavior is being assumed, the composition on a mole basis is the same as that on a volume basis. The volume percent data in Table 1.2 does not quite add up to 100% and it will be assumed that the unreported values correspond to noncombustible components.

The contribution of each component i to the LHV and the HHV are given by $y_i * \text{LHV}$ and $y_i * \text{HHV}$ as tabulated in Table 1.3, resulting in a mixture LHV = 35,290 kJ/Sm^3 and HHV = 39,081 kJ/Sm^3.

[2] At 15°C and 1 atm pressure used by natural gas suppliers in Europe and South America while normal m^3 (Nm^3) that is measured at 0°C and 1 atm pressure.

TABLE 1.3 Summary of heating value calculations

	From volume %	Given		Calculated		
	Mole %	LHV	HHV	$y_i \times$ LHV	$y_i \times$ HHV	H_2O formed by combustion
Component	100 y_i	kJ/Sm^3		kJ/Sm^3		Sm^3/Sm^3 of natural gas
CH_4	93	34,064	37,799	31,680	35,153	1.860
C_2H_6	3	61,220	66,853	1,837	2,006	0.090
C_3H_8	1	88,975	96,623	890	966	0.040
n-C_4H_{10}	0.5	116,134	125,722	581	629	0.025
n-C_5H_{12}	0.1	138,369	149,822	138	150	0.006
n-C_6H_{14}	0.1	164,595	177,652	165	178	0.007
$N_2 + CO_2$ + other noncombustibles	2.3	—	—	0	0	—
Total	100.0			35,290	39,081	2.028

In order to calculate the LHV from the preceding value of HHV, the amount of water formed by the complete oxidation of the combustibles is required, and the volumetric or molar contribution of each component i to form H_2O is given by $y_i * m/2$ where m is the number of H atoms in component i as tabulated in the last column of Table 1.3, resulting in a total amount of 2.028 Sm^3 H_2O formed per standard cubic meter of natural gas. Next, using the latent heat of H_2O as 2464 kJ/kg obtained from steam tables (e.g., Keenan et al., 1969) at the standard temperature of 15°C where the molar volume of an ideal gas is 23.64 Sm^3/kmol at a pressure of 1 atm:

$$\text{LHV} = \text{HHV} - \frac{2.028\ \text{Sm}^3/\text{Sm}^3}{23.64\ \text{Sm}^3/\text{kmol}} \times 2{,}464\ \text{kJ/kg} \times 18.015\ \text{kg/kmol} = 35{,}373\text{kJ/Sm}^3$$

This value compares well with the value of 35, 290kJ/Sm^3 shown in Table 1.3.

1.1.1.2 Petroleum Although petroleum resources occur to some degree in most parts of the world, the major commercially valuable resources occur in relatively fewer locations where appropriate geological conditions prevailed for the formation and storage of these fuels underground. It is widely accepted that the formation of petroleum (the word derived from Latin: petra = rock, oleum = oil) was biogenic in nature since petroleum deposits are found almost exclusively in sedimentary rock formations laid down millions of years ago when life flourished on the planet.

Petroleum is a mixture of various hydrocarbons with some sulfur, nitrogen, and organometallic compounds also present. A number of processing steps are involved in a refinery to produce the various high-value salable fuel streams such as gasoline, diesel, and jet fuel from the petroleum, the major processing step being the distillation operation. The petroleum distillation operation does not separate out individual compounds but produces various fractions consisting of mixtures. For example, fuel oil typically contains some 300 individual compounds. In fact petroleum as well as the various fuel products derived from petroleum are characterized by their boiling point curve rather than their chemical composition. This curve is generated by placing a known volume (100 ml) of sample in a round-bottom flask with means to collect and condense the vapor formed when it is heated at a specified rate. The vapor temperature when the first drop of condensate is collected is recorded as the "initial boiling point" as well as when the volume of vapor collected is 5, 10, 15, 20 ml, after which at every subsequent 10-ml interval to 80 ml, and finally at every subsequent 5 ml, and at the end of the test, which is recorded as the "end point" where no more evaporation occurs. Note that the residue remaining in the flask can decompose if the temperature continues to increase.

The petroleum received at the refinery is treated in a desalter to wash out salts before it is processed in the distillation unit operating near atmospheric pressure. The residual bottoms from this distillation unit may be fed to a vacuum distillation unit to maximize recovery of the valuable lower boiling point components. Two primary classes of liquid fuels are produced in a refinery, distillates, and residuals. Distillates are composed entirely of vaporized material from the petroleum distillation operation (subsequently condensed) and are clean, free of sediment, relatively low in viscosity, and free of inorganic ash. Gasoline, kerosene, or jet fuel and diesel are all distillates. Residuals contain fractions that were not vaporized in the distillation operation and contain inorganic ash components (that originate from the petroleum) and have higher viscosity such as heavy fuel oil. In addition to the gasoline, kerosene, diesel, and fuel oil fractions, other fractions having lower boiling points separated out by distillation include the following:

- A gaseous stream that is a mixture of primarily straight chain hydrocarbons, with an average carbon number of about 4 (the number of C atoms in a molecule) and is marketed as liquefied petroleum gas (LPG) after desulfurization, which may be accomplished using an amine process similar to desulfurization of natural gas.
- Naphtha that is a mixture of primarily straight chain hydrocarbons, with 5–12 carbon atoms and is supplied to a catalytic reforming process to increase its octane rating by rearranging the structure of the hydrocarbons. The naphtha is first hydrotreated (catalytically reacted with H_2 at high pressure) to form H_2S from the sulfur compounds, which is then followed by desulfurization before it is fed to the reforming process to protect the catalyst in the reformer.

The LPG, gasoline, kerosene, diesel, and fuel oil fractions are also further processed, to meet the required product specifications before they leave the refinery, which include removal of sulfur, nitrogen (their oxides formed by combustion being pollutants), and oxygen. These processing steps may include a Merox unit that assists in the desulfurization process by deoxidizing the mercaptans to organic disulfides; a hydrocracker unit that uses hydrogen to upgrade heavier fractions into lighter, more valuable products; a visbreaking unit that upgrades heavy residual oils by thermally cracking them into lighter, more valuable reduced viscosity products; an alkylation unit that uses H_2SO_4 or HCl as a catalyst to produce high octane components suitable for gasoline blending; a dimerization unit to convert olefins into higher octane gasoline blending components; an isomerization unit to convert straight chained to higher octane branched chained molecules again suitable as gasoline components; and a coking unit that can process very heavy residual oils to make additional gasoline and diesel while forming petroleum coke as a residual product. Depending on the purity (i.e., its S and metals such Va and Ni content) and type of process producing the petroleum coke, it can either be a valuable product suitable for making electrodes or may be used as a solid fuel like coal. A boiler plant utilizing the coke, however, should be designed to take into account its impurities by installing suitable flue gas treatment processes as well as disposing the ash in an environmentally responsible manner.

Petroleum-derived liquid fuels have become the major sources of energy in many countries because of the availability and convenience of these fuels for transportation engines (gasoline, kerosene, and diesel) and also sometimes for stationary power plants (fuel oils).

The U.S. Energy Information Administration's (2014) estimates of remaining recoverable petroleum in 2012 were the highest for Saudi Arabia at 267×10^9 bbl followed by Venezuela and the United States both at about 211×10^9 bbl while on a worldwide basis, the total was 1526×10^9 bbl (1 bbl is equivalent to 5.8×10^6 BTU or 6.1 GJ on a HHV basis). To put these numbers in perspective, the consumption rate of petroleum in the year 2012 was 89×10^6 bbl/day.

In order to maximize the production of the more valuable products as well as desulfurized and denitrified products, a number of processes are included downstream of the petroleum distillation operation. An example is hydrotreating to desulfurize, denitrify, and deoxygenate.

Catalytic reforming is a chemical process used to convert petroleum refinery naphthas, typically having low octane ratings, into high-octane liquid products called reformates, which are components of high-octane gasoline (also known as petrol). Basically, the process rearranges or restructures the hydrocarbon molecules in the naphtha feedstocks as well as breaks some of the molecules into smaller molecules. The overall effect is that the product reformate contains hydrocarbons with more complex molecular shapes having higher octane values than the hydrocarbons in the naphtha feedstock. In so doing, the process separates hydrogen atoms from the hydrocarbon molecules and produces very significant amounts of byproduct hydrogen gas for use in a number of the other processes involved in a modern petroleum refinery. Other byproducts are small amounts of methane, ethane, propane, and butanes.

1.1.1.3 Coal Coal is the most heterogeneous among all the fossil fuels. It was formed from plant life under the action of immense pressures and temperatures prevailing within the earth's crust and in the absence of air over time frames encompassing millions of years. The major element present in the "organic portion" of coal is carbon, with lesser amounts of hydrogen, oxygen, nitrogen, and sulfur. Sulfur mostly as iron pyrite is also present as part of the "inorganic portion" or ash in the coal that includes compounds of aluminum, silicon, iron, alkaline earths, and alkalis. Coal also contains some chlorine, mercury, and other volatile metals. The composition of coal is expressed by its "ultimate analysis" and by its "proximate analysis." The ultimate analysis provides the elemental analysis of the organic portion of coal along with the moisture content (if reported on a wet basis), ash content, and the sulfur content. The sulfur forms may also be reported, that is, the fractions that are part of the organic and the inorganic portions of the coal. The inorganic portion may be further subdivided into that present as a sulfide ("pyritic" since it is typically present as FeS_2) and as a sulfate. The proximate analysis shows the "fixed carbon," "volatile matter," moisture, and ash contents. Fixed carbon and volatile matter together constitute the organic portion of the coal with the volatile matter being that fraction that evolves when the coal is heated in the absence of O_2, the residue remaining being carbon (fixed carbon) and the inorganic constituents. The volatile matter content is typically an indicator of the reactivity and ease of ignition and affects the furnace design, for example, higher volatile matter increases flame length while affecting the amount of secondary air requirement and its distribution. In the case of gasification (partial oxidation to produce a combustible gas as discussed in Chapter 9) under milder temperatures, the volatile matter content provides an indication of the amount of organic components produced such as tars, oils, and gaseous compounds. The ash content is also important in the design of the furnace or the gasifier, pollution control equipment, and ash handling systems. It also affects the efficiency of the furnace or gasifier. The volatile matter is determined by placing a weighed freshly crushed coal sample in a covered crucible and heating it in a furnace at 900 ± 15°C for a specified period of time as specified by the American Society for Testing and Materials (ASTM). The volatile matter is then calculated by subtracting the moisture content of the coal from the loss in weight measured from the preceding heating process. The ash may next be determined by heating the solid remaining in the crucible over a Bunsen burner till all the carbon is burned off (by repeatedly weighing the sample till a constant weight is obtained). The residue is then reported as ash while the fixed carbon is obtained by difference.

Coal is classified according to the degree of metamorphism into the following four types:

1. Anthracite that is low in volatile matter and consists of mostly carbon (fixed carbon).
2. Bituminous that contains significant amounts of the volatile matter and typically exhibits swelling or caking properties when heated.
3. Sub-bituminous that is a younger coal contains significant amounts of volatile matter as well as moisture bound within the remnants of the plant cellular structure ("inherent" moisture).
4. Lignite that is the youngest form of coal, that is, when peat moss is not included in the broader definition of coal types, is very high in inherent moisture content resulting in a much lower heating value than the other types of coals.

TABLE 1.4 Composition of a bituminous and a lignite coal (as received basis)

Coal type	Bituminous	Lignite
Name of coal	Illinois No. 6	N. Dakota Beulah-Zap
Mine	Old ben	Freedom
Proximate analysis (%)[1]		
Moisture	11.12	36.08
Ash	9.7	9.86
Volatile matter	34.99	26.52
Fixed carbon	44.19	27.54
Total	100	100
Ultimate analysis (%)		
Carbon	63.75	39.55
Hydrogen	4.5	2.74
Nitrogen	1.25	0.63
Sulfur	2.51	0.63
Chlorine	0.29	12 ppmw
Ash	9.7	9.86
Moisture	11.12	36.08
Oxygen	6.88	10.51
Total	100	100
Heating value		
HHV (kJ/kg)	27,135	15,391
HHV (Btu/lb)	11,666	6,617
LHV (kJ/kg)	26,172	14,803
LHV (Btu/lb)	11,252	6,364
Trace component(s)		
Mercury (ppm)	0.18	0.116

Table 1.4 presents the composition and heating value data for a bituminous and a lignite coal. As can be seen, the moisture content of the lignite is much higher while its heating value lower. Coal is a highly heterogeneous solid and significant variation in its composition and heating value may be expected for coal supplied from the same formation.

Almost all the coal consumed in the world is for electric power generation by combusting the coal in boilers and generating power through the Rankine steam cycle, that is, generate high pressure steam to power a turbine as described in Chapter 8. Coal is being used to a limited extent in gasification-based plants to produce gas also known as syngas to fuel gas turbine-based combined cycles in an integrated gasification combined cycle (IGCCs) as described in Chapter 9. It is expected that with the introduction of more advanced gas turbines in the future, coal-based IGCC will have a strong economic basis in addition to its superior environmental signature, to compete with boiler-based power plants. The synthesis of chemicals is also being pursued, a majority of such plants being built in China.

The method of coal mining depends on the depth and quality of the coal seams and the local geology. Coal is mined on the surface or underground depending on the depth (surface mining usually when depth is limited to 50 m) and thickness of the coal seam and nature of the overburden. The coal after it is mined, which is called "run of mine" coal, is typically delivered to a coal preparation plant to separate out any rocks as well as machine parts or tools that may be left behind from the mining operation and reduce the inorganic impurities (ash along with the accompanying pyritic sulfur) in order to upgrade its value. Cleaning is accomplished at the current time by physical means (mechanical) separation of the contaminants using differences in the physical properties such as density. Chemical cleaning where the impurities are leached out as well as microbial processes (especially for sulfur removal) are under development.

In physical cleaning, coal is crushed and separated into coarse and fine fractions by a screening operation, the streams thus generated being treated separately. Coal coarser than 12.5 mm may be cleaned in a "heavy medium vessel" by gravity separation. A slurry consisting of suspended fine particles of magnetite or ferrosilicon in water forms the medium with a specific gravity that allows low-density materials like coal particles to float and be separated at the top and inorganic higher density materials to sink. Coal particles in the size range from 1 to 12.5 mm may be cleaned in a heavy medium cyclone that uses the centrifugal force generated by the circular motion to cause separation again based on density. Particles in the size range from 150 μm to 1 mm may be cleaned in a spiral separator that uses in addition to the density, difference in the hydrodynamic properties such as drag. Particles smaller than 150 μm may be cleaned by flotation, which uses primarily the difference in their hydrophobicity enhanced by the use of surfactants and wetting agents. The coal particles adhere to air bubbles induced into the agent and rise to the surface, thereby causing the separation. In addition to frothing agent (e.g., aliphatic or aromatic alcohols, poly glycol ethers) and flocculent (e.g., the water soluble polymer, anionic polyacrylamide), other chemicals are added to assist in the separation such as collectors, activators, depressants, as well as reagents for pH control.

The U.S. Energy Information Administration's (2014) estimates in 2008 of remaining recoverable coal were the highest for the United States at 236×10^9 tonne followed by Russia at 157×10^9 tonne, while on a worldwide basis, the total was 860×10^9 tonne. The annual consumption rate of coal in the year 2011 was 7.4×10^6 tonne. It should be borne in mind, however, that coal from different regions has different energy content. These data indicate that the reserves can last a long time if the current consumption rate continues, which is alarming if the accompanying CO_2 emissions go unchecked and are allowed to build up in the atmosphere.

Example 1.2. Estimate the percentage CO_2 capture (separated from the flue gas for subsequent sequestration) required from a supercritical steam coal fired boiler to result in the same amount of CO_2 emission based on a net MW of electric power produced by the plant as a natural gas fired combined cycle (note that the CO_2 emissions being considered in this example do not represent the emissions on a complete life cycle basis, which can be significant, e.g., in case of coal, does not include mining, cleaning, and transportation). The following data is available for the coal-fired plant and the natural gas-fired plant (note that heat rate is the amount of the energy in the fuel required to produce a unit of electric power and is inversely proportional to the efficiency). Assume the plant net heat rate varies linearly with the percentage CO_2 captured and complete combustion of both the fuels.

Coal-fired boiler plant:

Net heat rate (HHV basis) with no CO_2 capture = 9165 kJ/kWh

Net heat rate (HHV basis) with 90% CO_2 capture = 12,663 kJ/kWh

Coal: Illinois No. 6 bituminous coal with characteristics given in Table 1.4

Natural gas-fired combined cycle plant:

Net heat rate (LHV basis) with no CO_2 capture = 5800 kJ/kWh

Natural gas: with characteristics as calculated in Example 1.1.

Solution

The amount of CO_2 formed by complete combustion of the natural gas is calculated by adding up that formed from each of its constituents as $y_i n$ kmol CO_2/kmol of natural gas, where y_i is the mole fraction of specie i and n is the number of C atoms in that specie. These values are tabulated in Table 1.5 giving a total of 1.056 kmol CO_2/kmol of natural gas.

Next, LHV of natural gas

$$= 35{,}290\,\frac{\text{kJ}}{\text{Sm}^3}\,(\text{from Example 1.1}) = 35{,}290\frac{\text{kJ}}{\text{Sm}^3}\times\frac{22.414\,\text{Nm}^3}{\text{kmol}}$$

$$\times\frac{(273.18+15)^\circ\text{K Sm}^3}{(273.18+0)^\circ\text{K Nm}^3} = 834{,}422\,\frac{\text{kJ}}{\text{kmol of natural gas}}.$$

Then CO_2 emitted by natural gas combustion

$$= 1.056\frac{\text{kmol}}{\text{kmol}}\times 5{,}800\frac{\text{kJ}}{\text{kWh}}\times 1{,}000\frac{\text{kW}}{\text{MW}}\Big/\left(834{,}422\,\frac{\text{kJ}}{\text{kmol}}\right) = 0.007340\frac{\text{kmol}}{\text{MWh}}.$$

TABLE 1.5 Calculated amount of CO_2 formed by complete combustion

Component	Mole % $100\ y_i$	CO_2 formed by combustion kmol/kmol of natural gas
CH_4	93	0.930
C_2H_6	3	0.060
C_3H_8	1	0.030
n-C_4H_{10}	0.5	0.020
n-C_5H_{12}	0.1	0.005
n-C_6H_{14}	0.1	0.006
CO_2	0.5	0.005
N_2 + other noncombustibles	1.8	0
Total	100.0	1.056

The amount of CO_2 formed by complete combustion of the coal is calculated from the C content of the coal (0.6375 kg/kg of coal).

CO_2 formed by coal combustion = (0.6375 kg/kg/12.011 kg/kmol) = 0.05308 kmol/kg of coal.

CO_2 emitted by coal combustion with 0% capture = 0.05308 kmol/kg of coal × 9,165 kJ/kWh × 1,000 kW/MW /(27,135 kJ/kg) = 0.01793 kmol/MWh.

CO_2 emitted by coal combustion with 90% capture = 0.05308 kmol/kg of coal × 12,663 kJ/kWh × 1,000 kW/MW /(27,135 kJ/kg) × (1−0.9) = 0.00248 kmol/MWh.

Then, by linear interpolation between these two values for the CO_2 emitted by coal combustion to obtain the same emission as the natural gas-fired plant, CO_2 capture required for the coal plant = 62%, again not on complete life cycle basis.

1.1.1.4 Oil Shale The organic solids in oil shale rock are a wax-like material called kerogen and the oil is extracted by heating the rock in retorts in the absence of air where the kerogen decomposes forming oil, gas, water, and some carbon residue. The source of the kerogen is again biogenic in nature since most extensive deposits of oil shale are found in what used to be large shallow lakes and seas millions of years ago where subtropical and stagnant conditions favored the growth and accumulation of algae, spores, and pollen. The nitrogen content of shale oil is typically higher than that of petroleum and if left in the fuels produced from shale oil such as gasoline or jet fuel would result in significant emissions of NO and NO_2 (collectively denoted as NO_x). Production of the saleable fuels from the shale oil requires more extensive processing than most petroleum feedstocks. The United States has significant deposits of oil shale concentrated in Colorado and Utah, followed by Russia and Brazil.

1.1.2 Nuclear

The commercial production of power from nuclear energy involves conversion of matter to energy by nuclear fission reactions. These reactions consist of the capture of a neutron by nuclei of fissionable isotopes. Today's commercial nuclear reactors utilize U-235, a fissionable isotope of uranium containing a total of 235 neutrons and protons in its nucleus. Most of today's reactors require a fuel that contains between 3.5 and 5% of U-235 while naturally occurring uranium (which exists in the oxide form within the ore) may contain as little as 0.1% uranium (mostly in the form of U-238, which is not fissile, i.e., does not undergo fission) with only 0.7% of it fissile. So an initial step for preparing a suitable fuel for nuclear reactors consists of producing a concentrated form of uranium, generally containing 80% uranium. The tailings from this concentration process contain long-lived radioactive materials in low concentrations and toxic materials such as heavy metals and must be isolated from the environment. A number of other processing steps are required before nuclear "fuel rods" are produced as explained in Chapter 12.

Safety of nuclear power generation in terms of accidental release of radioactive materials from the reactor to the environment and the safe handling and disposal of the nuclear waste including spent nuclear fuel rods have been major issues. These issues have been impediments to the widespread use of nuclear energy in many countries. France, however, has been an exception where a majority of its electricity is generated by nuclear fission.

Nuclear power generation has a significantly smaller carbon footprint when compared to fossil-fueled power generation on a total life cycle basis that has led to a renewed interest in building nuclear power plants in a number of countries while incorporating design improvements based on lessons learned from previous nuclear power plant accidents.

The Joint Report by the Organisation for Economic Co-operation and Development (OECD) Nuclear Energy Agency and the International Atomic Energy Agency (Uranium 2011: Resources, Production and Demand, 2012) shows current estimates of remaining recoverable uranium ore if priced less than $130/kg are the highest for Australia at 1,661,000 tonne followed by Kazakhstan at 629,000 tonne, while on a worldwide basis the total is 5,327,200 tonne. At the current consumption rate, the supply is sufficient for another 100 years.

1.1.3 Renewables

1.1.3.1 Biomass and Municipal Solid Waste All living plant matter as well as organic wastes derived from plants, humans, animals, and marine life represent sources of energy that when properly managed and utilized can make a significant contribution toward conserving our finite energy resources of fossil and nuclear fuels, and more importantly have an impact on the global greenhouse gas emissions.

Specific examples of biomass are agricultural residues, trees, and grasses specifically grown for harvesting as energy crops, forestry residues, urban wood waste and mill residues, paper industry waste sludges, sewage, and animal farm wastes. These wastes or residues as well as municipal solid wastes in many instances pose a disposal problem while having significant energy content. The main driving force for utilizing plant-derived biomass as an energy source is its near CO_2 neutrality. Complete CO_2 neutrality would entail the growth rate of such biomass exactly balances the release rate of CO_2 but consideration of the CO_2 emissions associated with any fossil fuels utilized to synthesize fertilizers and other farm chemicals required to grow the biomass, as well as collecting, and any drying and transporting of the biomass should be taken into account. Growth of biomass by photosynthesis is a natural solar energy-driven CO_2 sink but the relative time scales of growth and utilization should be taken into account, however, when considering energy crops. For example, harvesting old growth trees for fuel should not be considered for energy use while fast-growing fuel crops, agricultural residues, and waste wood should. Agricultural residues are generated after each harvesting cycle of commodity crops and a portion of remaining stalks and biomass material is left on the ground, which can be collected.

Residues of wheat straw and corn stover make up the majority of crop residues but their value as a fertilizer when ploughed back into the soil should be taken into account. Energy crops that are produced solely or primarily for use as feedstocks in energy generation processes include hybrid poplar, hybrid willow, and switchgrass but should be grown on land without competing with food production. Potential environmental issues of growing energy crops that should be taken into consideration include loss of biodiversity due to growing only one plant specie for multiple crop cycles as well as fertilizer contamination of ecosystems.

Forestry residues include biomass material remaining in forests that have been harvested for timber. Timber harvesting operations do not extract all biomass material because only timber of certain quality is usable in processing facilities and thus a significant amount of residual material after timber harvest is potentially available for energy generation purposes. These forestry residues are composed of logging residues, rough rotten salvageable dead wood, and excess small pole trees. Urban wood waste and mill residues include waste woods from manufacturing operations that would otherwise be landfilled such as mill residues, pallets, construction waste, and demolition debris.[3]

In the paper industry, wood or other cellulosic raw materials are digested with alkali chemicals under high temperature and pressure to separate cellulosic fibers of wood from binders (lignins, resins) and the spent (black) liquor is currently regenerated by incineration in special "recovery" boilers with the addition of $Na(SO_4)_2$ while also recovering its energy value. Gasification of the black liquor is another option being considered. In the paper recycling industry, waste paper is received and de-inked prior to recovery of the fiber generating a sludge that contains particles of ink and fibers too short to be converted to finished paper product. This waste stream can also be utilized for energy recovery.

Biomass may be combusted in specially designed boilers, or gasified to generate a gas that could be utilized to generate power in internal combustion engines or fuel cells. Co-utilization of biomass along with coal in larger scale power generation facilities to realize economies of scale is gaining much interest. Combustion of waste streams such as municipal solid waste, treated wood, and plastics, especially those containing chlorine, could, however, produce highly toxic pollutants such as dioxins and furans (described later in this chapter) and proper design measures should be taken to limit these emissions.

Biomass as a renewable energy source provided about 50 EJ/year or 10% of global total primary energy supply in 2008 while it accounted for less than 4% of the total energy usage in the United States in 2009. The technical potential for biomass-derived energy may be as high as 500 EJ/year by 2050 (Edenhofer et al., 2011). Table 1.6 shows the composition and heating value data for woody biomass and covered field dried switchgrass alongside those for a bituminous coal. As can be seen, the heating value of the biomass is significantly lower than that of the coal. The moisture content

[3] Chemically treated construction waste wood may give rise to pollutants during combustion, their source being the treatment chemicals used, and proper pollution abatement measures should be designed into the plant scheme.

TABLE 1.6 Biomass versus coal characteristics

	Woody biomass		Switchgrass		Bituminous coal	
Basis	Dry	As received	Dry	As received	Dry	As received
Ultimate analysis (%)						
Carbon	52.36	26.18	42.60	36.21	71.72	63.75
Hydrogen	5.60	2.80	6.55	5.57	5.06	4.5
Nitrogen	0.37	0.19	1.31	1.11	1.41	1.25
Sulfur	0.03	0.02	0.01	0.01	2.82	2.51
Chlorine	0.10	0.05	0.04	0.03	0.33	0.29
Ash	1.38	0.69	7.41	6.30	10.91	9.7
Moisture	0.00	50.00	0.00	15.00[a]	0	11.12
Oxygen	40.16	20.08	42.08	35.77	7.75	6.88
Total	100.00	100.00	100.00	100.00	100.00	100.00
Heating value						
HHV (kJ/kg)	19,627	9,813	18,113	15,396	30,531	27,135
HHV (Btu/lb)	8,438	4,219	7,787	6,619	13,126	11,666
LHV (kJ/kg)	18,464	9,232	16,659	14,161	29,568	26,172
LHV (Btu/lb)	7,938	3,969	7,162	6,088	12,712	11,252

[a] Covered field dried.

of biomass is typically much higher than bituminous coals unless the biomass has been dried as was the case with the switchgrass.

Example 1.3. Express the equivalent thermal energy required for drying as received woody biomass to the same moisture content as the as received bituminous coal with characteristics shown in Table 1.6, as a percentage of HHV of the as received biomass. Assume the equivalent biomass contained energy required for the drying operation is 2400 kJ/kg of water evaporated (a steam-heated dryer is assumed and includes the electric energy used in the drying operation converted to an equivalent thermal energy).

Solution

Biomass moisture on an as-received basis = 50%.

Biomass moisture after drying = 11.12%.

Moisture removed = $0.5-(0.1112/1-0.1112)\times(1-0.5) = 0.4374$ kg H_2O/kg as received biomass.

Total equivalent fuel for drying = (0.4374 kg H_2O/kg as received biomass)(2400 kJ/kg H_2O) = 1050 kJ/kg as received biomass.

HHV of biomass = 27,135 kJ/kg as received.

Total equivalent fuel for drying as percentage of HHV in biomass= $(1{,}050\ \text{kJ/kg}/27{,}135\ \text{kJ/kg})\times 100\% = 3.87\%$.

1.1.3.2 Hydroelectric This is the most widely used form of renewable energy and accounted for more than16% of global electricity usage in 2008 while the total resource potential is 14,576 TWh/year (Edenhofer et al., 2011). It accounted for 2.7% of the total energy usage in the United States in 2009. This resource became important in modern times with the development of efficient electric generators and transmission technology that allow the transmission of electricity over vast distances from remotely located hydroelectric plants to the energy users several hundred miles away.

The energy that may be recovered from flowing water depends on the quantity of flow of water and the hydrostatic head, which is the height through which the water can be made to flow from the reservoir created by a dam to a hydraulic turbine. The construction of diversion and storage dams for such hydroelectric power plants thus requires suitable topography and other site conditions, and a steep drop in the elevation of the river. The environmental effects of submerging large areas of land by the construction of the dam can be devastating unless such projects are properly planned to minimize the impact on local human population as well as the flora and fauna. Hydroelectric plants are capital intensive but the operating costs are low since there is no fuel cost associated with such plants.

1.1.3.3 Solar Approximately half of the solar energy reaching earth's outer atmosphere reaches the earth's surface, while the remainder consists of the fraction absorbed by the atmosphere and the fraction reflected back into space. A portion of the energy reaching the surface ends up in biomass as chemical bond energy, in wind and waves as kinetic energy, in rivers as potential energy, and in oceans to create thermal gradients. These all represent renewable energy resources, but direct use of solar energy is being implemented more and more in recent times for heating and for conversion into electricity directly using solar photovoltaic cells and through the Carnot cycle using heat engines described in Chapter 12.

The potential for producing electric power from solar energy is huge with estimates ranging from 1,580 to 49,800 EJ/year, roughly equivalent to as much 3–100 times the world's primary energy usage in 2008 (Edenhofer et al., 2011) while it accounted for less than 0.1% of the total energy usage in the United States in 2009.

1.1.3.4 Wind Wind, which is the movement of air across the surface of the earth, is caused by pressure gradients. These result from the surface of the earth being heated unevenly by the sun. Heating by the sun varies with (i) latitude because the angle of incidence of the sun's rays varies with latitude as well as (ii) the earth's surface properties such as reflectivity and absorptivity (e.g., whether it is covered by ocean or dry land and whether it has vegetation), and of course (iii) time of day. The distribution of wind is not uniform over the earth; wind resources are higher in polar and temperate zones than in tropical zones and are also higher in coastal areas than inland. Furthermore, the drag on the atmosphere caused by earth's rotation results in turbulence, giving rise to a varying pattern of wind across the earth's surface.

Only a small fraction of the available wind energy resource is currently being harnessed. Wind energy is harnessed by installing rotating machines that are typically propellers called "wind mills" that are connected to electric generators as discussed in Chapter 12. Winds only in certain speed ranges may be harnessed economically at the current time. Because of friction losses, wind power turbines usually do not operate at wind velocities much lower than approximately 16 km/h, while the rotors of wind turbines are usually feathered (designed to vary blade pitch) to prevent damage at very high wind speeds such as winds of gale force.

The global potential for generating electricity from wind is very large, with estimates ranging from 70 EJ/year (19,400 TWh/year) for onshore installations to 450 EJ/year (125,000 TWh/year) for onshore and near-shore installations, while those for shallow offshore installations range from 15 EJ/year to 130 EJ/year (4,000 to 37,000 TWh/year) when only considering relatively shallower and near-shore applications. Wind, however, accounted for only 0.72% of the total energy usage in the United States in 2009.

1.1.3.5 Geothermal Geothermal energy is the heat energy stored within the earth's crust. The earth's core maintains temperatures in excess of 5000°C due to the heat generated from gradual radioactive decay of elements. It was previously thought incorrectly that the heat was left over from planet's formation but calculations have shown that the earth's core would have cooled down a long time ago. The heat within the core flows continuously to the earth's surface by conductive transfer and convective transfer by molten mantle beneath the crust. The heat intensity tends to be strongest along tectonic plate boundaries. Volcanic activity transports hot material to near surface but only a small fraction of molten rock actually reaches the surface while most of it is left at depths of 5–20 km beneath the surface. The heat is then transported to the surface by hydrological convection by groundwater, hot springs and geysers being examples. Until the beginning of the last century, the utilization of geothermal heat was limited to use as warm water, Roman baths in England being early examples. Also, geothermal hot springs began to enjoy widespread use throughout the world as therapeutic treatment. Only in more recent times has there been more extensive use made of geothermal energy for both power and nonpower applications. Typically, the useful geothermal heat for electric power generation is available in the form of hot brine with temperatures ranging from 150 to 200°C and discussed in more detail in Chapter 12.

World potential for electricity generation from the more suitable higher temperature geothermal energy is 33,600 TWh/year and that for heating from the lower temperature geothermal energy is greater than 1400 TWh/year, while it accounted for only 0.2% of the total energy usage in the United States in 2009.

1.2 ENERGY AND THE ENVIRONMENT

Energy is used in almost every human activity in an industrialized society, examples being transportation (automobiles, busses, trucks, trains, aircraft, rockets), household uses (cooking, hot water, air conditioning), agriculture (production of fertilizers, pumping of water), and manufacturing (heating, electricity). Thus, energy is consumed by us directly as in transportation and household uses as well as indirectly by consuming goods that require energy in their production. Energy usage can be a direct measure of the economic well-being of a society since it plays such an important role in our lives. The per capita energy use is often correlated to the per capita gross national product of a country and the trend line based on data for several countries shows essentially a straight line. Strictly speaking, this correlation should be corrected for other factors such as the energy usage efficiency of a given country as well as its climatic conditions and the type of economy it has, that is, whether it is dominated by a service economy, agriculture, or industrial production. Energy conservation measures practiced by or instituted in a given country such as mass transportation and the severity of the climate, that is, whether heating and/or cooling are required, can all affect the energy usage. In any case, it should be the goal of every country, both developing and developed, to practice energy conservation in order to reduce the slope of the energy usage trend line so that we have a planet that can sustain itself.

Energy use over the next two decades is expected to increase significantly throughout the world, with highest growth rates in Asia. The world energy consumption is projected to be about 20% higher in 2020 and almost 60% higher by 2040 over the 2010 values. More than 85% of the increase in this energy demand is due the developing nations driven by strong economic growth and expanding populations (U.S. Energy Information Administration, 2013b). Fossil fuels (liquid fuels, natural gas, and coal) are projected to supply most of the world's energy, as much as three-fourths of total world energy consumption in 2040.

World use of petroleum and other liquid fuels is expected to grow by more than 30% by 2040 over the 2010 values. The world's total natural gas consumption is expected to increase by more than 60% by 2040 over the 2010 values due to increasing supplies of natural gas, particularly from shale formations. Coal is expected to continue playing a significant role in the world energy markets, especially in the developing countries of Asia with world coal consumption increasing by almost 50% by 2040 over the 2010 values. Due to these significant increases in fossil fuel usage worldwide, the greenhouse gas CO_2 emissions to the atmosphere are expected to increase by an average annual growth of 1.3% in the 2010–2040 period or by as much as 46% by 2040 over the 2010 level. In addition to this greenhouse gas, the harnessing of the energy contained within a fuel by combustion produces pollutants such as oxides of sulfur, oxides of nitrogen, and unburned hydrocarbons that are introduced into the atmosphere, the amount depending on the degree of pollution abatement measures incorporated. A conflict thus arises between increasing the standard of living (measured in terms of per capita gross national product) and per capita energy use that can have a direct impact on the environment unless sustainable energy management is pursued.

1.2.1 Criteria and Other Air Pollutants

Criteria pollutants as defined by the U.S. Environmental Protection Agency (U.S. EPA) consist of CO, SO_2 and SO_3, NO and NO_2, O_3, Pb, and particulate matter

(PM). These pollutants are commonly found all over the United States and thus may be considered as indicators of air quality.

1.2.1.1 Carbon Monoxide and Organic Compounds CO along with another set of harmful pollutants consisting of organic compounds are formed due to improper combustion of carbonaceous fuels and can be minimized by designing the combustion process with good mixing (turbulence) of the fuel with the oxidant (air), and providing enough residence time at high temperatures. The following describes the mechanism of CO formation during combustion of natural gas.

The CH_4 molecule is very stable and requires high energy atoms to break loose an H atom forming the CH_3 radical that plays a key role in propagating the combustion process. This process includes the partial oxidation of CH_4, oxidation of CO and OH reactions. CO formation involves a number of steps but is a fast overall reaction while the oxidation of CO to CO_2 is very slow and as a result, the reciprocating engine produces significant amounts of CO due to the short residence time. In a gas turbine however, more residence time is available within the combustor and the CO emissions are much lower. High temperatures and O_2 concentrations and large residence times are required for the CO oxidation that involves the reaction with the OH radical formed during the combustion process. High CO emission not only means more pollution but also lower overall plant efficiency.

Postcombustion control may include oxidation of the CO as well as the volatile organic compounds (VOCs) in the presence of a noble metal catalyst. The harmful health effects of CO include reduction of O_2 supply to the organs such as the heart and brain as well as to the tissues while extremely high levels of CO can cause death. VOCs are responsible for the formation of a number of secondary pollutants such as peroxyacyl nitrates (PANs), which are powerful respiratory and eye irritants. They are also responsible for the formation of a major portion of the tropospheric O_3 such as in less polluted rural areas where the direct NO emissions and concentrations may be low.

Another set of toxic organic compounds that are also emitted depending on the type of fuel (its Cl content) belong to the families of compounds known as dioxins and furans, and among all of these compounds, 2,3,7,8-tetrachloro-*p*-dibenzo-dioxin (2,3,7,8TCDD) is considered the most toxic. These compounds can enter the body through air and water but mostly through food and build up in the fatty tissues. The U.S. EPA has stated that 2,3,7,8TCDD is likely to be a cancer-causing substance to humans. In addition, exposure to dioxins and furans changes hormone levels while high doses of dioxin have caused a skin disease called chloracne. Animal studies have shown that exposure to dioxins and furans changes their hormone systems, development of the fetus, decreased ability to reproduce, and suppresses the immune system. The formation of these compounds downstream of the combustion process may be minimized by quickly cooling the flue gas and by limiting the use of certain metals that are known to promote or catalyze their formation in the downstream equipment.

1.2.1.2 Sulfur Oxides About 90% of the sulfur oxides emitted during the combustion of a sulfur-bearing fuel is in the form of SO_2 with the remainder being SO_3. These oxides are collectively denoted as SO_x. Postcombustion capture of the SO_x from stationary sources is accomplished at the present time by reaction with Ca or Mg containing compounds such as limestone (or lime) and dolomite to form the corresponding sulfites that are then oxidized using air to form the stable sulfates. In high temperature gasification process, the sulfur in the fuel is mostly converted into H_2S and some COS and only minor amounts of CS_2 and mercaptans. COS is hydrolyzed to H_2S by catalytically reacting the H_2S with H_2O after which the H_2S is removed using either a chemical or physical solvent to eventually produce byproduct elemental sulfur or H_2SO_4. These processes are discussed in more detail in Chapters 8 and 9. The environmental effects of SO_x include the formation of H_2SO_4 in atmosphere under action of light, which reduces the pH of the rain water giving rise to "acid rain." Acid rain reduces soil productivity by causing damage to vegetation. It also causes damage to aquatic life and gives rise to the formation of sulfate particulates, mostly H_2SO_4 and $(NH_4)_2SO_4$.

1.2.1.3 Nitrogen Oxides NO is the primary product of combustion and is formed via three routes. The NO is produced from (1) high temperature oxidation of molecular N_2 (thermal NO), (2) hydrocarbon radical attack on the molecular N_2 (prompt NO), and (3) oxidation of chemically bound nitrogen in the fuel (fuel NO). The thermal NO, that is, the formation of NO from the N_2 present in the combustion air, requires the breaking of the covalent triple bond in the N_2 and requires very high temperatures and forms by the action of the O radical produced at high temperatures during the combustion process. Increased temperature, residence time, and O_2 concentration increase NO emission, which is in direct contrast to the conditions required for CO formation. As much as 60–80% of the fuel-bound nitrogen in fuels such as oil and coal forms NO. The formation of NO_2 is not significant during the combustion process, however the NO oxidizes to NO_2 in the atmosphere and thus all NO is potential NO_2. Another oxide of nitrogen, N_2O, which is also formed during combustion, has become important in recent years due to its role in the stratosphere as a greenhouse gas. It is formed in significant concentrations (from an environmental impact standpoint) in fluidized bed combustion.

Postcombustion control consists of reducing the NO by reacting with NH_3 or urea as discussed in more detail in Chapters 8 and 9. Prompt NO formed by the mechanism of hydrocarbon radical (formed during combustion) attack on N_2 is typically less than 5%. NO which oxidizes to form NO_2 in the atmosphere, can oxidize O_2 to the pollutant, O_3 under the action of sunlight. NO can also react with hydrocarbon free radicals present in the atmosphere to form components of smog while NO_2 can react with H_2O to form acid rain. Photolysis (decomposition under the presence of sunlight) of HNO_2 produces the OH radical. The reactive OH radical plays a dominant role during the daytime (while the nitrate NO_3 radical plays a major role in the night time) in the formation of photochemical air pollutants, acid rain and fogs, and toxic organics. Possible control strategies for N_2O which is a more potent greenhouse gas than CO_2 include burning of a gaseous fuel in the freeboard or cyclone of the fluidized bed, increasing bed temperatures, decreasing excess air, or catalytic reduction with metal oxides.

The thermal NO formation reaction suddenly takes off around 1540°C and thus a window of opportunity exists to control NO by staying just below this temperature during the combustion process. Thermal NO formation shows an inverse relationship with respect to hydrocarbon and CO emissions when the air to fuel ratio is varied. Control strategies include (i) burning under lean (high excess air) conditions, (ii) staged combustion with rapid quenching of the flame by the secondary air, (iii) premixed burners (ideally, with variable geometry for varying load of the boiler or engine), and (iv) flue gas recycle. Postcombustion processes are also sometimes applied but have certain disadvantages such as transforming NO into other undesirable species. To meet the ultra-low NO_x emissions being mandated, the internal structure of the combustion process that is complex and combines fluid dynamics, turbulent mixing, high temperature chemistry, and heat transfer need to be understood to develop new solutions without compromising efficiency at full or partial combustor load.

1.2.1.4 Ozone O_3 by itself can cause breathing problems as well as eye, nose, and throat irritation aggravating chronic diseases such as asthma, bronchitis, and emphysema. Heart disease and speeding up the aging process of lung tissue as well as a reduction in the resistance to fight off colds and pneumonia may also be consequences. It is also a major source of the OH radical in the atmosphere; under the action of light O_3 decomposes to generate the atomic O that reacts with H_2O to produce the OH radical.

1.2.1.5 Lead Pb accumulates in the body and can cause damage to the brain as well as the nervous system damage. It can lead to mental retardation especially in children and also to digestive problems; some of the chemicals containing Pb are known to have caused cancer in test animals. A major source of Pb entering the atmosphere was leaded gasoline that has been phased out in many countries. An organo Pb compound was added as an anti-knocking agent. Paints utilized pigments made out of Pb compounds and old houses could contain such paints. Plants to recover energy from construction debris should take this into account in their design. Electrodes made out of Pb and PbO_2 are used in storage batteries and recycling of the batteries is extremely important to limit the introduction of Pb into the environment (note that at the present time, batteries are utilized to store the electricity generated by small-scale intermittent renewables, especially by solar photovoltaic cells). Furthermore, proper control methods should be instituted at Pb manufacturing facilities and in Pb smelting processes.

1.2.1.6 Particulate Matter PM can be in solid or droplet (aerosol) form. Increased levels of fine particles in air are linked to heart disease, altered lung function, and lung cancer. Particles of approximately 10 μm or less (PM10[4]) can penetrate deepest parts of lungs while particles smaller than 2.5 μm (PM2.5) tend to penetrate into gas

[4] The International Organization of Standardization (ISO), which is an international body for setting standards composed of representatives from various national standards organizations, defines PM10 as particulate matter in which 50% of particles have an aerodynamic diameter less than 10 μm. The aerodynamic diameter of an irregularly shaped particle is defined as the diameter of a spherical particle with a density of 1000 kg/m^3 having the same settling velocity as the irregularly shaped particle.

exchange regions of lungs, can cause hardening of arteries, and can lead to heart attacks and other cardiovascular problems. Particles less than 100 nanometers may pass through lungs to affect other organs. Depending on the chemical nature of the particulates, they can also cause cell mutations, reproductive problems, and certain cancers. Sources of particulates can be coal (and to a lesser extent oil and natural gas) fired power and industrial plants, vehicles, especially those burning diesel as well as agriculture during plowing and burning off fields. Various types of filtration devices are utilized to filter out particulates and in the case of power plants burning a fuel such as coal, electrostatic precipitators are sometimes utilized instead of filtration devices called baghouses, as discussed in more detail in Chapter 8.

1.2.1.7 Mercury Hg can enter the food chain especially through aquatic organisms and exposure to it can cause neurological disorders. Studies have shown that Hg exposure can also cause cardiovascular problems. A major portion of the anthropogenic (human-caused) emissions of Hg is from the use of coal as a fuel. Other sources of Hg emissions include cement manufacture and the minerals processing industry. Hg can be repeatedly released into the environment after being first emitted into the atmosphere by entering the food chain. Activated carbon has been found to be an excellent Hg sorbent but its application in coal combustion-based power plants is in its early stages; its effectiveness under different conditions of fuel properties, flue gas temperatures, and trace-gas constituents is being investigated. In coal gasification-based plants, however, a sulfided activated carbon has been proven to be very effective in removing greater than 90% of the Hg from the syngas.

1.2.2 Carbon Dioxide Emissions, Capture, and Storage

The greenhouse gas concentrations of CO_2, CH_4, and N_2O in the atmosphere at the present time far exceed preindustrial values of the mid-eighteenth century due to human activities. CO_2 due to its much higher concentration in the atmosphere plays a more significant role as a greenhouse gas although the CH_4, N_2O as well as halocarbons should not be ignored. Currently, fossil fuel usage for energy recovery is the major source of CO_2 emissions on a worldwide basis, especially power plants firing coal and automobiles burning petroleum-derived fuels while the supply of primary energy is expected to be dominated by fossil fuels until at least the middle of the century. According to the U.S. Energy Information Administration, under the scenario where no new legislation or policies are passed to affect energy markets, the global energy use by 2035 will have grown by more than 50% over the 2008 value. This growth in energy use is fueled by projected rise in income and population in the developing world and fossil fuels are expected to continue playing the major role. In 1992, international concern over global climate change led to the United Nations Framework Convention on Climate Change whose ultimate objective is to stabilize greenhouse gas concentrations in the atmosphere at such a level that dangerous anthropogenic interference with the climate system is prevented.

The United Nations Intergovernmental Panel on Climate Change (IPCC) in their 2007 Fourth Assessment Report indicate that during the twenty-first century, the global surface temperature is likely to rise by 1.1–2.9°C for the lowest greenhouse gas emissions scenario and 2.4–6.4°C for the highest greenhouse gas emissions scenario based on projections made by different climate models. These models differ in sensitivity to greenhouse gas concentrations but the effects such as increases in wind intensity leading to higher and more intense hurricanes, increases of both drought and heavy precipitation, and decline of permafrost coverage are quite concerning. In fact, on the average, mountain glaciers and snow cover have already declined. Sea levels will continue to rise from the melting of land-based ice sheets while ocean warming will magnify this problem since seawater will expand as its temperature increases. The effect of higher concentrations of another "greenhouse gas," H_2O vapor in the atmosphere resulting from warmer oceans and air further compounds the challenge.

Carbon capture and sequestration (CCS) is receiving interest as a means for reducing global warming. CCS technology consists of separating the CO_2 from large point sources, such as fossil fuel power plants, and storing it so that it is not introduced to the atmosphere. Carbon capture technologies have been established and utilized in the chemical industry, such as the ammonia and urea synthesis processes where the CO_2 is scrubbed out of the syngas that consists mostly of CO and H_2 initially and the CO is converted by the water gas shift reaction, $CO + H_2O = CO_2 + H_2$ followed by scrubbing the gas with a suitable solvent to absorb the CO_2 leaving a decarbonized syngas. In the case of an IGCC, this decarbonized gas can be combusted in the gas turbine. A chemical solvent such as an aqueous amine solution or a physical solvent such as Selexol™ may be utilized, depending on the pressure of the syngas. This "precombustion" carbon capture technology is discussed in more detail in Chapter 9. In the case of direct combustion such as in a coal-fired boiler or a natural gas-fired gas turbine where "postcombustion carbon capture" technology is utilized, scrubbing the flue gas with a suitable aqueous amine solution (proven on natural gas-fired combined cycle) may be employed as discussed in more detail in Chapter 8. Both overall (energy) efficiency and capital cost are compromised significantly however, when these commercially available technologies are utilized for carbon capture and considerable research funds are being expended to develop technologies that reduce these penalties.

Technically, the major challenge in CCS is not the capture of the CO_2; it is the storage of the captured CO_2 for sequestration. The various proposals for sequestration include storage in depleted oil or gas wells, unminable coal beds, and deep saline aquifers. The U.S. National Energy Technology Laboratory has reported that there is enough storage capacity in North America to sequester CO_2 generated over the next 900 years at its current generation rate but the real concern is the geologic stability and the associated risk that the CO_2 might leak into the atmosphere. Research is underway to investigate the mineralization of the CO_2 by reactions with reservoir rocks that would be a long-term solution.

Deep ocean storage had been proposed but is no longer being considered due to the accompanying increase in acidity of the water. Using the CO_2 for enhanced oil recovery is another venue that can actually generate a positive revenue stream but such oil fields are limited. The CO_2 serves the purpose of pressurizing the oil field and reducing the viscosity of the petroleum to make it more free flowing by going into solution with the petroleum.

A large-scale CCS demonstration project has been operated since September 2000 in Canada. In this project, CO_2 separated from syngas generated in a gasification facility in the United States at 95% purity is transported via pipeline and injected into the Weyburn oil field in Saskatchewan at a rate of about 6500 tonnes/day CO_2.

1.2.3 Water Usage

Another environmental concern is the large amount of water used for heat rejection in today's power plants that utilize steam as the working fluid (in the Rankine cycle discussed in more detail in later chapters). Such power plants utilize a heat engine to convert heat to power and the conversion efficiency is far less than 100% due to limitations imposed by nature (and annunciated by the second law of thermodynamics discussed in Chapter 2) in converting a highly disorganized form of energy, the thermal energy, into a highly organized form of energy, the mechanical energy in the rotating shaft of the heat engine. In 2012, the water withdrawal/use in the United States was as much as 1300–1400 gal/person/day or 4.9–5.3 m^3/person/day (which includes industry, agriculture, and domestic uses) and more than half of this was used in power generation. The amount of water used by the power industry has an inverse relationship to plant efficiency when wet cooling towers are employed, which underscores the importance of power plant efficiency. Wet cooling towers provide cooling of the cooling water that is utilized for absorbing the heat rejected by the plant so that the cooling water may be recirculated for cooling. The cooling process, however, depends upon evaporating a portion of the cooling water (the latent heat of evaporation provides the cooling) by contacting the cooling water with ambient air, and this evaporated water is lost to the atmosphere. There are alternatives for wet cooling towers as discussed in the following, but some compromise efficiency and/or plant cost while others are currently under development.

Once-through systems take water from nearby sources such as rivers, lakes, and the ocean after some initial treatment and circulate the water to absorb the heat rejected by the plant. The water thus heated up is then returned to its source while taking precautions not to recirculate the warm returned water, that is, avoid the discharge to the intake. This type of cooling was initially the most popular because of its simplicity and low cost as long as the plant could be sited in close proximity to the source of water. The drawback with this type of cooling is the disruption caused to local ecosystems from the large rates of water withdrawals and returning of warmer water. Some plants employ a combination of wet cooling towers and this once-through cooling system to reduce the impact on the local ecosystem, but this scheme again has the disadvantage of water loss through evaporation in the wet cooling towers.

Dry-cooling systems use air instead of water to absorb the heat rejected by the plant and have been commercially employed. The major compromise with this type of cooling in a power plant is that the heat rejection temperature is typically increased due to the poorer heat transfer coefficient of air as compared to liquid water, requiring larger temperature driving forces within the heat exchanger to limit its size.

As explained in Chapter 2, the increase in the heat rejection temperature results in a reduction in the power cycle efficiency. The other drawback is that the plant cost is increased primarily due to the requirement for a very large (air cooled) exchanger.

A more recently developed cooling system technology called the "Green Chiller" technology is aimed at reducing the water usage as is the case with an air-cooled exchanger but without the accompanying compromise in plant efficiency. It also utilizes air as the cooling agent except that during warm days, the air is precooled by direct contact with water. In this precooled mode of operation, a portion of the water evaporates into the directly contacted air stream under adiabatic conditions and the latent heat required for the evaporation is taken from the air, thereby cooling the air. In this mode of operation, water is required for the cooling operation but the annual average water use of the plant is reduced when compared to a wet cooling tower-based system.

1.3 HOLISTIC APPROACH

In addition to practicing the environmental control strategies as discussed in previous sections at the generation point of the energy supply chain, improvements made at the raw materials supply side as well as at the end user side can greatly improve the sustainability of our planet.

1.3.1 Supply Chain and Life Cycle Assessment

Consider for example the drilling for shale gas. There are various environmental concerns due to not only the use of the large amount of water but also due to the various chemicals added to the water. There is concern over the migration of fracture fluids to aquifers and surface spills of undiluted chemicals. Measuring and disclosing operations and environmental data, engaging with local communities, surveying of the well sites to minimize their impacts, implementing robust designs in the construction of the wells, properly managing water usage and disposal, and providing appropriate resources to handle permitting and compliance are all essential to increasing the sustainability of such operations. There is also concern over the migration of gas through unplugged abandoned oil and gas wells as well as water wells (risk of house explosions and gas contamination of water wells). CH_4 is a potent greenhouse gas, much more potent than CO_2. Howarth et al. (2011), based on their comparison of total greenhouse gas emissions from shale gas that includes fugitive CH_4 emissions and coal, argue that shale gas has significantly higher greenhouse gas emissions than coal for the 20-year time horizon. This conclusion, however, is based on energy released by combustion as if thermal energy were the end product and does not necessarily apply when the end product is electrical energy because natural gas-fired power plants have a significantly higher overall efficiency than coal-fired plants, but nevertheless such fugitive emissions should be minimized. Coal-fired plants themselves have major challenges at the supply side, in the coal-mining operations. Coal-mining operations can result in diminishing the utility of land and if the land is left barren, then there is potential for erosion and landslides. Health and safety of the miners is also a major concern, especially when the mine owners do not provide safe working conditions and adequate equipment. In addition to the environmental impact of drilling or mining for the energy resource such as natural gas, oil, or coal, the processing of the resource before the resource is utilized for energy recovery as was described previously in this chapter should be taken into account with respect to its environmental impact. Inclusion of sustainability concepts in the original design of the operation as well as in ongoing operations is essential. A life cycle assessment starting from the acquisition of the raw materials to producing the end product as well as disposal of the used end product is absolutely necessary in identifying areas where improvements should be made, and in identifying which of the options is more sustainable on a cradle to grave basis. Purely an economic sustainability analysis, that is, one in which a monetary value to the environmental impact is not assigned, is not sufficient.

Higher efficiency processes led to the conservation of resources as well as reduction of the environmental impact. Energy efficiency has the potential to be a vast and low-cost "energy resource" for our planet both at the current time and in the future and to achieve this we need a coordinated international strategy to bring into practice existing high efficiency technologies, and continue research and development activities to make improvements in such technologies. There is potential for significant increase in efficiency at both the generation side as well as at the end user side. The emphasis in the following chapters is energy efficiency on the generation side but the importance of efficiency at the end user side should not be ignored. Changing end user habits is one aspect of increasing the overall system (inclusive of generation, transmission, and usage of electricity) efficiency and another is from a technological standpoint. As an example, consider the light bulb. Compact fluorescent lamps (CFLs) and light-emitting diodes (LEDs) offer similar levels of performance as the incandescent lamps but use a fraction of the energy. Specifically, a study conducted by the U.S. Department of Energy found that energy-in-use of the life cycle is the dominant environmental impact, with 15-W CFL and 12.5-W LED lamps performing better than the 60-W incandescent lamp, all producing approximately the same light output.

The U.S. Department of Energy study on lighting also addressed the sustainability of the manufacture and disposal of the CFLs and LEDs. The presence of Hg in CFLs requires strict controls over their recycling, although relatively little is known about the full environmental impact of manufacturing and disposing of LEDs. Based on the data available, it was found that both CFLs and LEDs were far superior to incandescent lamps in this life cycle assessment with a slight advantage for LEDs over CFLs.

Thus, a life cycle assessment to include an assessment of the impact on the resources and the environment is required to get the complete picture. In a life cycle assessment, the air-related environmental impacts can include the potential for global warming, acidification, photochemical oxidation, stratospheric O_3 depletion, and human toxicity. Water-related environmental impacts can include potential for freshwater aquatic ecotoxicity, marine aquatic ecotoxicity, and eutrophication (e.g., increase of phytoplankton due to increased levels of nutrients causing depletion of O_2 in the water for aquatic life). Soil-related environmental impacts include land use, ecosystem damage potential, and terrestrial ecotoxicity. Resource-related environmental impacts can include abiotic (nonbiological such as inorganic salts, mineral soil particles) resource depletion, nonhazardous waste landfill, radioactive waste landfill, and hazardous waste landfill.

Going back to the example of lighting, in recent years the replacement of the incandescent lamps in homes and businesses with alternative CFL and LED-based products has been accelerated due to increased legislation banning the production of the incandescent light sources, stressing the important role that governments can play for the sustainability of the planet. Public awareness and acceptance of this type of lighting is equally important in a free society.

1.4 CONCLUSIONS

Processes that use less energy and resources are necessary for the long-term sustainability. It is sometimes argued that sustainable designs tend to be more complex and riskier requiring unproven technology and often greater capital expenditure but any risks and increases in capital expenditure should be compared to the indirect costs incurred by society on the long run due to disruption of the environment and the associated adverse health effects, and depletion of the finite natural resources.

There are both external and internal costs brought about by the finiteness of the earth's capacity to withstand the environmental impacts. External costs are associated with environmental disruptions on society but are not reflected by the monetary values assigned to energy. Internal costs are costs due to measures such as pollution control equipment to reduce the external costs. External costs have been rising despite a substantial increase in the internal costs. The external costs are difficult to quantify but civilization depends heavily on regulating water supply, controlling pests and pathogens, and maintaining a tolerable climate. Climate governs most of the environmental processes on which our well-being depends and global climatic changes brought about by greenhouse gases could be devastating. The CO_2 problem may be postponed by switching to natural gas, which buys us some time. Nuclear energy is less disruptive on the climate but requires development of much safer reactors and management of nuclear wastes. Solar energy is clearly the long-term solution from a standpoint of minimizing the external costs but at the present time remains expensive. The near-term solution is to transition into low-impact energy supply technologies (without disrupting economic growth) and increase end use efficiency. If the energy use in the developing countries increases to near that of the developed countries per capita, CO_2 and other pollutants would affect both locally and globally on a scale much larger than ever experienced. International cooperation in energy research in the areas of long-term alternatives is crucial.

REFERENCES

Edenhofer O, Madruga RP, Sokona Y, Seyboth K, Matschoss P, Kadner S, Zwickel T, Eickemeier P, Hansen G, Schlömer S, von Stechow C, editors. *IPCC Special Report on Renewable Energy Sources and Climate Change Mitigation*. Cambridge: Cambridge University Press; 2011.

García R, Pizarro C, Lavín AG, Bueno JL. Spanish biofuels heating value estimation. Part I: Ultimate analysis data. Fuel 2014;117:1130–1138.

Howarth RW, Santoro R, Ingraffea A. Methane and the greenhouse gas footprint of natural gas from shale formations. Clim Change Lett 2011;106 (4):679–690.

Keenan JH, Keyes FG, Hill PG, Moore JG. *Steam Tables: Thermodynamic Properties of Water Including Vapor, Liquid, and Solid Phases.* New York: John Wiley & Sons, Inc.; 1969.

Rao A, Sander M, Sculati J, Francuz V, West E. Evaluation of advanced Gas Turbine Cycles. Prepared by Fluor Daniel, Inc., Report August 1993. Gas Research Institute Report nr 5091-294-2346. Chicago, IL: Gas Research Institute; 1993.

Uranium 2011: Resources, Production and Demand. 2012. A Joint Report by the OECD Nuclear Energy Agency and the International Atomic Energy Agency. Paris: OECD Nuclear Energy Agency; Vienna: International Atomic Energy Agency.

U.S. Energy Information Administration, Office of Energy Analysis, U.S. Department of Energy. 2013a. Technically recoverable shale oil and shale gas resources: An assessment of 137 shale formations in 41 countries outside the United States. Report June 2013. Available at http://www.eia.gov/analysis/studies/worldshalegas/pdf/fullreport.pdf. Accessed February 17, 2014.

U.S. Energy Information Administration, Office of Energy Analysis, U.S. Department of Energy. 2013b. International energy outlook 2013 with projections to 2040. Report July 2013. Available at http://www.eia.gov/forecasts/ieo/pdf/0484%282013%29.pdf. Accessed February 18, 2014.

U.S. Energy Information Administration, Office of Energy Analysis, U.S. Department of Energy. 2014. International Energy Statistics. Available at http://www.eia.gov/cfapps/ipdbproject/IEDIndex3.cfm#. Accessed February 17, 2014.

2

THERMODYNAMICS

The particularly useful forms of energy to us are mechanical and electrical energy to operate machines, thermal energy to process materials and to heat living spaces, and radiant energy to provide light. Mechanical energy at the current time is predominantly derived from thermal energy using engines of various types, the thermal energy being obtained from combustion of a fossil fuel. Mechanical energy may be converted into electrical energy using a generator and vice versa using a motor. In addition to thermal energy for heating that can be obtained from combustion of a fossil fuel, it may also be recovered from unconverted energy from a heat engine, or indirectly by producing electricity and then converting it back into heat, although this last method is very inefficient from a sustainability viewpoint. Radiant energy for lighting is derived from the conversion of electrical energy. Electrical energy plays the very useful role of energy transfer over distances to be converted into useful forms such as mechanical, thermal, or radiant energy at the end user. Production of various material products that we utilize directly or indirectly also involves many energy conversion steps that accompany material transformation processes, especially those involving chemical reactions.

Thermodynamics is the field of science that deals with conversion of one from of energy into another. It was originally concerned with heat and its relation to other forms of energy and work but has evolved into playing a much broader role in defining the laws governing the conversion of other forms of energy as well, for example, chemically bound energy directly to electrical energy as in a battery or a fuel cell

Sustainable Energy Conversion for Electricity and Coproducts: Principles, Technologies, and Equipment, First Edition. Ashok Rao.

without the intermediate step of conversion to thermal energy. The principles of thermodynamics may also be used in establishing the upper limit for conversion in a chemical reaction or separation of components from a mixture. Thus, a thorough understanding of this subject is essential in making the various energy and material transformation processes that our modern lives are so dependent on more sustainable. The analysis of such processes may be systematically accomplished by applying the two laws of thermodynamics. The intent of this chapter is to provide a refresher to these topics in thermodynamics, which are essential for understanding and analyzing energy conversion plants.

2.1 FIRST LAW

The first law of thermodynamics (for non-nuclear processes) may be stated concisely as follows:

$$\text{Change in energy of the system} + \text{Change in energy of the surroundings} = 0 \quad (2.1)$$

System here refers to the region in which a process of interest is occurring while the surroundings refer to the remainder of the universe; for engineering purposes, however, we may only consider that portion of the world that is influenced by the process. Typically, the boundary that separates the system from the surroundings is selected such that only heat transfer and work interactions occur through it. Equation 2.1 may then be written as:

$$\Delta U + \Delta(\text{KE}) + \Delta(\text{PE}) = Q - W \quad (2.2)$$

where U is the internal energy of the system (sum of the potential and kinetic energy of the molecules[1] making up the system), KE and PE are the macroscopic kinetic and potential energies of the system as a whole (i.e., without the KE and PE of the individual atoms or molecules), Q is the heat transferred into the system from the surroundings, and W is work done by the system on the surroundings. Note that W is a form of energy; in the case of a heat engine, it is the rotational kinetic energy of the shaft that connects the system to the surroundings. If the value of Q is negative, then heat is transferred from the system to the surroundings. Similarly, if the value of W is negative, then work is done by the surroundings on the system. Equation 2.1 may be modified to include nuclear reactions by invoking the conservation of matter principle and written as:

[1] In the absence of external electric and magnetic fields, internal energy is the sum of sensible energy (due to kinetic energy of the individual molecules, that is, translational, rotational, and vibrational), latent energy (due to binding forces between molecules and associated with phase of the system), chemical energy (due to atomic bonds), and nuclear energy (due to bonds within the atomic nuclei) for the general case that includes nuclear reactions.

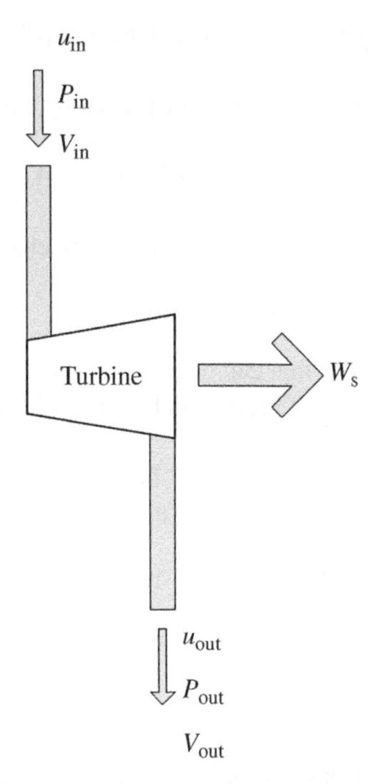

FIGURE 2.1 An open system

$$\Delta(\text{mass and energy of the system}) + \Delta(\text{mass and energy of the surroundings}) = 0 \quad (2.3)$$

As depicted in Figure 2.1, consider an open system with work interaction such as a mole of fluid entering a turbine with (specific) internal energy u_{in}, performing shaft work W_s, and then leaving the system with (specific) internal energy u_{out}, the entire process being at steady state, that is, without accumulation of either matter or energy within the system. As the fluid enters the system (the turbine), the surrounding (the fluid upstream of it) performs work W_{in} on the system, and as the fluid leaves the system, the system (the fluid upstream of it) performs work W_{out} on the surroundings. When these work interactions are taken into account, Equation 2.2 becomes:

$$\Delta u + \Delta(\text{KE}) + \Delta(\text{PE}) = Q - W_s + W_{in} - W_{out} \quad (2.4)$$

Expressing W_{in} and W_{out} in terms of the pressure P and specific volume V (on mole basis throughout this chapter) of the fluid,[2] Equation 2.4 becomes:

[2] Since $W = (\vec{F}) \cdot \Delta\vec{x} = (\vec{F}/A) \cdot (\Delta\vec{x}A) = PV$, where $\vec{F}$ is the force applied by the fluid entering (or to the fluid leaving), A is the cross-sectional area through which the fluid is entering (or leaving the system), and Δx is the distance a unit mole of fluid has moved while entering (or leaving the system).

$$u_{out} + P_{out}V_{out} - u_{in} - P_{in}V_{in} + \Delta(KE) + \Delta(PE) = Q - W_s$$

Defining (specific) enthalpy, $h = u + PV$, the preceding equation may be written as:

$$\Delta h + \Delta(KE) + \Delta(PE) = Q - W_s$$

For a system where the kinetic and potential energy changes may be neglected:

$$\Delta h = Q - W_s \quad (2.5)$$

There are other forms of work that can be produced such as electrochemical work, as in a fuel cell described in Chapter 11, in which case shaft work W_s in Equation 2.5 is replaced by the electrochemical work W_e.

For a closed system undergoing work and heat interactions over a time interval, since W_{in} and W_{out} are both = 0, the first law may be expressed as:

$$\Delta U = Q - W_s$$

2.1.1 Application to a Combustor

The combustion process plays an important role in many energy conversion devices or processes. In the case of a fuel-based heat engine such as a gas turbine, the fuel is first combusted in order to transform its internal energy associated with chemical bonds into thermal energy by exothermic reactions with O_2 followed by conversion to work. The first law may be applied in calculating the combustor effluent temperature, an essential step in the analysis of such energy conversion processes. Equation 2.5 when applied to a steady flow combustor where work is not being extracted during the combustion process (and since both Δ(KE) and Δ(PE) may be neglected, which are reasonable assumptions in most practical applications) reduces to:

$$\Delta H = H_{in} - H_{out} = Q \quad (2.6)$$

where the enthalpies are on a total molar flow basis, that is, $H = nh$ with n being the total number of moles and h the enthalpy per mole, Q is the heat loss from the combustor (which should always be minimized) and is often an empirical value taken as a percentage of the energy content (heating value) of the fuel but may be calculated using heat transfer equations described in Chapter 4 and a knowledge of the combustor geometry.

The enthalpy in our application needs to include the contributions due to sensible, latent, and chemical but not nuclear. ΔH can be calculated using tabulated data (either experimentally determined or estimated by empirical or other means) for heat released or absorbed by reactions at a certain set of "standard" conditions for the reactants and products and then adjusting it for conditions specific to our situation. A hypothetical process path to connect these specific conditions to the conditions corresponding to the tabulated data is defined so that the reaction occurs at those set of conditions at which the data is available. The results thus obtained will be valid for any real process with identical initial conditions and identical final conditions since H is a state function, that is, its value for a given amount of matter depends only on the state in which the matter is in (its temperature, pressure, composition and phase, which define completely its state).

Data has been tabulated for the heat of reaction of various compounds for special types of reactions occurring isothermally at 25°C, where the reactants are elements and the product is a single compound with both reactants and product in their "standard state." The corresponding heat of reaction expressed on a per g-mole of product basis is called the "standard heat of formation." The standard state of any element or compound is defined as a pure component, with gases in ideal gas state at 1 atm pressure, while with liquids and solids, the standard state pressure is usually also 1 atm. The standard heat of formation data can be used for calculating the heat of reaction for the set of reactants of interest, the fuel and oxidant in our case (the corresponding heat of reaction is then called the heat of combustion) and then adjusting this calculated heat of reaction to the actual conditions of interest using the hypothetical path mentioned previously. The methodology described in the following text for calculating the methane combustor exhaust temperature will make this procedure clearer.

For fuels such as coal and oil, the heat of formation data is typically not directly available and in such cases, the heating value of each of the products and each of the reactants (zero of course for combustion products in the case of complete combustion, and for air) may be utilized in calculating the heat of formation of such fuels. Alternately, the heat of reaction involving such fuels may be calculated directly using either the HHV or the LHV of each of the reactants and products in place of heat of formation. Note that either only the HHVs or the LHVs should be used consistently for each component. If HHVs are used, then the latent heat of water vapor in any reactant or product streams should be added to the enthalpy of the reactant or the product streams as the case may be.

2.1.1.1 Methane Combustor Exhaust Temperature As depicted in Figure 2.2, consider the combustion of pure CH_4 entering a combustor at the absolute temperature, T_{Fu}, and pressure, P_{Fu}, with a certain excess air (i.e., supplying more O_2 than its stoichiometric requirement to ensure complete combustion) at the absolute temperature, T_{Ox}, and pressure, P_{Ox}. Air, in addition to containing O_2 and N_2, also contains other components such as Ar and H_2O vapor (depending on its relative humidity) and

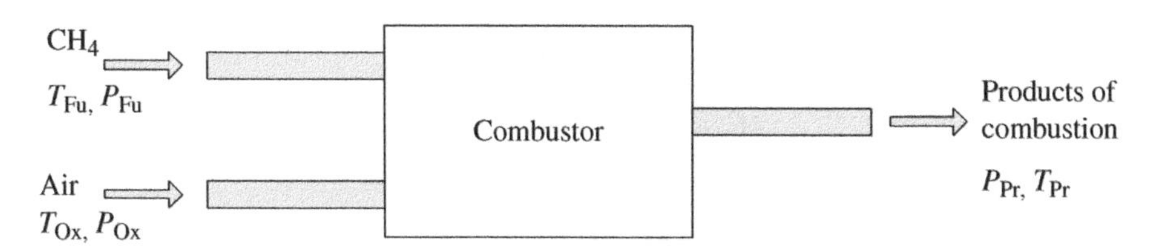

FIGURE 2.2 A simple combustor

are often neglected but will be considered in this exercise to lay the foundations for solving other reaction system problems that may involve many species. The products of combustion leave at pressure, P_{Pr} ($P_{Pr} < P_{Fu}$ and P_{Ox} due to frictional losses) and an unknown temperature, T_{Pr}, are to be calculated.

The first step consists of determining the composition of the products of combustion assuming complete combustion. The overall stoichiometric reaction for complete combustion[3] may be written as:

$$CH_4 + 2O_2 = 2H_2O + CO_2 \tag{2.7}$$

There are five elements in the system, namely, C, O, H, N, and Ar. These five elements will provide five elemental balance equations required to solve for the five unknowns, namely, the concentrations of CO_2, H_2O, O_2, N_2, and Ar in the combustor effluent as follows, with $\dot{M}_{in,i}$ representing the given moles of each of the species i entering the combustor, and $\dot{M}_{out,j}$ representing the moles of each of the species j leaving the combustor.

C balance:

$$\dot{M}_{in,CH_4} = \dot{M}_{out,CO_2}$$

O_2 balance:

$$\dot{M}_{in,O_2} + 0.5\dot{M}_{in,H_2O} = \dot{M}_{out,O_2} + \dot{M}_{out,CO_2} + 0.5\dot{M}_{out,H_2O}$$

H_2 balance:

$$2\dot{M}_{in,CH_4} + \dot{M}_{in,H_2O} = \dot{M}_{out,H_2O}$$

N_2 balance:

$$\dot{M}_{in,N_2} = \dot{M}_{out,N_2}$$

Ar balance:

$$\dot{M}_{in,Ar} = \dot{M}_{out,Ar}$$

Each $\dot{M}_{out,j}$ may now be determined by solving these five elemental balance equations.

If the fractional excess air is λ, then for a given $\dot{M}_{in,CH_4}$ and air composition, the other $\dot{M}_{in,i}$'s may be expressed as follows while utilizing the stoichiometry defined by Equation 2.7:

$$\dot{M}_{in,O_2} = 2(1 + \lambda)\,\dot{M}_{in,CH_4}$$

$$\dot{M}_{in,H_2O} = y_{H_2O}\dot{M}_{in,O_2}/y_{O_2}$$

[3] Formation of minor species such as CO, organic compounds, and NO_x may be neglected from an overall energy balance standpoint (but should not be when considering the environmental impact).

$$\dot{M}_{in,N_2} = y_{N_2}\dot{M}_{in,O_2}/y_{O_2}$$

$$\dot{M}_{in,Ar} = y_{Ar}\dot{M}_{in,O_2}/y_{Ar}$$

where $y_i^{\prime s}$ are the mole fractions of specie i in air.

The next step consists of calculating the combustor effluent temperature. The formation reactions (from the elements in their standard state to form each of the compounds i also in their standard state) and the corresponding standard heat of formations Δh_i^o (the superscript ^{o} is used to denote standard state) for the three compounds involved on a molar basis are:

$$C(s) + 2H_2(g) = CH_4(g); \Delta h^o_{CH_4} \quad (2.8)$$

$$H_2(g) + 0.5O_2(g) = H_2O(g); \Delta h^o_{H_2O} \quad (2.9)$$

$$C(s) + O_2(g) = CO_2(g); \Delta h^o_{CO_2} \quad (2.10)$$

The abbreviations s and g appearing within brackets in the preceding equations denote the phase of each of the components, that is, s for solid and g for gas. Since Equation 2.7 may be obtained by subtracting Equation 2.8 from the sum of Equation 2.9 multiplied by 2 and Equation 2.10, the standard heat of reaction (Δh^o_{reax}) on a molar basis for Equation 2.7 is then:

$$\Delta h^o_{reax} = 2\left(\Delta h^o_{H_2O}\right) + \Delta h^o_{CO_2} - \Delta h^o_{CH_4}$$

$$\Delta H^o_{reax} = \dot{M}_{in,CH_4}\left(\Delta h^o_{reax}\right)$$

Note that a hypothetical path is implied in the preceding calculations of the ΔH^o_{reax}, which consists of first decomposing the reactant CH_4 into its elements, C and H_2, and then oxidizing them into their respective oxides. The resulting ΔH^o_{reax} should also be valid for the path where CH_4 is directly oxidized since enthalpy is a state function.

Next, a path to account for the actual conditions of the reactants and products for the given combustor are defined. This path as shown in Figure 2.3 consists of the following:

1. Cooling (or heating as the case may be) the fuel stream, CH_4 at P_{Fu} and T_{Fu}, and the oxidant (air) stream at P_{Ox} and T_{Ox} at constant pressure to the standard state temperature of 25°C. The corresponding enthalpy changes for these processes = ΔH_{Fu1} and ΔH_{Ox1}.
2. Isothermally separating the required amount of O_2 for combustion from the air stream by passing the air through a hypothetical ideal membrane with its partial pressure driving the diffusion process through the ideal membrane (operating with zero pressure drop for driving force). The separated O_2 is now at a pressure equal to its partial pressure in the air. The corresponding enthalpy change for this process = ΔH_2.

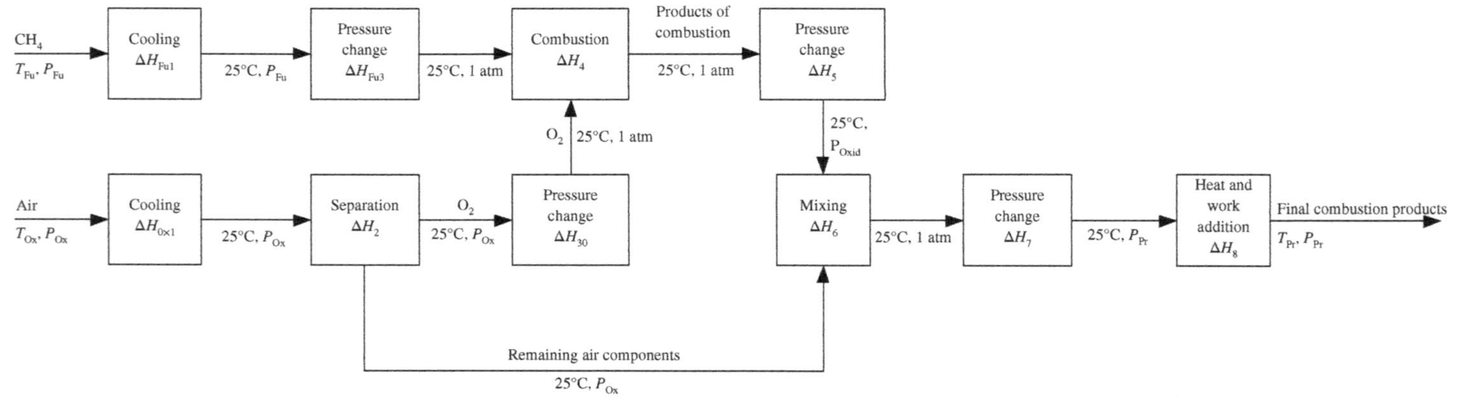

FIGURE 2.3 Hypothetical path for calculating ΔH_{Reax}

3. Isothermally decreasing (or increasing as the case may be) the pressure of the CH_4 and the separated O_2 to 1 atm by passing through ideal turbines (or compressors). The corresponding enthalpy change for these processes = ΔH_{Fu3} and ΔH_{Ox3}.
4. Isothermally reacting the CH_4 with the O_2 at 1 atm with all products including H_2O being in the gas phase. A hypothetical gas phase may be assumed for H_2O if not all of it may actually be in vapor form at 1 atm and 25°C. The corresponding enthalpy change for this process = ΔH_4.
5. Isothermally changing the pressure of the combustion products to the pressure of the remaining components of the air stream leaving the membrane using an ideal turbine (or compressor as the case may be). Again a hypothetical gas phase is assumed for H_2O if not all of it may actually be in vapor form at these conditions. The corresponding enthalpy change for this process = ΔH_5.
6. Isothermally mixing the combustion products with those remaining components of the air stream leaving the membrane. The corresponding enthalpy change for this process = ΔH_6.
7. Isothermally changing the pressure of this mixture to the specified combustor exit pressure of P_{Pr} using an ideal turbine (or compressor as the case may be). The corresponding enthalpy change for this process = ΔH_7.
8. Finally, transferring the heat or work added or removed in these previous seven isothermal processes into the mixture formed in step 7 as heat, resulting in a temperature T_{Pr} for the mixture. The corresponding enthalpy change for this process = ΔH_8.

By application of the first law, the sum of the enthalpy changes for the isothermal process defined in steps 1–7 may be written as:

$$\Delta H_{Fu1} + \Delta H_{Ox1} + \Delta H_2 + \Delta H_{Fu3} + \Delta H_{Ox3} + \Delta H_4 + \Delta H_5 + \Delta H_6 + \Delta H_7 = Q_{Net} - W_{S,Net} \quad (2.11)$$

While per step 8,

$$-Q_{Net} + W_{S,Net} = \Delta H_8 \quad (2.12)$$

Each of these enthalpy changes is given as follows while assuming that all the gas streams follow ideal gas behavior, which is a reasonable approximation with most of the practical combustors:

1. $\Delta H_{1Ox} + \Delta H_{Fu1} = \sum_i \dot{M}_{in,i} \int_{T_{Ox}}^{25+273} C_{pi} M_{wi}\, dT + \dot{M}_{in,CH_4} \int_{T_{Fu}}^{25+273} C_{pCH_4} M_{wCH_4}\, dT$

   where subscript i represents each of the species in the oxidant stream, and C_{pi} and M_{wi} are the mass-specific heat and molecular weight of component i.
2. $\Delta H_2 = 0$ for the separation process in the case of ideal solutions.

3. $\Delta H_3 = 0$ for ideal gases since ΔH is a function of temperature alone since $dH = dU + d(nPV) = M_w\,(C_v\,dT + n\,R\,dT)$, where M_w is the molecular weight, C_v the constant volume mass specific heat[4] (a function of T), n the number of moles, and R the individual ideal gas constant (on a mass basis).
4. $\Delta H_4 = \Delta H^\circ_{reax}$, the standard heat of reaction at 25°C and 1 atm with all reactants and products as vapors in our case.
5. $\Delta H_5 = 0$ for ideal gases since ΔH is a function of temperature alone.
6. $\Delta H_6 = 0$ for the mixing process in the case of ideal solutions.
7. $\Delta H_7 = 0$ for ideal gases since ΔH is a function of temperature alone.
8. $\Delta H_8 = \sum_j \dot{M}_{out,j} \int_{25+273}^{T_{Pr}} C_{pj} M_{wj}\, dT$

where subscript j represents each of the species in the combustor effluent, and C_{pj} and M_{wj} are the mass-specific heat and molecular weight of component j.

Substituting these preceding values or expressions for each of the ΔH_i corresponding to each of the eight steps into Equations 2.11 and 2.12 and combining the two equations, we get

$$\sum_i \dot{M}_{in,i} \int_{T_{Ox}}^{298^\circ} C_{pi} M_{wi}\, dT + \dot{M}_{in,CH_4} \int_{T_{Fu}}^{298^\circ} C_{pCH_4} M_{wCH_4}\, dT + \Delta H^o_{reax\ at\ 298^\circ} = -\sum_j \dot{M}_{out,j} \int_{298^\circ}^{T_{Pr}} C_{pj} M_{wj}\, dT \qquad (2.13)$$

T_{Pr} may now be determined by solving Equation 2.13 numerically, since the specific heats are functions (typically correlated empirically as polynomials) of temperature. Tabulated data are available in JANAF (Chase et al., 1985) for the heats of formation of various species corresponding to their standard state along with the enthalpies at various temperatures in which case Equation 2.14 may be used where the subscript l represents each component in case of a multicomponent fuel mixture:

$$\sum_i \dot{M}_{in,i}\left(\Delta h^o_i + \Delta h_i^{298^\circ \to T_{Ox}}\right) + \sum_l \dot{M}_{in,l}\left(\Delta h^o_l + \Delta h_l^{298^\circ \to T_F}\right) = \sum_j \dot{M}_{out,j}\left(\Delta h^o_j + \Delta h_j^{298^\circ \to T_{Pr}}\right) \qquad (2.14)$$

where Δh^o_i, Δh^o_l, Δh^o_j are the specific standard heats of formation (kJ/kmol) of each of the species in the oxidant, fuel, and product streams (=0 for elements in their

[4] Application of the first law to a constant volume process gives $\Delta U = Q = M_w\,(C_v\,\Delta T + R\,\Delta T)$. ΔU calculated in this manner should be the same for any other process with the same initial and final conditions; like H, U is a state function.

standard state) at the reference temperature of 298 K, and $\Delta h_i^{298°\rightarrow T_{Ox}°}$, $\Delta h_l^{298°\rightarrow T_F}$, $\Delta h_j^{298°\rightarrow T_{Pr}}$ are the specific enthalpies (kJ/kmol) of each of the species at their temperature above the reference temperature.

Recall that Equation 2.9 corresponded to the heat of formation of H_2O as a vapor. If instead the heat of formation of H_2O as a liquid were used in the preceding calculations, then a step for the phase change required to go from a liquid to a vapor has to be included and the corresponding ΔH.

Example 2.1. Assuming ideal gas behavior, calculate the exhaust composition and temperature (assuming complete combustion and ignoring pollutant formation) from a combustor burning a syngas (after cleanup) derived from an air-blown biomass gasifier with the following composition and conditions:

1. Syngas: $H_2 = 18\%$ by mole, $CO = 24\%$ by mole, $CH_4 = 3\%$ by mole, $CO_2 = 7\%$ by mole, $N_2 = 48\%$ by mole, available at 127°C and 50 bar pressure.
2. Air: $O_2 = 21\%$ by mole, $N_2 = 79\%$ by mole (ignoring Ar and H_2O), available at 25°C and 5 bar pressure.
3. Combustor: 10% excess air is required and has a pressure drop = 0.1 bar and heat losses are insignificant.

Solution

JANAF tables will be utilized for the enthalpies and heats of formation and thus Equation 2.14 will be utilized while using 100 moles of the fuel as a basis for the calculations. The calculation of the term $\sum_l \dot{M}_{lf}\left(\Delta h_l^o + \Delta h_l^{298°\rightarrow T_F}\right)$ is summarized in Table 2.1.

The oxidant (air) required (and its components) for 10% excess is summarized in Table 2.2.

The calculation of the term $\sum_i \dot{M}_{in,i}\left(\Delta h_i^o + \Delta h_i^{298°\rightarrow T_{Ox}}\right)$ is summarized in Table 2.3. Note that in this special case, the enthalpy is zero since the oxidant is at the reference temperature while containing only O_2 and N_2, which have zero heats of formation being elements. H_2O vapor that is typically present in air will have a nonzero heat of formation and should be included.

TABLE 2.1 Fuel enthalpy and heat of formation

Combustible component	Moles of fuel specie	Enthalpy at 127°C		Heat of formation		Enthalpy + heat of formation
		kJ/mole	kJ	kJ/mole	kJ	kJ
H_2	18	2.959	53.262	0	0	53.3
CO	24	2.976	71.424	−110.527	−2652.648	−2581.2
CH_4	3	3.861	11.583	−74.873	−224.619	−213.0
CO_2	7	4.003	28.021	−393.522	−2754.654	−2726.6
N_2	48	2.971	142.608	0	0	142.6
Total	100.0		306.9		−5631.9	−5325.0

TABLE 2.2 Oxidant (air) required

			Stoichiometric O_2 required		O_2 required for 10% excess air	N_2 in air
Combustible component	Chemical reaction	Moles of fuel specie	Mole per mole fuel specie	Total moles	Total moles	Mole N_2 = (79/21) × O_2
H_2	$H_2 + 0.5O_2 = H_2O$	18	0.5	9	9.9	37.24
CO	$CO + 0.5O_2 = CO_2$	24	0.5	12	13.2	49.66
CH_4	$CH_4 + 2O_2 = CO_2 + 2H_2O$	3	2	6	6.6	24.83
Total		45.0		27.0	29.7	111.7

TABLE 2.3 Oxidant (air) enthalpy and heat of formation

			Enthalpy 25°C		Heat of formation		Enthalpy + heat of formation
Oxidant component	Moles (calculated in Table 2.2)	Mole %	kJ/mole	kJ	kJ/mole	kJ	kJ
O_2	29.7	21	0	0	0	0	0
N_2	111.7	79	0	0	0	0	0
Total	141.4	100.0		0		0	0.0

TABLE 2.4 Products formed by combustion

			H_2O formed by combustion		CO_2 formed by combustion	
Combustible component	Chemical reaction	Moles of fuel specie	Mole per mole fuel component	Total moles	Mole per mole fuel component	Total moles
H_2	$H_2 + 0.5O_2 = H_2O$	18	1	18	0	0
CO	$CO + 0.5O_2 = CO_2$	24	0	0	1	24
CH_4	$CH_4 + 2O_2 = CO_2 + 2H_2O$	3	2	6	1	3
Total		45.0		24.0		27.0

Calculations for the amounts of products formed by combustion are summarized in Table 2.4 for the three combustibles present in the fuel.

Calculations for the amount of each component present in the flue gas is then determined as summarized in Table 2.5.

TABLE 2.5 Flue gas composition and heats of formation

Flue gas components	In fuel	From air, remaining after combustion	Formed by combustion	Total in flue gas		Heat of formation at 298 K	
	Mole	Mole	Mole	Mole	Mole %	kJ/mole	kJ
H_2O	0	0	24.0	24	10.89	−241.8	−5,803.8
CO_2	7	0	27.0	34	15.42	−393.5	−13,379.8
O_2	0	2.7	0	2.7	1.22	0	0
N_2	48	111.7	0	159.7	72.46	0	0
Total				220.4	100.00		−19,183.6

TABLE 2.6 Flue gas enthalpy and heat of formation at two trial temperatures

Flue gas components	Enthalpy at 1900 K		Enthalpy at 2000 K		Enthalpy + heat of formation at 1900 K	Enthalpy + heat of formation at 2000 K
	kJ/mole	kJ	kJ/mole	kJ	kJ	kJ
H_2O	67.7	1,624.9	72.8	1,747.0	−4,178.9	−4,056.9
CO_2	85.4	2,904.2	91.4	3,108.9	−10,475.5	−10,270.8
O_2	55.4	149.6	59.2	159.8	149.6	159.8
N_2	52.5	8,393.4	56.1	8,966.7	8,393.4	8,966.7
Total		13,072.2		13,982.3	−6,111.3	−5,201.2

The calculation of the term $\sum_j \dot{M}_{out,j}\left(\Delta h_j^o + \Delta h_j^{298^\circ \to T_{Pr}}\right)$ is summarized in Table 2.6 for two separate trial temperatures. By interpolation, the temperature of the flue gas is calculated to be 1986 K to give the same enthalpy + heat of formation as that of the fuel + oxidant = −5325.0 kJ.

2.1.2 Efficiency Based on First Law

The first law implies that the energy contained in the useful product(s) leaving an energy conversion system at steady state can be equal to or less than the energy supplied to the system, the remainder being rejected or wasted to the surroundings. The efficiency of a process or a plant based on this first law constraint also known as the "overall process or plant efficiency" measures how efficient a process is in terms of transforming the energy supplied to the system into the energy contained in the useful product(s) produced. This overall efficiency is defined as the ratio of energy contained in the product(s) to the energy supplied to the system, and provides a yard stick for comparing various alternate routes for converting the same type of energy supplied to the energy conversion system into the desired product(s). This type of efficiency, however, has its shortcomings as explained later in this chapter when investigating various alternate routes for energy conversion. As explained in Chapter 1, the overall efficiency for a fuel-based power plant is expressed by the ratio of the net power produced by the plant and the heating value of the fuel, either its HHV or its LHV as a measure of the energy supplied by the fuel. The first law efficiency is especially useful in determining the fuel cost (typically sold on a unit energy basis, either in terms of HHV or LHV) to produce a given amount of product, electrical energy in the case of a power plant.

2.2 SECOND LAW

The upper bound stipulated by the first law for the overall efficiency of a process is 100% while the efficiencies of real processes (e.g., fuel fired power plants) are far less than 100%. For example, a modern coal-fired power plant has an overall efficiency of about 40% when expressed on an HHV basis. In today's coal-fired power plants, the energy contained in the fuel is first released as thermal energy, which is then converted into work via a heat engine, the shaft of the engine being connected to an electric generator that converts this work into electrical energy. A significant amount of the coal-bound energy is rejected to the atmosphere (surroundings) and a significant fraction of this inefficiency is due primarily to the limit set by nature on how much of the thermal energy may be converted to work. This limit is a function of the temperature at which the thermal energy is utilized in the heat engine. It is found that higher this temperature, higher is the conversion efficiency. Another temperature that affects this efficiency is the temperature at which the unconverted thermal energy is rejected to the surroundings; lower this temperature, higher is the efficiency. These observations lead to a concept for an attribute for energy, quality in addition to quantity. Just as nature does not allow the creation of (quantity of) energy such that the quantity of energy leaving a system at steady-state cannot be greater than that entering the system, nature does not allow an increase in the total quality of energy leaving the system either. The total quality of energy, including energy in the form of work, leaving a system at steady-state cannot be greater than that entering the system, the degree of quality destruction depending upon how irreversible the process is. Defining θ_{in} and θ_{out} as the qualities of thermal energy entering (Q_{in}) and leaving (Q_{out}) the system on a unit energy basis (i.e., quality being defined as an intensive property), and θ_W as the quality of the net work (W_{net}) leaving the system, the second law may be written as:

$$Q_{in1}\theta_{in1} + Q_{in2}\theta_{in2} + \cdots \geq W_{net}\theta_W + (-Q_{out1})\theta_{out1} + (-Q_{out2})\theta_{out2} + \cdots \quad (2.15)$$

Note the quantities Q_{out1}, Q_{out2}, ... are negative quantities per our convention. The equality in the preceding expression applies to only a hypothetical reversible system. This quality of thermal energy (θ) may be related to the ideal gas temperature by considering a refrigeration cycle that uses heat to drive the process[5] utilizing an ideal gas

[5] In such a heat-driven refrigeration process where heat has to be transferred from a lower temperature to a higher temperature, additional heat at a higher quality is required so that there is no spontaneous increase in the overall quality of energy.

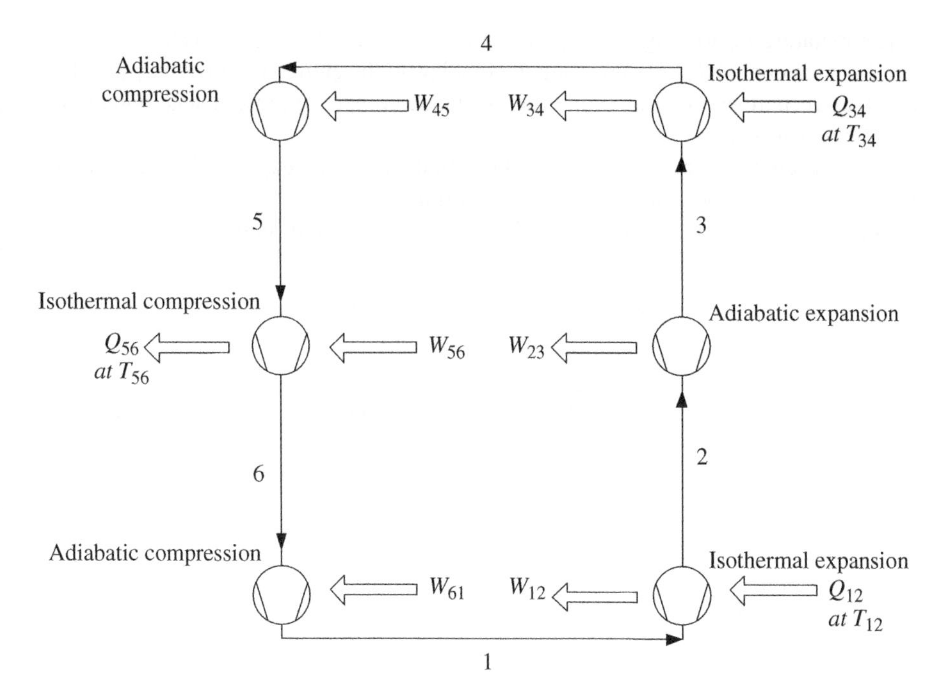

FIGURE 2.4 A heat-driven refrigeration cycle using an ideal gas

as a working fluid operating with the following steps, each step under reversible conditions (see Fig. 2.4):

1. Isothermal expansion of the gas at temperature T_{12} from state point 1 at pressure P_1 to state point 2 at pressure P_2 while absorbing heat Q_{12} and producing work W_{12}.
2. Adiabatic expansion from state point 2 at pressure P_2 to state point 3 at pressure P_3 while producing work W_{23} and in the process cooling down to temperature T_{34}.
3. Isothermal expansion of the gas at temperature T_{34} from state point 3 at pressure P_3 to state point 4 at pressure P_4 while absorbing heat Q_{34} and producing work W_{34}.
4. Adiabatic compression from state point 4 at pressure P_4 to state point 5 at pressure P_5 while consuming work W_{45} and in the process heating up to temperature T_{56}.
5. Isothermal compression at temperature T_{56} from state point 5 to state point 6 while releasing heat Q_{56} and consuming work W_{56}.
6. Adiabatic compression from state point 6 at pressure P_6 to state point 1 at pressure P_1 to complete the cycle while consuming work W_{61}.

The state point conditions are chosen such that there are no *other* work or heat interactions with the surroundings, that is, $W_{12}+W_{23}+W_{34}+W_{45}+W_{56}+W_{61}+Q_{12}+Q_{34}+Q_{56}=0$. Note that this refrigeration cycle transfers the heat stream Q_{34}

at a temperature T_{34} to a higher temperature and T_{56} by utilizing the heat stream Q_{12} at a higher temperature T_{12}. By applying the first law to the entire system, it may then be seen that the temperature of the gas after adiabatic compression from state point 6 to state point 1 is T_{12}.

The preceding work and heat inputs/outputs may be expressed in terms of pressure and/or temperature of the working fluid.

For an ideal gas (on a mole basis), $PV = R_u T$ and with constant T:

$$W_{12} = \int_{v_1}^{v_2} PdV = R_u T_{12} \ln(P_1/P_2) \tag{2.16}$$

$$W_{34} = \int_{v_3}^{v_4} PdV = R_u T_{34} \ln(P_3/P_4) \tag{2.17}$$

$$W_{56} = \int_{v_5}^{v_6} PdV = R_u \ln(P_5/P_6) \tag{2.18}$$

From the first law and with constant T:

$$Q_{12} = W_{12} = R_u T_{12} \ln(P_1/P_2) \tag{2.19}$$

$$Q_{34} = W_{34} = R_u T_{34} \ln(P_3/P_4) \tag{2.20}$$

$$Q_{56} = W_{56} = R_u T_{56} \ln(P_5/P_6) \tag{2.21}$$

Since $(T_{12}T_{34}T_{56})/(T_{12}T_{34}T_{56}) = 1$,

$$\ln[(T_{12}T_{34}T_{56})/(T_{12}T_{34}T_{56})] = 0, \text{ or}$$

$$\ln(T_{12}/T_{56}) + \ln(T_{34}/T_{12}) + \ln(T_{56}/T_{34}) = 0 \tag{2.22}$$

For an ideal gas with constant C_P/C_V undergoing reversible adiabatic expansion or compression (between state points 6 and 1, between state points 2 and 3, and between state points 4 and 5), Equation 2.22 may be written as:

$$\ln(P_1/P_6) + \ln(P_3/P_2) + \ln(P_5/P_4) = 0$$

Rearranging,

$$\ln(P_1/P_2) + \ln(P_3/P_4) + \ln(P_5/P_6) = 0$$

Substituting Equations 2.19–2.21 into the preceding:

$$\frac{Q_{12}}{T_{12}} + \frac{Q_{34}}{T_{34}} + \frac{Q_{56}}{T_{56}} = 0, \text{ or}$$
$$\frac{Q_{12}}{T_{12}} + \frac{Q_{34}}{T_{34}} = \frac{-Q_{56}}{T_{56}} \tag{2.23}$$

When irreversible processes are also considered, Equation 2.23 becomes:

$$\frac{Q_{12}}{T_{12}}+\frac{Q_{34}}{T_{34}}\leq\frac{-Q_{56}}{T_{56}} \tag{2.24}$$

Comparing expression (2.15) with expression (2.24), it may be seen that $\theta=-1/T$, that is, higher the temperature, higher is the quality of heat.

In terms of entropy that is defined by $\Delta S=\int\frac{dQ_{\text{reversible}}}{T}$ where $Q_{\text{reversible}}$ is the heat transferred reversibly, the expression (2.24) may be written for the reversible case as $\Delta S_{12}+\Delta S_{34}+\Delta S_{56}=0$, that is, entropy is conserved for a reversible process. Like ΔH, ΔS is a state function and may be calculated assuming a convenient reversible path connecting the initial state to the final state.

Since θ cannot be >0, the highest value θ can have is 0, which is the quality corresponding to work (θ_W). This may be verified by performing a similar analysis for a reversible heat engine employing an ideal gas for the working fluid as was done for the refrigeration cycle with the following steps (resulting in the Carnot cycle[6] and depicted in Fig. 2.5):

1. Isothermal expansion of the gas at temperature T_{12} from state point 1 at pressure P_1 to state point 2 at pressure P_2 while absorbing heat Q_{12} and producing work W_{12}.
2. Adiabatic expansion from state point 2 at pressure P_2 to state point 3 at pressure P_3 while producing work W_{23} and cooling down to temperature T_{34}.

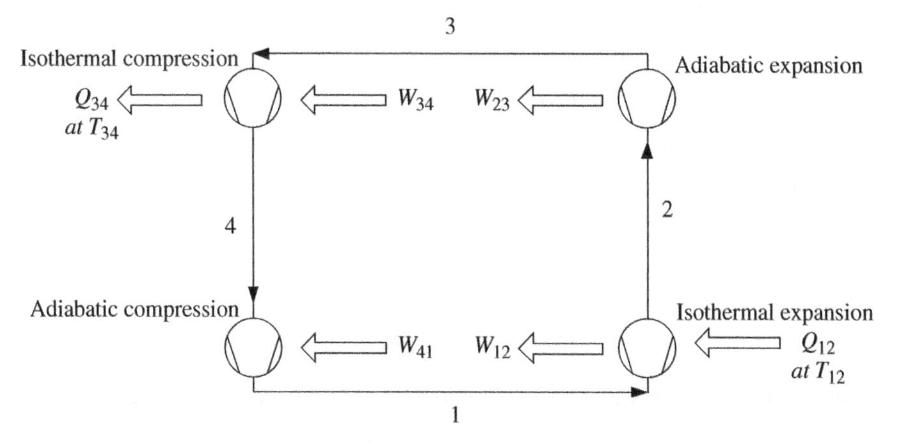

FIGURE 2.5 A cyclical reversible heat engine with an ideal gas

[6] This idealized cycle was proposed by Nicolas Léonard Sadi Carnot in 1824.

3. Isothermal compression at temperature T_{34} from state point 3 to state point 4 while releasing heat Q_{34} and consuming work W_{34}.
4. Adiabatic compression from state point 4 at pressure P_4 to state point 1 at pressure P_1 to complete the cycle while consuming work W_{41}.

The state point conditions are chosen such that there are no *other* work or heat interactions with the surroundings, that is, $W_{12} + W_{23} + W_{34} + W_{41} + Q_{12} + Q_{34} = 0$. By applying the first law to the entire system, it may then be seen that the temperature of the gas after adiabatic compression from state point 4 to state point 1 is T_{12}.

The preceding work terms and heat input may be expressed in terms of pressure and/or temperature of the working fluid. For an ideal gas (on a mole basis), $PV = R_u T$ and with constant T:

$$W_{12} = \int_{v_1}^{v_2} PdV = R_u T_{12} \ \ln(P_1/P_2)$$

$$W_{34} = \int_{v_3}^{v_4} PdV = R_u T_{34} \ \ln(P_3/P_4)$$

From the first law and with constant T:

$$Q_{12} = W_{12} = R_u T_{12} \ \ln(P_1/P_2) \tag{2.25}$$

$$Q_{34} = W_{34} = R_u T_{34} \ \ln(P_3/P_4) \tag{2.26}$$

The net work produced by the heat engine is given by $W_{net} = W_{12} + W_{23} + W_{34} + W_{41}$. Next, since $(T_{12}T_{34})/(T_{12}T_{34}) = 1$,

$$\begin{aligned} &\ln[(T_{12}T_{34})/(T_{12}T_{34})] = 0 \text{ or} \\ &\ln(T_{12}/T_{34}) + \ln(T_{34}/T_{12}) = 0 \end{aligned} \tag{2.27}$$

For an ideal gas with constant C_P/C_V undergoing reversible adiabatic expansion or compression (between state points 2 and 3, and between state points 4 and 1), Equation 2.27 may be written as:

$$\ln(P_3/P_2) + \ln(P_1/P_4) = 0$$

Rearranging,

$$\ln(P_1/P_2) + \ln(P_3/P_4) = 0$$

Substituting Equations 2.25 and 2.26 into the preceding equation, we get:

$$\begin{aligned} &Q_{12}/T_{12} + Q_{34}/T_{34} = 0 \text{ or} \\ &Q_{12}/T_{12} = -Q_{34}/T_{34} \end{aligned} \tag{2.28}$$

When irreversible processes are also considered, Equation 2.28 becomes:

$$Q_{12}/T_{12} \leq -Q_{34}/T_{34} \tag{2.29}$$

Note that Q_{34} is a negative quantity per our convention. Comparing this expression with expression (2.15), it may be seen that θ for work has the highest value, that is, $\theta_W = 0$, and in terms of entropy, for a reversible heat engine entropy is conserved. The analysis of real (irreversible) processes is conducted by assuming ideality, applying the entropy conservation principle, and then applying a correction (e.g., an empirically correlated isentropic efficiency in the case of compressors, turbines, and other heat engines), the actual work and outlet conditions from the process may be determined for a given set of inlet conditions.

The efficiency of a reversible heat engine (known as the Carnot efficiency), η_{Carnot}, may be calculated by applying the first law to express the net work produced in terms of the heat supplied and rejected by the engine as $W_{net} = Q_{12} + Q_{34}$ and Equation 2.28:

$$\eta_{Carnot} = W_{net}/Q_{12} = \frac{(Q_{12} + Q_{34})}{Q_{12}} = 1 - T_{34}/T_{12} \tag{2.30}$$

From the preceding examples, it may be seen that:

1. Heat may be transferred from a lower temperature to a higher temperature but then heat has to be also supplied at a higher temperature[7] so that the overall quality of energy is not upgraded, the amount of quality destruction depending on how irreversible the process is.
2. Work can be produced from heat by converting a portion of the heat while the remaining heat has to be rejected at a downgraded quality such that the overall quality of energy is not upgraded, again the amount of quality destruction depending on how irreversible the heat engine is.
3. The efficiency of a reversible heat engine as seen from Equation 2.30 is independent of the working fluid and only a function of the temperature at which heat is supplied to the engine and the temperature at which the heat is rejected by the engine. This efficiency, the Carnot cycle efficiency, represents the upper limit for the efficiency of a heat engine as long as it operates at the maximum temperature at which heat can be absorbed (i.e., same temperature as the heat source) and the minimum temperature at which heat may be rejected (i.e., same temperature as the heat sink).

2.2.1 Quality Destruction and Entropy Generation

Lost potential to do work by an irreversible process results in the destruction of quality of energy which can be shown to be equivalent to entropy generation. Consider the "free expansion" of an ideal gas from a pressure P_1 to a pressure P_2. This expansion process as depicted in Figure 2.6 may be modeled as the gas first performing isothermal work in a reversible turbine while absorbing heat $Q_{reversible}$ at the constant temperature T and producing work W_S. The work W_S is then degraded into heat by

[7] Or work (as in the case of mechanical refrigeration), which is of course higher quality energy.

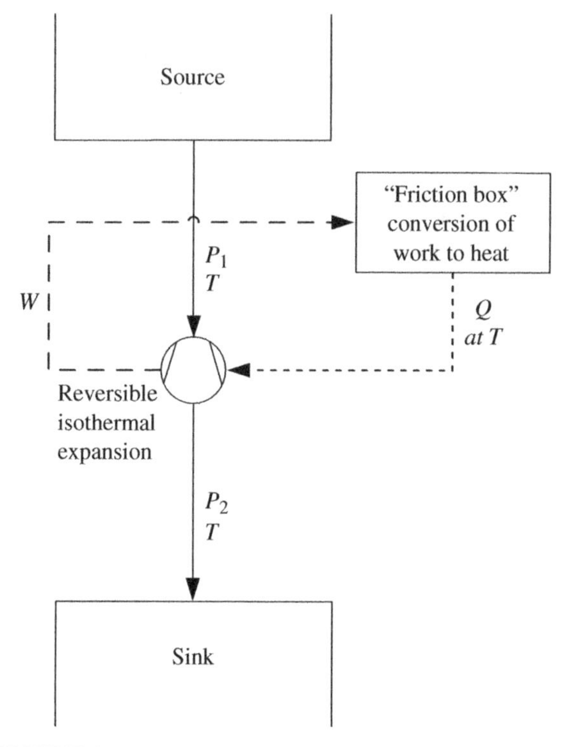

FIGURE 2.6 Entropy generation in free expansion of a gas

friction and the corresponding heat has to be equal to $Q_{\text{reversible}}$. This heat is then supplied back to the reversible turbine. The net effect is that no work is produced by the process, that is, free expansion. The corresponding quality destruction by friction (in the "friction box" shown in the figure), while noting that $\theta_W = 0$, $\theta_T = -1/T$, and $\Delta S = Q_{\text{reversible}}/T$, is

$$W_S\,\theta_W - Q_{\text{reversible}}\theta_T = -Q_{\text{reversible}}\theta_T = \frac{Q_{\text{reversible}}}{T} = \Delta S$$

Thus, entropy generation is a measure of the destruction of the quality of energy. Next, for this example $Q_{\text{reversible}} = W_S = R_u T\ \ln(P_1/P_2)$, and the entropy generation for this irreversible process $\Delta S = R_u\ \ln(P_1/P_2)$. This is an example of how ΔS may be calculated, that is, by devising a reversible step for the heat transfer, even when there are no heat interactions with the surroundings. The lost work is given by $T\Delta S$.

Another example is the irreversible heat transfer from a higher temperature T_1 to a lower temperature T_2, and the loss of quality of energy may again be shown to be the entropy generated, $\Delta S = Q(\theta_1 - \theta_2) = Q\left(\frac{1}{T_2} - \frac{1}{T_1}\right)$.

2.2.2 Second Law Analysis

The second law of thermodynamics may be applied to assess how useful an energy source or a material stream can be to produce work. It can also be applied in the analysis of systems to identify areas of inefficiency quantitatively and thus, areas where system modifications will make significant improvement in the overall system efficiency. The analysis consists of calculating the maximum work potential of each of the streams (entering, leaving, and within the system) when it is brought in thermodynamic equilibrium with the environment. In this manner, the amount of work potential destroyed across each major equipment within the system is also determined.

Suppose we want to produce work from the thermal energy Q provided by a heat source at a constant temperature T_1 and a heat sink is available at a constant temperature T_S. A Carnot heat engine is available for this task with a maximum temperature for its working fluid T_2 ($T_2 < T_1$) due to limits placed by the materials used in the heat engine. The efficiency of converting heat Q to work W produced by this engine is then

$$W = Q\left(1 - \frac{T_S}{T_2}\right)$$

Lower this temperature T_2, lower is the amount of heat converted to work and is due to the loss in quality of the thermal energy associated with the reduction in temperature from T_1 to T_2. By hypothesizing an additional heat engine that can accept the heat at T_1 while rejecting the unconverted heat at T_2 to the previous heat engine, additional work may be theoretically produced and the work lost by utilizing the heat at the lower temperature T_2, which is the area of inefficiency due to material limitations in the heat engine, is given by

$$W_{\text{lost}} = Q\,T_S\left(\frac{1}{T_2} - \frac{1}{T_1}\right)$$

A heat source or a heat sink in real world situations exist as material streams except heat sources in the case of solar or nuclear energy. The heat source stream has to contain internal energy at a quality higher than that of the heat sink stream to extract work from the heat source. Unless the material stream is a single component saturated vapor or saturated liquid, the internal energy of the material stream measured by its temperature will decrease or increase depending on whether heat is being removed or added (i.e., whether it is a source or a sink). Thus, heat sources or sinks seldom remain at constant temperature, and this aspect should be accounted for in the thermodynamic analysis of processes. Current technology allows us to produce work directly from the following types of internal energy: sensible and latent, as well as chemical using a fuel cell as discussed in Chapter 11. When a source stream is at a pressure greater than the sink stream, work can also be produced by expansion of the source stream itself (to the sink stream pressure) while its internal energy is converted to work with a corresponding decrease in its temperature depending upon whether or how much thermal energy is supplied to the stream during expansion. In addition to utilizing the total pressure gradient between the source and the sink streams, partial pressure of components may also be utilized for producing work, and again it is the internal energy that is converted to work.

A property that measures the work potential of a stream is called exergy but unlike the conventional definition of exergy, the maximum work potential in this book is being defined to also include the work that may be produced from a stream when its temperature is lower than the environment (ambient) temperature. In such cases, a heat engine is hypothesized using the environment as its heat source and the stream as its heat sink to quantify the work potential. In this manner, the refrigeration potential of streams such as liquefied natural gas (when supplied to a power plant as fuel) may be taken into account. A possible use of the refrigeration potential of liquefied natural gas may be to cool the inlet air of a gas turbine in order to reduce the compressor work and thus increase the thermal efficiency (and the net power output) of the engine.

Exergy (χ) of a steady-flow material stream with a given composition at temperature T and pressure P (when the kinetic and potential energy effects may be neglected) relative to a dead state at temperature T_o and pressure P_o (the environment) may be defined as:

$$= W_{heat} + W_{expan} + W_{chem} + W_{conc} \quad (2.31)$$

where W_{heat} is the reversible work that may be obtained by a heat engine operating between the temperature of the stream (at temperature T) and that of the environment (at temperature T_o), W_{expan} is the reversible work that may be obtained from the stream at temperature T_o and pressure P by isothermal expansion (after equilibrating to the temperature T_o in the previous step) to the pressure of the environment P_o, W_{chem} is the work (after equilibrating to the temperature T_o and expansion to pressure P_o in the previous step) that may be produced by reversible isothermal oxidation of the combustibles that may be present in the stream with enough ambient air to fully oxidize the combustibles, and W_{conc} is the additional work that may be obtained by reversibly diffusing each component i within the stream at partial pressure P_i through a membrane to the environment where it may be at a partial pressure of P_{oi}.

The first term on the left hand side of Equation 2.31 defines the maximum work that may be obtained based on Equation 2.30, the Carnot cycle efficiency, and should include the work potential of streams that contain vapors that could undergo a phase change when equilibrated with the environment. This term (W_{heat}) may be written as:

$$W_{heat} = -\int_{T}^{T_o} \left(1 - \frac{T_o}{T}\right) dH(T)$$

where the enthalpy as a function of temperature $H(T)$ may include both sensible as well as latent energy. When $T < T_o$, it may be shown that the preceding expression for W_{heat} still applies where dH is now the heat rejected by the reversible heat engine to the stream and accounts for the refrigeration potential of a stream while absorbing heat from the environment.

The stream after equilibrating to T_o may undergo further phase change during isothermal pressure reduction and so the second term on the left hand side of Equation 2.31, using the first law, may be written as:

$$W_{expan} = -\Delta H_{expan} + Q_{reversible}$$

Since $\Delta S = Q_{reversible}/T_o$, we have:

$$W_{expan} = -\Delta H_{expan} + T_o\,\Delta S_{expan}$$

where ΔH_{expan} and ΔS_{expan} are the corresponding enthalpy and entropy changes for this isothermal expansion of the stream as its pressure is reversibly reduced to that of the environment. ΔH_{expan} and ΔS_{expan} being state functions may be determined by:

$$H_{expan} = H_{Po} - H_P$$

$$S_{expan} = S_{Po} - S_P$$

where H_{Po} and S_{Po} are the enthalpy and entropy after expansion to pressure P_o, and H_P and S_P are the enthalpy and entropy before expansion, that is, at the initial pressure P, all at temperature T_o. Note that for an ideal gas without phase change, on a molar flow basis,

$$W_{expan} = -R_u\,T_o\ \ln\left(\frac{P_o}{P}\right)$$

The next step in the path for the stream to equilibrate with the environment is the reversible isothermal oxidation reaction for complete combustion at T_o and P_o of the combustibles present in the stream utilizing stoichiometric amount of the ambient air. The corresponding work that may be produced by this reaction is given by the following expression (may be related to H and S in a similar manner as W_{expan} or see Chapter 11 under the section on fuel cells):

$$W_{chem} = -\Delta H_{chem} + T_o \Delta S_{chem}$$

Defining the combination of properties $G = H - T\,S$ as the Gibbs free energy G (a state function),

$$W_{chem} = -\Delta G_{chem}$$

where ΔG_{chem} is the Gibbs free energy change for the isothermal oxidation reaction of the combustibles (say by reaction, $aA + bB \rightarrow cC + dD$) at T_o and P_o. ΔG_{chem} may be determined from the relationship on a molar flow basis:

$$\Delta G_{chem} = \Delta G^o_{chem} + R_u\,T_o\ \ln\left(\frac{f_C^c f_D^d}{f_A^a f_B^b}\right)$$

where f_i = fugacity of specie i, with the standard state fugacity for each of the species being taken as unity. Tabulated data are available in JANAF (Chase et al., 1985) for the Gibbs free energy of formation of various species corresponding to their standard state at various temperatures that can be used in calculating ΔG^{o}_{chem} in a similar manner as ΔH^{o}_{reax} described previously in this chapter. The second term on the right corrects for the actual state of each of the components, which is different from the standard state where the fugacity = 1. For an ideal gas, where P_i is the partial pressure of component i,

$$\Delta G_{chem} = \Delta G^{o}_{chem} + R_u T_o \ \ln\left(\frac{P^c_C P^d_D}{P^a_A P^b_B}\right)$$

The final step in the path for the stream to equilibrate with the environment is the reversible isothermal expansion at temperature, T_o, and total pressure, P_o, of component i through the hypothetical reversible membrane to its partial pressure in the ambient air, P_{io}, that is, as depicted in Figure 2.7 by allowing the component to reversibly exchange through a selective membrane between the stream and a chamber, and similarly between the ambient air and a second chamber, the two chambers being connected by a turbo-expander operating reversibly between the two pressures:

$$W_{conc} = -\Delta G_{conc}$$

Note that for an ideal gas containing i components without phase change on a molar flow basis,

$$W_{conc} = -\sum_i R_u T_o \ \ln\left(\frac{P_{io}}{P_i}\right)$$

The major components considered for this type of expansion are the two components, H_2O vapor and CO_2. The O_2 and N_2 could also be expanded but in the reverse direction while producing work, since a concentration gradient may exist for these components between the system and the environment (the concentration in the environment being typically higher). Such considerations would, however, lead to misleading results or provide impractical guidance for cycle improvements.

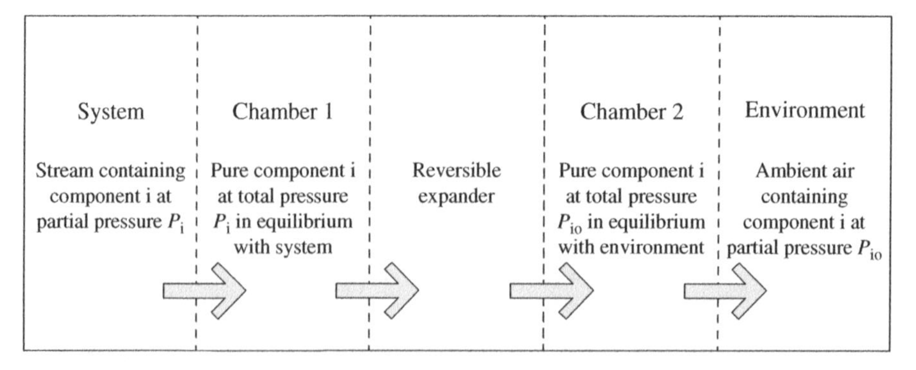

FIGURE 2.7 Reversible expansion of component after diffusion through membrane

The destruction of exergy or the maximum work potential across each equipment within the system and the maximum work potential carried away from the system by streams leaving it identify the sources of inefficiency and provide guidance for making improvements to arrive at an improved system configuration. With each of the four components of maximum work potential known separately, identification of the specific type of change or modification required of the system configuration may be assisted. In this manner, new configurations may be identified aimed at maximizing the overall system efficiency. Simplicity of the configuration is an important factor to maintain in developing these improved designs, however, in order to not compromise plant capital cost and process controllability.

2.2.3 First and Second Law Efficiencies

The efficiency referred to so far is actually the first law efficiency since it is defined on the basis of the first law of thermodynamics being simply a ratio of the rate of useful energy obtained (in the form of electric power for a power plant) from a system to the rate of energy supplied to the system. Shortcomings of this type of efficiency are that it does not account for the quality of the energy input to the system, does not provide a true yard stick for comparing different types (in terms of quality) of energy input, nor provides a realistic upper limit for the efficiency. The second law of efficiency, which is the ratio of the rate of useful energy obtained (in the form of electric power for a power plant) to that if the system were operated reversibly, is more appropriate especially when investigating alternate routes for energy conversion. It may be defined as

$$\eta_{\text{II}} = \frac{\dot{W}_{\text{actual}}}{\dot{W}_{\text{heat}} + \dot{W}_{\text{expan}} + \dot{W}_{\text{chem}} + \dot{W}_{\text{conc}}} = \frac{\dot{W}_{\text{actual}}}{\dot{\chi}} \tag{2.32}$$

Example 2.2. Calculate the first and the second law efficiencies for a process that generates 100 MW of electric power by converting 300 MJ/s of thermal energy available at a constant temperature of 500°C while the heat sink temperature remains constant at 15°C (i.e., assuming both the heat source and sink are infinite reservoirs supplying or absorbing heat at constant temperature).

Solution

Heat source $\dot{Q} = 300$ MJ/s

Source temperature $T_{\text{souce}} = 500°\text{C} = 773.15$ K

Sink temperature $T_{\text{sink}} = 15°\text{C} = 288.15$ K

Power $\dot{W}_{\text{actual}} = 100$ MW

The maximum amount of work that may be produced in this case is given by an ideal heat engine operating between the heat source and sink temperatures given by the Carnot cycle efficiency, that is, referring to Equation 2.32, $\dot{W}_{\text{expan}} = \dot{W}_{\text{chem}} = \dot{W}_{\text{conc}} = 0$ for this case: $\eta_{\text{carnot}} = 1 - T_{\text{sink}}/T_{\text{souce}} = 0.6273$.

First law of efficiency: $\eta_I = \dot{W}_{\text{actual}} \dot{Q} = \frac{100}{300} = 33.33\%$

Second law of efficiency: $\eta_{\text{II}} = \frac{\dot{W}_{\text{actual}}}{\dot{Q}\eta_{\text{carnot}}} = \eta_{1\text{st law}}/\eta_{\text{carnot}} = 33.33\%/0.6273 = 53.14\%$

2.3 COMBUSTION AND GIBBS FREE ENERGY MINIMIZATION

Thermodynamics may also be applied in estimating the composition of products of reactions if it can be assumed that they are in thermodynamic equilibrium. Important reactions in the energy field are the combustion reactions. In most current applications, the energy contained as the chemical bond energy in fossil fuels and biomass is harnessed by combustion. These combustion-based sources provide as much as 90% of the current energy supply in the United States. The thermal energy released by the combustion process, in addition to providing domestic energy needs such as heat for cooking and making hot water, is utilized in industrial energy and manufacturing applications such as in boilers and furnaces. It may also be transformed into shaft power using a heat engine, for example, by generating steam for turning a turbine, producing mechanical motion as in an auto engine, thrust as in an aircraft jet engine, or shaft power as in a land-based gas turbine.

The chemical energy contained in fossil and biomass fuels that is released by the combustion process consists of exothermic reactions with O_2 contained in the air. The combustion process involves some 1000 reactions to complete the oxidation process forming CO_2 and H_2O, the ultimate products of combustion. However, pollutants such as CO, hydrocarbons, oxygenated organic compounds, soot, NO_x, and SO_x are also formed during the combustion process as a result of the various reactions. The quantity of these pollutants is small and negligible from most efficiency calculations standpoint but very significant from the environmental standpoint. Models based on reaction kinetics have been or are being developed to predict the concentration of some of these pollutants. The formation mechanism involves a number of elementary reactions and many intermediate species. In cases where assuming the various reactions reach chemical equilibrium may be justified, a good approximation when reaction rates are fast and there is sufficient residence time, the second law may be applied in predicting the concentration of the various species including the pollutant concentrations. The first step is to identify the various chemical species (compounds as well as free radicals) that may be present in the combustor effluent. The next steps include performing elemental and energy balances (application of the first law) as was done in Section 2.1 along with the application of the second law as explained in the following since the products of combustion are not simply CO_2 and H_2O.

Consider the following reversible reaction occurring in the combustor as representative of all species within the combustor:

$$aA + bB \leftrightarrow cC + dD \tag{2.33}$$

where the uppercase letters refer to the chemical specie and the lowercase letters refer to the number of moles of the specie taking part in the reaction. Chemical equilibrium in the combustor exhaust is reached when both the forward and the reverse reaction rates are the same. The reaction can be moved to the left or to the right by introducing a small amount of species C and/or D, or A and/or B, that is, reversible conditions exist in the combustor exhaust. A criterion for equilibrium can be derived as follows for this reversible process occurring at constant pressure and temperature as follows.

Applying the first law:

$$dH = \delta G_{\text{reversible}} - \delta W_S$$

Applying the second law, that is, replacing $\delta Q_{\text{reversible}}$ by $T\,dS$ in the preceding equation, we get:

$$dH = T\,dS - \delta W_S \tag{2.34}$$

Writing the expression for G in differential form for constant T,

$$dG = dH - T\,dS \tag{2.35}$$

Combining Equations 2.34 and 2.35, $dG = -W_S$. With no work interactions for the reaction at constant temperature T and pressure P,

$$dG = 0 \tag{2.36}$$

Thus, the combustor effluent composition that minimizes the Gibb's free energy will correspond to equilibrium conditions in the effluent for the particular temperature, pressure, and assumed species.

Gibb's free energy may be thought of as a potential that drives chemical reactions just as temperature drives heat transfer. When this potential is zero, the reaction ceases and reaches equilibrium.

Alternately, equilibrium constants may also be utilized in determining the equilibrium composition (or composition with a specified approach to equilibrium) as described in Chapter 5 where the system can be specified with a limited number of thermodynamically independent reactions. This approach is typically used in the modeling of many of the reactors encountered in coproduction facilities described in Chapter 10. In such cases, this method of using equilibrium constants may be simpler. The equilibrium constants are actually functions of temperature (only in the case of ideal gases) but can be related to it by equations that are not too cumbersome. The concept of equilibrium constant is discussed next.

According to Equation 2.36 while representing the molar Gibbs free energy of component i by g_i:

$$dG = cg_C + dg_D - ag_A - bg_B = 0 \tag{2.37}$$

The Gibbs free energy of formation data is available at different temperatures but at a fixed pressure (Chase et al., 1985). To calculate the Gibbs free energy for each of the components for a different pressure, an adjustment for pressure is required. For an ideal gas, the variation of Gibbs free energy with pressure at constant temperature may be written as:

$$\Delta G = -\int_{v_1}^{v_2} P\,dV = \int_{P_1}^{P_2} V dP = RT \ln(P_2/P_1) \tag{2.38}$$

If the "standard-state Gibbs free energy" for component i corresponding to the reference pressure P^o is g_i^o, then Gibbs free energy at pressure P_i is

$$gi = g_i^o + RT \ln(P_i/P^o) \tag{2.39}$$

Typically, $P^o = 1$ atm. Then from Equation 2.37

$$\begin{aligned} &c\left[g_C^o + RT \ln(P_C)\right] + d\left[g_D^o + RT \ln(P_D)\right] \\ &-a\left[g_A^o + RT \ln(P_A)\right] - b\left[g_B^o + RT \ln(P_B)\right] = 0 \end{aligned} \tag{2.40}$$

where P_i is the partial pressure of component i. Equation 2.40 may be written in terms of the "standard state Gibbs free energy change" (a function of temperature alone) for the reaction as $\Delta G^o = cg_C^o + dg_D^o - ag_A^o - bg_B^o$,

$$\Delta G^o = -RT \ln\frac{P_C^c P_D^d}{P_A^a P_B^b} = -RT \ln K_P$$

$K_P = \left(P_C^c P_D^d / P_A^a P_B^b\right)$, which is a function of temperature alone, is known as the equilibrium constant and may be calculated from the standard state Gibbs free energy change for the reaction. It may then be utilized in determining the molar composition of the reaction products utilizing the relationship $P_i = \left(M_i \big/ \sum_i M_i\right) P$, where P is the total pressure as explained in Chapter 5. If K_P is known at a certain temperature, its value may be calculated at the temperature of interest empirically by the relationship

$$\ln K_P = \frac{C_{-1}}{T} + C_0 + C_1 T + C_2 T^2 + C_3 T^3 + \cdots \tag{2.41}$$

where C_i^s is the empirical constant. Equation 2.41 is actually based on the relationship between ΔG^o, ΔH^o, and ΔS^o; when specific heats are independent of temperature, $C_{-1} = (-\Delta H^o/R)$ and $C_0 = (\Delta S^o/R)$ with the remaining $C_i^s = 0$.

2.4 NONIDEAL BEHAVIOR

2.4.1 Gas Phase

The nonideal behavior of gases manifests itself when the pressure is high as compared to its critical pressure P_C and the temperature is low as compared to it critical temperature T_C. The nonideal behavior may be captured by introducing the compressibility factor $\mathcal{Z}$ into the ideal gas equation ($\mathcal{Z} = 1$ for an ideal gas):

$$PV = \mathcal{Z} R_u T \tag{2.42}$$

Generalized charts are available in most textbooks of thermodynamics correlating $\mathcal{Z}$ to the reduced pressure P/P_C and reduced temperature T/T_C. The critical property data may be obtained from appendix A of Poling et al. (2001). Other equations of state that account for nonideality have been developed to specific types of gases and may be used in predicting $\mathcal{Z}$. These equations of state may also be used in determining the mixture properties as well as applied to estimate their vapor–liquid equilibrium compositions. Mixing rules based on mole fractions of the different pure components present in the mixture, their individual properties, and the binary interaction[8] parameters for each molecular pair (published values obtained either from experimental data or molecular simulation) have been devised to calculate the pseudo-critical temperature and pressure and other parameters required by the equation of state. An example is the Peng–Robinson equation of state (Peng and Robinson, 1976), which is a modification of the van der Waals equation. The attractive pressure term in the van der Waals equation was modified by Peng and Robinson with a significant improvement in the accuracy of property predictions. A dimensionless eccentric factor was introduced in the calculation of the attractive pressure term, the functional dependence of this factor determined by using vapor pressure data. The Peng–Robinson equation of state is not recommended for highly polar systems.

The behavior of most gas streams in power cycles such as the Brayton and the Diesel cycles may be approximated by assuming ideality. In the Rankine cycle, however, the working fluid is much closer to its saturated vapor conditions and hence nonideal behavior has to be taken into account. For steam, correlations to reproduce the ASME steam tables are available (Keenan et al., 1969).The REFPROP database published by the National Institute of Standards and Technology has thermodynamic properties for many of the fluids used in low-temperature Rankine cycles as well in mechanical refrigeration.

Nonideality affects the various thermodynamic properties of substances such as specific heat, internal energy, enthalpy, entropy, and Gibbs free energy. The corrections for pressure may be obtained by the application of the suitable equation of state utilizing "departure functions." General expressions for these functions are given on page 6.5 in Poling et al. (2001) with $\lim_{V\to\infty} \mathcal{Z} = 1$; the departure functions for enthalpy, entropy, and Gibbs free energy are given by the following Equations 2.43–2.45:

$$\frac{h - h_{\text{ideal}}}{RT} = \int_{V}^{\infty} \left[T \left(\frac{d\mathcal{Z}}{dT} \right)_V \frac{dV}{V} \right] + 1 - \mathcal{Z} \tag{2.43}$$

[8] Binary interactions are between just two molecules at a time. The probability of a three or higher molecular interaction is quite low.

$$\frac{s-s_{\text{ideal}}}{R}=\int_{V}^{\infty}\left[T\left(\frac{dZ}{dT}\right)_{V}-1+Z\right]\frac{dV}{V}-\ln Z \tag{2.44}$$

$$\frac{g-g_{\text{ideal}}}{RT}=\int_{V}^{\infty}(1-Z)\frac{dV}{V}+\ln Z+1-Z \tag{2.45}$$

where the h_{ideal}, s_{ideal}, and g_{ideal} are the ideal properties. The preceding expressions may be evaluated using the appropriate equation of state.

The enthalpy and entropy of nonideal mixtures are given by:

$$h_{\text{mix}}=\sum_{i} y_i h_i+\Delta h$$

$$s_{\text{mix}}=\sum_{i} y_i s_i+R_u\sum_{i} y_i \ \ln\frac{1}{y_i}+\Delta s$$

where y_i is the mole fraction, h_i and s_i are the enthalpy and entropy on a molar basis of each of the pure components forming the mixture with the last term in each of the preceding two expressions, and Δh and Δs accounting for the deviation from ideal solution behavior, disappearing for ideal solutions. One approach to estimating Δh and Δs is by applying Equations 2.43 and 2.44 to the mixture, that is, in an appropriate equation of state applied to the mixture.

In the previous section, the equilibrium constant for ideal gases was defined as $K_P=\left(P_C^c P_D^d / P_A^a P_B^b\right)$. When nonideality cannot be neglected,

$$K_P=\frac{(\gamma_C \varnothing_C P_C)^c(\gamma_D \varnothing_D P_D)^d}{(\gamma_A \varnothing_A P_A)^a(\gamma_B \varnothing_B P_B)^b}$$

where γ_i and $\varnothing_i$ are the activity and the fugacity coefficients for component i. Activity coefficient takes into account the nonideal nature of a mixture (typically $\gamma_i=1$ for gases) while the fugacity coefficient ($\varnothing_i=f_i/P_i$) corrects for nonideal behavior of the gas. These coefficients may be obtained from experimentally correlated data or estimated theoretically.

2.4.2 Vapor–Liquid Phases

Rankine-type cycles, especially those used in the conversion of low-temperature thermal energy (e.g., waste heat and geothermal) to electrical energy, may employ a working fluid that contains more than a single component to increase the conversion efficiency as explained in Chapter 8. A binary mixture (mixture of just two components) of propane and pentane is sometimes used in organic Rankine cycles (ORCs) while the Kalina cycle (Kalina, 1984) uses a mixture of NH_3 and H_2O.

The composition of the vapor phase and the liquid phase in such cases is not the same and the two compositions vary with temperature and pressure. Often, the two phases may be considered to be in equilibrium, which allows determining the composition of the two phases. Vapor–liquid phase equilibrium data is also utilized in the design and analysis of mass transfer operations discussed in Chapter 5. If the solutions behave as ideal solutions, then Raoult's law may be used in calculating the composition of the two phases. If P'_i is the vapor pressure of component i, then its partial pressure in the vapor phase is related to the mole fraction x_i of this component in the liquid phase at equilibrium by the Raoult's law as

$$P_i = P'_i x_i \tag{2.46}$$

To account for nonideal behavior, the activity and the fugacity coefficients, obtained from empirical or semi-empirical methods, may be introduced as

$$\emptyset_i P_i = \emptyset'_i P'_i \gamma_i x_i$$

Or in terms of vapor mole fraction y_i and total pressure P

$$\emptyset_i y_i P = \emptyset'_i \gamma_i P'_i x_i$$

The ratio of the vapor and liquid mole fractions is sometimes referred to as the "K-value," which is a function of pressure, temperature, and composition of the vapor and liquid phases:

$$\frac{y_i}{x_i} = \frac{\emptyset'_i \gamma_i P'_i}{\emptyset_i P} = K_i$$

An empirical relation that is usually valid for dilute solutions is Henry's law, which may be considered a modification of the Raoult's law and is given by

$$P_i = \mathcal{H}_i x_i$$

where $\mathcal{H}_i$ is an empirically determined constant for a given system.

Example 2.3. Tabulated in Table 2.7 are experimental data collected at 300°F (149°C) for the vapor–liquid equilibrium data for the H_2O–CO_2 system (Gillespie and Wilson, 1982). Compare the moisture content of the vapor phase with that calculated using Raoult's law for each of the pressures.

TABLE 2.7 Experimental data at 300°F (149°C) for vapor–liquid equilibrium data for H_2O–CO_2 system

Total pressure, bar (P_{Total})	6.89	25.34	50.68	101.35	202.70
Mole % CO_2 in liquid phase	0.0499	0.316	0.646	1.241	2.05
Mole % CO_2 in vapor phase	32.6	79.91	89.02	93.42	95.03

TABLE 2.8 Calculated versus experimental data for moisture content of the vapor phase

	From experimental data				Calculated using Raoult's law	
Total pressure bar	Mole % CO_2 in liquid phase	Mole % H_2O in liquid phase	Mole % CO_2 in vapor phase	Mole % H_2O in vapor phase	Mole % H_2O in vapor phase	Percent error
6.89	0.0499	99.9501	32.6	67.40	66.97	−0.6
25.34	0.316	99.684	79.91	20.09	18.18	−9.5
50.68	0.646	99.354	89.02	10.98	9.06	−17.5
101.35	1.241	98.759	93.42	6.58	4.50	−31.6
202.70	2.05	97.95	95.03	4.97	2.23	−55.1

Solution

Equation 2.46 may be written in terms of mole fraction of the water vapor as:

$$y_{H_2O} = \frac{P'_{H_2O} x_{H_2O}}{P_{Total}}$$

From steam tables (Keenan et al., 1969), the saturated vapor pressure of water (P'_{H_2O}) at 300°F (149°C) is 4.62 bar. Substituting $P'_{H_2O} = 4.62$ bar, and $x_{H_2O} = (1-0.0499/100)$ corresponding to $P_{Total} = 6.89$ bar from Table 2.7 into the preceding expression, we have:

$$y_{H_2O} = \frac{4.62 \times (1-0.0499/100)}{6.89} = 0.670$$

The experimentally determined value is $1-32.6/100 = 0.674$ and the % error is $(0.67-0.674)/0.674 \times 100 = -0.6\%$

Table 2.8 shows the results calculated in a similar manner for all the data points. As expected, the error becomes larger as the total system pressure increases, that is, the system deviates more from ideal behavior as the total pressure increases for a given temperature. At a total pressure of 202.7 bar, the error is as much 55%.

REFERENCES

Chase MW Jr, Davies CA, Downey JR Jr, Frurip DJ, McDonald RA, Syverud AN. JANAF Thermochemical Tables, 3rd ed. J Phys Chem Ref Data 1985;14(1):1985.

Gillespie, PC, Wilson, GM. Vapor–liquid and liquid–liquid equilibria: water–methane, water–carbon dioxide, water–hydrogen sulphide, water–*n*-pentane, water–methane–*n*-pentane. Research Report. Gas Processors Association Research Report; 1982. Report nr. RR-48. Tulsa, OK: Gas Processors Association.

Kalina IA. Combined cycle system with novel bottoming cycle. ASME J Eng Gas Turbines Power 1984;106:737–742.

Keenan JH, Keyes FG, Hill PG, Moore JG. *Steam Tables: Thermodynamic Properties of Water Including Vapor, Liquid, and Solid Phases*. New York: John Wiley & Sons, Inc.; 1969. p 24–97.

Peng D, Robinson DB. A new two-constant equation of state. Ind Eng Chem Fundam 1976; 116(1):59–64.

Poling BE, Prausnitz JM, O'Connell JP. *The Properties of Gases and Liquids*. 4th ed. New York: McGraw Hill; 2001.

3

FLUID FLOW EQUIPMENT

3.1 FUNDAMENTALS OF FLUID FLOW

Most power plants involve a working fluid which is either a gas or a liquid, for example, steam and water in the case of a plant utilizing a Rankine cycle, as well as gaseous combustion products that provide the thermal energy for the Rankine cycle in the case of fossil- and biomass-fired power plants. In the case of an oil-fired power plant utilizing the Brayton cycle, the fluids involved within the cycle are oil, air, and combustion products. A coproduction plant can involve significantly more gaseous and liquid streams. A fluid stream is also required for the rejection of the energy that is not usefully converted (say, into electricity or chemical energy in the case of a coproduct) and has to be transferred to the atmosphere. This heat rejection process may consist of circulating cooling water or air involving again fluid flow. Thus, a thorough understanding of the nature of fluid flow is essential in configuring, analyzing, or designing such energy conversion plants as well as understanding how to account for losses associated with fluid flow and possibly finding solutions to minimize these losses. The intent of this chapter is to provide an introduction to certain topics in fluid mechanics that are essential for the analysis and design of energy conversion plants.

Solution of fluid flow problems requires knowledge of the physical properties of the fluid such as viscosity and density. Viscosity is a measure of its resistance to internal deformation or shear and is highly influenced by temperature. For liquids, viscosity reduces, that is, liquids flow more easily as the temperature increases, while

Sustainable Energy Conversion for Electricity and Coproducts: Principles, Technologies, and Equipment, First Edition. Ashok Rao.

for gases, viscosity increases. Effect of pressure on the viscosity of a liquid is quite insignificant while for gases, it can be substantial, especially when close to the critical point. Viscosity of many common fluids may be found in *Perry's Chemical Engineers' Handbook* (Green, 2007). For Newtonian fluids, the shear force per unit area is proportional to the negative of the local velocity gradient with the proportionality constant being the viscosity. Not all fluids follow this type of proportionality. Most fluids encountered in power cycles are Newtonian fluids, however, and this chapter will be restricted to flow of such fluids.

Density in the oil industry is often expressed in degrees API (American Petroleum Institute) and is related to the specific gravity (SG), the ratio of density of a liquid at a given temperature to that of water at 15.6°C (60°F) by:

$$^{\circ}\text{API} = 141.5/\text{SG} - 131.5$$

For gases using Equation 2.42, density is given by:

$$\rho_{\text{G}} = \frac{P M_{\text{w}}}{\mathcal{Z} R_{\text{u}} T} \tag{3.1}$$

where $\mathcal{Z}$ is the compressibility factor to account for nonideality of the gas, P is the pressure, M_{w} the molecular weight, R_{u} the universal gas constant, and T the absolute temperature.

3.1.1 Flow Regimes

There are two distinct types of fluid flow, laminar and turbulent. Laminar flow is characterized by smooth motion in "layers" such that there is no macroscopic mixing of the adjacent layers of the fluid. On the other hand, turbulent flow is characterized by mixing of the fluid particles from adjacent layers. Osborne Reynolds conducted experiments to determine the conditions that govern the fluid flow regime by observing the motion of a dye introduced into a fluid flowing through a pipe. The type of flow was found to depend on the following dimensionless grouping of variables known as the Reynolds number (Re):

$$\text{Re} = \frac{D v \rho}{\mu} = \frac{4\dot{m}}{D\mu}$$

where D is the pipe diameter, v magnitude of the fluid velocity, ρ the fluid density, μ the fluid viscosity, and $\dot{m}$ the fluid mass flow rate. In the case of a fluid flowing through a pipe, the flow may be considered laminar when $\text{Re} < 2100$ and turbulent when $\text{Re} > 4000$, with a transition region existing in-between which is rather unpredictable. For engineering purposes, the safest practice may be to assume that the flow is turbulent when $\text{Re} > 2100$. When dealing with flow cross section that is noncircular such as a pipe not flowing full, or a duct that is oval or rectangular, a hydraulic radius R_{H} is defined for calculating Re (with $D = 4R_{\text{H}}$):

$$R_{\mathrm{H}}=\frac{\text{cross-sectional area of flowing fluid}}{\text{wetted perimeter}}$$

This method is applicable to turbulent flow and should not be applied where the cross-sectional area of the flowing fluid is very narrow, that is, width is small relative to height. In such cases,

$$R_{\mathrm{H}}\approx 1/2(\text{width of passage})$$

R_{H} for various cross sections can be found in handbooks such as *Marks' Standard Handbook for Mechanical Engineers* (Avallone et al., 2006) and *Perry's Chemical Engineers' Handbook* (Green, 2007).

3.1.2 Extended Bernoulli Equation

A fundamental equation of fluid flow that may be derived by the application of Newton's second law of motion is the extended Bernoulli equation or the total mechanical energy balance that relates the flow parameters of fluid velocity v, pressure p, and elevation[1] z. In differential form it may be written as:

$$\frac{dp}{\rho}+g\,dz+v\,dv+dW_{\mathrm{S}}+dW_{\mathrm{f}}=0$$

And in English units as:

$$\frac{dp}{\rho}+\frac{g}{g_{\mathrm{c}}\,dz}+\frac{v\,dv}{g_{\mathrm{c}}}+dW_{\mathrm{S}}+dW_{\mathrm{f}}=0$$

where g is the local gravitational acceleration, g_{c} the conversion factor in Newton's law of motion, $-W_{\mathrm{S}}$ is the shaft work input to the system, and W_{f} the mechanical energy loss due to friction, both on a unit mass of fluid basis. When the velocity v is not constant across the flow cross section as in flow through a pipe, a correction factor α_{c} is required in the aforementioned Bernoulli equation as follows where v is now the average linear velocity:

$$\frac{dp}{\rho}+g\,dz+\frac{v\,dv}{\alpha_{\mathrm{c}}}+dW_{\mathrm{S}}+dW_{\mathrm{f}}=0 \tag{3.2}$$

For turbulent flow, where the velocity profile is nearly flat, α_{c} is nearly unity while in fully developed laminar flow, $\alpha_{\mathrm{c}}=0.5$.

[1] Effect of elevation becomes important when a dense fluid such as a liquid is being pumped from one elevation to another.

Friction may include that occurring at the interface of the fluid and the walls of the pipe or channel ("skin friction"), or internally within the fluid due to changes in the flow pattern caused by changes in the cross section of the pipe or change in direction as in a pipe bend, or due to "obstructions" in the fluid path ("form friction") such as in a catalyst bed. Correlations in the form of charts or equations have been developed that relate the frictional loss to fluid properties such as fluid density and viscosity, fluid velocity, surface roughness in the case of turbulent flow through a pipe, and a characteristic length (e.g., pipe diameter) defining the geometry of the system.

3.2 SINGLE-PHASE INCOMPRESSIBLE FLOW

Integration of Equation 3.2 for constant density and velocity (i.e., constant cross-sectional flow area), and no shaft work yields:

$$\frac{\Delta p}{\rho} + g\,\Delta z + W_{\mathrm{f}} = 0 \tag{3.3}$$

3.2.1 Pressure Drop in Pipes

For fully developed laminar flow in a long straight constant diameter pipe at constant temperature such that density and viscosity μ do not vary, the mechanical energy loss W_{F} due to friction in Equation 3.3 based on the Hagen–Poiseuille relation (derived from first principles) is given by:

$$W_{\mathrm{f}} = \frac{32 L v \mu}{D^2 \rho} \tag{3.4}$$

where L is the length of pipe and D the diameter of pipe.

In turbulent flow (which is typically encountered in industrial practice in order to reduce piping costs), the dependence of W_{f} on the various flow conditions and fluid properties is too complicated to be derived from first principles. In addition to the variables appearing in Equation 3.4, W_{f} also depends on the pipe roughness ε (which has the dimensions of length and varies with the type of pipe material). Dimensional analysis and experimental data have been used to develop useful correlations. It is convenient to correlate the dependence of W_{f} on the flow conditions and fluid properties by introducing a dimensionless friction factor f known as the Fanning friction factor:

$$dW_{\mathrm{f}} = \frac{2 f v^2}{D}\, dL \tag{3.5}$$

where f is found to be a function of Re and ε. For laminar flow, it may be shown that

$$f = \frac{16}{\mathrm{Re}}$$

Note that f is independent of ε for laminar flow while for turbulent flow, charts showing the dependence of the friction factor on Re and the relative roughness ε/D (as well as tabulated values for ε for various pipe materials) are available in handbooks such as *Perry's Chemical Engineers' Handbook* (Green, 2007) and *Marks' Standard Handbook for Mechanical Engineers* (Avallone et al., 2006). Marks' handbook as well as the Technical Paper No. 410 by Crane Co. (1988) which is an excellent guide to flow of fluids through valves, pipes, and fittings utilize Darcy's (or Moody's) friction factor f' which is four times the Fanning friction factor f. The defining equation for f', equivalent to Equation 3.5 is:

$$dW_{\mathrm{f}} = \frac{f' v^2}{2D}\, dL \tag{3.6}$$

Empirical correlations such as the Colebrook–White equation may also be utilized for estimating the friction factor for turbulent flow for $4000 < \mathrm{Re} < 10^8$ and $0 < \varepsilon/D < 0.05$ but the solution requires an iterative procedure such as the Newton–Raphson technique:

$$\frac{1}{\sqrt{f}} = -4\log\left[\frac{\varepsilon/D}{3.7} + \frac{1.256}{\mathrm{Re}\sqrt{f}}\right]$$

Equation 3.7 provides f explicitly and was developed by Romeo et al. (2002) from a statistical analysis of the various correlations found in literature (their original expression is for f') and is valid for $3000 < \mathrm{Re} < 1.5 \times 10^8$ and $0 < \varepsilon/D < 0.05$.

$$\frac{1}{\sqrt{f}} = -4\log\left[\frac{\varepsilon/D}{3.7065} - \frac{5.0272}{\mathrm{Re}}\log\left\{\frac{\varepsilon/D}{3.827} - \frac{4.567}{\mathrm{Re}} \times \log\left(\left(\frac{\varepsilon/D}{7.7918}\right)^{0.9924} + \left(\frac{5.3326}{208.815 + \mathrm{Re}}\right)^{0.9345}\right)\right\}\right] \tag{3.7}$$

3.2.2 Pressure Drop in Fittings

A significant fraction of the pressure loss in a piping system is typically due to friction in valves and fittings. Fittings can include elbows, return bends, tee junctions, as well as expansions or contractions to accommodate changes in diameter. It is important that the associated losses be properly accounted for in the design and analysis of power systems. The resistance in valves and fittings is calculated using empirical methods and a common method consists of estimating the friction from each fitting in terms of equivalent straight pipe. A dimensionless characteristic K (resistance coefficient) for a particular type and size of fitting is defined by:

$$W_{\mathrm{f}} = K\frac{V^2}{2} \tag{3.8}$$

Values for K or expressed in the form of $K = \text{constant} \times f'_T$ where f'_T is the friction factor corresponding to flow in a zone of complete turbulence, are available in the Technical Paper No. 410 by Crane Co. (1988) including the values of f'_T for clean commercial steel pipes of various diameters.

Comparing Equation 3.8 with the integrated form of Equation 3.5 or Equation 3.6 when the other variables appearing in Equation 3.5 or Equation 3.6 are constant, it may be seen that

$$K = 4 f_T \frac{L_e}{D} = f'_T \frac{L_e}{D} \tag{3.9}$$

with f replaced by f_T or f'_T, and L replaced by L_e, the equivalent length of straight pipe. When $K = \text{constant} \times f'_T$, then the ratio L_e/D, the equivalent length of straight pipe in pipe diameters may be expected to be also a constant, and has been tabulated for many valves and fittings, and can be found in *Marks' Standard Handbook for Mechanical Engineers* (Avallone et al., 2006) as well as *Perry's Chemical Engineers' Handbook* (Green, 2007). For a given type of fitting and diameter D, the equivalent length L_e may be calculated from the tabulated data, added to the straight pipe length (when the diameter is the same), and then the friction factor f or f' determined as if the total equivalent length is for a straight pipe to calculate the friction loss using the integrated form of Equation 3.5 or Equation 3.7. For greater accuracy however, use of f_T or f'_T may be preferred for fittings.

3.3 SINGLE-PHASE COMPRESSIBLE FLOW

Combining Equations 3.2 and 3.5 while replacing v with $\bar{m}/\rho$, where $\bar{m}$ is the mass velocity (mass flow rate divided by the cross-sectional flow area) yields:

$$\frac{dp}{\rho} + g dz + \frac{\bar{m}^2}{2\alpha c} d\left(\frac{1}{\rho^2}\right) + 2\frac{\bar{m}^2 f}{\rho^2 D} dL = 0 \tag{3.10}$$

Compressible flow of a gas at isothermal conditions that follows Equation 3.1 will be considered. Integration of Equation 3.10 for flow through a constant cross-sectional pipe from point 1 to 2 (with no shaft work and omitting the dz term since the effect of elevation for gases is negligible, their density being quite low) yields the following expression to calculate the pressure drop:

$$p_1^2 - p_2^2 = \frac{G^2 Z_{mean} R T}{M}\left[\frac{4 f L}{D} + \frac{2}{\alpha c} \ln\left(\frac{p_1}{p_2}\right)\right] \tag{3.11}$$

where the compressibility factor Z_{mean} may be approximated by the arithmetic average $(Z_1 + Z_2)/2$. Note that f is a function of $\text{Re} = (4\dot{m}/D\mu)$, which remains constant for constant $\dot{m}$ and D.

Equations are available for the case of adiabatic flow through a pipe (Lapple, 1943) but for long pipes in practical applications, it may be assumed that the flow is isothermal and Equation 3.11 may be utilized.

3.3.1 Pressure Drop in Pipes and Fittings

Equation 3.8 may be used after accounting for the change in velocity caused by the change in pressure as the gas moves through the piping system while the topic of pressure drop due to expansions or contractions to accommodate changes in diameter has been pursued by a number of investigators such as Marjanovic and Djordjevic (1994), and Benedict et al. (1966).

3.3.2 Choked Flow

The velocity of a compressible fluid as it discharges adiabatically into a reservoir attains a maximum value as the pressure within the reservoir is decreased. Any further decrease in pressure below this critical pressure has no effect on the flow rate. The reason for this phenomenon is that a point is reached when the upstream velocity becomes equal to the sonic velocity, the velocity at which disturbances travel through a fluid medium and thus, the information regarding any reduction in the pressure within the reservoir below the critical pressure does not travel upstream. The flow is said to be "choked." It can be shown that this maximum velocity for an ideal gas is given by:

$$v_{\mathrm{sonic}} = \sqrt{\frac{C_{\mathrm{p}} P}{C_{\mathrm{v}} \rho}} \tag{3.12}$$

where v_{sonic}, P, and ρ are the upstream magnitude of (sonic) velocity, pressure, and density of the gas, while C_{p} and C_{v} are the gas specific heats at constant pressure and volume, respectively.

Flow restrictions in gas systems such as orifices and control valves can also lead to choked flow if the pressure differential across them is large enough such that sonic velocity prevails in the flow path. "Restriction orifices" take advantage of this phenomenon to supply a fixed flow rate of gas in cases where the downstream pressure decreases (or increases but by a limited amount so that sonic conditions still prevail in the flow path) over time.

3.4 TWO-PHASE FLUID FLOW

Two-phase flow with one phase as a vapor and the other a liquid may be encountered in Rankine cycles as the working fluid changes phase from a liquid to a vapor. Note that unless a once-through system is used, the liquid stream entering an evaporator of a Rankine cycle is not entirely vaporized. The vapor phase is separated from the liquid phase and the liquid phase returned to the evaporator (along with additional liquid equal in mass to the net vapor leaving the evaporator). Two-phase flow in general should be avoided however, unless absolutely necessary (as in an evaporator to avoid depositing the dissolved solids that may be present in the liquid being evaporated) and it is often prudent to separate the two phases as quickly as possible and have separate pipes for each phase. Pressure drop for two-phase flow depends on the type of flow regime; there being various regimes as described next and are broadly classified according to the continuity of the two flowing phases. The flow regime can vary depending on a number of system variables as well as whether the pipe is horizontal, vertical, or inclined. Furthermore, the flow regime may change as the operating conditions change and certain flow regimes can structurally strain the piping network and should be avoided.

3.4.1 Gas–Liquid Flow Regimes

The orientation of the pipe and the relative gas and liquid volumetric flow rates largely govern the flow regime. Other factors include liquid density, viscosity, and surface tension. Described in the following are the various flow regimes encountered in horizontal pipe flow:

- Bubble flow. This type of flow occurs when the volumetric liquid flow rate is relatively high compared to the gas flow rate such that the liquid occupies most of the pipe cross section while the gas flows as small bubbles. Both the gas and liquid velocities are similar. The bubbles are typically dispersed throughout the pipe cross section but due to buoyancy, they tend to be more concentrated in the top portion of the pipe. A special case of this type of flow is called froth flow in when the bubbles predominate and get dispersed throughout.
- Plug flow. This type of flow occurs when the volumetric liquid flow rate is relatively lower compared to the gas flow rate than in bubble flow. The bubbles coalesce resulting in alternating plugs of gas and of liquid flowing along the top cross section of the pipe while the bottom cross section has a continuous phase of the liquid.
- Stratified flow. When the volumetric liquid flow rate is further reduced compared to the gas flow rate, the plugs become separated and the gas flows along the top cross section of the pipe while the liquid flows along the bottom section with the interface between the two phases remaining relatively smooth. The friction due to each phase remains relatively constant along the length of the pipe.
- Wavy flow. When the gas flow rate is increased in stratified flow, waves form over the liquid surface due to friction between the two phases and the wave amplitude increases as the gas flow rate increases.
- Slug flow. If the liquid flow rate is now increased, the higher liquid flow rate causes the wave crests to touch the top of the pipe and form frothy slugs resulting in a segregated intermittent flow. The velocity of the liquid slugs and the alternating large gas bubbles will be greater than the average velocity of the liquid

phase and the large gas bubbles end up occupying almost the entire cross section of the pipe. The flow is similar to plug flow described previously and the distinction between these two types of flow is rather arbitrary.

- Annular flow. At higher velocities, the gas causes the liquid to flow as an annular film of varying thickness along the pipe wall with the gas flowing through the central core. A significant difference in the velocity of the two phases exists and the gas can contain entrained liquid droplets sheared from the liquid phase.
- Mist or spray flow. At higher gas velocity, the amount of entrained liquid phase increases while the annular liquid ring decreases.

The various flow regimes formed in vertical pipe flow are bubble flow, slug flow, froth or churn flow, annular flow, and mist or spray flow, depending on the velocity of each of the two phases. Slug flow can transition to froth or churn flow when the liquid flow rate is decreased or gas flow rate is increased such that the gas–liquid interface becomes more turbulent. The smaller gas bubbles mix with the liquid forming a turbulent disordered pattern of short liquid slugs separated by successively large gas bubbles.

In inclined pipes which may be used to allow the drainage of the liquid-phase into a vessel used for separating the two phases (a "knock-out KO drum"), a small deviation in the inclination of the pipe can significantly affect the onset of each flow regime described for horizontal pipes. As the pipe orientation deviates more than 20° from the horizontal, the occurrence of the flow regimes resemble more those of a vertical pipe.

3.4.2 Pressure Drop in Pipes and Fittings

The Beggs and Brill method for calculating the pressure drop in a pipe consists of first establishing the flow regime (either segregated, intermittent, or distributed) by applying their criteria and then calculating the pressure drop using correlations specific to each of these flow regimes (Beggs and Brill, 1973). For pressure drop in fittings, a calculation procedure described by Chisholm (Collier, 1972) may be used.

3.4.3 Droplet Separation

Liquid separation from vapor streams is one of the most common operations in process plants (performed in KO drums) and often encountered in power plants, especially those employing the Rankine cycle.

KO drums are either vertically or horizontally oriented vessels and the choice depends on factors such as the relative economics and plot space requirements. These devices may also be designed with a mist elimination device (such as a wire mesh pad) to ensure the vapor is essentially free of the liquid. The separation in such equipment is accomplished in the following three steps: (1) primary separation using an inlet diverter to impinge the larger droplets carried by the vapor onto the diverter such that they drop by gravity; (2) secondary separation, which occurs in the disengagement area of the vessel for the smaller droplets to drop by gravity; and (3) final separation where the mist is eliminated by coalescing the smallest droplets into larger droplets, which again separate by gravity. The allowable velocity for the vapor ν_V should be low enough for the secondary separation to occur, that is, the velocity of the vapor should be less than the terminal velocity of the droplets ν_T, which is determined by equating the difference in gravitational force and the buoyancy force to the drag force acting on the droplet:

$$\nu_T = \sqrt{\frac{4\,g\,D_P\,(\rho_L - \rho_V)}{3\,C_D\,\rho_V}} \tag{3.13}$$

where D_P is the droplet diameter, ρ_L and ρ_V are the densities of the liquid and vapor, and C_D is the drag coefficient and may be estimated from correlations developed for rigid spheres (*GPSA Engineering Data Book*, 2013). C_D is a function of the droplet diameter and thus the smallest droplet to be separated from the vapor phase has to be specified.

When mist eliminators are included, an empirical form of Equation 3.13 is utilized:

$$\nu_V = K\sqrt{\frac{(\rho_L - \rho_V)}{\rho_V}} \tag{3.14}$$

where K is empirically derived from past experience. Tabulated values of K are available in the *GPSA Engineering Data Book* (2013) for vertically as well as horizontally oriented vessels. Adjustments to K for certain specific liquid–vapor systems as well as for certain types of special services are also recommended. The velocity ν_V in a vertical vessel is based on the horizontal cross-sectional area while that in a horizontal vessel is based on the vertical cross-sectional area above the highest liquid level (e.g., which sets off an alarm). In this manner the diameter of the vessel may be calculated.

After the diameter is calculated, the seam-to-seam (where the heads[2] are welded to the cylindrical portion of the vessel) length, also known as tangent-to-tangent length of the vessel, is specified in the process industry for KO drums based on the surge time and holdup time requirements. The surge time is the time duration that the vessel can accommodate the incoming liquid flow while the outgoing flow is cut off and is used in setting the distance between the normal liquid level (NLL) and the high liquid level (HLL) where the alarm is set off. The surge time provides a buffer between variations in the incoming and the outgoing flow rates, as well as a means to prevent a sudden

[2] Head may be either 2:1 elliptical, hemispherical, or dished. Typically, elliptical heads are used when vessel diameter is less than 4.5 m and pressure is greater than 7 barg; hemispherical heads are used when vessel diameter is greater than 4.5 m and pressure is greater than 7 barg; and dished heads with "knuckle" radius (at which the cylindrical shape ends) of 0.6 times the diameter are used when vessel diameter is less than 4.5 m and pressure is less than 7 barg.

shutdown of the upstream facility when a reduction in flow rate by the downstream equipment is required. Holdup time, on the other hand, is the time duration that the vessel can supply the liquid feed to the downstream equipment while the vessel's incoming flow is shut off and is used in setting the distance between the NLL and the low liquid level (LLL). Holdup time is based on considerations such as how stable the incoming flow is and the sensitivity of the downstream equipment to changes in flow rate. An increase in level to the high–high liquid level (HHLL) above HLL would cause the inlet flow to be shut off and the distance between the HLL and the HHLL is based on providing enough time for operator intervention. A decrease in level to the low–low liquid level (LLLL) would cause the outlet flow to be shut off and the distance between the LLL and the LLLL is again based on providing enough time for operator intervention. The vessel dimensions have to accommodate the distance between the HLL and HHLL, and between the LLL and LLLL. The LLLL itself is kept a certain distance above the bottom to prevent the vapor entering the bottom discharge, the distance dependent on the time it takes from when the valve closure command is triggered and the time at which the valve is fully closed. A vortex breaker that may consist of radial vanes or baffles around the liquid exit to reduce the angular velocity of the liquid discharging is often installed over the bottom discharge to avoid vapor being carried with the liquid unless a siphon-type drain is used. In addition to the height required for surge and holdup, a vertical vessel height needs to accommodate the mist eliminator if included, while typically an aspect ratio (the seam-to-seam length to diameter ratio) can range from 1.5 to as high as 6 depending on the design pressure (this ratio increasing with higher pressure) and is selected based on cost, while the aspect ratios for a horizontal vessels typically range from 2.5 to 6.

Steam drums are typically horizontal vessels but vertical vessels are also being built. For smaller horizontal steam drums where the residence is lower to depend on gravity for the separation, internal cyclones are installed for the primary separation of the two phases. The centrifugal force created within the cyclone by the circular motion of the fluid mixture pushes the heavier water droplets further toward its walls and away from the steam. Vane-type separators and/or coalescing pads inside the vessel cause the final separation by coalescing the fine droplets.

In vertical steam drums (often called "steam separator"), the steam–water mixture enters the drum tangentially from the evaporator creating a centrifugal force to cause the primary separation of the two phases and thus internal cyclones are not required as in horizontal drums. A vane-type separator located at the top of the drum can cause the final separation. These types of separators are being offered for heat recovery steam generators (HRSGs) for fast startup to complement the intermittent renewables such as wind and solar discussed in Chapter 12. HRSGs are used for recovery of heat from the exhaust of gas turbines and sometimes reciprocating internal combustion engines and such power plants will have to come online as the power output of these intermittent renewables goes down. These vertical drums are configured with smaller diameters to reduce the hoop stress allowing thinner walls, which reduces the thermal stresses induced during the startup process. Multiple units may be also installed, while maintaining the small diameters.

Example 3.1. A horizontal KO drum has been designed without a mist elimination device to separate out liquid droplets of diameter $D_P > 150\ \mu m$ from a stream with a vapor flow rate $\dot{m}_V$ of 20 kg/s. Given that the drag coefficient C_D on a 150 μm diameter sphere is 1.4 corresponding to the operating conditions within the vessel, liquid density ρ_L of 500 kg/m^3, and vapor density ρ_V of 33 kg/m^3, determine if a vessel with inside diameter D of 1.2 m and tangent-to-tangent (T/T) length L of 4.6 m suffices in separating out these droplets if the liquid level in the vessel is maintained such that it occupies ½ the total volume of the vessel.

Solution

Using Equation 3.13, the terminal velocity of the droplets is

$$v_T = \sqrt{\frac{4\times 9.81\times 150\times 10^{-6}(500-33)}{3\times 1.4\times 33}} = 0.14\ \text{m/s}$$

Assuming that the time required for the droplet to fall from the top of the vessel to the liquid surface maintained in the vessel (a distance of $D/2$) = the time required by the vapor to flow from the inlet to the outlet of the vessel (a distance of L), we have

$$\frac{D/2}{v_T} = \frac{L((\pi/4)D^2)/2}{\dot{m}_V/\rho_V}$$

The given value of $D = 1.2$ m, the minimum T/T length of the vessel is $L = \frac{4\dot{m}_V/\rho_V}{\pi v_T D} = \frac{4\times 20/33}{\pi \times 0.14\times 1.2} = 4.57$ m, which is less than the given value of 4.6 m. Thus, the sizing of the vessel is adequate assuming that the liquid level maintained at a height of ½ the diameter is also adequate from the instrumentation and controls standpoint, which includes the surge and holdup requirements.

3.5 SOLID FLUID SYSTEMS

3.5.1 Flow Regimes

When a fluid flows through a porous bed of solids at low velocities, the solid particles may remain stationary but as the velocity is increased, the particles do not remain in contact with each other anymore but "fluidize" due to the higher forces imparted by the fluid to the solid particles. When the solid particles are stationary, the bed is called a "fixed bed," while under fluidizing conditions, the bed is called a "fluidized bed" and the fluid velocity at which fluidization begins is called the "critical velocity." The fluid undergoes a pressure drop as it flows through the bed and the relationship between the superficial fluid velocity (velocity calculated using the entire cross-sectional area of the bed) and fluid pressure drop across the bed shows the distinct flow regimes as presented in Figure 3.1. In batch fluidization, the particles move vigorously, swirling and moving in random directions and the bed resembles a boiling liquid and hence it is sometimes

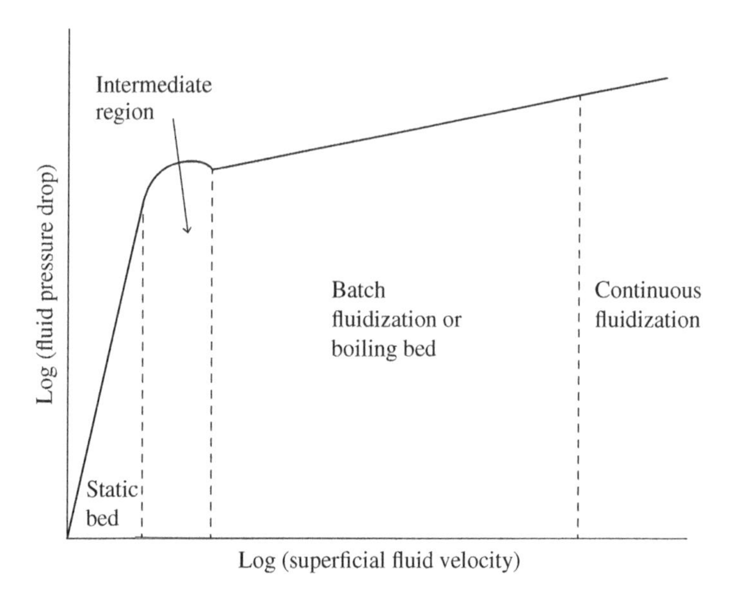

FIGURE 3.1 Pressure drop and flow regimes in solid-fluid flow

called a "boiling bed." The particles are not carried away by the fluid, however, as the velocity of the fluid between the particles is much greater than the velocity in the space above the fluidized bed, resulting in essentially all the particles dropping out of the fluid. As the fluid velocity is further increased such that "continuous fluidization" occurs, the particles are carried away by the fluid, the fraction and size of the particles depending on the fluid velocity and the density of the particles. When the entire bed material is carried away by the fluid, the bed is said to be an "entrained bed."

Flow of a gas or a liquid through a bed of solids is often encountered in "energy plants" (throughout this book, plants in which the principal product is energy such as electrical or thermal will be referred to as energy plants). For example, in the treatment of water to make it suitable for the generation of steam, water is passed through a bed of ion-exchanging solids. Another example is in a cooling tower for rejecting heat from say, a power plant, where a solid "packing" material is used to enhance the heat and mass transfer between the ambient air and cooling water. These are a few examples of fixed bed operations and both batch fluidization and continuous fluidization are also often encountered. For example, both types of these fluidization operations are used in the combustion of coal, and in gasification of coal and biomass as discussed in more detail in Chapters 8 and 9, respectively. Pneumatic transport of solid particles by a gas media, which is an example of continuous fluidization, is also often encountered in energy plants.

3.5.2 Pressure Drop

The pressure drop Δp in a static bed in addition to depending on the superficial fluid velocity $\bar{v}_s$ as seen in Figure 3.1 also depends on the bed height L; the bed porosity ε, which is the ratio of volume of voids to total volume of bed; density ρ_F and viscosity μ_F of the fluid; and the average hydraulic radius R_H of the channels formed between the solid particles. The pressure drop is actually controlled by the velocity of the fluid between these channels ($\bar{v} = \bar{v}_S/\varepsilon$) and may be correlated to a friction factor $f_{\text{packed bed}}$, which in turn may be correlated to a Reynolds number R_e as:

$$\frac{\Delta p\, R_H}{L \rho_F \bar{v}^2} = f_{\text{packed bed}} = \phi(\text{Re}) = \phi\left(\frac{\bar{v}\, R_H\, \rho_F}{\mu_F}\right) \tag{3.15}$$

For N solid particles in the bed, the mean hydraulic radius may be approximated by the ratio of the total volume of the voids $[= (N\, V_P\, \varepsilon/(1-\varepsilon))]$ in the bed to the total surface area (NS_P) of the solid particles where V_P and S_P are the mean individual particle volume and surface area. Also, by utilizing the superficial velocity ($\bar{v}_s$) of the fluid, Equation 3.15 may be written as:

$$\frac{\Delta p\, \varepsilon^3\, V_P}{L \rho_F \bar{v}_S^2\, (1-\varepsilon)\, S_P} = f_{\text{packed bed}} = \phi(\text{Re}) = \phi\left[\frac{\bar{v}_s\, V_P\, \rho_F}{\mu_F\, (1-\varepsilon)\, S_P}\right] \tag{3.16}$$

Equation 3.16 is a form of the Kozeny–Carman equation. Although the two authors never published together, Kozeny (1927) was the first to propose such an equation and was later modified by Carman (1937, 1956). Plots of this Reynolds number versus the friction factor may be found in McCabe and Smith (1956) for granular and for hollow particles. At low Reynolds number, the relationship $f_{\text{packed bed}} = 5/\text{Re}$ may be substituted into Equation 3.16 to make the equation computer-friendly. With the shape factor λ_P also introduced, the resulting equation is the more popular form of the Kozeny–Carman equation:

$$\frac{\Delta p}{L} = 150 \frac{(1-\varepsilon)^2\, \lambda_P^2\, \mu_F\, \bar{v}_s}{\varepsilon^3\, D_P^2} \tag{3.17}$$

The shape factor, λ_P, accounts for the nonsphericity of the particles and is defined as the ratio of the surface area of the particle to that of an equivalent volume sphere or $\frac{(1/6\,\pi)^{2/3}}{\pi} S_P/V_P^{2/3} = 0.2068\, S_P/V_P^{2/3}$ and is unity for spheres and greater than 1 for irregular-shaped particles. Tabulated values of shape factor are available in McCabe and Smith (1956).

Ergun's equation (1952) for a packed bed with solids of diameter D_P which is also computer-friendly and valid also for larger values of the Reynolds number is given by Equation 3.18. It may be considered an extension of Equation 3.17 to higher Reynolds number; the first term on the right-hand side being that given by Equation 3.17:

$$\frac{\Delta p}{L} = 150 \frac{(1-\varepsilon)^2\, \lambda_P^2\, \mu_F\, \bar{v}_s}{\varepsilon^3\, D_P^2} + 1.75 \frac{(1-\varepsilon)\, \lambda_P\, \rho_F\, \bar{v}_s{}^2}{\varepsilon^3\, D_P} \tag{3.18}$$

For fluidized beds, Leva et al. have correlated a modified friction factor with the Reynolds number while utilizing the particle shape factor λ_P and plots are presented in McCabe and Smith (1956) for the following relationship:

$$\frac{\Delta p\, \varepsilon^3 D_P}{2\, L \rho_F \bar{v}_S^2 (1-\varepsilon)^2 \lambda_P^2} = f_{\text{fluid bed}} = \phi(\text{Re}) = \phi\left(\frac{\bar{v}_s D_P \rho_F}{\mu}\right) \quad (3.19)$$

The minimum fluidization velocity may be calculated by equating the upward force exerted to the bed while it is still static, that is, on the onset of fluidization ($\Delta p \times$ Bed cross-sectional area) given by Equation 3.18 to the net downward force (weight of solid particles – buoyancy of the solid particles in the fluid) resulting in the following expression with ρ_P being the density of the solid particles and g the acceleration due to gravity:

$$(\rho_P - \rho_F)\, g = 150 \frac{(1-\varepsilon)^2 \lambda_P^2 \mu \bar{v}_s}{\varepsilon^3 D_P^2} + 1.75 \frac{(1-\varepsilon)\, \lambda_P \rho_F\, \bar{v}_s^{\,2}}{\varepsilon^3 D_P}$$

3.5.3 Pneumatic Conveying

This type of conveying which is widely used for transporting dry bulk solid materials in powdered, granular, or pelletized form has the advantage over other means of transport by having fewer moving parts, has a compact layout, can have multiple solids entry and discharge points, is completely enclosed and heating, cooling, drying, or blending can be accomplished while conveying. The transport and flow characteristics of pneumatic conveying for vertical flow are distinctly different from horizontal flow. Furthermore, there are two broad types of pneumatic conveying: dilute phase and dense phase. In vertical dilute phase conveying where the solids-to-gas ratio is low, the particles are carried as a uniform suspension, while in vertical dense phase conveying where the solids-to-gas ratio is higher, heavier or coarser particles could be carried as slugs but the lighter or smaller particles could flow without slugging. The flow regimes for horizontal pneumatic conveying are more complex because it is dependent on the deaeration characteristics of the solids being conveyed. Vogt and White (1948) had developed a correlation for estimating the frictional pressure drop through both vertical and horizontal pneumatic conveying systems involving a relative pressure drop defined as the ratio of the pressure drop for the flow with the solids to the pressure drop for fluid flow alone (i.e., without the solids) through the same pipe and at the same velocity. McCabe and Smith (1956) describe a methodology to estimate the pressure drop by adding the energy required to lift and move the solid particles while taking into account the slip velocity between the solids and the gas.

3.6 FLUID VELOCITY IN PIPES

The fluid flow velocities should not exceed certain limits to avoid noise/vibrations and damaging wear and tear of pipes and fittings. Following are some guidelines:

- Liquids: maximum of 15 ft/s (4.6 m/s) if pressure drop is not limiting; minimum of 3 ft/s (0.9 m/s) to avoid possible solid or liquid accumulation in pipe bottoms. However, on the pump suction side, the velocity should be limited to 8 ft/s (2.4 m/s) if subcooled, and to 4 ft/s (1.2 m/s) if at bubble point (the bubble point of a liquid mixture is the point where the liquid just starts to evaporate) to reduce suction line losses and consequently the available net positive suction head which is discussed in the following section.
- Gases or vapors: maximum of 150 ft/s (46 m/s) for single phase but limited to 60 ft/s (18 m/s) when in-plant noise is a concern; minimum of 15 ft/s (4.6 m/s) to avoid liquid accumulation in pipe bottoms.
- Cross-country pipelines: velocity determined based on economics (based on trade-off between pressure drop which dictates the energy required to move the fluid as well as the cost of the associated equipment and pipe size which affects its cost).
- Two-phase flow: maximum erosional velocity determined by $v_e = C/\sqrt{\rho_M}$, where v_e is the erosional velocity, ρ_M is the density of the gas–liquid mixture at the operating conditions while C is a constant = 100 in imperial units[3] (v_e in feet per second and ρ_M in pounds per cubic feet) for continuous service with periodic surveys to assess pipe wall thickness, $C = 125$ in imperial units for non-continuous service, and $C = 150$–200 in imperial units for solid-free fluids in noncorrosive service or is controlled by additives or by employing corrosion-resistant alloys.

3.7 TURBOMACHINERY

3.7.1 Pumps

Pumps are used to transport liquids through piping systems by adding mechanical energy to overcome the frictional forces in the piping system, elevation and meet the pressure requirement at the discharge end of the piping system. Sometimes the term "pump" is also used in devices for pressurizing a gas such as an "air pump" and a "vacuum pump." There are various types of pumps that may be classified as either positive displacement or kinetic pumps. Both reciprocating and rotary pumps come under the category of positive displacement pumps, while kinetic pumps may be divided into centrifugal, axial, or peripheral. These categories may be further subdivided into various subcategories. In centrifugal pumps, work is performed on the liquid primarily by the action of centrifugal force while in axial pumps, work is performed by the aerodynamic lifting or propelling action of the impeller vanes. A radial flow pump is an example of a centrifugal pump where the liquid exits the pump radially, that is, perpendicular to the pump shaft. An axial flow pump consists of an impeller in the shape of a propeller installed within a pipe that imparts aerodynamic lift to

[3] Corresponding value of C in SI units = 7.62 (V_e in meter per second and ρ_M in kilogram per cubic meter).

the fluid with the fluid entering and leaving nearly in the axial direction. In mixed flow pumps the work is performed partially by centrifugal force and partly by the aerodynamic lifting action of the vane. The main types that may be most often encountered in energy plants are described in the following.

3.7.1.1 Centrifugal Pumps The liquid enters the pump near the center of the rotating impeller contained within a casing and is flung outward by the centrifugal action of the vanes. The kinetic energy of the fluid increases from the center of the impeller to its tips and as the fluid exits the vanes, the kinetic energy is converted into pressure of the liquid. This conversion occurs within the diffuser, which may be fashioned by making the casing spiral in shape to gradually increase the cross-sectional flow area and thus decrease gradually the liquid velocity in which case it is called a volute casing design (see Fig. 3.2). The spiral casing is also designed to maintain a constant liquid velocity around the impeller when operating at the best efficiency point. An alternate design for the diffuser may consist of surrounding the impeller by multiple diffuser vanes discharging into a circular casing. In multistage centrifugal pumps, a series of impellers are mounted on a shaft and the liquid leaving the first impeller enters the second and so on through internal channels inside the casing. In this manner, the liquid can be pumped to a much higher pressure without compromising efficiency. Boiler feed water (BFW) pumps in a high pressure Rankine steam cycle utilize such pumps.

There are three types of impellers: open, semi-open, and shrouded (fully enclosed, i.e., with side walls on both sides) impellers. An open impeller has blades attached to a central hub and no side walls and is lighter than the other types of impellers for a given diameter resulting in less force applied to the shaft and consequently reduced shaft diameter and cost. A disadvantage of an open impeller is that it has to be carefully positioned within the casing while minimizing the gap between the impeller and the casing to minimize the pump efficiency loss. Minimizing the gap, however, leads to high fluid velocities and formation of vortices close to the casing, which can cause excessive wear, especially when the fluid contains abrasives. Typically, open impellers have lower efficiency due to leakage within the clearances in the front and back of

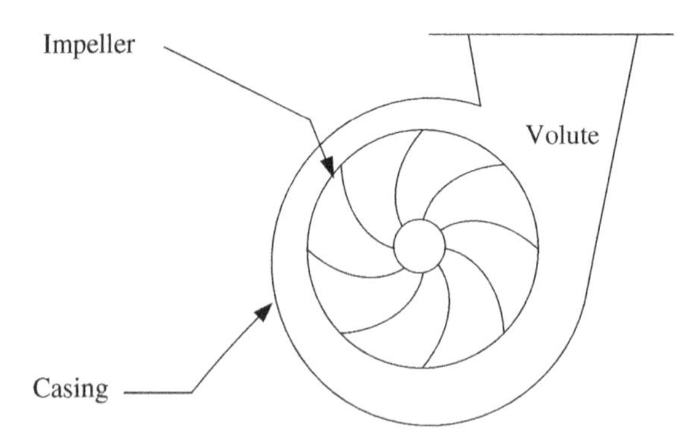

FIGURE 3.2 Centrifugal pump with volute casing

the impeller but are more suitable for viscous liquids or liquids containing suspended solids. In shrouded impellers, the friction caused by the rotating shroud close to the casing causes some loss in pump efficiency. Wear rings in combination with impeller balance holes may be used to reduce the leakage between the shroud and the casing by minimizing the fluid flowing back to the suction side but the tight clearance required between the rotating and the stationary wear rings can cause high wear rates, especially when the fluid contains abrasives. Semi-open impellers consisting of only one shroud at the back of the impeller are a compromise between the open and shrouded impellers. Impellers may be single-suction type where the liquid enters from only one side or double-suction type where the liquid enters from both sides.

3.7.1.2 Axial Pumps These pumps consist of an axial impeller to move the liquid in the axial direction rather than the radial direction. These pumps are especially suitable when a relatively high discharge flow rate is required at relative low pressure increases but some designs can be operated efficiently at low flow rates and high pressure increases by changing the pitch on the impeller blades (i.e., changing the angle of attack of the fluid relative to the blade chord). In energy plants, axial pumps are typically used for cooling water supply systems and storm water pumping stations. Sewage treatment plants also use such pumps; such plants in a sustainable world will need to be brought under the energy plant umbrella.

3.7.1.3 Rotary Pumps Gear, screw, vane, and cam pumps are all classified as rotary pumps. In gear pumps, the liquid entering the pump is trapped in the spaces between the teeth of slowly turning (intermeshing) gears with close clearance and the casing and transported to the discharge side of the pump. In the case of screw pumps, the intermeshing gears have screw threads but the seal between the screws is not highly effective and are not suitable for low viscosity liquids. In the sliding vane pump, vanes are mounted to a rotor located inside a chamber. In some cases, these vanes can be variable length and can be tensioned to maintain contact with the walls as the pump rotates to avoid leakage from higher-pressure to lower-pressure regions. In its simplest form, the vane pump has a circular rotor within an eccentric circular casing to create a chamber between two vanes with increasing volume on the inlet side and decreasing volume on the discharge side. Some uses of vane pumps are high-pressure hydraulic pumps as well as in vacuum pump service. In the case of a cam pump, the impeller oscillates axially on an eccentric mounted rotating shaft. The entering liquid flows into the space between the impeller and the casing while discharging on the other side of a divider vane. Such pumps are used for highly corrosive liquids. The gear, sliding vane, and cam pumps are usually limited to applications with clear, nonabrasive liquids but especially suitable for high viscosity or low vapor pressure liquids. They are also used as metering pumps since they have a constant displacement at a set speed.

3.7.1.4 Reciprocating Pumps Piston type, plunger type, and diaphragm type are all examples of these positive displacement pumps. In the case of the piston type, liquid being pumped is drawn through a check valve into the cylinder by a reciprocating piston. Each time the piston travels the length of a full stroke within the cylinder, a volume of liquid equivalent to the volume displacement of the piston is discharged from the pump resulting in a flow that is pulsating. Most commercially available piston pumps have multiple cylinders, however, to smooth out the pulsating flow. In double-acting piston pumps, the cylinder has inlet and outlet ports at both ends of the cylinder. Liquid is drawn into the cylinder at the back end as the piston moves forward, while liquid is discharged at the front end. The sequence is reversed when the piston direction is reversed. With double-acting pumps, the pulsations are lower than in the single-acting pumps. Plunger-type pumps are used for higher-pressure service and consist of a heavy walled cylinder of small diameter containing the reciprocation plunger, which is basically an extension of the driven shaft. Piston and plunger type pumps have been replaced by centrifugal or rotary pumps to a large extent but they are still used in applications such as for pumping volatile or highly viscous liquids and where the need for variable speed and stroke such as metering a specified liquid flow rate are important, say for the addition of chemicals to BFW. In the case of a diaphragm pump, a flexible disk fixed at the periphery but movable at the center (in the axial direction) performs the plunging action. The diaphragm may be operated mechanically or by a pressurized fluid and these pumps are typically used in pumping slurries while the piston and plunger types are not suitable where solids, abrasives, or dirt are present in the liquid.

3.7.1.5 Specific Speed A dimensionless parameter used in obtaining generalized correlations for turbomachinery (in our case, kinetic pump) behavior is the specific speed N_S defined as the rotational speed at which a geometrically similar impeller would operate if sized to deliver 1 m³/s of fluid against 1 m of head (head is the height of the liquid column that the pump can create from the kinetic energy imparted to the liquid by the impellers). Mathematically, N_S is defined as:

$$N_S = \frac{\dot{N}\,\dot{V}^{0.5}}{\left(g\,H_{pump}\right)^{0.75}} \tag{3.20}$$

where $\dot{N}$ is the actual rotational speed, $\dot{V}$ the volumetric flow rate (for double suction pumps, half of the total flow rate should be used), g the gravitational acceleration, and H_{pump} the head. Specific speed may be used to assist in the selection of the type of pump required for a design flow rate and head. Guides for narrowing down the types of suitable pumps for a given service are also available in the form of charts showing regions representing the suitable pump with the required volumetric flow rate plotted against the total head developed. Such a plot for example, may be found in *Perry's Chemical Engineers' Handbook* (Green, 2007).

3.7.1.6 Net Positive Suction Head Unlike positive displacement pumps that can draw a liquid to the inlet, centrifugal pumps require "priming," that is, to be filled up by the liquid.[4] The liquid must be fed to the pump under pressure unless the pump is

[4] Pressure head has units of length and is defined as $p/g\rho$ where p is the pressure of the fluid, g the local gravitational acceleration, and ρ the density of the fluid.

equipped with a self-priming element. A centrifugal pump once primed can draw the liquid but the height to which it can draw is limited by the vapor pressure of the liquid. Even when the liquid level is above the pump center line, the losses due to friction as well the pressure above the liquid level affect the static pressure of the liquid entering the pump. The static pressure of the liquid entering the pump minus its vapor pressure, which is defined as the "net positive suction head" (NPSH), should be greater than zero by a significant margin to avoid vaporization within the pump. Without enough NPSH, cavities or bubbles form within the liquid, which collapse when they pass into regions of higher pressure within the pump, causing noise, vibration, and damage to the pump, while a loss in capacity, a reduction in the head developed, and pump efficiency may be experienced. For a centrifugal pump, the minimum NPSH requirement or $NPSH_R$ is related to the suction specific speed N_{suct}, which is the specific speed obtained by replacing H_{pump} in Equation 3.20 by the $NPSH_R$. Sometimes, g is left out in this relationship resulting in a quasi nondimensional number for N_{suct}. If N_{suct} and rotational speed are known for a given type of pump (defined by the pump manufacturer), then $NPSH_R$ may be estimated for the required volumetric flow rate. This can then allow specifying the elevation to which a tank or vessel supplying the liquid to a given type of pump has to be raised to, or for a given elevation or available NPSH, possibly finding a suitable pump. Plots of volumetric flow rate against the $NPSH_R$ showing regions of availability of different type of pumps/speeds are sometimes available from the pump manufacturer.

In the case of reciprocating pumps, sufficient $NPSH_R$ is required to completely fill the cylinder during the suction stroke. In its calculation, the "acceleration head" $H_{pump,acc}$, which allows for the inertia of the fluid mass in the suction line as its velocity varies during the pump stroke, should be taken into account. The acceleration head reduces the available NPSH and is calculated by:

$$H_{pump,\,acc} = K_{pump}\frac{L v \dot{N}}{g\,\mathcal{Z}_L}$$

where K_{pump} is a constant specific to a pump and should be available from the pump manufacturer, L is the length of the suction line, v is the mean liquid velocity, $\dot{N}$ is the pump rotational speed, $\mathcal{Z}_L$ is a measure of the compressibility of the liquid (1.4 for water; 2.4 for hot oil), and g is the acceleration due to gravity.

3.7.1.7 Pumping Power The power required to operate a pump has to account not only for the hydraulic inefficiency of the pump but also any mechanical inefficiency associated with the pump and the inefficiency of the pump driver (a motor or a turbine). The shaft work required by a pump may be calculated assuming incompressibility for turbulent flow by integrating Equation 3.2 to yield:

$$-W_S = \frac{\Delta p}{\rho} + g\,\Delta z + \frac{\Delta v^2}{2} + W_f$$

where Δ represents the difference in magnitude from a station 2 just downstream of the pump to a station 1 just upstream of the pump in a piping system (i.e.,

$\Delta X = X_2 - X_1$). The entire right-hand term may be replaced by $\Delta P_{\text{pump}}/\rho$, where ΔP_{pump} is the static pressure rise across the pump since typically the changes in the kinetic and the gravitational potential energy are quite small between the inlet and outlet of the pump. The hydraulic power required by the pump $\dot{W}_{\text{Hy}}$ may then be written as:

$$\dot{W}_{\text{Hy}} = \frac{\Delta P_{\text{pump}}\,\dot{V}}{\eta_{\text{Hy}}} \tag{3.21}$$

where $\dot{V}$ is the volumetric flow rate, and η_{Hy} is the pump hydraulic efficiency. In the case of a reciprocating pump, the volumetric efficiency which accounts for leakage of the liquid should also be taken into account. Any mechanical losses associated with running the pump such as friction in bearings may be accounted for by the mechanical efficiency. These various efficiencies appear multiplicatively in the denominator of Equation 3.21 to result in the total power required by the pump. Additionally, the total power required to operate the pump should account for the efficiency of the driver.

3.7.1.8 System Requirements and Pump Characteristics Process and power systems do not necessarily operate at constant flow rates. For example, in a Rankine power cycle, the demand for power (load) may decrease, which in turn will require less heat being input to the cycle and a correspondingly lower flow rate of the working fluid. The decrease in flow rate causes a corresponding decrease in the system pressure drop. In order to match a pump that operates satisfactorily in the entire range of flow rates and corresponding pressures, the system curve with the total head difference across the system versus the flow rate is plotted on to the pump characteristics plot. The pump characteristics as depicted in Figure 3.3 are represented in the form of a curve of flow rate against head for a given set of impeller sizes φ_i or rotational speeds. Also included in this plot are the efficiency η_i contours. The NPSH_R and pumping

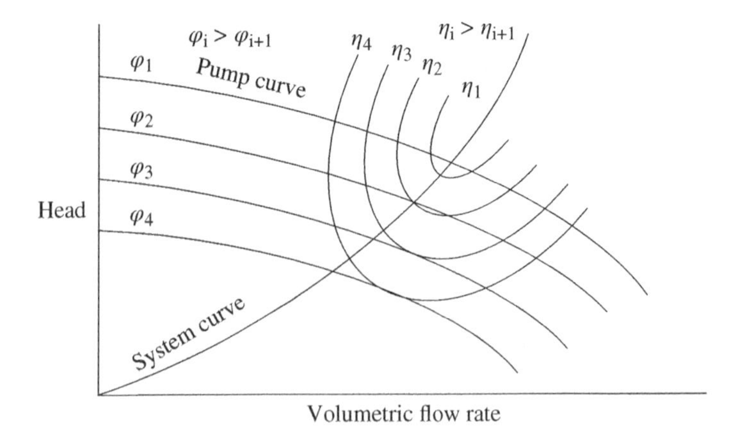

FIGURE 3.3 Centrifugal pump characteristics and system requirements

power are also often included. These characteristic curves are typically available for water at 60°F in the United States or about 15°C and corrections have to be applied to correspond to the particular liquid being pumped and its temperature. For kinetic pumps, the plot of flow rate versus head shows a rapid decrease in the head with increasing flow rate while the system curve shows the opposite trend, that is, the head decreases with decreasing flow rate. For positive displacement pumps it appears as a line with a slightly negative slope, which indicates that the pump capacity is nearly independent of downstream resistance at a given speed.

The pump operating point will be at the intersection of the system curve and the pump characteristic curve and the desired operating point should be close to the best efficiency point. The selection of a suitable control valve located downstream of the pump to regulate the flow rate is also facilitated. When a reduction in the flow rate is required, the control valve has to introduce additional pressure drop to "use up" the increase in the pump discharge pressure (typically, the control valve is sized to have a pressure drop equal to half of the variable system pressure drop exclusive of the control valve at the normal design flow rate). Recirculation of a portion of the liquid back to the pump suction may be used to reduce the net flow rate but increase in liquid temperature which could lead to cavitation should be guarded against. This type of control is used especially with positive displacement pumps since they are not throttled. Speed control of the pump is another approach to control flow rate and could be used when the driver is a turbine, variable speed motors being rather expensive.

Example 3.2. As depicted in Figure 3.4, an electric motor driven pump draws 70 m^3/h of water at 25°C from a tank open to the atmosphere (at 1.014 bar) with the level in the tank 2 m below the pump center line. The pump discharges the water into a vessel at a pressure of 20 bar and the liquid level in the vessel is at a height of 10 m above its center line (the discharge pipe is assumed full of water and thus the extra height to which the water is raised above the vessel does not have to be considered due to the final downcoming "siphon leg"). The entrance to the pipe is well-rounded to minimize the entrance frictional loss, and a strainer is installed in the pump suction piping to filter out any debris (suspended solids). A check valve (swing type) to avoid back flow through the pump, a globe valve to control the level in the pressure vessel, and an orifice meter to measure the flow rate (which is used to open the valve on the recycle line when flow rate reaches a certain minimum value for the pump) are installed in the pump discharge piping. The pressure drop measured by the orifice meter is 0.25 bar but it may be assumed that 60% of this pressure drop is recovered downstream. The total piping lengths are 10 m on the suction side and 30 m on the discharge side. The isentropic efficiency of the pump is 60%, while the losses due to shaft bearings, seals, and the motor are 5% (or $\eta_{mech+elec} = 95\%$). Calculate the required pipe size assuming schedule 40 steel pipe[5] and the electric power required by the motor under these conditions.

[5] The dimensionless schedule number specifies the pipe wall thickness and is given by: 1000 (internal working pressure)/allowable stress.

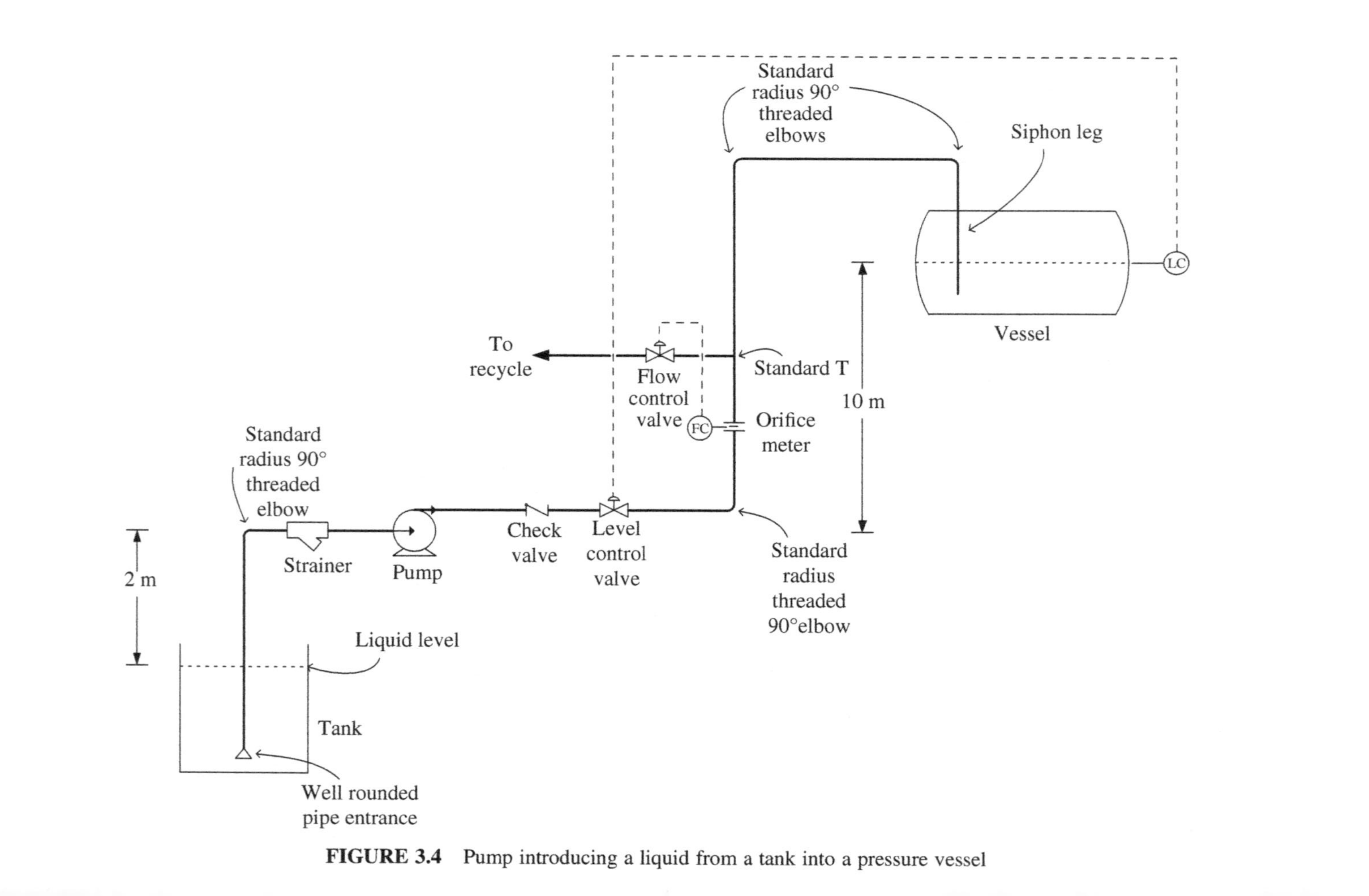

FIGURE 3.4 Pump introducing a liquid from a tank into a pressure vessel

TABLE 3.1 Selected piping parameters

	Units	Suction	Discharge
Fluid flow rate ($\dot{V}$)	m^3/h	70	70
	m^3/s	0.01944	0.01944
Maximum velocity allowable	m/s	2.4	4.6
Corresponding minimum diameter	mm	102	73
Closest schedule 40 nominal pipe size or DN (diameter nominal) (Crane Co., 1988)		100	80
Outside diameter (D_o)	mm	114.3	88.9
Wall thickness (t)	mm	6.02	5.486
Inside diameter ($D_i = D_o - 2t$)	mm	102.26	77.93
Fluid velocity corresponding to selected pipe diameter (v)	m/s	2.37	4.08
Fluid density (ρ)	kg/m^3	997.0 at 25°C	997.0 at 25°C
Fluid viscosity (μ)	kg/m/s	0.00091 at 25°C	0.00091 at 25°C
Reynolds number ($Re = \frac{D_i v \rho}{\mu}$)		265,236	348,052
Pipe roughness for clean steel pipe (ε) (Crane Co., 1988)	m	4.57×10^{-5}	4.57×10^{-5}
Relative roughness (ε/D)		0.000447	0.00059
Fanning's friction factor from Equation 3.7 (f)		0.004652	0.00479
Moody's friction factor ($f' = 4 \times f$)		0.01861	0.01916
Friction factor for flow in zone of complete turbulence, (f'_T) (Crane Co., 1988)		0.01700	0.01800

Solution

Table 3.1 summarizes the given data and selected piping parameters using maximum allowable fluid velocity criteria discussed previously in this chapter, while Table 3.2 summarizes the pressure losses for the suction and discharge sides of the pump.

Source pressure (P_{source}) = 1.014 bar; destination pressure ($P_{destination}$) = 20 bar.

Source elevation (Z_{source}) = −2 m; discharge elevation ($Z_{destination}$) = 10 m.

Pressure rise required across pump,

$$\Delta P_{pump} = \Delta P_{suct} + \Delta P_{disch} + P_{destination} - P_{source} + Z_{destination}\, g\, \rho_{disch} - Z_{source}\, g\, \rho_{suct}$$
$$= 0.12 + 1.4 + 20 - 1.014 + 10 \times 9.81 \times 997.0/100{,}000 - (-2 \times 9.81 \times 997.0/100{,}000)$$
$$= 21.68 \text{ bar or } 2{,}168{,}000 \text{ Pa}.$$

Then electrical power required by motor:

$$\dot{W}_{elec} = \frac{\dot{W}_{Hy}}{\eta_{mech+elec}} = \frac{\Delta P_{pump} \dot{V}}{\eta_{Hy}\eta_{mech+elec}} = \frac{2{,}168{,}00 \times 0.01944}{0.6 \times 0.95} = 73{,}940 \text{ W} = 74 \text{ kW}.$$

TABLE 3.2 Pressure losses (most K values are from Technical Paper No. 410 by Crane Co., 1988)

	L (m)	K/f'_T or $K/f'=$ L_e/D	K	Quantity	$\Delta P = W_f\rho$ $=KV^2\rho/2$ (kg/m/s² or N/m² or Pa)	ΔP (bar)
Suction						
Inlet loss (well-rounded)			0.04	1	112	0.00
Elbows (standard threaded)		30.00	0.51	1	1,425	0.01
Strainer (Y-type, 1–10 depending on how clean)			2.00	1	5,588	0.06
Straight pipe (calculated $K=f'\ L/D$)	10	97.79	1.82	1	5,084	0.05
Total ΔP_{suct}					12,209	0.12
Discharge						
Elbows (standard threaded)		30	0.54	3	13,421	0.13
Check valve (swing type)		100	1.7	1	14,084	0.14
Control valve (globe, fully open)		340	5.78	1	47,886	0.48
Orifice meter (40% of indicated ΔP)				1	10,000	0.10
T junction (standard with no flow through branch)		5.4	0.092	1	761	0.01
Exit loss			1	1	8,285	0.08
Straight pipe (calculated $K=f'L/D$)	30	293.37	5.46	1	45,229	0.45
Total ΔP_{disch}					139,666	1.40

3.7.2 Compressors

Compressors are used to transport gases through piping systems by adding mechanical energy to the gas such that its pressure is increased to overcome the frictional forces in the piping system and meet the pressure requirement at the discharge end of the piping system. The three main types of compressors are: the dynamic compressors, positive displacement compressors, and ejectors. These categories may be further subdivided into various subcategories. In dynamic compressors, work is performed by applying inertial forces to the gas with revolving bladed impellers or rotors. In positive displacement compressors, a trapped amount of gas is reduced in volume with a subsequent increase in pressure (and temperature) and then released at the higher pressure. Ejectors increase the pressure of a gas or vapor at the expense of another fluid's pressure. The main types of compressors that may be most often encountered in energy plants are described in the following.

3.7.2.1 Centrifugal Compressors The gas enters this dynamic compressor near the center of the rotating impeller contained within a casing and is accelerated in the outward direction by the centrifugal action of the vanes or blades. The impeller consists of radial vanes fitted to a hub and may be open, semi-closed, or shrouded as in centrifugal pumps. The kinetic energy of the fluid increases from the center of the impeller to its tips and as the fluid exits the vanes, the kinetic energy is converted into pressure by decelerating the gas within the radial diffusers and further by the discharge diffusers. In multistage centrifugal compressors, a series of impellers are mounted on a shaft and the liquid leaving the first impeller enters the second and so on. In this manner, the gas can be compressed to a much higher pressure without compromising efficiency.

Three types of blades may be configured: the radial, the backward curved, and the forward curved, with respect to the direction of motion. The dependence of the head developed or pressure ratio on flow rate is a function of the type of blade; with the radial blades, the head developed remains essentially flat, while with the backward curved blades, it decreases with increase in flow rate and with the forward curved blades it increases. Most compressors used in process plants consist of either radial or backward curved blades. Backward curved blades provide higher efficiency and stability over a wider range of flow rates while radial blades have simplicity in design.

3.7.2.2 Axial Compressors This dynamic compressor also consists of a rotor mounted within a casing but the blades attached to the rotor move between rows of fixed blades attached to the casing. The fixed blades direct the gas at the proper angle of attack to the rotating blades while the rotating blades impart aerodynamic lift to the gas. A multistage compressor contains series of alternating rotating and stationary blades. The gas enters and exits the blades in nearly the axial direction. This type of compressor is used in larger scale gas turbines and for supplying the large quantity of pressurized air required in an air separation unit to provide O_2 for an IGCC.

3.7.2.3 Reciprocating Compressors Compression in this positive displacement compressor is achieved by drawing the gas into a cylinder and pressurizing it by the movement of a piston within the cylinder. The thermodynamic cycle for this compressor is depicted in Figure 3.5. During the intake stroke, gas is drawn into the cylinder at lower pressure P_{in} while the cylinder volume increases from V_1 to V_2 with the suction valve open and the discharge valve closed. During compression, both the suction and the discharge valves are closed while at the completion of the compression stroke when pressure is increased to P_{disch} and the cylinder volume is reduced to V_3, the discharge valve is opened releasing the higher pressure gas from the cylinder. At the end of this delivery stroke, the cylinder volume is reduced to V_4 and the discharge valve closes while the remaining gas contained within the cylinder clearance remains at the discharge pressure. This small remaining volume V_4 known as the clearance volume caused by the space remaining within the cylinder when the piston is at this most advanced position in its travel is required to prevent contact between the piston and the head of the cylinder. Next, with both valves closed, the piston continues to move and the gas expands to the suction pressure P_{in} and then additional gas is drawn in after the intake vale opens while the piston moves continuing the cyclic operation. Because the suction and the

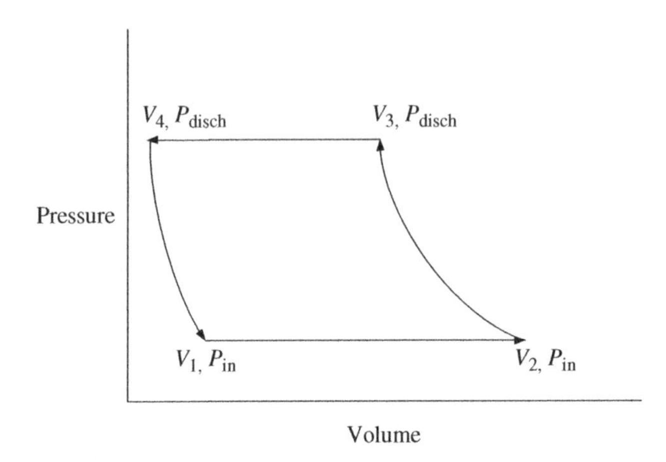

FIGURE 3.5 Pressure-volume diagram for a reciprocating compressor

discharge valves are open during only a part of the piston stroke, the flow is pulsating and it should be dampened to provide a smooth flow into and out of the compressor as well as to reduce vibration and prevent the compressor from overload or underload conditions. The simplest approach for dampening the pulsations is to include a chamber for the gas close to the cylinder inlet or outlet. Pulsation dampeners which are internally baffled devices reacting to the pulsation pressure signals are also used. Reciprocating compressors are selected over centrifugal and axial compressors when significantly higher pressures at lower flow rates are required. In double-acting reciprocating compressors, the cylinder has inlet and outlet ports at both ends of the cylinder. Gas is drawn into the cylinder at the back end as the piston moves forward, while gas is discharged at the front end. The sequence is reversed when the piston direction is reversed. With double-acting compressors, the pulsations are lower than in the single-acting compressors.

3.7.2.4 Rotary Screw Compressors These positive displacement compressors use two rotating helical screws to compress the gas. In dry rotary compressors, the two screws maintain precise alignment using timing gears, while in oil-flooded rotary compressors, lubricating oil bridges the space between the screws to provide a hydraulic seal as well as transfers mechanical energy between the driving and driven screws. Gas moves through the threads as the screws rotate and the meshing screws force the gas to the high-pressure side of the compressor. The effectiveness of this compression method depends on precisely fitting clearances between the two screws, and between the screws and the chamber containing the screws. This type of compressor finds its use in applications where the gas flow rate is much lower than that suitable for centrifugal compressors such as for compressing the fuel gas for small power plant applications.

3.7.2.5 System Requirements and Compressor Characteristics As mentioned previously under sections addressing pumps, process systems do not necessarily operate at constant flow rates. For example, in a gas-fueled power plant, the demand for

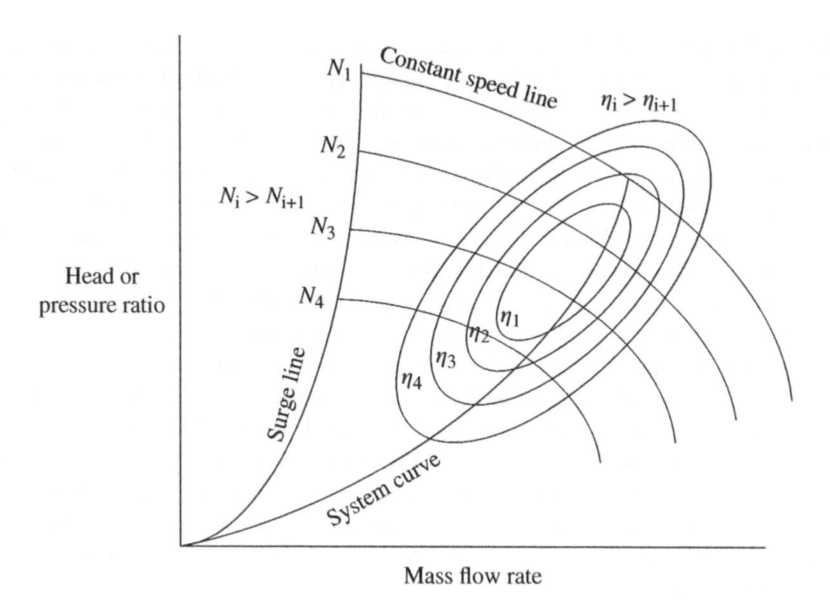

FIGURE 3.6 Dynamic compressor characteristics and system requirements

power (load) may decrease, which in turn will require less energy being input to the cycle and a correspondingly lower flow rate of the gaseous fuel. The decrease in flow rate causes a corresponding decrease in the system pressure drop. In order to match a compressor that operates satisfactorily in the entire range of flow rates and corresponding pressures, the system curve with the total head difference across the system versus the flow rate is plotted on to the compressor characteristics plot. The compressor characteristics as depicted in Figure 3.6 are represented in the form of a curve of flow rate against head for a given rotational speed N_i. Also included in this plot are the efficiency η_i contours, and in the case of dynamic compressor, the operating stability line is also shown which represents the lower flow limit for a given head under which surge may occur. At surge, flow reversal can occur causing severe damage to the machine. For both centrifugal and axial compressors, the plot of flow rate versus head at a constant speed shows a decrease in the head with increasing flow rate but the steepness of this curve is much higher for axial compressors. The system curve shows the opposite trend, that is, the head decreases with decreasing flow rate. When the system flow rate decreases, the lower pressure requirement per the system curve can be met by either speed control if the compressor is driven by a turbine or a variable speed motor (such motors tend to be more expensive than constant speed motors, however) or the less efficient use of a valve is required. For reciprocating compressors at constant speed, the characteristic curve appears as an almost vertical line (with a negative slope) which indicates that the compressor flow capacity is nearly independent of downstream resistance at a given speed.

The compressor operating point will be at the intersection of the system curve and the compressor characteristic curve and the desired operating point should be close to the best efficiency point. Only in dynamic compressors where suction or discharge throttling may be used, the selection of a suitable control valve to regulate the flow rate is also facilitated. When a reduction in the flow rate is required, the control valve has to introduce additional pressure drop to "use up" the increase in the compressor discharge pressure (typically the control valve is sized to have a pressure drop equal to half of the variable system pressure drop exclusive of the control valve at the normal design flow rate). Moveable inlet guide vanes may be used for large centrifugal and axial compressors to throttle the gas flow rate by changing the angle of these guide vanes but without resulting in a significant pressure drop. The effect is: altering the characteristic of the first stage of compression and lowering the surge limit; their effect becomes less pronounced as the total number of stages is increased. Recirculation of a portion of the gas back to the compressor suction may be used to reduce the net flow rate and a heat exchanger to cool the recycle gas should be included since the compression work performed raises the temperature of the gas. This type of control is especially used for positive displacement compressors since they are not throttled. Speed control of the compressor is another approach as mentioned previously to control flow rate and could be used when the driver is a turbine, variable speed motors being rather expensive.

3.7.2.6 Compression Power and Intercooling For an ideal gas undergoing adiabatic compression, the compression power is inversely proportional to the absolute inlet temperature as will be seen from Equation 3.26. Since work is performed on the gas, its temperature increases; cooling the gas during compression reduces the compression power since it lowers the average gas temperature within the compressor. A practical means of achieving this is to cool the gas between the compressor stages in a heat exchanger against a cooling medium such as water or air and readmitting it to the compressor for further compression. The pressure ratio split π to locate a given number of intercoolers to minimize the compression power when the gas returned for compression after intercooling is at the same temperature as the gas entering the compressor and the intercooler pressure drop is small, is given by Equation 3.22:

$$\pi = \left(\frac{P_{out}}{P_{in}}\right)^{1/(n+1)} \tag{3.22}$$

where P_{in} and P_{out} are the inlet and outlet pressures, and n is the number of intercoolers. For example, in the case of a single intercooler, the overall pressure ratio of the stages upstream and downstream of the intercooler must be the same, assuming the individual stage pressure ratios are identical. A large number of intercooled stages do not necessarily mean an overall reduction in compression power since a point of diminishing returns is reached where the additional pressure drop associated with the intercoolers may negate the reduction in power associated with a reduction in temperature. Cost of the intercoolers should also be taken into account while determining the suitable number of intercooled stages. In some cases, intercooling may be achieved by directly injecting a suitable liquid into the gas instead of using a heat exchanger where the gas and the cooling medium do not come into direct contact. In addition to decreasing the compression power, temperature may also affect the compressor materials of construction as well as the design of the compressor. Typically, the upper limit for gas temperature in centrifugal compressors used in the process industry is around 150°C especially when the gas has high concentration of oxygen since O_2 at high partial pressures is very reactive. For gases with high concentrations of chlorine or acetylene, the temperature may have to be limited to as low 90°C for safety reasons. Axial compressors for air (e.g., in gas turbine applications) or for an inert gas such as nitrogen, however, may be designed for much higher temperatures, in excess of 600°C. The outlet conditions from a compressor may be calculated assuming an isentropic path and then applying the appropriate efficiency to determine the actual outlet conditions for a given inlet set of conditions. For a given gas (with composition defined) at inlet set of conditions of temperature and pressure, the inlet enthalpy and entropy are determined. For isentropic compression, $\Delta S_{\text{in}} = \Delta S_{\text{out,ideal}}$, where $\Delta S_{\text{out,ideal}}$ is the outlet entropy. With $\Delta S_{\text{out,ideal}}$ known for a given outlet pressure, the enthalpy is determined. An adiabatic efficiency η_A is calculated from the polytropic efficiency η_P and then utilized in establishing the actual enthalpy of the outlet stream. The polytropic efficiency if available from the equipment manufacturer or derived from empirical relationships that take into account the gas flow rate (and hence the size of the compressor) may be used. The relationship between these two types of efficiencies for an ideal gas is given by Equation 3.23:

$$\eta_A = \frac{\left[(P_{\text{out}}/P_{\text{in}})^{(\kappa-1)/\kappa} - 1\right]}{\left[(P_{\text{out}}/P_{\text{in}})^{(\kappa-1)/(\eta_P\kappa)} - 1\right]} \tag{3.23}$$

where κ is the ratio of the specific heats, C_p/C_v. The polytropic efficiency is the isentropic efficiency of an infinitesimal adiabatic compression step whereas the adiabatic efficiency represents the isentropic efficiency of the entire unintercooled compression step (i.e., for a process with no heat interactions with the surroundings). The polytropic efficiency by definition is independent of the overall pressure ratio of the compression step and allows for a more accurate comparison between compressors of different overall pressure ratios. Once the adiabatic efficiency is calculated from the polytropic efficiency, the outlet enthalpy may be determined by Equation 3.24:

$$\Delta H_{\text{out}} = \Delta H_{\text{in}} + \left(\Delta H_{\text{out,ideal}} - \Delta H_{\text{in}}\right)/\eta_A \tag{3.24}$$

where ΔH_{in} and ΔH_{out} are the inlet and outlet enthalpies, $\Delta H_{\text{out,ideal}}$ is the outlet enthalpy corresponding to the isentropic path. With the outlet enthalpy established, the temperature is determined. The work required or produced is given by the difference in the actual outlet and inlet enthalpies. For an ideal gas, the power required for adiabatic compression $\dot{W}_{\text{Gas}}$ at an inlet volumetric flow rate of $\dot{V}_{\text{in}}$ may be calculated directly from Equation 3.25 or Equation 3.26:

$$\dot{W}_{\text{Gas}} = \frac{\kappa P_{\text{in}} \dot{V}_{\text{in}}}{(\kappa - 1)\eta_{\text{A}}} \left[\left(\frac{P_{\text{out}}}{P_{\text{in}}} \right)^{(\kappa-1)/\kappa} - 1 \right] \tag{3.25}$$

Or in terms of temperature T_{in}, molar flow rate $\dot{n}_{\text{in}}$, and the universal gas constant R_{u}:

$$\dot{W}_{\text{Gas}} = \frac{\kappa R_{\text{u}} T_{\text{in}} \dot{n}_{\text{in}}}{(\kappa - 1)\eta_{\text{A}}} \left[\left(\frac{P_{\text{out}}}{P_{\text{in}}} \right)^{(\kappa-1)/\kappa} - 1 \right] \tag{3.26}$$

It may be seen from Equation 3.26 that the power required by the compressor is directly proportional to the absolute temperature of the gas entering the compressor.

The above equation may be extended to cover nonideal gases to a certain extent by introducing the compressibility factor in front of T_{in}. The total power required to operate the compressor should account for the mechanical efficiency due to losses such as friction in bearings and gear if used, as well as the efficiency of the driver.

Example 3.3. 2500 kmol/h of an oxygen stream (assume 98% O_2 purity with the rest all Ar, that is, neglecting the typically accompanying N_2 content) at a temperature of 25°C and a pressure of 2 bar is to be compressed to a pressure of 45 bar. The polytropic efficiency of the compressor is 88%, while the motor driving the compressor has an efficiency of 97%. A gear is required to match the compressor and the motor speeds with a corresponding power loss of 2%. It is necessary to limit the temperature of the oxygen to 150°C due to safety considerations, O_2 at high partial pressures being very reactive. Determine the minimum number of intercoolers required to limit the temperature to the stated 150°C and the electric power required by the motor, assuming the oxygen stream behaves as an ideal gas. The oxygen stream may be cooled to 25°C in the intercooler. Pressure drop for the intercooler (along with the associated piping) in bar may be estimated by $\Delta P_{\text{intercooler}} = 0.09 (P_{\text{in,intercooler}})^{0.7}$, where the intercooler inlet pressure $P_{\text{in,intercooler}}$ is in bar (this correlation should be used with caution when the inlet pressure is very high as the resulting large pressure drop calculated may not be necessary), while the compressor bearing and seal losses in kilowatts may be estimated by Scheel's (1961) equation (converted from the original gas horsepower to kilowatts), $\dot{W}_{\text{loss, bearing + seal}} = 0.8386\,(\text{kW})^{0.4}$.

Solution

The first step is to get an initial guess for the number of "process stages" of compression (with one intercooler, the resulting number of process stages is two while each stage will typically have many mechanical stages of compression). For isentropic–adiabatic compression of an ideal gas, $(P_{\text{out}}/P_{\text{in}}) = (T_{\text{out}}/T_{\text{in}})^{\kappa/(\kappa-1)}$, where the subscripts out and in refer to the outlet and inlet streams of a given stage.[6] Substituting the numerical values, $T_{in} = 25 + 273 = 298K$, $T_{out} = 150 + 273 = 423$ K, an average $\kappa = 1.4$ for this mixture which is essentially made up of a diatomic gas (when the change in temperature and pressure over a process stage is high enough to have an effect on κ, a trial-and-error procedure is required to calculate the average κ), we obtain $(P_{out}/P_{in}) = 3.4$. For nonisentropic compression, each of the pressure ratios will have to be < 3.4 to keep the discharge temperature less than or equal to150°C or 423 K. With two process stages, while neglecting $\Delta P_{intercooler}$, the pressure ratio will be $(45/2)^{1/2} = 4.7$, which is greater than 3.4 and not acceptable. With three stages, the pressure ratio will be $(45/2)^{1/3} = 2.8$, which is less than 3.4 and thus three process stages may suffice. The actual pressure ratio per stage will have to be slightly higher to compensate for the $\Delta P_{intercooler}$. A trial-and-error procedure may be used by assuming a slightly larger per stage pressure ratio, calculating the outlet pressure from the first stage and the $\Delta P_{intercooler}$ between the first and the second stages, doing the same for the second stage and so on and then comparing the discharge pressure from the final stage with the required value.

Once the pressure profile is established, Equation 3.23 may be utilized to calculate η_A, and Equation 3.26 to calculate $\dot{W}_{Gas}$. Finally, the electric power required by the motor may be calculated after accounting for the bearing and seal losses (by Scheel's correlation), gear losses of 2%, and motor efficiency of 97% as:

$$\dot{W}_{\text{power to motor}} = \left(\dot{W}_{\text{Gas}} + \dot{W}_{\text{loss, bearing + seal}}\right)/(1-0.02)/0.97$$

The results of these calculations are summarized in Table 3.3. The total electric power required by the motor is $2650 + 2650 + 2650 = 7950$ kW. The power required by each of the stages is the same because the inlet temperature, molar flow rate, pressure ratio, and adiabatic efficiency are identical.

3.7.3 Fans and Blowers

Fans and blowers have much lower pressure ratios than those developed by compressors while their discharge pressures are also significantly lower. The American Society of Mechanical Engineers (ASME) defines fans as having maximum pressure ratio of 1.11 and discharge pressure of 1136 mm Wg (gauge pressure measured in terms of height of a water column), and defines blowers as having higher pressure ratios with a maximum of 1.2 and higher discharge pressure with a maximum of 2066 mm Wg. Fans are designed as axial as well as centrifugal machines while the main type of blowers are centrifugal and positive displacement machines.

[6] The corresponding expression for nonideal compression by incorporating the adiabatic efficiency is: $\left(\frac{P_{out}}{P_{in}}\right) = \left[\eta_A\left(\frac{T_{out}}{T_{in}} - 1\right) + 1\right]^{\kappa/(\kappa-1)}$.

TABLE 3.3 Compressor power requirement

Process stage	Unit	1	2	3
T_{in}	K	298	298	298
P_{in}	Bar	2	5.52	15.45
$\Delta P_{intercooler}$	Bar	0.31	0.63[a]	
P_{out}	Bar	5.83	16.08	45.00
η_P	%	88.00	88.00	88.00
η_A	%	86.09	86.09	86.09
T_{out}	K	422	422	422
$\dot{n}_{in}$	kmol/s	0.6944	0.6944	0.6944
R_u	kJ/kmol/K		8.314	
$\dot{W}_{Gas}$	kW	2500	2500	2500
$\dot{W}_{loss,\ bearing+seal}$	kW	19	19	19
$\dot{W}_{loss,\ gear}$	%	2	2	2
η_{motor}	%	97	97	97
$\dot{W}_{electrical}$	kW	2650	2650	2650

[a] Such a high pressure drop may not be necessary and should be determined by a trade-off analysis between equipment cost and electrical power.

3.7.4 Expansion Turbines

Turbines are used for producing work from a pressurized gas or vapor by expansion. These machines may be considered as compressors operating in reverse. The radial inflow turbine is analogous to the centrifugal compressor while the axial flow turbine is analogous to the axial compressor. Like their compressor counterparts, the radial inflow turbine is suitable for flow rates that are smaller than those suitable for an axial turbine. The first application of the axial turbine was for the production of work from steam and this technology was later applied for the expansion of pressurized gases. Axial turbines also contain alternating rows of stationary and rotating blades like axial compressors.

3.7.4.1 Expansion Power and Reheat The work that may be extracted from a pressurized gas or vapor increases as its temperature is increased; for an ideal gas the work produced is in direct proportion to the absolute temperature of the gas being expanded.

Reheating the gas after partial expansion, analogous to intercooling in the case of compression may also be utilized to increase the amount of work extracted. Chapter 6 discusses the various types of expansion cycles in detail. The outlet conditions from an expander may be calculated assuming an isentropic path and then applying the appropriate efficiency to determine the actual outlet conditions for a given inlet set of conditions. For a given gas composition and inlet set of conditions of temperature and pressure, the inlet enthalpy and entropy are determined. For isentropic compression, $\Delta S_{in} = \Delta S_{out,ideal}$, where $\Delta S_{out,ideal}$ is the outlet entropy. With $\Delta S_{out,ideal}$ known for a given outlet pressure, the enthalpy is determined. The enthalpy of the gas or vapor leaving the expander may be determined by:

$$\Delta H_{out} = \Delta H_{in} + \left(\Delta H_{out,ideal} - \Delta H_{in}\right) \eta_A \tag{3.27}$$

where ΔH_{in} and ΔH_{out} are the inlet and the actual outlet enthalpies, $\Delta H_{out,ideal}$ is the outlet enthalpy corresponding to the isentropic path. With the actual outlet enthalpy established, the temperature is determined. The work required or produced is given by the difference in the actual outlet and inlet enthalpies. Adiabatic efficiencies are typically dependent upon the volumetric flow rate of the gas being expanded as well as the inlet conditions of temperature, pressure and the expansion pressure ratio (i.e., ratio of inlet pressure to outlet pressure). For steam turbines greater than 16.5 MW, empirical correlations developed by Spencer et al. (1974) are available.

An equation similar to Equation 3.25 may be used to calculate the power developed by the turbine in the case of an ideal gas:

$$\dot{W}_{Gas} = \frac{\kappa P_{in} \dot{V}_{in} \eta_A}{(\kappa - 1)} \left[1 - \left(\frac{P_{out}}{P_{in}}\right)^{(\kappa-1)/\kappa}\right] \tag{3.28}$$

Or in terms of temperature T_{in}, molar flow rate $\dot{n}_{in}$ and the universal gas constant R_u:

$$\dot{W}_{Gas} = \frac{\kappa R_u T_{in} \dot{n}_{in} \eta_A}{(\kappa - 1)} \left[1 - \left(\frac{P_{out}}{P_{in}}\right)^{(\kappa-1)/\kappa}\right] \tag{3.29}$$

It may be seen from Equation 3.29 that the power extracted by the turbine is directly proportional to the absolute temperature of the gas entering the turbine.

REFERENCES

Avallone E, Baumeister T, Sadegh A, editors. *Marks' Standard Handbook for Mechanical Engineers*. 7th ed. New York: McGraw-Hill; 2006.

Beggs HD, Brill JP. A study of two-phase flow in inclined pipes. *J Petrol Tech* 1973; May:607–617.

Benedict RP, Carlucci NA, Swetz SD. Flow losses in abrupt enlargements and contractions. *J Eng Power Trans ASME* 1966;88:73–81.

Carman PC. Fluid flow through granular beds. *Trans Inst Inst Chem Eng Lond* 1937;15:150–166.

Carman PC. *Flow of Gases through Porous Media*. London: Butterworths; 1956.

Collier JG. *Convective Boiling and Condensation*. 2nd ed. Berkshire: McGraw-Hill; 1972.

Crane Co. Flow of fluids through valves, fittings, and pipe. Technical Paper No. 410; Lacey, WA: Crane Co.; 1988; p 410.

Ergun S. Fluid flow through packed columns. *Chem Eng Progress* 1952;48(2):89.

GPSA Engineering Data Book. Vol. 1. Section 7. 13th ed. Tulsa, OK: Gas Processors Suppliers Association; 2013.

Green DW, editor. *Perry's Chemical Engineers' Handbook*. 8th ed. New York: McGraw-Hill; 2007.

Kozeny J. Ueber kapillare Leitung des Wassers im Boden. *Sitzungsber Wien Akad* 1927;136:271–306.

Lapple CE. Isothermal and adiabatic flow of compressible fluids. *Trans Am Inst Chem Eng* 1943;39:385.

Marjanovic P, Djordjevic V. On the compressible flow losses through abrupt enlargements and contractions. *J Fluids Eng* 1994;116:756–762.

McCabe WL, Smith JC. *Unit Operations of Chemical Engineering*. New York: McGraw-Hill; 1956.

Romeo E, Royo C, Monzón A. Improved explicit equations for estimation of the friction factor in rough and smooth pipes. *Chem Eng J* 2002;86(3):369–374.

Scheel LF. *Gas and Air Compression Machinery*. New York: McGraw-Hill; 1961.

Spencer RC, Cotton KC, Cannon CN. A method for predicting the performance of steam turbine-generators, 16,500 kW and larger. Paper no. 62-WA-209; ASME Winter Annual Meeting; Revised July 1974 version; New York.

Vogt EG, White RR. Friction in the flow of suspensions—granular solids in gases through pipes. *Ind Eng Chem* 1948;40:1731–1738.

4

HEAT TRANSFER EQUIPMENT

4.1 FUNDAMENTALS OF HEAT TRANSFER

Heat transfer plays an essential role in power plants that utilize heat engines. For example, in plants employing the Rankine cycle, heat has to be transferred from the heat source to the working fluid using heat exchangers to heat up the working fluid in the liquid state to near its boiling point, then to vaporize it, and then typically to superheat the vapor. Next, due to constraints placed by the second law of thermodynamics, a significant fraction of the heat transferred to the working fluid which is not converted to work has to be rejected to the surroundings (atmosphere). This is another example of a heat transfer operation requiring a heat exchanger to transfer the heat from the working fluid to a cooling medium such as cooling water or directly to ambient air depending on the mode of heat rejection used. The working fluid here undergoes phase change again but from the vapor to the liquid phase to complete the cycle. Thus, heat transfer between fluids may or may not involve phase change, while the corresponding heat transfer rates can differ significantly.

There are three modes of heat transfer: radiation, conduction, and convection. Consider combusting a fuel such as coal or biomass in a furnace and transferring the heat thus released into a Rankine cycle working fluid consisting of water/steam flowing within the tubes exposed to the hot combustion products. The principle mode of heat transfer from the combustion products to the tubes is by radiation. In this mode of heat transfer, thermal energy is transferred by photons emitted in the infrared and

Sustainable Energy Conversion for Electricity and Coproducts: Principles, Technologies, and Equipment, First Edition. Ashok Rao.

visible portions of the electromagnetic spectrum. Radiant heat being exchanged between two surfaces is a "two way street" since both surfaces emit radiation, the amount depending upon the surface temperatures and properties, and it is the net direction and net amount of heat transfer that is of interest from an engineering standpoint.

Next, the heat is transferred through the walls of the tubes from the outer surface to the inner surface by conduction. In this mode of heat transfer, thermal energy is transferred from regions with greater molecular kinetic energy to regions with lower molecular kinetic energy through direct collision of molecules. In addition, in the case of metals, a significant fraction of the thermal energy transported is carried by conduction-band electrons.

Finally, the heat is then transferred from the inner surface to the water by convection. In this mode of heat transfer, thermal energy is transferred both by diffusion, and by advection which involves motion of the fluid at larger scales.

A thorough understanding of the nature of heat transfer is essential in configuring, analyzing, or designing energy conversion plants as well as understanding how to account for losses associated with heat transfer and possibly finding solutions to minimize these losses. The intent of this chapter is to provide an introduction to some of the heat transfer equipment, which is essential for the analysis and design of energy conversion plants.

Temperature is the driving force for heat transfer but several material properties and operating conditions also control the heat transfer rate between two regions at differing temperatures. Depending on the mode of heat transfer, these could include thermal conductivity, specific heat, mass density, fluid velocity and viscosity, and surface emissivity in the case of radiant heat transfer.

4.1.1 Conduction

The governing equation for steady-state conduction is Fourier's law shown for one dimensional heat flow in Cartesian coordinates (with temperature within the material T a function of distance x alone) and for axially symmetrical heat flow in cylindrical coordinates (with temperature within the material T a function of the radial distance r alone):

$$\frac{dQ}{dt} = -kA\frac{dT}{dx} = -k\,(2\pi\,r\,L)\,\frac{dT}{dr} \tag{4.1}$$

where dQ/dt is the rate of heat transferred, k is a physical property of the material called the thermal conductivity through which conduction is occurring whose units depend on the units used for the other variables appearing in Equation 4.1, A is the heat transfer surface area normal to the direction of heat flow, and L is the axial length. The thermal conductivity of various materials may be found in handbooks such as *Marks' Standard Handbook for Mechanical Engineers* (Avallone et al., 2006) and *Perry's Chemical Engineers' Handbook* (Green, 2007). It varies with temperature and either the valid temperature range is specified or sometimes correlations are available to adjust its value for the required temperature.

4.1.2 Convection

There are two types of convection: natural and forced. Motion of the fluid in the case of natural convection is enabled by density differences in the different regions of the fluid caused by differences in temperature, while in the case of forced convection motion is imparted to the fluid by external forces. The governing equation for convection which is similar to Newton's law of cooling is given by:

$$\frac{dQ}{dt} = \hbar A\,\Delta T \tag{4.2}$$

where dQ/dt is the rate of heat transferred, $\hbar$ is the proportionality constant known as the heat transfer coefficient whose units depend on the units used for the other variables appearing in Equation 4.2, A is the heat transfer surface area normal to the direction of heat flow, and ΔT is the temperature difference. The heat transfer coefficient has been correlated to certain dimensionless groupings characterizing the type of flow (laminar or turbulent), direction of flow in relation to the solid surface or whether phase change is occurring, and certain physical properties of the fluid in dimensionless groupings. Such correlations are available in standard textbooks on heat transfer such as *Process Heat Transfer* (Kern, 1950) as well as in handbooks such as *Marks' Standard Handbook for Mechanical Engineers* (Avallone et al., 2006) and *Perry's Chemical Engineers' Handbook* (Green, 2007). The following are some examples for flow situations most encountered in surface type heat exchangers used in energy plants, that is, heat exchangers employing a solid surface (such as tubes) to keep the two fluids exchanging heat separate. Another type of heat exchangers involve direct physical contact between the fluids exchanging heat and are introduced in Chapter 5 as they also typically involve simultaneous mass transfer (e.g., cooling towers and humidifiers).

4.1.2.1 Heat Transfer by Free Convection from Vertical and Horizontal Flat Surfaces Heat loss calculations from the outside surfaces of storage tanks as well as furnace walls requires knowledge of the heat transfer coefficient for free convection from such surfaces. For vertical surfaces (being cooled or heated), it may be calculated by the following correlation (Drew et al., 1936):

$$\mathrm{Nu} = c_1\,[(\mathrm{Gr})\,(\mathrm{Pr})]^{c_2} \tag{4.3}$$

where Nu is the Nusselt number given by the dimensionless grouping $\hbar L/k$, Gr is the Grashof number given by the dimensionless grouping $L^3\,\rho^2\,g\,\beta\,\Delta T/\mu^2$, and Pr is the Prandtl number given by the dimensionless grouping $C_\mathrm{p}\,\mu/k$. The constants $c_1 = 0.13$

and constant $c_2 = {}^1/_3$ for turbulent range as characterized by the product of the Grashof and the Prandtl numbers (also known as the Rayleigh number) being in the range of 10^9–10^{12}. The constants $c_1 = 0.59$ and $c_2 = {}^1/_4$, for laminar range characterized by the Rayleigh number being in the range of 10^4–10^9. The various variables appearing in the dimensionless numbers are:

$\hbar$ = heat transfer coefficient

L = height of the surface

g = acceleration due to gravity

ΔT = temperature difference between the surface and the fluid

k, ρ, β, μ, and C_p = thermal conductivity, density, coefficient of thermal expansion, viscosity, and constant pressure mass specific heat of the fluid surrounding the surface, respectively.

For horizontal surfaces being cooled and facing upward (or heated and facing downward), the constants appearing in Equation 4.3, $c_1 = 0.27$ and $c_2 = 1/4$, for laminar range characterized by the product of the Rayleigh number being in the range of 10^5–10^{10} (Incropera and DeWitt, 1996).

For horizontal surfaces being cooled and facing downward (or heated and facing upward), the constant c appearing in Equation 4.3 has again different values (Incropera and DeWitt, 1996); $c_1 = 0.15$ and $c_2 = 1/3$, for turbulent range characterized by the Rayleigh number being in the range of 10^7–10^{11}. The constants $c_1 = 0.54$ and $c_2 = 1/4$, for laminar range characterized by the product of the Rayleigh number being in the range of 10^4–10^7.

4.1.2.2 Heat Transfer by Free Convection from Horizontal Pipes For estimating the heat loss from the outside surfaces of pipes, Equation 4.3 may be utilized for calculating the heat transfer coefficient for free convection using constants, c_1 and c_2 provided by Morgan (1975) for different ranges of the Rayleigh number. The physical properties of the fluid surrounding the pipe should, however, be evaluated at the film temperature defined as the arithmetic mean of the surface temperature and the fluid bulk temperature. Churchill and Chu (1975) provide a single correlation valid for Rayleigh number up to about 10^{12}:

$$\mathrm{Nu} = \left[0.6 + \frac{0.387\{(\mathrm{Gr})(\mathrm{Pr})\}^{1/6}}{\left\{1 + (0.559/\mathrm{Pr})^{9/16}\right\}^{8/27}}\right]^2$$

4.1.2.3 Heat Transfer by Forced Convection through a Tube This type of flow is encountered in equipment such as heat exchangers with the flow regime being typically in the turbulent region. The inside heat transfer coefficient is required in the calculation of the overall heat transfer coefficient that accounts for heat transfer from the bulk fluid flowing through the tube to the inside tube wall surface, the heat transfer through the wall of the tube and finally from the outside tube wall to the bulk fluid flowing outside the tube. The inside heat transfer coefficient may be calculated by the following Dittus–Boelter correlation (McAdams, 1942):

$$\mathrm{Nu} = c_1 \mathrm{Re}^{0.8}\,\mathrm{Pr}^{c_2} \tag{4.4}$$

where the Nusselt number $\mathrm{Nu} = \hbar D_i/k$, the Reynolds number $\mathrm{Re} = D_i v\rho/\mu$, and the Prandtl number $\mathrm{Pr} = C_p\mu/k$. The constants $c_1 = 0.023$ and $c_2 = 0.4$ for heating, while for cooling, $c_2 = 0.3$. The various variables appearing in these dimensionless numbers are:

$\hbar$ = inside heat transfer coefficient

D_i = inside tube diameter

v = velocity of the fluid inside the tube

k, ρ, μ, and C_p = thermal conductivity, density, viscosity, and constant pressure mass specific heat of the fluid flowing through the tube, all evaluated at the arithmetic mean of the fluid bulk temperature entering and leaving the tube.

For liquids of high viscosity, a correction factor of $(\mu/\mu_S)^{0.14}$ is applied to the heat transfer coefficient as calculated by Equation 4.4 with μ_S evaluated at the tube inside surface temperature, which may be calculated as explained later in Section 4.1.2.9.

4.1.2.4 Heat Transfer by Forced Convection over a Bank of Tubes This type of flow is again encountered in equipment such as heat exchangers with the flow regime being typically in the turbulent region. Just as the inside heat transfer coefficient is required in the calculation of the overall heat transfer coefficient, so is the outside heat transfer coefficient required. This outside heat transfer coefficient may be calculated by the following correlation (Colburn, 1933) which works well for 10 or more rows of tubes in a staggered arrangement and for $10 < \mathrm{Re} < 40{,}000$:

$$\mathrm{Nu} = 0.33(\mathrm{Re})^{0.6}(\mathrm{Pr})^{1/3} \tag{4.5}$$

The various variables appearing in the dimensionless groupings $\mathrm{Nu} = \hbar D_o/k$, $\mathrm{Re} = D_o v\rho/\mu$ and $\mathrm{Pr} = C_p\mu/k$ are:

$\hbar$ = heat transfer coefficient

D_o = outside tube diameter

v = mean velocity of fluid in the smallest cross-sectional area between tubes through which the fluid flows

k, ρ, μ, and C_p = thermal conductivity, density, viscosity, and constant pressure mass specific heat of the fluid flowing through the tube. The thermal conductivity and viscosity are evaluated at the average film temperature.

A more generalized correlation for 20 or more rows of tubes in staggered and aligned arrangements, for $1000 < \text{Re} < 2 \times 10^6$, and $0.7 < \text{Pr} < 500$, provided by Zukauskas (1987) takes the form:

$$\text{Nu} = c_1 (\text{Re})^{c_2} (\text{Pr})^{0.36} \left(\frac{\text{Pr}}{\text{Pr}_s}\right)^{1/4}$$

where values for c_1 and c_2 have been provided for different geometrical parameters. All the properties appearing in the dimensionless numbers are calculated at the arithmetic mean of the inlet and outlet conditions except Pr_s which is based on the properties at the outside surface temperature of the tube. Correction factors for number of tubes less than 20 are also provided.

4.1.2.5 Heat Transfer by Condensation outside a Tube A vapor can condense on a colder surface by forming fine drops, in which case it is called "dropwise condensation" or as a liquid film on the surface in which case it is called "film condensation." Film condensation is more common in heat exchangers in condensing service and the corresponding heat transfer coefficient may be calculated by the following correlation developed by W. Nusselt (Jacob, 1936) with $\text{Nu} = \hbar L/k$:

$$\text{Nu} = c_1 \left(\frac{g \rho_L^2 \lambda_{\text{con}} L^3}{\mu_L k_L \Delta T}\right)^{1/4} \tag{4.6}$$

The constant $c_1 = 0.943$ for horizontal tubes and 0.725 for vertical tubes. The various variables appearing in the dimensionless groupings are:

$\hbar$ = heat transfer coefficient

L = height of a vertical tube or outside tube diameter of a horizontal tube

g = acceleration due to gravity

k_L, ρ_L, μ_L = thermal conductivity, density, and viscosity of the condensate (liquid formed from the condensing vapor) calculated at its average temperature (arithmetic mean of the saturation temperature and the outside wall temperature)

λ_{con} = latent heat of condensation

ΔT = difference between the vapor temperature (its saturation temperature) and the outside tube surface temperature

Incropera and DeWitt (1996) present expressions similar to Equation 4.6 but with ρ_L^2 replaced by $\rho_L(\rho_L - \rho_V)$, where ρ_V is the density of the vapor phase, and the constant $c_1 = 0.729$ for vertical tubes. At high vapor densities, it would be preferable to use these modified expressions.

4.1.2.6 Heat Transfer by Boiling outside a Tube Analogous to condensation, a liquid can boil on a hotter surface by forming a vapor film on the surface in which case it is called "film boiling" (heat being transferred from the surface through the film to the bulk liquid) or by forming bubbles at active nuclei sites on the surface itself in which case it is called "nucleate boiling." Film boiling occurs when the temperature difference between the surface and the fluid is very high (above a critical temperature) and the resulting vapor film on the surface forms an insulating layer reducing the heat transfer coefficient. This type of boiling also increases the fouling of the tubes. Fouling consists of deposits of extraneous material that can appear on heat transfer surfaces during the operation of the heat exchanger. In the case of steam generation, the BFW will contain a certain amount of dissolved solids depending on the degree of treatment of the BFW and these solids can give rise to fouling as the water evaporates and leaves behind the solids on the heat transfer surface.

Heat exchange equipment in boiling service is typically designed to avoid film boiling unless it cannot be avoided due to very high temperature driving forces. The maximum heat flux obtained with nucleate boiling outside the tube surface of a bank of tubes may be estimated by the following relationship (Palen and Small, 1964):

$$\left(\frac{\dot{Q}}{A}\right)_{\text{max}} = c_1 \frac{p \rho_V \lambda_{\text{vap}}}{D_o \sqrt{\mathcal{N}}} \left(\frac{g \sigma (\rho_L - \rho_V)}{\rho_V^2}\right)^{1/4} \tag{4.7}$$

The constant $c_1 = 0.43$ in SI units, while the various variables appearing in Equation 4.7 along with their corresponding SI units are:

$\dot{Q}$ = heat transfer rate (J/s)
A = heat transfer surface area (m^2)
p = tube pitch, defined as the shortest distance between two adjacent tubes (m)
ρ_V and ρ_L = density of vapor and liquid phases (kg/m^3)
λ_{vap} = latent heat of evaporation (J/kg)
D_o = outside tube diameter (m)
$\mathcal{N}$ = number of tubes
g = acceleration due to gravity (m/s^2)
σ = surface tension between liquid and vapor phases (N/m).

For film boiling, the heat transfer coefficient may be estimated by the following correlation (Bromley et al., 1953):

$$\hbar = c_1 \left(\frac{k_V^3 \rho_V g \lambda' (\rho_L - \rho_V)}{\mu_V D_o \Delta T}\right)^{1/4} \tag{4.8}$$

with

$$\lambda' = \lambda_{vap}\left(1 + \frac{0.4\,\Delta T\,C_{pV}}{\lambda_{vap}}\right)^2 \tag{4.9}$$

The constant $c_1 = 4.306$ in SI units, while the various variables appearing in Equations 4.8 and 4.9 along with their corresponding SI units are:

$\hbar$ = heat transfer coefficient (J/s/m^2/K)

k_V, μ_V, and C_{pV} = thermal conductivity (J/s/m/K), density (kg/m^3), viscosity (Pa s) and constant pressure specific heat (J/kg/K) of the vapor phase

ρ_L = density of the liquid phase (kg/m^3)

g = acceleration due to gravity (m/s^2)

λ' = effective difference in heat content between vapor at its average temperature and liquid at its boiling point (J/kg)

D_o = outside tube diameter (m)

ΔT = difference between tube surface temperature and liquid at its boiling point (K).

4.1.2.7 Heat Transfer by Boiling inside a Tube Kandlikar (1990) presents a generalized correlation for both vertical and for horizontal tubes that takes into account contributions due to nucleate boiling and convective mechanisms. It incorporates a fluid-dependent parameter in the nucleate boiling term to make it applicable to different fluids. The predictive ability of this generalized correlation for water and different refrigerants (which may also be considered for low-temperature Rankine cycles) was tested and gave a mean deviation of about 16% for water data, and about 19% for the refrigerants. Other correlations cited by Kandlikar (1990) may show greater accuracy but with limited applicability to the number of fluids.

4.1.2.8 Heat Transfer from Tubes with Fins When the outside tube heat transfer coefficient is low as may be the case with a gas at low pressure, externally finned tubes may be utilized to increase the heat transfer rate per unit tube length substantially by increasing the heat transfer surface area. On the other hand, the fin itself increases the conductive heat transfer resistance and as a general rule, the fin effectiveness, defined as the ratio of the fined heat transfer rate to that without any fins should be higher than 2 to justify the use of fins. As a simplifying assumption, the surface heat transfer coefficient is sometimes assumed to be the same as an unfinned tube and then only accounting for the higher heat transfer surface area obtained by the addition of fins. For more detailed analysis, plots presented in some of the standard textbooks on heat transfer such as Process Heat Transfer (Kern, 1950) may be utilized for determining the heat transfer coefficient (as well as pressure drop) for certain types of finned tubes.

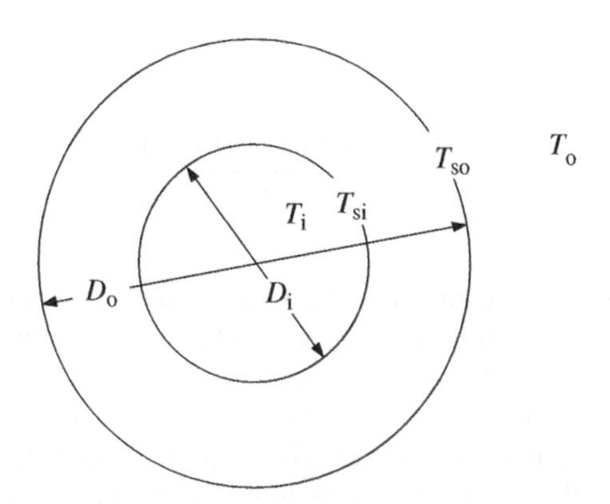

FIGURE 4.1 Heat transfer between fluids separated by tube wall

4.1.2.9 Overall Heat Transfer Coefficient for Heat Transfer between Fluids Separated by Tube Wall Consider the steady-state heat conduction from the inside to the outside surface of cylindrical wall with convection on both sides of the wall as shown in Figure 4.1. An overall heat transfer coefficient $\mathcal{U}_o$ based on the outside tube surface area A_o that accounts for the resistance to heat flow from the inside bulk fluid at temperature T_i to the tube inner surface, through the tube wall and then from the tube outer surface to the outside bulk fluid at temperature T_o, may be defined by:

$$\frac{dQ}{dt} = \mathcal{U}_o A_o (T_i - T_o) \tag{4.10}$$

Equation 4.10 may be utilized to calculate the heat transfer rate dQ/dt for a given area A_o or the area required for a given heat transfer rate if $\mathcal{U}_o$ is known. $\mathcal{U}_o$ may be expressed in terms of known quantities by applying Equation 4.1 in cylindrical coordinates for conduction through the tube wall after integrating over the entire thickness of the tube (having inside and outside diameters of D_i and D_o), and by applying Equation 4.2 for convection at the inner and outer surfaces (at temperatures T_{si} and T_{so}). At steady state,

$$\frac{dQ}{dt} = \mathcal{U}_o A_o (T_i - T_o) = \hbar_i A_i (T_i - T_{si}) = k[A_o/(D_o/2)]\frac{(T_{si} - T_{so})}{ln(D_o/D_i)}$$
$$= \hbar_o A_o (T_{so} - T_o)$$

where $\hbar_i$ and $\hbar_o$ are the inside and outside convective heat transfer coefficients between the respective fluid and the tube surface, and A_i is the inside surface area of the tube. Eliminating T_{si} and T_{so} from the above set of equations while expressing the tube outside and inside surface areas per unit length in terms of their corresponding diameters,

$$\mathcal{U}_o A_o (T_i - T_o) = \frac{1}{\dfrac{D_o}{\hbar_i D_i} + \dfrac{D_o \ln(D_o/D_i)}{2k} + \dfrac{1}{\hbar_o}} A_o (T_i - T_o)$$

or,

$$U_o = \frac{1}{\dfrac{D_o}{h_i D_i} + \dfrac{D_o \ln(D_o/D_i)}{2k} + \dfrac{1}{h_o}} \quad (4.11)$$

4.1.2.10 Cocurrent, Countercurrent, and Cross Flow The two fluids can flow either in the same direction within the heat (or mass) transfer equipment, in which case it is called a "coflow" or "parallel flow" or "cocurrentflow" configuration, or in opposite directions, in which case it is called a "countercurrent flow" configuration, or normal to each other, in which case it is called a "cross flow" configuration. The advantage with countercurrent flow is that the mean driving force for the transfer process (involving heat or mass transfer or chemical reactions) between the two fluids is the highest while the irreversibility is the lowest. Consider the example where a fluid A is to be heated to the highest temperature possible using a second fluid B available at temperature T_{hot}. Only in countercurrent flow can fluid A be heated to a temperature approaching T_{hot}, whereas in coflow arrangement, the fluid A can be heated only to a temperature approaching the final (lower) temperature of the fluid B after heat transfer. Note that the amount of thermal energy transferred is also limited with cocurrent flow.

4.1.2.11 Log Mean Temperature Difference In most cases where the hot and cold fluids are not both undergoing phase changes, the local temperature difference between the hot stream and the cold stream will not be constant throughout the heat exchanger. An effective mean temperature difference is required for use in Equation 4.10, which depends on the configuration of the heat exchanger. For simple countercurrent exchangers (where the hot and cold fluids flow in opposite directions) and cocurrent (or parallel flow) exchangers (where the hot and cold fluids flow in the same direction), it may be shown that this effective mean temperature difference is the log mean temperature difference LMTD under the following conditions: (i) specific heats of the two fluids remain constant; (ii) overall heat transfer coefficient is constant; (iii) heat transfer occurs only in the radial direction; (iv) heat losses are negligible; and (v) changes in potential and kinetic energy may be ignored. For a true countercurrent exchanger, if T_{Ci} and T_{Co} are the cold fluid inlet and outlet temperatures, and if T_{Hi} and T_{Ho} are the hot fluid inlet and outlet temperatures, then

$$\text{LMTD} = \Delta T_{ln} = \frac{(T_{Hi} - T_{Co}) - (T_{Ho} - T_{Ci})}{\ln[(T_{Hi} - T_{Co})/(T_{Ho} - T_{Ci})]} \quad (4.12)$$

For exchanger configurations with flow arrangements that are partially countercurrent and partially cocurrent as in shell and tube exchangers, which are described later in this chapter, it is common practice to calculate the LMTD as though the exchanger were in true countercurrent flow, and then to apply a correction factor F in Equation 4.10 as:

$$\frac{dQ}{dt} = U_o A_o F \Delta T_{ln} \quad (4.13)$$

The magnitude of the correction factor F, depends on the exchanger configuration and the stream temperatures. Values of F for various exchanger arrangements can be found in the form of plots in standard textbooks on heat transfer such as *Process Heat Transfer* (Kern, 1950), as well as in handbooks such as *Marks' Standard Handbook for Mechanical Engineers* (Avallone et al., 2006) and *Perry's Chemical Engineers' Handbook* (Green, 2007). Typically, if the value obtained for F is less than 0.8, it indicates that the initially selected exchanger configuration is not appropriate and a configuration that more closely approaches a countercurrent flow consisting of more shell passes or multiple shells in series should be chosen. Heat exchanger shells are described in Section 4.2. For now, consider having multiple shells as "breaking up" the exchanger into multiple smaller exchangers in series. When nonlinear heat release or absorption with respect to the fluid temperature (e.g., a fluid mixture of two or more components undergoing a phase change) is encountered, an average weighted mean temperature difference must be used and a short cut method is described by Gulley (1966). Note that in such situations, the "pinch temperature (minimum temperature difference between the two fluids at any point in the exchanger) may not occur at the inlet or at the outlet of the exchanger." In fact, "temperature cross-over" (temperature of the cold fluid being higher than that of the hot fluid) should be checked for during the plant configuration/process design phase and not simply base the design on having reasonable terminal temperature differences between the two fluids at the two ends of the exchanger.

Example 4.1. A water-cooled steam surface condenser (often operates under subatmospheric pressure or a "vacuum") to condense 100,000 kg/h of exhaust steam from a turbine is designed with a pinch temperature of 5°C (difference in temperature of the cooling water leaving the condenser and temperature of the condensing steam). The cooling water flowing through the tubes of the exchanger is supplied at 30°C and is returned to the cooling tower at 40°C (i.e., a 10°C rise or "range"), at the design point operation of the turbine. If the steam flow rate to the turbine is reduced by 25%, calculate the condenser pressure at this turned down flow rate, assuming the steam entering the condenser remains at its saturation temperature with quality (vapor fraction) of 1.0 while the cooling water flow rate and its supply temperature remain constant (a reduction in this temperature can actually be expected for a given set of ambient conditions unless the cooling tower air supply is reduced at the lower loads). Neglect any changes to the LMTD correction factor.

Solution

Dividing the rate of heat transfer for the off-design case by that for the design case:

$$\frac{Q_{\text{offdesign}}}{Q_{\text{design}}} = \frac{U_{\text{offdesign}}\, A_{\text{condenser}}\, \Delta T_{\text{ln, offdesign}}}{U_{\text{design}}\, A_{\text{condenser}}\, \Delta T_{\text{ln, design}}}$$

The ratio of the heat transfer rates in this case is simply proportional to the steam flow rates, $\left(Q_{\text{offdesign}}/Q_{\text{design}}\right) = 1 - 0.25 = 0.75$.

The heat transfer coefficient on the cooling water side as seen from Equation 4.4 may be assumed to remain constant since the cooling water flow rate is being held the same at the off-design point while the small change in its exit temperature at the off-design point having an insignificant effect on its thermal properties. The heat transfer coefficient on the steam side as seen from Equation 4.6 may also be assumed to remain constant since the change in its thermal properties will be quite small due to the small change in the condensing pressure/temperature. The change in the ΔT term appearing in Equation 4.6 will also be quite small again due to the small change in the condensing pressure/temperature while the change in the cooling water exit temperature at the off-design point will have a small reduction in the outer wall surface temperature of the tubes. The effect of these changes is further dampened by the exponent 0.25 in Equation 4.6.

With these justifications we have, $\mathcal{U}_{\text{offdesign}}/\mathcal{U}_{\text{design}} = 1$.

Next, the steam condensing temperature at the design point is calculated by adding the given pinch temperature of 5°C to the given cooling water exit temperature of 40°C giving 45°C. If $T_{\text{H,out}}$ represents the steam condensing temperature and $T_{\text{C,out}}$ the cooling water exit temperature at the off-design point, we have:

$$\frac{Q_{\text{offdesign}}}{Q_{\text{design}}} = 0.75 = \frac{\Delta T_{\text{ln, offdesign}}}{\Delta T_{\text{ln,design}}} = \frac{\dfrac{(T_{\text{H,in}} - T_{\text{C,out}}) - (T_{\text{H,out}} - 30)}{\ln[(T_{\text{H,in}} - T_{\text{C,out}})/(T_{\text{H,out}} - 30)]}}{\dfrac{(45-40)-(45-30)}{\ln[(45-40)/(45-30)]}}$$

We also need the energy balance equation to solve for the two unknowns. This may be written for the cooling water side while noting that its flow rate $\dot{m}$ remains constant and neglecting the slight variation in the specific heat C_p of water with temperature in this case, as:

$$\frac{Q_{\text{offdesign}}}{Q_{\text{design}}} = 0.75 = \frac{\dot{m}_{\text{offdesign}}\ Cp_{\text{offdesign}}\ \Delta T_{\text{offdesign}}}{\dot{m}_{\text{design}}\ Cp_{\text{design}}\ \Delta T_{\text{design}}} = \frac{T_{\text{C,out}} - 30}{(40-30)}$$

Solving the earlier two equations with $T_{\text{H,in}} = T_{\text{H,out}}$, $T_{\text{C,out}} = 37.5°\text{C}$ and $T_{\text{H,out}} = 41.5°\text{C}$. The saturated pressure of steam corresponding to 41.5°C gives the condenser pressure of 0.08 bar at this off-design point of operation.[1]

4.1.3 Radiation

When radiant energy falls on a surface, a fraction α may be absorbed, another fraction ρ reflected while the remaining τ transmitted through the surface. Then by definition,

$$\alpha + \rho + \tau = 1 \tag{4.14}$$

[1] As noted previously, steam condensers operate under subatmospheric pressure; lower the pressure, higher is the power output of the steam turbine, and lower is the heat rejection temperature consistent with the second law efficiency. One way of looking at the creation of the vacuum within the condenser is to consider the significant decrease in volume of the fluid as it goes from the vapor state to the liquid state.

Most solid materials encountered in engineering are opaque, in which case, the transmissivity $\tau = 0$ and Equation 4.14 becomes:

$$\alpha + \rho = 1 \tag{4.15}$$

For a perfect black body, the absorptivity $\alpha = 1$, while the reflectivity $\rho = 0$. Emissivity ε for a given wavelength (and temperature), which is defined as the ratio of radiant energy of that wavelength emitted (at that temperature) by a plane surface per unit area and unit time to that by a perfect black body can be shown to be identical with the absorptivity. When the emissivity is independent of the wavelength and temperature, the body is called a grey radiator. The governing equation for heat radiated from a grey surface is given by the generalization of the Stefan–Boltzmann equation:

$$\frac{dQ}{dt} = -\sigma_{SB}\,\varepsilon_S\,A\,T^4 \tag{4.16}$$

where dQ/dt is the rate of heat transferred, σ_{SB} ($= 5.67 \times 10^{-8}$ J/s/m^2/K^4) is the Stephan–Boltzmann constant, ε_S is the emissivity of the surface, A is the exposed heat transfer surface area, and T is the absolute temperature of the surface.

The net heat transfer rate between two black surfaces, one with an area A_1 at temperature T_1 and the other with an area A_2 at temperature T_2 is given by:

$$\frac{dQ_{net}}{dt} = \sigma_{SB}\,A_1\,F_{12}\left(T_1^4 - T_2^4\right) = \sigma_{SB}\,A_2\,F_{21}\left(T_1^4 - T_2^4\right) \tag{4.17}$$

where F_{12} (or F_{21}) represents a configuration or view factor based on area A_1 (or A_2) to take into account the geometrical arrangement of the two surfaces (i.e., "how much one surface views the other"). Values of the configuration factor for different geometrical arrangements may be obtained from graphs or equations found in handbooks such as *Marks' Standard Handbook for Mechanical Engineers* (Avallone et al., 2006) and *Perry's Chemical Engineers' Handbook* (Green, 2007). Overall factors that not only take the geometrical arrangement of the two surfaces into account but also account for nonblack surfaces with different emissivities (and geometries that involve reradiating surface connecting the two surfaces such as within a furnace) are also available in these handbooks.

4.1.3.1 Gas Radiation Certain gases such as the triatomic species H_2O, CO_2, and SO_2 (typically found in products of combustion) and to a lesser extent the diatomic gas CO are not transparent to radiation. They emit and absorb radiation within a narrow band of wavelengths with the emission and absorption taking place through the entire body of the gas which is dependent on the number of its molecules that at a given temperature is proportional to the gas specie partial pressure and a measure of the size of the body called the mean beam length L. Thus, the mean beam length accounts for the effect of geometry defining an isothermal gas volume and its boundary and may be found for various geometries in handbooks such as *Marks' Standard Handbook for*

Mechanical Engineers (Avallone et al., 2006) and *Perry's Chemical Engineers' Handbook* (Green, 2007). Approximate values may be estimated by:

$$L \approx 3.5 \frac{\text{Volume of body}}{\text{Its surface area}}$$

The net radiant energy exchange per unit area q_{net} between a gas at temperature T_{G} and a black surface enclosing it at temperature T_{S} is given by:

$$\frac{dq_{\text{net}}}{dt} = \sigma_{\text{SB}} \left(\varepsilon_{\text{G}} T_{\text{G}}^4 - \alpha_{\text{G}} T_{\text{S}}^4 \right) \qquad (4.18)$$

where ε_{G} and α_{G} represent the gas emissivity (defined as the ratio of the radiation from the gas to the surface to the radiation from a black body at the same temperature as the gas) and gas absorptivity (of black body radiation from the surface). Since the gas and the surface may not necessarily be at the same temperature, ε_{G} and α_{G} will not always be identical. The gas emissivity which depends on its temperature T_{G}, the product of the partial pressure of each of the gas species and the mean beam length may be determined from charts and tables presented in handbooks such as *Marks' Standard Handbook for Mechanical Engineers* (Avallone et al., 2006) and *Perry's Chemical Engineers' Handbook* (Green, 2007). The absorptivity may be assumed equal to the gas emissivity but corresponding to the surface temperature T_{S}. The emissivity of any suspended particles such as soot, coal, biomass, or ash particles should be accounted for and is discussed in the aforementioned handbooks.

4.1.3.2 Heat Loss from Insulated Pipe by Conduction, Convection, and Radiation Consider an insulated pipe carrying a hot fluid at a bulk temperature T_{i} and exposed to the ambient (still) air at temperature T_{o} as shown in Figure 4.2. Heat transfer from the hot fluid to the inner pipe wall at temperature T_{Pi} occurs by convection, followed by conduction to the outside surface of pipe (of thermal conductivity k_{P}) and at temperature T_{Po} and then through the insulating layer (of thermal

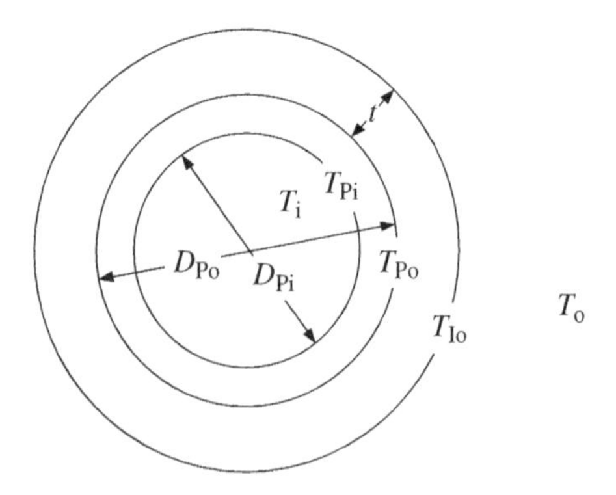

FIGURE 4.2 Heat loss from insulated pipe by conduction, convection, and radiation

conductivity k_I) at outside surface temperature of T_{Io}, and finally by combined convection and radiation from the outside surface of the insulation to the ambient air.

Radiation due to its free passage through the air can be simply added to the convective heat loss from the insulation and thus the combined heat transfer rate from the insulation surface may be written as:

$$\frac{dQ}{dt} = h_o A_{Io}\left(T_{Io} - T_o\right) + \varepsilon_I \sigma_{SB}\left(T_{Io}^4 - T_o^4\right) \tag{4.19}$$

where h_o is the convective heat transfer coefficient between the outer surface of the insulation and the ambient air, A_{Io} and ε_I are the surface area and emissivity of the outer surface of the insulation. Sometimes, it is convenient to replace the radiant heat transfer term in Equation 4.19 by a form similar to the convective heat transfer term:

$$\frac{dQ}{dt} = h_o A_{Io}\left(T_{Io} - T_o\right) + h_r A_{Io}\left(T_{Io} - T_o\right) \tag{4.20}$$

where

$$h_r = \varepsilon_I \sigma_{SB}\left(T_{Io}^2 + T_o^2\right)\left(T_{Io} + T_o\right) \tag{4.21}$$

Next, an overall heat transfer coefficient U_o based on the outside insulation surface area A_{Io} may be defined by:

$$\frac{dQ}{dt} = U_o A_{Io}\left(T_i - T_o\right) \tag{4.22}$$

Equation 4.22 may be used to calculate the heat loss from the insulated pipe if U_o is known. U_o may be expressed in terms of known quantities by solving a set of equations obtained by equating the heat transfer rates through the various sections at steady state:

$$\begin{aligned}\frac{dQ}{dt} &= U_o A_{Io}\left(T_i - T_o\right) = h_i A_{Pi}\left(T_i - T_{Pi}\right) = k_P A_{Io}\frac{\left(T_{Pi} - T_{Po}\right)}{\left(D_{Io}/2\right) ln\left(D_{Po}/D_{Pi}\right)} \\ &= k_I A_{Io}\frac{\left(T_{Po} - T_{Io}\right)}{\left(D_{Io}/2\right)\ln\left(D_{Io}/D_{Po}\right)} = \left(h_o + h_r\right) A_{Io}\left(T_{Io} - T_o\right)\end{aligned} \tag{4.23}$$

where D_{Pi} and D_{Po} are the inside and outside diameters of the pipe and $D_{Io} = D_{Po} + 2t$, t being the thickness of the insulation.

The expression for U_o may then be derived in a similar manner as was done previously for steady-state heat transfer between fluids separated by a tube wall resulting in:

$$U_o = \frac{1}{\dfrac{D_{Io}}{h_i D_{Pi}} + \dfrac{D_{Io}\ln\left(D_{Po}/D_{Pi}\right)}{2k_P} + \dfrac{D_{Io}\, ln\left(D_{Io}/D_{Po}\right)}{2k_I} + \dfrac{1}{h_o + h_r}} \tag{4.24}$$

The "radiant heat transfer coefficient" h_r, however, involves the unknown T_{Io}, which can be calculated by an iterative process by assuming a value for T_{Io}, then calculating $\mathcal{h}_r$ using Equation 4.21, substituting its value into Equation 4.24 to calculate $\mathcal{U}_o$, then calculating dQ/dt using Equations 4.20 and 4.22, and finally checking if the two values of dQ/dt match. This calculation procedure may be made less tedious, facilitated by using values for $\mathcal{h}_o + \mathcal{h}_r$ tabulated for outside diameters, D_{Io} and temperature differences, $T_{Io} - T_o$ (Jakob and Hawkins, 1957).

Example 4.2. Calculate the heat loss from a horizontal 6 in. nominal steel pipe (corresponding outside diameter $D_{Po} = 6.625$ in. or 16.83 cm) and 6 m long with 2.5 cm thickness insulation (or $D_{Io} = 21.83$ cm) with a thermal conductivity for the insulating material $k_I = 0.061$ J/s/m/°C. The pipe surface is at a constant temperature $T_{Po} = 225°C$ and the ambient air temperature is at $T_o = 20°C$.

Solution
From Equation 4.23,

$$\frac{dQ}{dt} = k_I A_{Io} \frac{(T_{Po} - T_{Io})}{(D_{Io}/2)\ln(D_{Io}/D_{Po})} = (\mathcal{h}_o + \mathcal{h}_r) A_{Io} (T_{Io} - T_o)$$

and

$$\mathcal{h}_o + \mathcal{h}_r = k_I \frac{(T_{Po} - T_{Io})}{(D_{Io}/2)\ln(D_{Io}/D_{Po})(T_{Io} - T_o)} \tag{4.25}$$

In the preceding equation, the combined convection and radiation heat transfer coefficient $\mathcal{h}_o + \mathcal{h}_r$ and the outside surface temperature of the insulation T_{Io} are not known, and an iterative procedure will be used as follows:

1. Assuming a value $T_{Io} = 50°C$, then $T_{Io} - T_o = 50 - 20°C = 30°C$, and by substituting into Equation 4.25 along with the given values, yields $\mathcal{h}_o + \mathcal{h}_r = \frac{(0.061)(225-45)}{\left(\frac{21.83}{2\times 100}\right) ln(21.83/16.83)(30)} = 12.9$ J/s/m^2/°C. But, by interpolating tabulated values given for various pipe diameters, D_{Io} and temperature differences, $T_{Io} - T_o$ (Jakob and Hawkins, 1957), $\mathcal{h}_o + \mathcal{h}_r = 10.9$ J/s/m^2/°C.
2. Assuming a value $T_{Io} = 55°C$, then $T_{Io} - T_o = 55 - 20°C = 35°C$, and by substituting into Equation 4.25 along with the given values yields, $\mathcal{h}_o + \mathcal{h}_r = \frac{(0.061)(225-45)}{\left(\frac{21.83}{2\times 100}\right) ln(21.83/16.83)(35)} = 11.1$ J/s/m^2/°C. Next, by interpolating tabulated values given for various pipe diameters, D_{Io} and temperature differences, $T_{Io} - T_o$ (Jakob and Hawkins, 1957), $\mathcal{h}_o + \mathcal{h}_r = 11.1$ J/s/m^2/°C which matches the value calculated earlier.

Then the heat loss,

$$\frac{dQ}{dt} = (\mathcal{h}_o + \mathcal{h}_r) A_{Io} (T_{Io} - T_o) = (11.1)\left(\pi \times \frac{21.83}{100} \times 6\right)(35) = 1.6 \text{ kJ/s}$$

4.2 HEAT EXCHANGE EQUIPMENT

Heat exchange equipment for transfer of thermal energy between two fluids without the two fluids coming into direct contact with each other will be considered in this section. The heat transfer in such exchangers then occurs through a barrier that keeps the two fluids separate; tubes in the case of shell and tube exchangers, which are discussed in the following. Air-cooled exchangers which also have tubes but no shell are discussed next. Finally, other equipment that utilize tubes for the transfer of heat are discussed such as HRSGs. As stated previously, heat exchange between fluids with direct contact typically also involves mass transfer between the two fluids and these types of unit operations[2] and the corresponding equipment utilized are addressed in Chapter 5.

Heat exchangers must be designed by taking into consideration the fouling characteristics of the fluids. Factors that can affect the rate of fouling are temperature, materials of construction, and surface roughness. An additional resistance to heat transfer is introduced which reduces the capacity of the heat exchanger. The deposit can be large enough to reduce the cross-sectional fluid flow area requiring additional pressure drop to maintain the original flow through the exchanger. The overall heat transfer coefficient must account for this additional fouling resistance and must also consider the necessary mechanical arrangements to permit easy cleaning of the fouled surfaces.

Allowance for fouling is largely based on experience but tables for typical resistances for various fluids are available in the Standards of the Tubular Exchanger Manufacturers Association (TEMA) (2007). TEMA is a trade association of shell and tube heat exchanger manufacturers who have pioneered research and development of heat exchangers for over 60 years and have established rules and guidelines for the design of these heat exchangers.

Example 4.3. A steam generator is being designed for heat recovery from flue gas derived from combustion of natural gas. The gas will flow outside the tubes while steam will be generated within the tubes. The outside diameter of the tubes is 50.8 mm and thickness is 7.75 mm. Calculate the increase in the heat transfer surface area of the exchanger when fouling is taken into account. Assume that the fouling factor (defined as the resistance to heat transfer caused by the layer of deposited material or inverse of the corresponding heat transfer coefficient) on the gas side is 0.00025 m^2 °C h/kJ and that on the steam side is 0.000025 m^2 °C h/kJ, and the overall heat transfer coefficient without fouling is 250 kJ/m^2/°C/h.

Solution

Including the additional resistance due to fouling on the inside surface of the tubes h_{fi} and outside surface of the tubes h_{fo} to the clean overall heat transfer coefficient

[2] A unit operation is a basic step (or subsystem) within an overall process involving physical change such as heating or cooling (heat transfer), separation (mass transfer). A unit process on the other hand involves chemical change.

given by Equation 4.11 results in the following for the overall heat transfer coefficient:

$$\mathcal{U}_{fo} = \frac{1}{\dfrac{D_o}{h_i D_i} + \dfrac{D_o \ln(D_o/D_i)}{2k} + \dfrac{1}{h_o} + \dfrac{D_o}{h_{fi} D_i} + \dfrac{1}{h_{fo}}}$$

The increase in heat transfer surface area is then:

$$\frac{A_{fo}}{A_o} - 1 = \frac{\mathcal{U}_o}{\mathcal{U}_{fo}} - 1 = \frac{\dfrac{D_o}{h_i D_i} + \dfrac{D_o \ln(D_o/D_i)}{2k} + \dfrac{1}{h_o} + \dfrac{D_o}{h_{fi} D_i} + \dfrac{1}{h_{fo}}}{\dfrac{D_o}{h_i D_i} + \dfrac{D_o \ln(D_o/D_i)}{2k} + \dfrac{1}{h_o}} - 1$$

$$= \frac{\dfrac{D_o}{h_{fi} D_i} + \dfrac{1}{h_{fo}}}{\dfrac{D_o}{h_i D_i} + \dfrac{D_o \ln(D_o/D_i)}{2k} + \dfrac{1}{h_o}}$$

$$= \mathcal{U}_o \left(\frac{D_o}{h_{fi} D_i} + \frac{1}{h_{fo}} \right)$$

Substituting the given values for the inside fouling factor $\dfrac{1}{h_{fi}} = 0.000025\ \text{m}^2\ {}^\circ\text{C h/kJ}$, outside fouling factor $\dfrac{1}{h_{fo}} = 0.00025\ \text{m}^2\ {}^\circ\text{C h/kJ}$, and inside tube diameter $D_i = 50.8 - 2 \times 7.75 = 35.3$ mm, and outside tube diameter $D_o = 50.8$ mm into the preceding expression, $\dfrac{A_{fo}}{A_o} - 1 = 250 \left(\dfrac{0.000025 \times 50.8}{35.3} + 0.00025 \right) = 0.07$ or a 7% increase.

4.2.1 Shell and Tube Heat Exchangers

For larger-scale applications, a shell and tube exchanger with multiple tubes contained within the shell are used, and often the exchanger may include a multiple of such shells.

Each shell consists of a large cylindrical vessel containing a bundle of tubes with one fluid flowing through the tubes (tube-side fluid) while the other flowing over the tube bundle (shell-side fluid). The tubes may be finned (longitudinally) to increase the heat transfer surface area. The tube bundle may be removable or nonremovable depending on the type of application.

Transverse baffles are used to direct the shell-side fluid to flow normal to the tube bank which increases the turbulence and thus the heat transfer rate, and also to provide mechanical support for the tubes. Different types of baffles are available such as the segmental baffles which are the most commonly used in shell and tube exchangers, and disc and donut baffles. The baffle spacing and the distance between baffles is governed by the required mechanical support for the tubes and the pressure drop available on the shell side. Tie rods and spacers are used to retain the baffles and tube support plates securely in position. A longitudinal baffle may be installed to subdivide the shell to create two passes for the shell-side fluid. Impingement baffles are plates located adjacent to the shell inlet nozzles to protect the tube bundle from cutting, pitting, or eroding from suspended particles or high velocity fluids and also help distribute the entering fluid into the tube bundle.

4.2.1.1 Removable Bundles There are two types of removable bundles: the U-tube and the floating head. U-tube exchangers are most economical requiring the simplest form of construction for removable bundles. Straight tubes are bent in the form of a U and are welded to a tubesheet at one end and rolled at the other end. The U-bend section of the tube bundle is free to expand in the shell eliminating the need for an expansion joint to relieve stresses from thermal expansion. This type of exchanger can be used in applications requiring extremely high pressures on one side. The design allows the shell inlet nozzle to be located beyond the U-bends eliminating the need for an impingement plate. This type of exchanger is suitable only for clean service, however, and individual tubes are difficult to replace in case of tube rupture; only the tubes in outer rows can easily be replaced. Only an even number of tube passes are possible but most industry practice is to have exchangers with an even number of passes. Figure 4.3 shows a schematic for this type of shell and tube exchanger with two shell passes made possible with the longitudinal baffle (a “2–4 exchanger” where 2 represents the number of shell passes and 4 represents the number of tubes passes). The two fluids are entering the exchanger in countercurrent

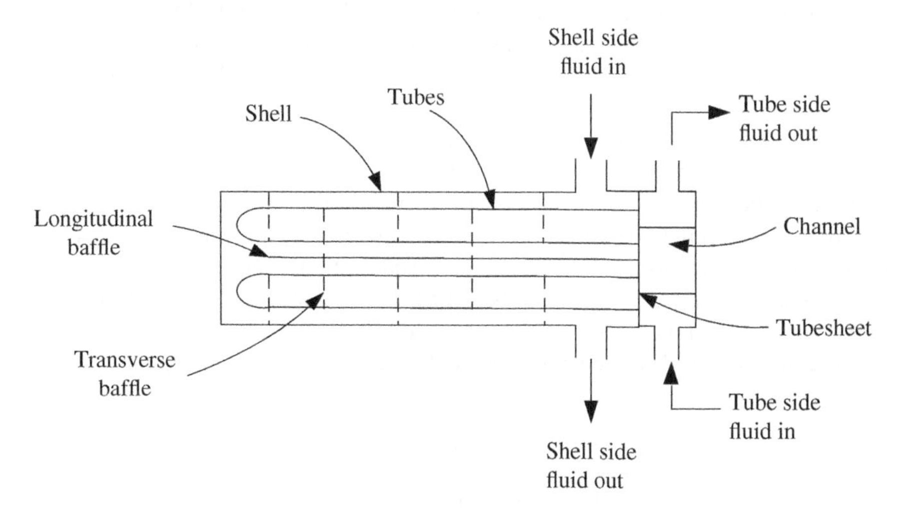

FIGURE 4.3 Shell and tube exchanger with U-tubes

directions but the shell with its transverse baffles directs the flow of the shell-side fluid normal to the tube banks for greater heat transfer coefficient.

The floating head type of heat exchanger is used where regular maintenance is an integral part of plant operation. These exchangers have straight tubes secured at both ends in tubesheets. One tubesheet is free to move, thereby providing for differential thermal expansion between the tube bundle and the shell. Tube bundles may be removed for inspection, replacement, and external cleaning of the tubes. Tube-side headers, channel covers, gaskets are also accessible for maintenance and replacement, and tubes may be cleaned internally. In the "outside-packed stuffing box" type floating head exchanger (typically TEMA[3] Type AEP), shell-side fluid is sealed by rings of packing compressed within a stuffing box by a packing-follower ring. The packing allows the floating tubesheet to move back and forth. Since the stuffing box only contacts shell-side fluid, shell-side and tube-side fluids do not mix, should leakage occur through the packing. Application is limited for shell-side services up to 600 psig and 600°F (4140 kPa gauge and 316°C) and to nonhazardous services as leakage of shell-side fluid to the environment is possible. In the "outside-packed lantern ring" type floating head exchanger (typically TEMA Type AJW), the shell-side and tube-side fluids are each sealed by separate rings of packing (or O-rings) separated by lantern rings provided with weep holes, so that leakage through either packing will be to the outside. The width of the tubesheet must be sufficient to allow for the two packings, the lantern ring, and for differential thermal expansion. A small skirt is sometimes attached to the floating tubesheet to provide a bearing surface for packings and lantern rings. Application is limited to 150 psig and 500°F (1030 kPa gauge and 260°C) and not applicable in services where leakage of either tube-side or shell-side fluid is unacceptable, or where possible mixing of tube-side and shell-side fluid cannot be tolerated. Furthermore, this design is limited to one or two tube passes only. The "pull-through bundle" type floating head exchanger (typically TEMA Type AKT) has a separate head bolted directly to the floating tubesheet with both the assembled tubesheet and head designed to slide through the shell, and the tube bundle that can be removed without breaking any joints at the floating end. Clearance requirements between the outermost tubes and the inside of the shell must provide for both the gasket and the bolting at the floating tubesheet. For an odd number of passes, the nozzle must extend from the floating head cover through the shell cover. Packed joints or internal bellows are required for both differential thermal expansion and tube-bundle removal. Application is restricted to services where mixing of shell-side and tube-side fluids does not present a hazardous situation if a failure of the internal gasket between floating tubesheet and its head occurs. In the "inside split backing-ring" type floating head exchanger (typically TEMA Type AES), the floating head cover is secured against the floating tubesheet by bolting to a split backing ring. This floating head cover is contained inside a shell cover and to dismantle the heat exchanger to remove

[3] The TEMA has designations for standard heat exchanger designs. For shell and tube heat exchangers, it consists of a series of three letters with the first letter representing the front-end stationary head of the exchanger, the second letter representing the shell type, and the third letter indicating the rear end head of the exchanger.

the tube bundle from the stationary end requires the shell cover to be removed first, followed by the split backing ring, and then the floating head cover. The nozzle must extend from the floating head cover through the shell cover for odd number of tube passes. Typically, this type of shell and tube exchanger tends to be the most expensive.

Floating head exchangers should not be used where tube-side and shell-side differential pressure exceeds about 500–1000 psi (about 3400–6900 kPa) because of potential leakage (e.g., via the internal gasket) and U-type bundles instead should be used.

4.2.1.2 Nonremovable Bundles (Fixed Tubesheet) This is the most popular type of heat exchanger in smaller-scale applications and consists of straight tubes secured at both ends in tubesheets welded to the shell. The exchanger can be designed with removable channel covers (TEMA Type AEL), bonnet type channels (TEMA Type BEM), and integral tubesheets (TEMA Type NEN) and provides maximum protection against leakage of shell-side fluid to the environment, and the shell diameter is lower than other types of shell and tube exchangers for a given heat transfer surface area, number of tubes, their diameter and length, and the number of tube passes making them relatively inexpensive. Not having a removable bundle makes the shell side not accessible for mechanical cleaning and hence, its applications are limited to clean service on the shell side. Stresses due to differential thermal expansion are relieved by including an expansion joint (bellows).

4.2.1.3 Shell Types Various types of shell designs are available, which can broadly be classified as single pass or two pass, single-pass shell (TEMA Type E) being the most common design but its use is restricted to applications where there is no temperature cross. The two-pass shell (TEMA Type F shell) is used in services where temperature cross is unavoidable due to process considerations, and where space limitations exclude the use of two or more shells in series. This type of shell is available with removable and nonremovable longitudinal baffles. A major problem with removable longitudinal baffles is fluid leakage through the clearance between the longitudinal baffle and shell, which can make the exchanger ineffective depending on the amount of leakage. Longitudinal seal strips are required to prevent the fluid leakage. In applications where shell-side heat transfer is not controlling and low shell-side pressure drop is required, then split flow and divided flow shells may be utilized (TEMA Shell Types G, H, J, and X).

4.2.1.4 Tube side Typical tube diameters used in the process industries in the United States are ¾, 1, 1½, and 2 in., with standard tube wall thicknesses ranging from 18 to 10 BWG (tube-wall thickness is frequently measured in Birmingham Wire Gauge or BWG units). Tubes of 1 in. diameter are normally used when fouling is expected because smaller sizes are difficult to clean mechanically. The maximum tube length for most manufacturers in the United States is about 40 ft (12.2 m) but typical tube lengths are 8 ft (2.4 m), 10 ft (3 m), 12 ft (3.7 m), 16 ft (4.9 m), and 20 ft (6.1 m), with the longer tubes in most applications being less costly since a smaller number of tubes, reduced size of tubesheets and flanges are required for the same heat transfer surface area. For clean streams or when chemical cleaning of the tube is possible and minimizing tube-side pressure drop is not critical, smaller diameter tubes are preferred since the heat transfer surface area is greater for a given bundle diameter.

4.2.1.5 Tube Pitch and Pattern Tube pitch is the shortest center-to-center distance between adjacent tubes while the clearance is the shortest distance between two tubes. In most shell and tube exchangers, the pitch is in the range of 1.25–1.5 times the tube diameter, and 3/16 in. (4.76 mm) is usually considered to be a minimum clearance. Either a square or triangular pattern is used for the tube layout.

4.2.1.6 Materials Tubes, shell, and shell cover are usually made from carbon steel unless process conditions and mechanical design consideration require more exotic metallurgy. Cladding or lining consisting of the more exotic metallurgy may be used for the shell with carbon steel as the base material to reduce cost. Materials of construction of baffles, tie rods, and spacers are similar to that of the shell.

4.2.1.7 Fluid Allocation Selecting which fluid flows on the tube side and which on the shell side is important from an operating and capital cost standpoint. Listed in the following are some of the factors to be considered in this selection process. Engineering judgment may have to be used when one factor favors tube-side flow for a given fluid while another factor favors shell-side flow:

1. Viscous Fluid—Generally, higher heat transfer rates are obtained when the fluid with a higher viscosity is on the shell side.
2. Toxic and Lethal Fluids—Generally, the toxic fluid should be on the tube side while utilizing a double tubesheet to minimize the possibility of leakage.
3. Flow Rate of Fluid—A more economic design results when the fluid with the lower flow rate flows on the shell side since turbulent flow can be achieved at much lower velocities with the use of transverse baffles.
4. Mixed-Phase Fluid—The fluid should be on the shell side.
5. Corrosive Fluid—A more economic design results when the corrosive fluid is on the tube side since only the tubes have to be constructed with the more expensive alloy.
6. Cooling Water—In water-cooled heat exchangers, the cooling water is typically placed on the tube side to avoid corrosion on the shell side that may occur due to pocketing and deposits at stagnation points caused by baffles. Also, cleaning of the deposits is easier on the tube side.
7. Fouling Fluid—Increased velocities tend to reduce fouling, and thus the fluid with the higher tendency to foul is placed on the tube side. Furthermore, chemical cleaning to remove the fouling can usually be more effective on the tube side while straight tubes can be physically cleaned without having to remove the tube bundle.

8. Temperature—High-temperature fluids are placed on the tube side since only the tubes have to be constructed with the more expensive alloy to withstand the high temperature.
9. Pressure—Higher-pressure fluids are placed on the tube side since the tubes with their significantly lower diameters have significantly lower hoop stresses. Furthermore, since the tube-side fluid undergoes a higher pressure drop, the absolute pressure drop penalty for a higher pressure fluid is less severe.
10. Pressure Drop—A fluid with a lower available pressure drop is generally placed on the shell side.

4.2.1.8 Double Pipe Heat Exchangers These types of exchangers can be considered special types of shell and tube exchangers with a larger pipe used to form the shell and a smaller pipe or pipes contained inside it as the tube or tubes which may or may not be finned. The exchanger can be built in a U-shape, in which case it is referred to as a hairpin heat exchanger. Double pipe exchangers are used when the required heat transfer surface area is small, approximately less than 50 m^2, or in high-pressure applications. Standard designs are typically available for pressures up to 5000 psig (345 barg) on the tube side and up to 500 psig (34.5 barg) on the shell side with special designs being available for even higher pressures. Because true countercurrent flow is achieved within this type of exchanger, the "approach temperature" (how closely one fluid approaches the other fluid temperature) can be much lower than that with conventional shell and tube exchangers, the log–mean correction factor being unity. Furthermore, cleaning, inspection, and tube replacement are easier due to the simple construction of this type of exchanger. An advantage with the hairpin design is that expansion joints are not required because of the U-tube construction.

4.2.1.9 Surface Condenser This is a special type of water-cooled shell and tube exchanger and is an important part of a steam Rankine-cycle-based plant. It maximizes the power recovery from a steam turbine by providing a constant turbine discharge vacuum while allowing the steam condensate to be recovered. Turbine exhaust steam is condensed on the shell side while cooling water on the tube side removes heat. The unit includes equipment (steam jet ejector or the more efficient mechanical vacuum "pumps") to bleed out noncondensables (O_2 and CO_2) that may otherwise accumulate on the shell side and increase the pressure. Condensate is collected in a "hotwell" and pumped back to the deaerator (for removal of dissolved gases such as O_2 and CO_2 in the water, which can cause corrosion). The shell may be constructed in many different forms, such as round, oval, rectangular, or a combination, while the hotwell may be a cylindrical pot type or bathtub design.

4.2.1.10 Reboilers These types of heat exchangers are used in mass transfer equipment such as distillation and stripping columns which are used for the separation of components from a stream containing two or more components. These types of columns provide intimate contact between a vapor flowing upward and a liquid flowing downward through packing or multiple trays to enhance the mass transfer of components between the two streams while the reboiler provides the thermal energy required for separation. It provides the required amount of vapor traffic through the column while boiling out a greater fraction of the more volatile components to assist in the overall separation process. These types of operations are discussed in Chapter 5. Most reboilers are of the shell and tube design and normally condensing steam provides the thermal energy but other heat sources such as hot oil or Dowtherm™ may be used, and in some cases fuel-fired furnaces may also be used as reboilers. There are various types of reboilers which are described in the following.

In a once-through thermosyphon reboiler, the column bottom tray liquid flows directly into the reboiler while producing the liquid bottoms product. Fluid flow through the reboiler depends upon the hydrostatic driving force and hence the name thermosyphon. The hydrostatic driving force must overcome the total pressure drop in the circuit. Once-through reboilers can either be horizontal or vertical and can have the boiling liquid either on the shell or tube side. In the recirculating thermosyphon reboiler, the liquid from the bottom tray is mixed with the liquid fraction of the reboiler return stream and recirculated back as reboiler feed. This type of reboiler (with baffle in the tower bottom) is usually preferred over the once-through type since the hydraulic design is less critical. Another type of reboiler is the kettle reboiler, which is commonly used in applications where a wide range in turndown capability or when a high vapor quality stream leaving the reboiler is required. These reboilers tend to be generally more expensive than the previously described reboilers due to the larger shell size and product surge volume requirement. The pump-through or forced circulation reboilers are generally the highest cost and used when handling a viscous liquid or in heavily fouling applications and may or may not include recirculated liquid. Table 4.1 summarizes the advantages and disadvantages of different types of reboilers that are installed outside the column.

4.2.2 Plate Heat Exchangers

A plate heat exchanger uses metal corrugated plates to transfer heat between two fluids and has a major advantage over conventional heat exchangers in that the fluids are exposed to a much larger heat transfer surface area since the fluids are spread out over the plates. These exchangers are primarily used for liquid-to-liquid heat transfer but can also be used for liquid-to-gas heat transfer. The plates are pressed together to form narrow channels (typically <1.5 mm wide) through which alternating hot and cold fluids flow. The plates may be gasketed, welded, or brazed together depending on the type of application and are compressed together in a rigid frame to form the parallel flow channels. Making each channel narrow allows a greater fraction of the fluid to come into contact with plate surface and hence better heat transfer rate. The corrugations also create turbulent flow in the fluid which increases the heat transfer coefficient. These heat exchangers can be designed economically for a very small temperature approach, as low as 1°C whereas in shell and tube heat exchangers it is typically limited to an approach of 10°C or more since the overall heat transfer coefficient can be as high as three to four times that of a shell and tube exchanger. These exchangers are useful in cryogenic operations such as in air separation plants

TABLE 4.1 Advantages and disadvantages of different types of external reboilers

Type	Advantage	Disadvantage
Kettle	• Equivalent to a theoretical stage • Stable operation and easier to control • Good turndown • Lower column skirt height	• Low to moderate heat transfer coefficient • Higher fouling tendency • Low surge time • Extra piping • Higher capital cost • Larger plot space • Higher residence time in heated zone which can cause degradation of temperature sensitive fluids • May not be suitable for corrosive process fluid since it will be on shell side and thus both shell and tube side metallurgy are effected
Vertical thermosyphon	• Once-through types may be considered equivalent to a theoretical stage • Less piping and simpler layout • Heat transfer rates can be high due to higher circulation rate • Fouling tendency low due to higher circulation rate • Low residence time in heated zone which avoids degradation of temperature sensitive fluids • Lower capital cost than kettle type • Less plot space • Suitable for corrosive process fluid since it will be on tube side and thus only tube side metallurgy is effected	• Maintenance and cleaning can be difficult • Higher column skirt height • Limited turndown characteristics and susceptible to instability • May not be suitable for operation at critical conditions as the difference in liquid and vapor densities provides the driving force for recirculation
Horizontal thermosyphon	• Once-through types may be considered equivalent to a theoretical stage • Heat transfer rates can be moderately high	• Higher column skirt height but less than vertical type • Limited turndown and susceptible to instability • Extra piping than vertical type

(continued overleaf)

TABLE 4.1 *(continued)*

Type	Advantage	Disadvantage
	• Low residence time in heated zone which avoids degradation of temperature sensitive fluids • Lower pressure drop than vertical type • Longer tubes can be utilized to reduce cost • Maintenance easier than vertical type; removable tube bundle may be utilized • Lower capital cost than kettle type	• Larger plot space than vertical type • Fouling tendency can be higher but not as high as in kettle type • Both shell and tube side metallurgy are effected with corrosive process fluid since it will be on shell side
Vertical forced circulation	• Heat transfer rates high due to high circulation rate • Fouling tendency very low due to higher circulation rate • Low residence time in heated zone which avoids degradation of temperature sensitive fluids • Lower column skirt height than vertical thermosyphon unless pump NPSH is high • Less plot space but additional space for pumps • Suitable for corrosive process fluid since it will be on tube side and thus only tube side metallurgy is effected • Suitable for high viscosity liquids	• Not equivalent to theoretical stage • Maintenance and cleaning can be difficult • Extra piping due to recirculation pumps • Higher operating cost due to pump power • Higher capital cost than corresponding thermosyphon type due to additional piping of pumps but lower than kettle type
Horizontal forced circulation	• Heat transfer rates high due to high circulation rate • Fouling tendency very low due to high circulation rate • Low residence time in heated zone which avoids degradation of temperature sensitive fluids • Lower column skirt height than horizontal thermosyphon unless pump NPSH is high • Longer tubes can be utilized to reduce cost • Maintenance easier than vertical type; removable tube bundle may be utilized • Suitable for high viscosity liquids	• Not equivalent to theoretical stage • Extra piping due to recirculation pumps • Larger plot space than vertical type • Higher operating cost due to pump power • Higher capital cost than corresponding thermosyphon type due to additional piping of pumps but lower than kettle type • Both shell and tube side metallurgy are effected with corrosive process fluid since it will be on shell side

(say, to provide oxygen for gasification) and in natural gas liquefaction plants, to maximize "cold recovery" and thus the process efficiency. They are also becoming commonplace in district heating networks supplying hot water from a CHP plant which is discussed in Chapter 10. Heat recovery from a geothermal fluid is often accomplished with this type of heat exchanger. The smaller physical size of this type of exchanger is also an advantage in reducing the plot space requirement, requiring less than 50% of that required by a shell and tube exchanger. Limitations for this type of exchanger are due to constraints on the highest allowable fluid temperature which is determined by the gasket material and pressure; the maximum design temperature and pressure being about 400°F (200°C) and 300 psig (21 barg). The pressure drop across this type of exchanger is higher than that for a shell and tube exchanger which may also place limitations on its use.

4.2.3 Air-Cooled Exchangers

This type of heat exchanger is used to cool fluids with ambient air. The air flows over a bank of finned tubes through which the fluid being cooled flows. Fins typically made out of aluminum are used to provide an extended surface to compensate for the relatively low heat transfer coefficient on the air side. Air-cooled condensers are used in place of water-cooled surface condensers in geographic locations where water is in short supply. Either a forced or an induced draft fan forces or draws the air through the exchanger. In a forced draft design, the tube bundles are located on the discharge side of the fan while in an induced draft design, tube bundles are located on the suction side of the fan. A car radiator is an example of a forced draft air-cooled exchanger installed vertically. In industry, however, horizontally oriented air-coolers are the most commonly used. Most installations utilize forced draft units; however, there are many advantages to the induced design that should be considered. The advantages of forced draft designs are: slightly lower power use since the air entering the fans is cooler, better accessibility of mechanical components for maintenance, and easily adaptable for warm air recirculation in cold climates to avoid overcooling. The disadvantages are poor distribution of air, greatly increased possibility of hot air recirculation due to lower discharge air velocity and absence of an exhaust stack (air recirculation may be limited to some extent by the use of "fences" between bays), and low natural draft capability when fan failure occurs. The advantages of induced draft designs are: better distribution of air; less possibility of the hot air recirculating to the intake; and more resilient to fan failure since the natural draft stack effect is much higher. The disadvantages are higher fan power use since warm air enters the fan and fan drive components are less accessible for maintenance. Coolers are regularly manufactured in the United States in tube lengths from 6 to 50 ft (1.8–15.2 m) and in bay (a bay constitutes one or more tube bundles) widths from 4 to 30 ft (1.2–9.1 m). Due to shipping limitations, 14 ft (4.3 m) width is the most commonly used. Use of longer tubes usually results in a less costly design; 30 or 40 ft (9.1–12.2 m) tube lengths being commonly used.

4.2.4 Heat Recovery Steam Generators (HRSGs)

An HRSG is a heat exchanger specially designed to recover heat from a near atmospheric pressure hot gas stream such as the exhaust from an internal combustion engine or a gas turbine to generate steam that can be used in a process as in a CHP plant or used to drive a steam turbine as in a combined cycle as discussed in Chapter 9 and it is recommended that this section on HRSGs be revisited after studying Chapter 9. It consists of heat exchange tubes contained within a casing of a rectangular cross section, and without any refractory lining due to significantly lower temperatures as compared to a fired boiler. The hot gas stream flows outside the tubes and the principle mode of heat transfer is by convection. Depending on the type of steam cycle, coils or tubes may be provided to heat up the BFW (in the "economizer" coils), evaporate the water (in the "evaporator" coils), superheat the generated steam (in the "superheater" coils), as well as reheat the steam (in the "reheater" coils) coming out of the high-pressure section of a steam turbine before readmitting it back into the steam turbine.

The smaller HRSGs are often designed for the gas to flow vertically (over horizontal tubes) in order to save plot space but the larger ones have to be designed for horizontal gas flow (over vertical tubes). Unless the HRSG is a once-through unit, that is, where the BFW is completely evaporated within the evaporator coils, most of the units are designed for natural circulation of the BFW from the steam drum through the evaporator coils. The steam drum also serves the purpose of separating the steam from the water returning from the evaporator coils. Single-pressure HRSGs produce steam at a single pressure and hence have only one steam drum. A triple-pressure HRSG with reheat steam cycle consists of three sections: a lower temperature section, which produces low-pressure saturated steam which is then superheated and preheats the BFW; an intermediate temperature section, which produces intermediate-pressure saturated steam which is then superheated and further preheats the BFW; and a high temperature section, which produces high-pressure saturated steam which is also superheated and a reheater that reheats the steam from the high-pressure section of the steam turbine before it is admitted into the intermediate pressure section of the steam turbine. Each of these sections has a steam drum connected to the corresponding evaporator coil. The high-pressure steam is superheated by a significant amount not only to avoid water droplets entering the steam turbine but also to increase the Rankine cycle efficiency as discussed in Chapters 8 and 9.

The tubes within an HRSG are typically finned because of the poor outside heat transfer coefficient. The high-pressure superheater coils may not be finned however, if a possibility exists for the oxidation of the fins due to the much higher temperature gas flowing over these tubes while the inside heat transfer coefficient is not as high as in the evaporator. The inside heat transfer coefficient for the economizer is about 10 times, for the evaporator about 50–200 times and for the high pressure superheater about 20 times the outside heat transfer coefficient.

Some HRSGs include supplemental or duct firing which consists of burners to provide heat for producing more steam which may be required in CHP applications, or sometimes to generate additional power in the steam turbine (when the steam turbine and other associated equipment and piping are designed for this larger steam flow).

Smaller HRSGs associated with smaller (<20 MW) gas turbines are shipped as fully assembled units from the factory ("packaged" units) resulting in cost savings.

In the once-through HRSG design, the BFW flows in a continuous path without separate sections for the economizer, evaporator, and superheater, which provides a higher degree of flexibility as the functionality of a portion of each section can change based on the heat available in the gas turbine exhaust. Furthermore, quick changes in steam production are possible without the steam drums. However, the BFW has to be ultrapure to minimize fouling of the coils where phase change occurs.

4.2.5 Boilers and Fired Heaters

Boilers and fired heaters employ combustion of a fuel such as natural gas, refinery gas, liquefied petroleum gas, landfill gas, distillate oil, residual oil, or coal to provide the heat to generate steam or a heated process fluid. When the fuel is a liquid, an atomizer is used to avoid simply pouring the fuel onto the furnace floor creating a hazardous situation, while the finer the droplet size, greater is the combustion efficiency. Natural gas is typically used as the pilot fuel which produces a small flame to light the main fuel for units designed with a pilot lighter which in turn is lit by an ignition transformer. Boilers range in sizes from commercial units typically rated less than approximately 10 GJ/h to industrial units rated at approximately 10 to as high as approximately 250 GJ/h, to utility boilers which are generally rated above approximately 250 GJ/h and are discussed in more detail in Chapter 8.

4.2.5.1 Fire Tube Design In boilers of the fire tube design, hot combustion products produced from the combustion of either a gaseous, liquid, or a solid fuel pass through the tubes while steam is produced from the BFW on the shell side. Fire tube boilers are generally used for relatively small steam capacities of up to 12,000 kg/h and low-to-medium steam pressures of up to 18 bar. Most fire tube boilers due to their smaller size can be shipped as "packaged" units resulting in cost savings.

4.2.5.2 Water Tube Design In subcritical boilers of the water tube design, BFW flows through the tubes where a fraction of the water changes phase and enters the boiler stream drum for separation of the steam. The required heat is supplied by products of combustion from outside the tubes. These boilers are selected when the steam flow rate as well as its pressure are high as in the case of large process and power plant boilers. Many of these boilers are also packaged when they are oil and/or gas fired while packaged units are less common if they are solid-fuel fired. Water tube boilers include a radiant section where the tubes receive almost all of their heat by radiation from the flame of the combusting fuel followed by a cooler convection section where heat transfer takes place by convection. Tubes in the radiant section are placed either vertically or horizontally along the refractory wall but a distance away from the insulation such that radiant energy can be reflected to the back of the tubes to maintain uniform tube wall temperatures. The tubes in the convection section are typically finned for enhanced heat transfer except the first few tube rows at the bottom of the convection section since they are exposed to significant amount of radiation from the flame. These tubes provide shielding from the radiation to the remaining tubes in the convection section. Large boilers are equipped with both forced and induced draft blowers. The combustion air is forced into the boiler by the forced draft fan while the induced draft fan assists the flow of the flue gas to the stack, the fans being sized to make the pressure within the boiler itself near atmospheric (a slight vacuum to avoid the hot gases flowing out as a safety measure).

Fired heaters are similar to water tube design boilers except the process fluid to be heated flows through the tubes. Furnace designs vary depending on their functionality, amount of thermal energy to be transferred (heating duty), type of fuel, and method of introducing the combustion air. Burners can be arranged in cells which heat a particular set of tubes and can be floor mounted, wall mounted, or roof mounted. The burners in a vertical, cylindrical furnace are located in the floor and fire upward. The fuel is typically natural gas or a fuel oil.

REFERENCES

Avallone E, Baumeister T, Sadegh A, editors. *Marks' Standard Handbook for Mechanical Engineers*. 11th ed. New York: McGraw-Hill; 2006.

Bromley LA, Leroy NR, Robbers JA. Heat transfer in forced convection film boiling. Ind Eng Chem 1953;44(12):2639–2646.

Churchill SW, Chu HH. Correlating equations for laminar and turbulent free convection from a horizontal cylinder. Int J Heat Mass Transf 1975;18(9):1049–1053.

Colburn AP. A method of correlating forced convection heat transfer data and a comparison with fluid friction. Trans Am Inst Chem Eng 1933;29:174.

Drew TB, Hottel HC, McAdams WH. Heat transmission. Trans Am Inst Chem Eng 1936;32:271.

Green DW, editor. *Perry's Chemical Engineers' Handbook*. 8th ed. New York: McGraw-Hill; 2007.

Gulley DL. How to calculate weighted MTD's. Hydrocarb Process 1966;45(6):116–122.

Incropera FP, DeWitt DP. *Fundamentals of Heat and Mass Transfer*. New York: John Wiley & Sons, Inc.; 1996.

Jacob M. Heat transfer in evaporation and condensation. Mech Eng 1936;58:643–729.

Jakob M, Hawkins GA. *Elements of Heat Transfer*. 3rd ed. New York: John Wiley & Sons, Inc.; 1957.

Kandlikar SG. A general correlation for saturated two-phase flow boiling heat transfer inside horizontal and vertical tubes. J Heat Transf 1990;112(1):219–228.

Kern DQ. *Process Heat Transfer*. New York: McGraw-Hill; 1950.

McAdams WH. *Heat Transmission*. 2nd ed. New York: McGraw-Hill; 1942.

Morgan VT. The overall convective heat transfer from smooth circular cylinders. Adv Heat Transf 1975;11:199–264.

Palen JW, Small W. A new way to design kettle and internal reboilers. Hydrocarb Process 1964;46(11):199–208.

Tubular Exchanger Manufacturers Association. *Standards of the Tubular Exchanger Manufacturers Association*. 9th ed. New York: Tubular Exchanger Manufacturers Association; 2007.

Zukauskas A. Heat transfer from tubes in cross-flow. Adv Heat Transf 1987;18:87.

5

MASS TRANSFER AND CHEMICAL REACTION EQUIPMENT

5.1 FUNDAMENTALS OF MASS TRANSFER

Mass transfer plays an essential role in energy conversion plants such as coproduction plants as well as power plants utilizing a "dirty fuel" or heat engines. Gas cleanup operations using a liquid solvent or reagent wash such as desulfurizing the syngas in the case of a coproduction plant or the flue gas in the case of a power plant burning a dirty fuel involve transfer of mass (the pollutant in this case) from the gas phase to the liquid phase. Another example is the cooling tower used for plant heat rejection in which the cooling water after being heated up in the plant is cooled by contacting it with ambient air. A fraction of the cooling water transfers into the air by evaporation. In addition to gas–liquid mass transfer, mass transfer between other phases is also of industrial importance. Liquid–liquid mass transfer is used in extraction processes, an example being the dehydration of ethanol although these days ethanol dehydration is accomplished using adsorption which is again a mass transfer process involving a fluid and a solid. Adsorption is used in drying of gas streams as well as in separating H_2 from syngas, O_2 from air in smaller scale air separation applications and in the removal of contaminants both from gases and liquids. For example, desulfurization of natural gas is accomplished using activated carbon particles. Drying of biomass such as wood is again a mass (and heat) transfer operation in which the liquid water held within the wood has to ultimately transfer out as a vapor. Mass transfer also plays a very important role in heterogeneous chemical reactions. In catalytic reactors, depending

Sustainable Energy Conversion for Electricity and Coproducts: Principles, Technologies, and Equipment, First Edition. Ashok Rao.

on the operating conditions, the reactor sizing may either be controlled by reaction kinetics or mass transfer or combination of the two. Mass transfer allows the bringing of the reactants together and then removal of the products away from the reaction zone to make room for fresh reactants so that the process may continue.

Within fluids, there are two types of mass transfer processes, diffusive mass transfer or molecular diffusion that occurs when fluids are stagnant or in laminar flow, and convective mass transfer that occurs when fluids are in turbulent flow. In the case of wood drying, the diffusional mechanism is involved in the migration of the bound water through the cell-wall matrix. In an industrial wood dryer where a drying medium such as flue gas is used, the external mass transfer of the moisture into the flue gas phase occurs by the convective mechanism.

Chemical potential is the true driving force for mass transfer but concentration in the case of ideal liquids and partial pressure in the case of ideal gases may be used to represent this driving force. In the case of adsorbents the driving force may be taken as the fraction of active sites (where adsorption occurs) occupied on the solid surface when the direction of mass transfer is from the solid to the fluid while the concentration of the fraction of active sites unoccupied on the solid surface when the direction of mass transfer is from the fluid to the solid. There are also several material properties and operating conditions that effect the mass transfer rate and in the case of gas–liquid mass transfer, properties such as diffusivity, viscosity, mass density, surface tension, and fluid velocity. Other important factors influencing the rate of mass transfer are the available fluid interfacial area and the degree of dispersion of one fluid in the other. In the case of mass transfer between two immiscible fluids, the mass transfer area may be enhanced by including packing in the flow path while the degree of dispersion may be increased by agitation or by forming bubbles (in the case of a gas) or droplets. In the case of mass transfer between a fluid and a solid, the mass transfer rate may be increased by decreasing the solid particle size and more importantly by increasing the surface area within the pores, that is, making the solid highly porous where most of the surface resides.

The intent of this chapter is to provide an introduction to the governing equations for mass transfer followed by basic design methods for major types of the mass transfer equipment as well as reactors that may be encountered in energy conversion plants which is essential for the analysis and design of such systems. More details on these subject may be found in books such as *Mass Transfer Operations* (Treybal, 1980), *Separation Processes* (King, 2013), *Chemical Reaction Engineering* (Levenspiel, 1972), and *Chemical and Catalytic Reaction Engineering* (Carberry, 1976).

5.1.1 Molecular Diffusion

Molecular diffusion is governed by a random walk process involving the transport of molecules from a region of higher concentration to a region of lower concentration and this diffusion of individual molecules also brings about bulk motion of the molecules. The governing equation for steady state molecular diffusion which is analogous to conductive heat transfer is Fick's first law shown for one dimensional flow in Cartesian coordinates moving at molar average velocity of the mixture:

$$J_A = -\mathcal{D}_{AB}\frac{dC_A}{dz} = -C_{TOT}\,\mathcal{D}_{AB}\frac{dy_A}{dz} \tag{5.1}$$

where J_A is the molar flow rate per unit area normal to the direction of diffusion or molar flux of component A per unit area diffusing into medium B, $\mathcal{D}_{AB}$ a physical property of the binary system called the diffusivity or diffusion coefficient whose units depend on the units used for the other variables appearing in Equation 5.1, C_A the molar concentration of component A, C_{TOT} the (total) molar concentration of the mixture, y_A the mole fraction of component A, and z the length along the direction of diffusion. The diffusivity of various binary systems may be found in Poling et al. (2001) along with methods for estimating the diffusivity. *Perry's Chemical Engineers' Handbook* (Green, 2007) also describes methods for estimating the diffusivity. It varies with temperature and in the case of gases with pressure and correlations are available to adjust its value for the required set of conditions. In the case of diffusion in multicomponent mixtures, an effective diffusivity is used which is derived from the binary diffusivities of the diffusing component and each of the other species. The equivalent form of Equation 5.1 for stationary coordinates in terms of mole fraction of component A is:

$$N_A = -C_{TOT}\,\mathcal{D}_{AB}\frac{dy_A}{dz} + y_A\,(N_A + N_B) \tag{5.2}$$

where N_A and N_B are the molar fluxes of components A and B with respect to stationary coordinates. The second term on the right hand side of Equation 5.2 is due to the net molar bulk flow.

When steady state diffusion occurs in a unidirection, and linear concentration profiles may be assumed over a thickness δ, Equation 5.2 may be written as:

$$N_A = \frac{C_{TOT}\mathcal{D}_{AB}}{\delta}\Delta y_A + y_A(N_A + N_B) \tag{5.3}$$

where Δy_A is the change in mole fraction of component A over the thickness δ.

5.1.2 Convective Transport

The mechanism of mass transfer in industrial processes involving gas (vapor)–liquid contact as in a syngas decarbonizing scrubber where a soluble gas species (CO_2) transfers from the vapor to the liquid phase includes typically convective transport. The transfer of the vapor through the vapor–liquid interface may be modeled as passing through stagnant vapor and liquid films present on each side of the interface. The resistance to mass transfer is primarily controlled by these stagnant films. The vapor–liquid interface offers essentially no resistance to mass transfer while the mechanism of transfer from the bulk vapor to the stagnant vapor film and then from the stagnant liquid film to the bulk liquid is accomplished by eddies and the corresponding transfer processes occur at much faster rates than through the films.

Since the films are stagnant, the mass transfer mechanism within the films is by diffusion and at steady state, Equation 5.3 may be applied to the transfer process in each of the films as:

$$N_A = k_V\left(y_A - y_A^{int}\right) + y_A\left(N_A + N_B\right) = k_L\left(x_A^{int} - x_A\right) + x_A\left(N_A + N_B\right) \tag{5.4}$$

where $k_V = (C_{TOT,V}\,\mathcal{D}_{AB}/\delta_V)$ and $k_L = (C_{TOT,L}D_{AB}/\delta_L)$ for the vapor and liquid films and are called the gas and the liquid mass transfer coefficients, y_A and x_A are the bulk concentrations of component A in the vapor and liquid phases, while y_A^{int} and x_A^{int} are the respective concentrations at the interface. Another form of mass transfer coefficients, $k_{C,V}$ and $k_{C,L}$ is also used based on concentrations and defined by:

$$N_A = C_{TOT,V}\,k_{C,V}\left(y_A - y_A^{int}\right) + y_A\left(N_A + N_B\right) = C_{TOT,L}\,k_{C,L}\left(x_A^{int} - x_A\right) + x_A\left(N_A + N_B\right) \tag{5.5}$$

5.1.3 Adsorption

In fluid–solid adsorbent systems, the adsorption phenomenon is the binding of molecules to the solid surface. In absorption, the species being absorbed (absorbate) enters the absorbing phase by forming a solution and the whole volume of the absorbing phase is involved whereas adsorption is limited to the surface of the adsorbent ("sorption" is a generalized term for both absorption and adsorption). Depending on the nature of the fluid component and the solid surface, the binding in adsorption may occur by weak van der Waals forces (physisorption) or by covalent bonding (chemisorption) or by electrostatic attraction. Adsorption is usually weak and reversible and increasing the temperature or reducing the partial pressure of the species adsorbed reduces the amount adsorbed. Thus, regeneration of the adsorbent while releasing the species adsorbed as a separate stream from the feed stream may be accomplished by increasing the temperature in which case the process is called temperature swing adsorption or by decreasing the partial pressure of the spice in which case it is called pressure swing adsorption. Since it is a surface phenomenon, a good adsorbent should have a very high surface area available per unit mass of adsorbent as is the case with common industrial adsorbents such as activated carbon. Most of its surface is within its pores (typically macroscopic pores from the surface to the interior of the solid branch out into several microscopic pores). Catalysts for chemical reactions also involve surface phenomena and include adsorption of the reactants and desorption of the products, and the amount of catalyst required by a chemical reactor can be reduced significantly by increasing its internal surface area, that is, within the pores.

A semiempirical model expressed by Equation 5.6 was derived by Langmuir and is commonly used due to its simplicity and its applicability to a vast number of systems.

$$\Theta = \frac{K\,C_i}{1 + K\,C_i} \tag{5.6}$$

Where Θ represents the fraction of surface sites covered with adsorbed molecules at equilibrium, K a coefficient which is a function of temperature for a given system of adsorbate and adsorbent, and C_i is the molar concentration of the adsorbate i in the fluid (or may be substituted for the partial pressure in the case of an ideal gas). The model is based on simplifying assumptions that all of the adsorption sites are equivalent, each site can accommodate only one molecule which does not interact with the other adsorbed molecules nor do they undergo phase transitions, the surface is homogeneous, and only a monolayer is formed at the maximum adsorption level. Equation 5.6 may be derived by assuming that the adsorption rate is proportional to the driving force C_i and the fraction of surface sites available for adsorption $(1 - \Theta)$ while the desorption rate is proportional to the fraction of surface sites occupied (Θ). At equilibrium, since adsorption rate = desorption rate:

$$k_{\text{adsorp}}\,C_i\,(1 - \Theta) = k_{\text{desorp}}\,\Theta \tag{5.7}$$

where k_{adsorp} and k_{desorp} are the rate constants. Next, by replacing $k_{\text{adsorp}}/k_{\text{desorp}}$ with K in Equation 5.7 and rearranging provides Equation 5.6 which may be utilized in determining the amount of adsorbent required for a given service as explained later in this chapter under Fluid–Solid Systems. Other models have been developed for systems where any of these assumptions are not valid such as the Langmuir–Freundlich model given by Equation 5.8 where ξ is the index (or measure) of heterogeneity, and the BET model for systems where the adsorbed molecules do form multilayers (Brunauer et al., 1938).

$$\Theta = \frac{K\,C_i^{\xi}}{1 + K\,C_i^{\xi}} \tag{5.8}$$

5.2 GAS–LIQUID SYSTEMS

5.2.1 Types of Mass Transfer Operations

5.2.1.1 Absorption Gas absorption involves the transfer of a component or components from the gas phase to the liquid phase. The gas is contacted with a liquid to preferentially dissolve one or more components of the gas mixture into the liquid. The absorption operation is typically accomplished in a column with the liquid introduced near the top and the vapor introduced near the bottom with the liquid flowing downward by gravity. The column is equipped with internals to enhance the surface area available for mass transfer between the two phases as described later in this chapter. Agitated vessels are also used for the gas absorption operation in applications where the gas to liquid ratio is low or when the liquid contains a suspended solid such as a catalyst, or a microorganism as in a fermentation process (e.g., to make C_2H_5OH).

The gas is dispersed as bubbles through the liquid contained within the agitated vessel. An example of the use of agitated vessels is in the oxidation of the $CaSO_3$ formed during the flue gas desulfurization process to the stable $CaSO_4$ (gypsum) which is more stable for disposal or for use in the manufacture of dry walls. Although this process is not purely a physical process since a chemical reaction occurs, mass transfer plays a very important role since the reactants have to be brought together. Depending on the type of components and the liquid to gas ratio, the heat released during absorption (heat of solution) may be significant enough to increase the temperature of the liquid phase (in the case of a column, as it descends through it) and should be accounted for in the design of the absorber. Intercooling may be included in the design wherein a cooling coil is inserted into the column or by installing an external exchanger wherein the liquid is withdrawn from the column, cooled, and then reintroduced into the column to continue the absorption process.

An example of gas absorption is in the desulfurization and/or decarbonization of syngas using the Selexol^{TM} solvent as illustrated in Figure 5.1 (Palla, 2009). This process is described in Chapter 9 but it suffices for now to observe the two absorbers shown in series, one to desulfurize the syngas and the other to decarbonize the syngas (the syngas after further processing such as humidification and preheating may be combusted in a gas turbine as in an IGCC described in Chapter 9). The gas is contacted countercurrently with a solvent consisting of a mixture of dimethyl ethers of polyethylene glycol to transfer primarily the H_2S in the first absorber to the liquid phase and to transfer the CO_2 in the second absorber to the liquid phase. Another example of gas absorption is the decarbonization of the flue gas in a coal fired boiler plant as illustrated in Figure 5.2 (Reddy et al., 2008) where the flue gas is contacted with a solvent consisting of aqueous monoethanolamine (MEA) to transfer the CO_2 into the liquid phase. This process is described in Chapter 8 but it suffices for now to observe the two sections in series within the column, one to absorb the CO_2 and the other to wash the flue gas before it enters the atmosphere. This second section in addition to washing out the liquid amine droplets from the flue gas may also absorb polluting vapors into the water phase.

5.2.1.2 ***Stripping*** Stripping is the reverse of absorption involving the transfer of a component or components dissolved in the liquid phase into the gas phase. The liquid is contacted with a vapor (steam) or gas (such as N_2 depending on what is done with the components removed from the liquid phase) to preferentially transfer one or more components in the liquid phase into the vapor phase. The stripping operation is typically accomplished in a column with the liquid introduced near the top and the stripping agent (vapor or gas) introduced near the bottom with the liquid flowing downward by gravity. When steam is used as the stripping medium, a fraction of it condenses as it travels up the column providing the heat required for separation or the stripping operation. An example of this type of operation is in conjunction with the previously mentioned desulfurization and/or decarbonization of syngas using the Selexol^{TM} solvent as illustrated in Figure 5.1 or flue gas decarbonization using the amine process illustrated in Figure 5.2. The rich solvent leaving the absorber is stripped using steam to transfer the dissolved gases (H_2S and/or the CO_2) into the vapor phase and thus regenerate the solvent so that it may be reused in the absorption

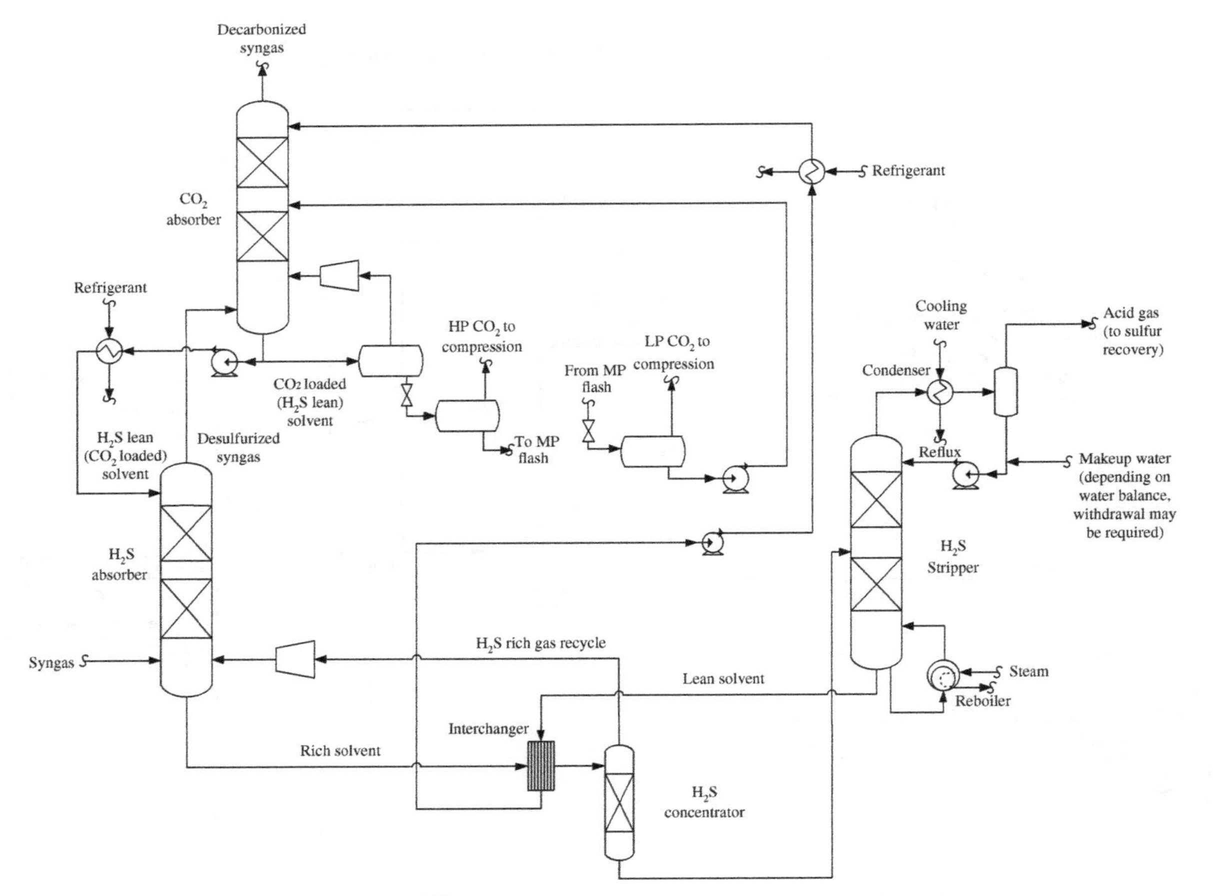

FIGURE 5.1 Selexol^{TM} process for desulfurization and decarbonization of syngas

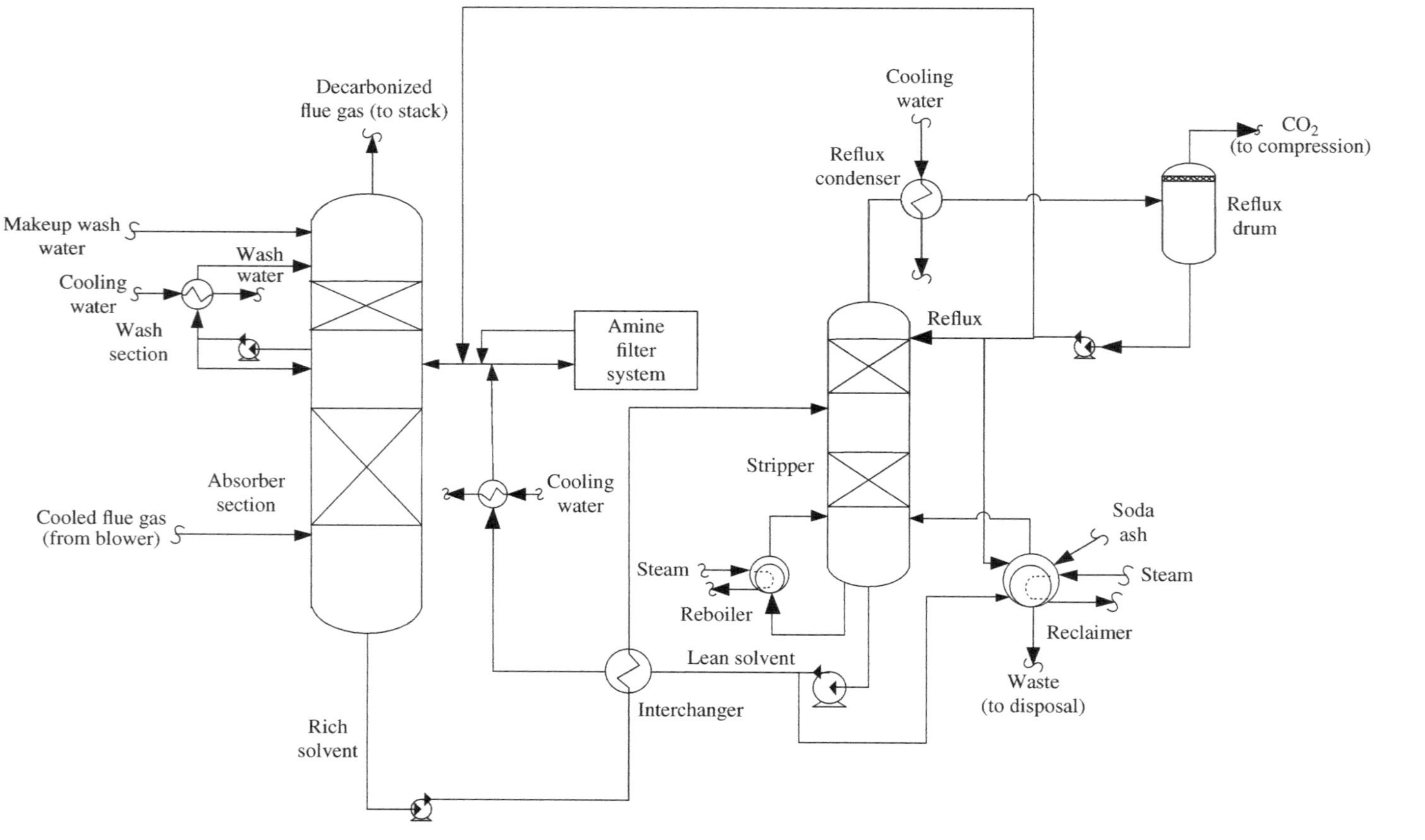

FIGURE 5.2 MEA process for decarbonization of flue gas

operation. Another example of the stripping operation is the removal of contaminants such as NH_3 and H_2S from waste (“sour”) water produced in a gasification plant. When steam is used as the stripping medium, either a reboiler (type of heat exchanger described in Chapter 4) is used where the solvent with the dissolved water (formed by the condensation of the stripping steam) is heated to vaporize the water and thus generate the stripping steam as is done in the Selexol™ process for syngas treatment as well as the amine process for flue gas treatment, or “live” steam is injected into the column as is typically done in a sour water stripper. This second option eliminates the cost associated with having a reboiler (operating in a corrosive environment) but the treatment cost of providing the BFW to generate the steam (which now becomes part of the stripped water) should be accounted for in the trade-off analysis. The vapor stream leaving the column at the top consisting of steam and the components stripped off from the liquid phase are cooled in a condenser[1] (a horizontal shell and tube heat exchanger with the cooling medium such as cooling water flowing through the tubes or an air cooled exchanger) to condense out the steam and any vaporized solvent which are then reintroduced near the top of the column as “reflux.” The moisture content of the vapor stream leaving the condenser depends on its temperature and pressure. The condenser may be a partial condenser in which case the liquid stream is fed back to the column as reflux while the vapor stream constitutes the distillate product. In the case of a total condenser, a portion of the liquid stream is fed back as reflux while the remaining liquid constitutes the distillate product.

5.2.1.3 ***Distillation*** Distillation combines both absorption and stripping to separate a component or components from a liquid mixture based on differences in the relative volatility of the components. Consider a continuous “single stage” distillation process in which a binary liquid mixture (mixture of two components) enters a still heated to vaporize a portion of the liquid. Assume the idealized case where the heating is even, each of the phases is well mixed and there is enough residence time for equilibrium to prevail between the two phases. In most cases (except azeotropic mixtures discussed later), the vapor phase in equilibrium with the liquid phase will contain higher concentration of the more volatile component while the liquid phase will contain less of the more volatile component as compared to the liquid mixture entering. Thus, by partially vaporizing a portion of the liquid, two streams may be produced, a vapor phase enriched with the more volatile component and a liquid phase enriched with the less volatile component and it is the difference in the composition of the vapor phase and the liquid phase that allows separation to occur. For a binary mixture (of components i and j where component i is more volatile), the relative volatility α_{ij} at a given temperature and pressure is defined by:

[1] In the case of a sour water stripper, there is an advantage in withdrawing the liquid from a tray close to the top (rather than cooling the vapor leaving the top tray), cooling it and then feeding it back to the top of the column to cool the vapor stream by direct contact while also scrubbing it before the vapor leaves the column. This type of configuration avoids fouling of the condenser by precipitating salts.

$$\alpha_{ij} = \frac{(y_i/x_i)}{(y_j/x_j)} = \frac{K_i}{K_j} \tag{5.9}$$

where the equilibrium molar fractions of the vapor and the liquid phases are represented by the y^s and the x^s while the K^s are known as the vapor–liquid distribution ratios or sometimes also referred to simply as the "K values." Thus, higher the relative volatility (over unity), easier it is to cause separation by distillation.

The above single stage distillation process may be made into a multistage operation by having a series of such equilibrium stages as illustrated in Figure 5.3 so that a high degree of separation may be accomplished. Such multistage distillation is performed in a column with the bottom section functioning as a stripper while the top section functioning as an enriching or rectification section. The absorption and the stripping operations conducted in columns as described previously may also be visualized as being comprised of such multiple equilibrium stages.

Distillation columns require reboilers to provide the energy required for the separation while a reflux condenser is used to provide the liquid (reflux) in the top section of the column for the mass transfer to occur while the feed is introduced toward the middle section of the column. The overhead product or that leaving the condenser is called the "distillate" while the product leaving the bottom of the column or the reboiler is called the "bottoms." The reflux stream also functions as a recycle stream to increase the purity of the distillate.

5.2.1.4 ***Energy Saving Measures*** The feed to the column may be preheated to reduce the heat load of the reboiler as well as match the temperature within the column in the region where the feed is introduced. It may enter the column either as a subcooled liquid below its bubble point (as stated in a previous chapter, the bubble point of a liquid mixture is the point where the liquid just starts to evaporate and is the equivalent of "boiling point" for a single component liquid), or as a saturated liquid at its bubble point, or as a two-phase mixture, or as a saturated vapor at its dew point[2] (the dew point of a vapor mixture is the point where the vapor just starts to condense), or a superheated vapor above its dew point. The feed is introduced into the column such that the composition is close to the composition of the corresponding phase within the column (requires trial and error procedure or estimating method to locate the feed).

Figure 5.4 illustrates the methanol distillation operation to purify methanol to meet the product specifications discussed in Chapter 10. The distillation unit consists of a set of energy saving heat integrated distillation columns with one column operating at a pressure higher than a second column with the condenser of the higher pressure column providing heat for the reboiler of the lower pressure column. The dew point of the overhead vapor leaving the higher pressure column is increased by the higher operating pressure. The heat of condensation can now be transferred with reasonable

[2] Note that for nonazeotropic mixture, the dew point temperature of a vapor mixture is higher than the bubble point temperature of a liquid mixture for a given composition, whereas in the case of a single component, the boiling point and dew point temperatures are identical. For azeotropic mixture, the dew point and bubble point temperatures become identical at a certain composition.

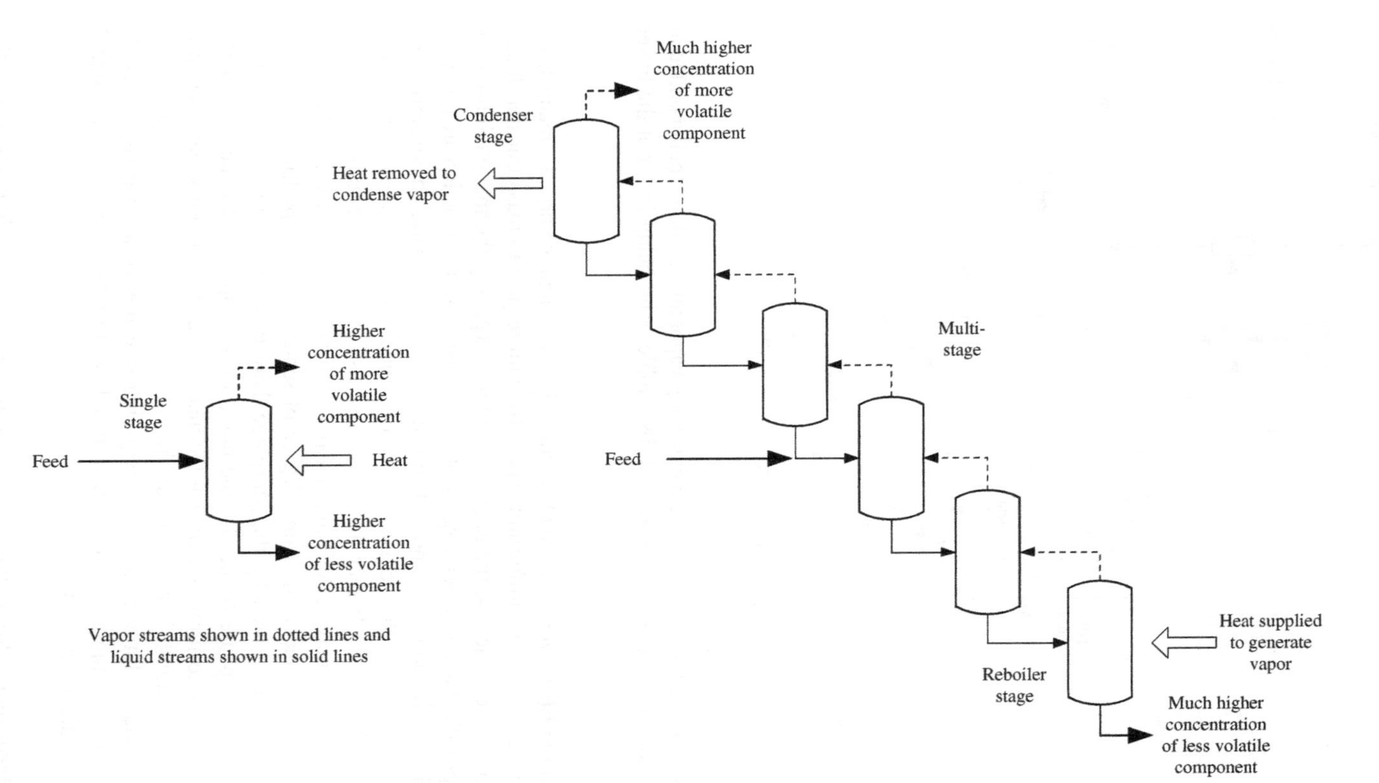

FIGURE 5.3 Single stage versus multi-stage distillation

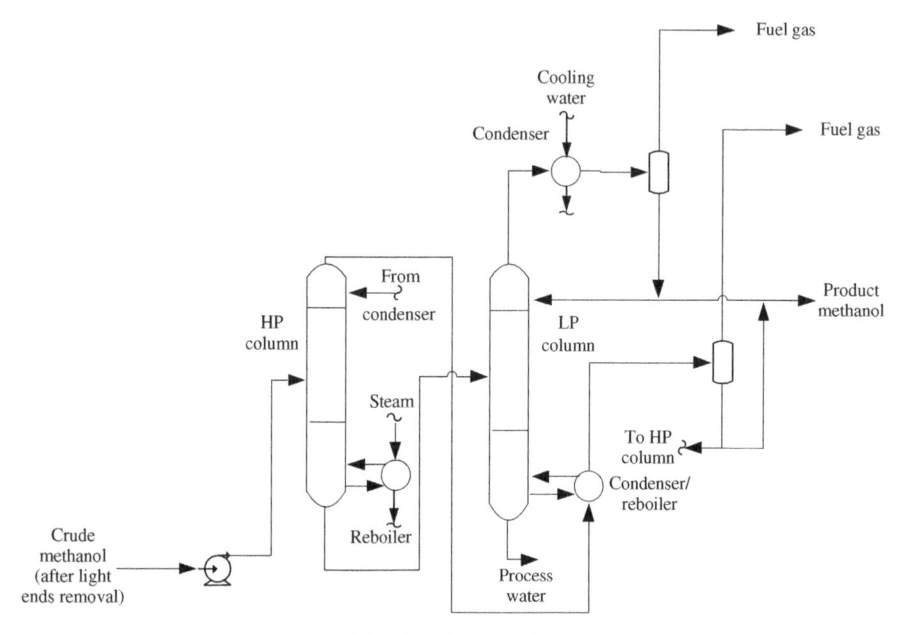

FIGURE 5.4 Methanol distillation

temperature driving force between the condensing vapor and the liquid being boiled in the reboiler of the lower pressure column, its bubble point being lower at the lower pressure.

5.2.1.5 Stage Efficiency In actual practice, the vapor and the liquid streams leaving each stage are not in equilibrium especially with respect to mass transfer due to insufficient contact time and the degree of mixing (typically thermal equilibrium is closely approached unless when large heat effects due to heat of reaction are involved as in reactive distillation where both chemical reaction and separation are carried out simultaneously). Thus, the actual number of contact stages required for a given separation will be much higher than calculated using the equilibrium assumption. An overall efficiency may be defined as the ratio of the equilibrium stages to the actual stages required to achieve the same degree of separation. The performance of an individual tray is measured by the Murphree Tray Efficiency and is specific to a component and the phase (either for the liquid or the vapor) and is defined as the change in actual concentration between the incoming and outgoing streams divided by the change if equilibrium prevailed. Tray efficiencies are affected by a number of factors including fouling, corrosion requiring selection of appropriate materials of construction, foaming, and entrainment which lead to the carryover of the liquid by the vapor back to the stage above.

5.2.1.6 Azeotropes Some liquid mixtures as mentioned previously (such as the binary mixture of ethanol and water) form azeotropes under certain conditions of composition and pressure having identical compositions in both the vapor and the liquid phases. In such cases, the degree of separation is limited by the corresponding azeotropic composition and to achieve a higher degree of separation, the distillation operation has to be followed by an extraction operation or use of extractive distillation described in the following.

5.2.1.7 Extraction This unit operation consists of contacting two immiscible liquids (feed and solvent streams) to transfer preferentially a component i (solute) from the feed stream (containing components i and j) into the solvent stream. The effectiveness of a solvent is measured by the "selectivity" which is analogous to relative volatility in distillation and is defined as:

$$\beta_{ij} = \frac{(\mathfrak{X}_i \text{ in solvent leaving or extract}/\mathfrak{X}_i \text{ in depleted feed leaving or raffinate})}{(\mathfrak{X}_j \text{ in solvent leaving or extract}/\mathfrak{X}_j \text{ in depleted feed leaving or raffinate})} \tag{5.10}$$

where the $\mathfrak{X}^s$ represent the mass fractions of the two components at equilibrium. Thus, a high value for the selectivity (over unity) is desirable. Other desirable characteristics for a solvent include being nontoxic, nonflammable, and chemically stable, having low cost, low viscosity, low vapor pressure, low freezing point, high solubility for the component being extracted to minimize the solvent flow rate, and being recoverable in terms of ease of separation of the solvent and the component(s) extracted since the recovered solvent is recycled for extraction.

5.2.1.8 Extractive Distillation Extractive distillation uses a solvent with low volatility to assist in the separation by distillation by increasing the relative volatility. The less volatile component enters into solution with the solvent introduced into the upper part of the column and leaves as the bottoms stream from the distillation unit. This mixture is then separated in a second distillation unit and the solvent recycled to the first distillation unit. The insoluble components in the original feed leave the column as the distillate. Desirable characteristics for a solvent include high selectivity to alter the vapor–liquid equilibria of the components in the original feed stream and increase the relative volatility, being nontoxic, nonflammable and chemically stable, having low cost, low viscosity, low vapor pressure, low freezing point, high solubility for the component(s) being extracted to minimize the solvent flow rate, and being recoverable in terms of ease of separation of the solvent and the component(s) extracted since the recovered solvent is recycled.

5.2.1.9 Humidification and Cooling Towers Countercurrent humidification in a column to introduce large quantities of water vapor into a gas stream is practiced in energy plants instead of generating steam and then mixing the steam with the gas which is thermally less efficient. An example of such a humidification operation is in steam methane reforming where a mixture of methane and steam is required to produce syngas by the reaction: $CH_4 + H_2O = 3H_2 + CO$. Another example is in IGCC plants where H_2O vapor is introduced as a thermal diluent into the gas turbine fuel gas (syngas) to reduce the flame temperature in a gas turbine and thus limit the amount of NO_x formed. Because of the large latent heat of water, a significant amount of heat transfer occurs within the column as the water evaporates and thus the rate of heat transfer occurring simultaneously with the mass transfer has to be taken into consideration. The types of humidifiers described above operate at high pressures (25–30 bar) while a cooling tower may be considered as an ambient air humidifier operating near atmospheric pressure.

5.2.2 Types of Columns

The most common types of equipment used in mass transfer (including simultaneous heat and mass transfer) operations are either tray or packed columns consisting of vertical cylindrical vessels in which the vapor and the liquid are contacted countercurrently in a series of "trays" or beds of packing material. The liquid introduced at the top of the column and flowing down by gravity is opposed by the pressure exerted by the vapor flowing up through the column.

5.2.2.1 Tray Columns There are various types of trays also referred to as "plates" and the most common ones are discussed in the following.

The bubble cap tray consists of risers covered by caps with gap between the riser and the cap forming an annular space. The vapor entering from a tray below it flows up through the riser and then through the annular space by reversing its flow and bubbles out through slots into the surrounding liquid contained on the tray. Bubble cap trays have moderately high capacities and maintain their effectiveness in terms of mass transfer rates over a wide range of flow rates. Entrainment of the liquid by the vapor which reduces the overall stage efficiency and the effectiveness of a countercurrent operation is much higher than with other types of trays such as the perforated and the sieve trays described next due to the jetting action of the bubbles. The liquid level on each tray is maintained by weirs and the liquid flowing over the weir enters the tray below by gravity through a conduit or "downcomer," their number per tray dependent on the diameter of the column. This type of tray is used in applications where the flow rates can become extremely low, the tray remaining wet and maintaining a vapor seal under such conditions. Typically, the trays are spaced 18 in (46 cm) apart but when the operating pressure is low such as below atmospheric, the spacing is increased to 24 in (61 cm) and sometimes as high as 36 in (91 cm) for extreme low pressure.

A sieve tray or perforated tray has small holes in the tray through which the vapor bubbles through the liquid fairly uniformly. The liquid level in the tray is again maintained by weirs and the liquid flowing over the weir enters the tray below through downcomer(s), their number again dependent on the diameter of the column. The performance of these trays remains as good or better than bubble cap trays down to 60% of the design rate but the performance falls off rapidly at lower flow rates. Entrainment is much lower than bubble cap trays, about a third but at low vapor load, tray weeping can be an issue. Tray weeping (known as dumping in the extreme case) is caused by insufficient pressure being exerted by the vapor at the lower flow rates to hold back the liquid from flowing down through the holes in the tray. Excessive weeping leads to buildup of this effect in each successive tray and eventually the column may have to be shutdown and restarted. Sieve trays are used in applications with high operating capacities that remain close to the design rate and are also suitable for applications where suspended solid particles are present. Tray spacing can be less than that for bubble cap trays due to lower entrainment with 15 in. (38 cm) being an average although 9 in. (23 cm) to 12 in. (30 cm) spacing is sometimes used. Again, spacing is increased for low pressure or vacuum service to 20 in. (51 cm) to 30 in. (76 cm).

Valve trays are similar to sieve trays (with downcomers for the liquid flow) but the holes in the tray are covered by caps that can move up and down (within limits) depending on the vapor flow rate. This provides a much better turndown capability since the tendency to weep at low flow rates is reduced by the caps. Valve trays have replaced bubble cap trays in a number of applications due to their wider operating range, ease of maintenance, and cost.

Perforated trays without downcomers are sometimes used in which the vapor flows up through the holes as in sieve trays but the liquid also flows through these same holes countercurrently to the tray below. The capacity and entrainment characteristics of these trays are similar to sieve trays but the performance deteriorates rapidly at lower flow rates to unacceptable values below 60% of the design flow rate. Like sieve trays, these trays are used in applications with high operating capacities but remaining close to the design rate. These trays are also suitable when suspended solid particles are present. Twelve inches (30 cm) tray spacing is typical although 9 in. (23 cm) to 18 in. (46 cm) spacing is sometimes used. For low pressure or vacuum service, the spacing is increased to 18 in. (46 cm) to 30 in. (76 cm).

5.2.2.2 Packed Columns Packed columns are continuous contact columns in contrast to trayed columns which are staged contact columns. The liquid is distributed over a packed bed to flow down through the packing by gravity while the vapor flows upward. There are various types of geometric shapes for the packing optimized to provide large interfacial surface between the vapor and the liquid phases for efficient mass transfer. Thus, the packing affectivity which is expressed in terms of the surface area per unit volume of packed space should be high. Other desirable attributes for packing include large fractional void space to permit large flow of fluid with low pressure drop, be chemically inert to the fluids being processed, have structural strength to withstand the weight of the packing above it, have low density and low cost. Typically, packing is chosen over trays for applications where the column diameter is small or is in corrosive service since there are more choices available for the packing material such as plastics, ceramics, as well as metal alloys, and where lower pressure drop is required. Packed columns also have less liquid entrainment and cause less foaming due to less agitation of the liquid by the vapor. On the other hand, packed columns are less suitable for services where there is a wide variation in the liquid or the vapor flow, fouling fluids are present, the liquid to vapor ratio is large, the number of required stages is large, large temperature gradients can occur which can cause cracking of the packing, or if a cooling coil has to be inserted into the column as an intercooler to limit the temperature.

Packing can be either random or structured type. Random packing is made up of individual shaped pieces that are dumped into the column in sections as packed beds with the maximum height of each section dependent on the strength of the packing material (its crushing strength) and the liquid flow distribution across the cross section. These individual packing pieces come in various shapes with names such as pall rings, raschig rings, intelox saddles as illustrated in *Perry's Chemical Engineers' Handbook* (Green, 2007), and some consisting of proprietary designs and different sizes. Typically, the height of a packed section is limited to 15 ft (4.6 m) to 18 ft (5.5 m). Note that the packing at the bottom of a packed section has to bear the weight of the entire packing material in that section. The bed is held on a grate and sometimes the bed may be loaded with larger packing material followed by smaller sized packing to avoid the packing from dropping through the holes in the grate. Smaller the diameter of the column, smaller should be the characteristic length of the random packing. It is typically limited to 1/8 of the column diameter. Structured packing typically consist of thin corrugated metal plates or gauzes stacked within the column in an orderly arrangement and has the advantage of high efficiency with low pressure drop but is more expensive than random packing.

Liquid distributors are installed to obtain uniform flow distribution across the cross-section of the packed section below. Redistribution may also be required especially for small diameter columns to avoid channeling of the liquid flow by breaking up the bed into smaller sections and installing redistribution trays between the beds. Various types of distributors are illustrated in *Perry's Chemical Engineers' Handbook* (Green, 2007). When a portion of the liquid has to be withdrawn from the column, a chimney tray is installed between the packed beds and consists of risers for the vapor flow with hats to avoid the liquid raining down into the riser, while the liquid is collected for draw-off.

5.2.2.3 Spray Columns Spray columns are typically used when limited pressure drop on the vapor side is available and consist of liquid spray nozzles installed inside a vessel. The inlet gas stream usually enters the bottom of the tower and moves upward, while liquid is sprayed downward for countercurrent contact from one or more levels. Carryover of droplet by the vapor can be significant, however, to obtain true countercurrent contact.

5.2.3 Column Sizing

An introduction to determining the column height and its diameter is presented next for distillation and more details on this subject may be found in books on mass transfer operations such as the one by Treybal (1980). Distillation can be considered as the more general case compared to absorption and stripping since distillation columns typically have a condenser and a reboiler while the feed to the column is not typically introduced at the top nor at the bottom tray but at a tray in between. The more general case would be when feed streams of different compositions are introduced into the column, each at the appropriate stage and multiple "side-draw" product streams (in addition to the distillate and the bottoms streams) are withdrawn, again each from a different stage. Both graphical (for binary mixtures) and numerical methods (that may be used for multicomponent mixtures) have been developed. The following discussion will be limited to the numerical methods which are used in process plant simulators widely accessible to energy engineers, made possible by the availability of the powerful desk top computer. A column with a single feed stream and no side-draws is considered for this illustration. The numerical methods described here consist of the equilibrium stage approach for trayed columns and the rate-based approach which may be more appropriate for packed columns, after a few concepts are first introduced. The equilibrium stage approach may be applied to packed columns using the "height equivalent to a theoretical plate" (HETP) by using experimentally determined HETP data but the HETP in addition to being specific to a particular type of packing also depends on the system conditions including the vapor and liquid flow rates, the species concentrations, and type of species involved. Enormous amount of experimental data is required and a paper by Piché et al. (2003) summarizes the various sources of data and correlations available. Use of HETP entails multiplying the ideal number of stages (i.e., without applying an efficiency) by the HETP and then applying a design allowance, typically 10%.

5.2.3.1 Key Components In multicomponent distillation, the more volatile components are referred to as "light" while the less volatile components as "heavy," the relative volatility being the gauge. When there is a single component whose maximum allowable concentration in the bottoms is a design specification, this component is called the "light key." Similarly, when there is a single component whose maximum allowable concentration in the distillate is a design specification, this component is called the "heavy key."

5.2.3.2 Column Specifications The following specifications are typically known or specified:

1. Feed temperature, pressure, composition and rate, and the stage where it is introduced.
2. Column operating pressure which is typically selected based on the condenser cooling medium temperature.
3. Heat losses which are typically neglected for large-scale units due to the relatively smaller surface per unit capacity of the system.

Additional specifications required to completely define the system may be chosen from the following:

1. Number of stages
2. Reflux flow rate or ratio or condenser duty
3. Flow rate of vapor generated in reboiler (boilup rate) or ratio of vapor generated in reboiler to bottoms product (reboil or boilup ratio) or reboiler duty
4. Concentration of a particular component in the distillate or in the bottoms or in both

5. Ratio of flow rate of a particular component in the distillate or in the bottoms to that in the feed (fractional recovery)
6. Ratio of total distillate or total bottoms to total feed.

5.2.3.3 Reflux Ratio and Number of Stages As the reflux ratio (defined as the ratio of the amount of liquid distillate fed back to the column to that withdrawn as the overhead product) is increased, more of the liquid (rich in the more volatile components) gets recycled back to the column, and for a specified degree of separation, the required number of stages within the column decreases while the heat duty of the condenser and the reboiler increase since more vapor has to be condensed within the condenser and consequently more liquid has to be vaporized in the reboiler. In the limit, when the entire overhead stream is condensed and refluxed back to the column, that is, without withdrawing any amount of the distillate as product, the resulting number of stages is the minimum under this "total reflux" condition. Knowledge of the minimum number of stages is required to pick a reasonable initial value when this parameter is one of the chosen specifications for the column design. A method for estimating the minimum number of stages under total reflux was proposed by Fenske (1932) and may be used to provide a reasonable value without performing tedious calculations.

The largest value for the reflux ratio that results in an infinite number of stages to separate the key components is called the minimum reflux ratio and under these conditions, it is possible to exclude all components heavier than the heavy key from the distillate, and all components lighter than the light key from the bottoms. Knowledge of the minimum reflux ratio is required to pick a reasonable reflux ratio when this parameter is one of the chosen specifications for the column design. Typically, the reflux ratio chosen is 1.2–1.5 times the minimum reflux ratio. A number of methods for estimating the minimum reflux ratio have been proposed but the one by Underwood (1948) which uses simplifying assumptions may be used to provide a reasonable value without performing tedious calculations.

5.2.3.4 Feed Tray Location The optimum tray location may be determined by running a number of cases on the process plant simulator but a reasonable initial value may be estimated by the method described in Treybal (1980) which considers only the key components while making certain simplifying assumptions such as (i) constant molar overflow (i.e., the net molar exchange between the two phases is zero, the molar heats of vaporization of all components are equal, and heats of solution, variations in heat capacity and heat losses are negligible) and (ii) the system follows ideal solution behavior.

5.2.3.5 Equilibrium Stage Approach The column is modeled as comprised of a series of equilibrium stages, that is, the vapor and the liquid streams leaving such a stage are assumed to be in mass and heat transfer equilibrium (see Fig. 5.3). A partial condenser and a partial reboiler may each be also counted as an equilibrium stage. Larger the number of stages required for a specified degree of separation, greater is the height of the column. The actual number of stages or trays may then be determined by using an appropriate tray efficiency discussed previously in this chapter.

The tray efficiency is typically derived from empirical data but semiempirical and first principle models have been developed for certain types of trays to predict the tray efficiency (Burgess and Calderbank, 1975; Garcia and Fair, 2000; Geddes, 1946; Hughmark, 1971; Neuberg and Chuang, 1982; Prado and Fair, 1990; Zuiderweg, 1982) and require a knowledge of the mass transfer coefficients, and the heat transfer coefficients when heat effects are significant.

A system of equations for each stage is solved consisting of a molar balance for each species, an equilibrium relationship for each species relating the fractions in the two streams leaving the stage, and an overall energy balance with identical temperatures for the two streams leaving the stage, the two leaving streams being assumed to be in equilibrium. The column design can be optimized by varying the values for the specifications and running a number of cases on the computer using the process plant simulator.

The following set of equations is written for a "general stage" involving external heat transfer $\dot{Q}$, a feed stream and a liquid side-draw stream. Typically, molar fractions are used in the specification of product streams as well as equilibrium data and thus it is more convenient to use molar flow rates in the mass and energy balances. Referring to Figure 5.5, the equations for Stage k are as follows.

- Species i molar balance with the x^s, y^s, and z representing the liquid, vapor, and feed molar fractions of species i:

$$\dot{F}_k z_{k,i} + \dot{L}_{k-1} x_{k-1,i} + \dot{V}_{k+1} y_{k+1,i} = \left(\dot{L}_k + \dot{S}_k\right) x_{k,i} + \dot{V}_k y_{k,i}$$

- Summation of the molar fractions

$$\sum_i y_{k,i} = 1 \quad \text{and} \quad \sum_i x_{k,i} = 1$$

- Overall molar balance with the $\dot{L}^s$, $\dot{V}^s$ representing the liquid and vapor molar flow rates while $\dot{F}_k$ and $\dot{S}_k$ representing the feed and liquid side-draw molar flow rates. Note that this equation may actually be obtained by combining the above two sets of equations, that is, by summing up the species molar balances and then substituting the summation of the molar fractions. Thus, it does not represent an independent equation but might be helpful in simplifying the solution procedure.

$$\dot{F}_k + \dot{L}_{k-1} + \dot{V}_{k+1} = \dot{L}_k + \dot{V}_k + \dot{S}_k$$

- Species i equilibrium relationship (refer to Chapter 2):

$$y_{k,i} = f\left(x_{k,i}\right)$$

  $i-1$ such relationships are required in light of $\sum_i y_{k,i} = 1$.

- Overall energy balance with the h^s representing the feed, liquid, and vapor enthalpies on a mole basis with $\dot{Q}_{\text{in}} = 0$ and $\dot{Q}_{\text{out}} = \dot{Q}_{\text{cond}}$ for the condenser,

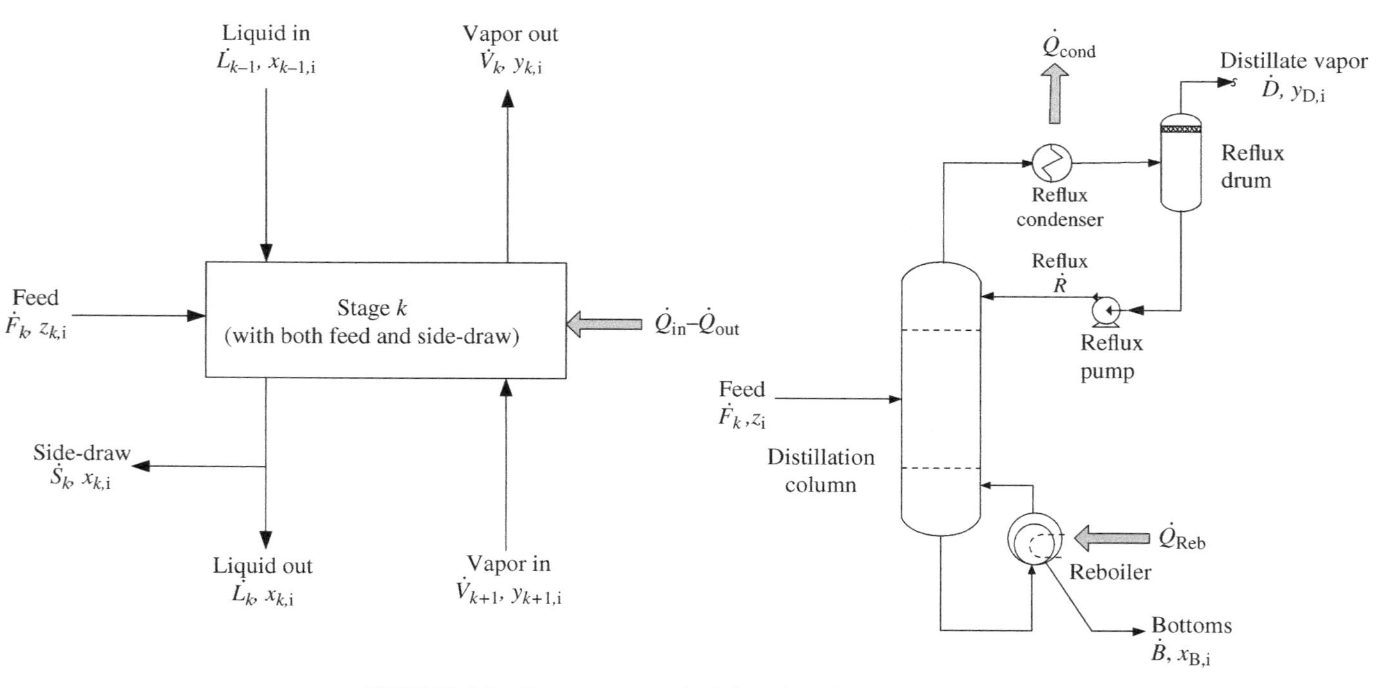

FIGURE 5.5 Stage to stage calculations in a distillation column

$\dot{Q}_{\text{in}} = \dot{Q}_{\text{reb}}$ and $\dot{Q}_{\text{out}} = 0$ for the reboiler, and $\dot{Q}_{\text{in}} = 0$ and $\dot{Q}_{\text{out}} = 0$ for any other stage when heat losses are negligible:

$$\dot{F}_k h_{k,\text{F}} + \dot{L}_{k-1} h_{k-1,\text{L}} + \dot{V}_{k+1} h_{k+1,\text{V}} + \dot{Q}_{\text{in}} = \left(\dot{L}_k + \dot{S}_k\right) h_{k,\text{L}} + \dot{V}_k h_{k,\text{V}} + \dot{Q}_{\text{out}}$$

Equations for the overall column (including the condenser and the reboiler within the control volume) may be helpful in simplifying the solution procedure. For example, if the distillate flow rate is specified and not the bottoms flow rate, the bottoms flow rate may be determined by an overall column mass balance, or if the condenser duty is specified and not the reboiler duty, the reboiler duty may be determined from the overall column energy balance (for the case where there are no other heat exchangers such as intercoolers or side-draw heaters).

- Overall molar balance with $\dot{F}^{\text{s}}$, $\dot{D}$, $\dot{B}$, and $\dot{S}^{\text{s}}$ representing the molar flow rates of the feeds, distillate, bottoms, and side-draws:

$$\sum_k \dot{F}_k = \dot{D} + \dot{B} + \sum_k \dot{S}_k$$

- Species i molar balance with the z_{i}, $y_{\text{D,i}}$, and $x_{\text{B,i}}$ representing the molar fractions in the feed, the distillate (for a case where the entire distillate is in the form of a vapor), and the bottoms:

$$\sum_k \dot{F}_k z_{k,\text{i}} = \dot{D}\, y_{\text{D,i}} + \dot{B} x_{\text{B,i}} + \sum_k \dot{S}_k\, x_{k,\text{i}}$$

- Overall energy balance with the h^{s} representing the feed, distillate, bottoms, and side-draw enthalpies on a mole basis:

$$\sum_k \dot{F}_k h_{k,\text{F}} + \dot{Q}_{\text{reb}} = \dot{D}\, h_{\text{D}} + \dot{B}\, h_{\text{B}} + \dot{Q}_{\text{cond}} + \sum_k \dot{S}_k\, h_{k,\text{L}}$$

When the light and the heavy key fractions are specified, a simplifying assumption would be to assume that all components lighter than the light key would be present only in the distillate while all components heavier than the heavy key would be present only in the bottoms.

Next, a simple application of the equilibrium stage approach to calculate the composition of the vapor and liquid phases formed by cooling a multicomponent vapor is illustrated in Example 5.1.

Example 5.1. A multicomponent vapor stream at a molar rate of $\dot{F}$ of a give composition z_{i} is cooled to a given temperature below its dew point temperature at a given pressure. Derive the working equations to calculate the molar rates $\dot{V}$ and $\dot{L}$ and compositions y_{i} and x_{i} of the vapor and liquid phases formed using the equilibrium stage approach given the K-values (which can be derived from vapor–liquid equilibrium data and discussed in Chapter 2).

Solution

From an overall molar balance, $\dot{F} = \dot{L} + \dot{V}$

From species molar balances, $\dot{F} z_i = \dot{L} x_i + \dot{V} y_i$

Species equilibrium relationships with the K-values, $y_i = K_i x_i$

Combining the above equations, the following working equation is obtained

$$x_i = \frac{z_i}{1 + (\dot{V}/\dot{F})(K_i - 1)}$$

which may be solved by trial and error by assuming $\dot{V}/\dot{F}$ and checking if the resulting x_i satisfy $\sum_i x_i = 1$. Then y_i may be calculated using $y_i = K_i x_i$ while $\dot{V}$ may be obtained from $\dot{F} \times (\dot{V}/\dot{F})$ and $\dot{L}$ from $\dot{L} = \dot{F} - \dot{V}$. The following working equation which is numerically well behaved may be used instead (Rachford and Rice, 1952) obtained by subtracting the above working equation from $y_i = K_i x_i$ and summing over all i^s to solve for *V*/*F*:

$$\sum_i (y_i - x_i) = \sum_i \frac{z_i (K_i - 1)}{1 + (\dot{V}/\dot{F})(K_i - 1)} = 0$$

5.2.3.6 Rate-based Approach Since in packed columns, the vapor and the liquid phases are contacted continuously rather than step-wise as in tray columns, the computation procedure consists of dividing the packing height into differential elements of height dz as illustrated in Figure 5.6 instead of using the stage to stage computational procedure (unless the HETP method is used).

The equilibrium stage approach may still be appropriate for a partial reboiler or a partial condenser, however. Referring to Figure 5.6, the differential equations for an

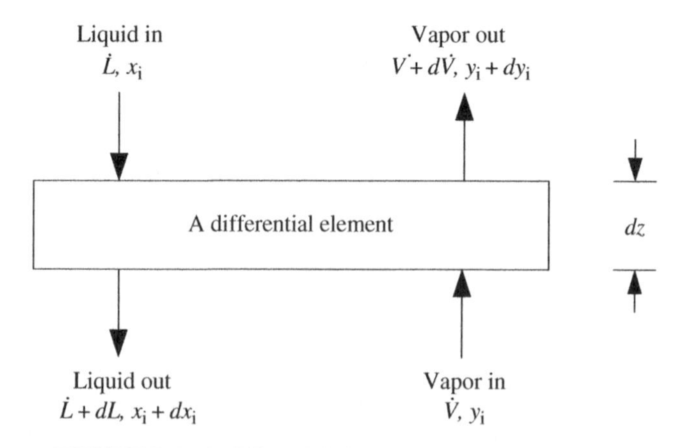

FIGURE 5.6 A differential element in a distillation column

element are as follows, assuming plug flow for both the liquid and vapor phases, and that each of the bulk phases is homogeneous:

- Species i molar balance with x_i and y_i representing the bulk liquid and vapor molar fractions of species i:

$$\frac{d(\dot{L}x_i)}{dz} = \frac{d(\dot{V}y_i)}{dz}$$

- Summation of the molar fractions for species i with x_i^{int} and y_i^{int} representing the molar fractions of species i at the liquid–vapor interface:

$$\sum_i y_i = 1, \sum_i x_i = 1, \sum_i y_i^{int} = 1, \text{ and } \sum_i x_i^{int} = 1$$

- Overall molar balance with $\dot{L}$ and $\dot{V}$ representing the liquid and vapor molar flow rates. As noted previously, the overall molar balance may actually be obtained by combining the above two sets of equations, that is, by summing up the species molar balances and then substituting the summation of the molar fractions. Thus, it does not represent an independent equation but might be helpful in simplifying the solution procedure.

$$\frac{d\dot{L}}{dz} = \sum_i \frac{d(\dot{L}x_i)}{dz} = \frac{d\dot{V}}{dz} = \sum_i \frac{d(\dot{V}y_i)}{dz}$$

- Species i mass transfer rate relationship with a representing the interfacial area per unit volume of packing which is specific to the type of packing but can vary with both the vapor and liquid mass flow rates (determined experimentally and provided by the packing manufacturer), k_L and k_V representing the mass transfer coefficients in the liquid and the vapor phases on a mole basis while applying Equation 5.4 or Equation 5.5 gives:

$$\frac{d(\dot{L}x_i)}{dz} = aA\,k_L\left(x_i^{int} - x_i\right) + x_i \sum_i \frac{d(\dot{L}x_i)}{dz}$$

$$= \frac{d(\dot{V}y_i)}{dz} = aAk_V\left(y_i - y_i^{int}\right) + y_i \sum_i \frac{d(\dot{V}y_i)}{dz}$$

or

$$\frac{d(\dot{L}x_i)}{dz} = aA\,C_{T,L}\,k_L^*\left(x_i^{int} - x_i\right) + x_i \sum_i \frac{d(\dot{L}x_i)}{dz}$$

$$= \frac{d(\dot{V}y_i)}{dz} = aA\,C_{T,V}\,k_V^*\left(y_i - y_i^{int}\right) + y_i \sum_i \frac{d(\dot{V}y_i)}{dz}$$

For equimolar counterdiffusion of components between the vapor and liquid phases which is the case frequently encountered in distillation operations,[3] the above simplify to:

$$\frac{\dot{L}\,dx_{\mathrm{i}}}{dz} = a\,A\,k_{\mathrm{L}}\left(x_{\mathrm{i}}^{\mathrm{int}} - x_{\mathrm{i}}\right) = \frac{\dot{V}\,dy_{\mathrm{i}}}{dz} = a\,A\,k_{\mathrm{V}}\left(y_{\mathrm{i}} - y_{\mathrm{i}}^{\mathrm{int}}\right) \tag{5.11}$$

or

$$\frac{\dot{L}\,dx_{\mathrm{i}}}{dz} = a\,A\,C_{\mathrm{T,L}}\,k_{\mathrm{L}}^{*}\left(x_{\mathrm{i}}^{\mathrm{int}} - x_{\mathrm{i}}\right) = \frac{\dot{V}\,dy_{\mathrm{i}}}{dz} = a\,A\,C_{\mathrm{T,V}}\,k_{\mathrm{V}}^{*}\left(y_{\mathrm{i}} - y_{\mathrm{i}}^{\mathrm{int}}\right)$$

- Species i equilibrium relationship (refer to Chapter 2) at interphase (but not between the bulk phases):

  $$y_{\mathrm{i}}^{\mathrm{int}} = f\left(x_{\mathrm{i}}^{\mathrm{int}}\right)$$

  $i-1$ such relationships are required in light of $\sum_{\mathrm{i}} y_{\mathrm{i}}^{\mathrm{int}} = 1$.

- Overall energy balance and heat transfer rates with the h^{s} representing the liquid and vapor molar specific enthalpies at their respective fluid bulk temperatures, $\hbar_{\mathrm{L}}$ and $\hbar_{\mathrm{V}}$ the liquid and the vapor heat transfer coefficients, and T^{s} representing the liquid and vapor temperatures:

$$\frac{d(\dot{L}h_{\mathrm{L}})}{dz} = aA\hbar_{\mathrm{L}}\,(T_{\mathrm{int}} - T_{\mathrm{L}}) + \sum_{\mathrm{i}} h_{\mathrm{L,i}}\frac{d(\dot{L}x_{\mathrm{i}})}{dz} = \frac{d(\dot{V}h_{\mathrm{V}})}{dz}$$

$$= aA\hbar_{\mathrm{V}}\,(T_{\mathrm{V}} - T_{\mathrm{int}}) + \sum_{\mathrm{i}} h_{\mathrm{V,i}}\frac{d(\dot{V}y_{\mathrm{i}})}{dz}$$

The above shows that the heat released due to condensation of the transferring vapor mass at the interface has to be dissipated into the liquid and vapor phases. Similarly, the heat required for evaporation has to be supplied by the two phases.

Equations for the overall column (including the condenser and the reboiler within this control volume) listed under the Equilibrium Stage Approach may be helpful in simplifying the solution procedure.

Mass transfer coefficients which are specific to the type of packing are typically available for binary systems but those required for a multicomponent system may be estimated from the binary coefficients (Krishna and Standart, 1979). Wang et al. (2005) compiled correlations for mass transfer coefficients developed by various

[3] Another situation is the net transfer of a single specie through another that does not transfer as is the case frequently encountered in absorbers. In such cases, a simplification is made by redefining mass transfer coefficients as $\mathfrak{k}_{\mathrm{L}}$, $\mathfrak{k}_{\mathrm{V}}$, $\mathfrak{k}_{\mathrm{L}}^{*}$ and $\mathfrak{k}_{\mathrm{V}}^{*}$ empirically by $\left(d(\dot{L}x_{\mathrm{i}})/dz\right) = aA\mathfrak{k}_{\mathrm{L}}\left(x_{\mathrm{i}}^{\mathrm{int}} - x_{\mathrm{i}}\right)$, $\left(d(\dot{V}y_{\mathrm{i}})/dz\right) = aA\mathfrak{k}_{\mathrm{V}}\left(y_{\mathrm{i}} - y_{\mathrm{i}}^{\mathrm{int}}\right)$, $\left(d(\dot{L}x_{\mathrm{i}})/dz\right) = aA\,C_{\mathrm{T,L}}\mathfrak{k}_{\mathrm{L}}^{*}\left(x_{\mathrm{i}}^{\mathrm{int}} - x_{\mathrm{i}}\right)$ and $\left(d(\dot{V}y_{\mathrm{i}})/dz\right) = aA\,C_{\mathrm{T,V}}\mathfrak{k}_{\mathrm{V}}^{*}\left(y_{\mathrm{i}} - y_{\mathrm{i}}^{\mathrm{int}}\right)$.

researchers for both random and structured packings. *Perry's Chemical Engineers' Handbook* (Green, 2007) also provides correlations for various packings. When heat transfer coefficients are not available, then these may be estimated for most cases from the mass transfer coefficients using the Chilton–Colburn analogy (Chilton and Colburn, 1934) which was actually developed to estimate the mass transfer coefficients from heat transfer and fluid friction (momentum transfer) data. These transport phenomena are similar and the analogy is valid when they are all governed by similar set of dimensionless equations and when they have similar boundary conditions. Heat and binary mass transfer are related by the Chilton–Colburn factors j_{H} and j_{M}:

$$j_{\mathrm{H}} = \frac{\mathrm{Nu}}{\mathrm{Re\,Pr}}\mathrm{Pr}^{2/3} = \frac{\hbar}{C_{\mathrm{p}}\rho\bar{v}}\mathrm{Pr}^{2/3} = j_{\mathrm{M}} = \frac{\mathrm{Sh}}{\mathrm{Re\,Sc}}\mathrm{Sc}^{2/3} = \frac{k_{\mathrm{x}}}{\rho\bar{v}/M_{\mathrm{w}}}\mathrm{Sc}^{2/3}$$

where the Sherwood number $\mathrm{Sh} = (k_{\mathrm{x}} D / C_T \mathcal{D}_{\mathrm{AB}})$ and the Schmidt number $\mathrm{Sc} = (\mu / \rho \mathcal{D}_{\mathrm{AB}})$ play the same role in mass transfer as the Nusselt number Nu and the Prandtl number Pr do in heat transfer. The various variables appearing in the dimensionless numbers are:

$\hbar$ = heat transfer coefficient

k_{x} = binary mass transfer coefficient

$\mathcal{D}_{\mathrm{AB}}$ = binary diffusion coefficient

C_{T} = concentration of mixture which appears in Sh but does not appear in j_{M}.

D = characteristic length of packing which appears in Nu, Sh, and Re but does not appear in j_{H} or j_{M}.

$\bar{v}$ = superficial velocity (volumetric flow rate per unit cross-sectional area of column) of fluid

M_{w} = molecular weight of fluid

k, ρ, μ, and C_{p} = thermal conductivity, density, viscosity, and constant pressure mass specific heat of the fluid.

Countercurrent humidifiers used in IGCC and advanced power cycle applications such as the humid air turbine (HAT) cycle (Rao, 1989) may be modeled using the above rate- based approach. Cooling towers as well as countercurrent dehumidifiers may also be similarly modeled. The equations take much simpler forms in these applications since only one component (H_2O) is being transferred and there is no reboiler or condenser. When very high mass transfer rates are encountered in such application (Parente et al., 2003), the available data on mass transfer coefficients may have to be adjusted using Ackerman's factor (Bird et al., 2007).

Example 5.2. For a packed column with countercurrent mass transfer and equimolar counterdiffusion between a liquid and a vapor, derive an expression for the height of packing assuming the entire resistance to mass transfer is in the vapor phase, in terms of the vapor phase mass transfer coefficient. Also assume the interfacial area per unit volume of packing a, and the mass transfer coefficient k_V remain constant, while the inlet compositions $x_{i,in}$ and $y_{i,in}$, outlet compositions $x_{i,out}$ and $y_{i,out}$, and liquid and vapor flow rates $\dot{L}$ and $\dot{V}$ are given.

Solution
From Equation 5.11,

$$\left(\frac{\dot{V}}{a\,k_V}\right)\frac{dy_i}{\left(y_i - y_i^{int}\right)} = dz \tag{5.12}$$

$$\dot{L}\,dx_i = \dot{V}\,dy_i \tag{5.13}$$

Integrating Equation 5.12 over the entire height of the column, and Equation 5.13 to any point within the column starting at the bottom (where the vapor enters and the liquid leaves the column) with $\dot{L}$ and $\dot{V}$ constant for equimolar counter diffusion, we have

$$Z = \left(\frac{\dot{V}}{a\,k_V}\right)\int_{y_{i,in}}^{y_{i,out}} \frac{dy_i}{\left(y_i - y_i^{int}\right)} \tag{5.14}$$

$$\left(y_i - y_{i,in}\right) = \left(\frac{\dot{L}}{\dot{V}}\right)\left(x_i - x_{i,out}\right) \tag{5.15}$$

Since the two phases may be assumed to be in equilibrium at the interface and the entire resistance to mass transfer is in the vapor phase (and none in the liquid phase, i.e., $x_i^{int} = x_i$), we have $y_i^{int} = f\left(x_i^{int}\right) = f(x_i)$. Substituting into Equation 5.14,

$$Z = \left(\frac{\dot{V}}{a\,k_V}\right)\int_{y_{i,in}}^{y_{i,out}} \frac{dy_i}{\left[y_i - f(x_i)\right]} \tag{5.16}$$

Equation 5.16 may then be integrated with the help of Equation 5.15 which provides the required relationship between x_i and y_i, either graphically or numerically, using vapor–liquid equilibrium data generated either experimentally or derived by semiempirical methods. A similar expression as Equation 5.16 may be derived for cases where the entire resistance to mass transfer lies in the liquid phase. When both the phases contribute to the resistance, an overall mass transfer coefficient may be utilized with the proper choice of the driving force.

The first term (within brackets) on the right hand side of Equation 5.14 has the unit of length and is called the height of a transfer unit (HTU) which when multiplied by the second term (the integral) which is dimensionless and called the number of transfer units (NTU), gives the total height required for the mass transfer operation. Empirical correlations have been developed for estimating HTUs. It should be noted that the

HTU may not be a constant as $(a\, k_V)$ may vary since the vapor and liquid mass flow rates may not remain constant due to differences in molecular weights of the diffusing species even when the molar flow rates remain constant, and thus the constancy of the height of a transfer unit should be checked before applying the above methodology.

5.2.3.7 ***Overall Column Height*** In addition to the height of the column required to accommodate the trays or packing (and the space between trays or packing sections), height is required both at the top and at the bottom of the column. Space at the top is required (typically 1.5–3 m) to accommodate the reflux nozzle, any manway, any demister and enough space for disengagement of the liquid droplets from the vapor. The height at the bottom is required to accommodate the nozzle for the vapor from the reboiler and to provide enough surge capacity for the liquid which is set by the downstream operation which is typically 5 min. Adequate height should be provided between the HHLL and the bottom tray or packing section (see the criteria discussed for KO drums under Droplet Separation in Chapter 3 for setting the various liquid levels).

5.2.4 Column Diameter and Pressure Drop

If the column diameter is made too large to minimize pressure drop (on the vapor side), a tray column may weep, that is, the liquid starts flowing through the openings in the trays as explained previously, while in the case of a packed column, the mass and heat transfer coefficients will be diminished. In either case, separation efficiency is severely compromised. On the other hand, if the column diameter is made too small to reduce equipment cost (at the expense of pressure drop and thus operating cost), the column may flood. The liquid which flows by gravity has to overcome the increased pressure exerted by the vapor and a point is reached where the liquid cannot flow and starts backing up (in the downcomer in the case of a trayed column). Excessive vapor velocity also causes the liquid to be entrained up the column as droplets or froth, or froth to build up and reach the tray above, which not only reduces the advantages of countercurrent operation and the effective tray efficiency but also leads to flooding. The criterion used in determining the column diameter is to first determine the minimum cross-sectional area at which flooding occurs and then increasing this area by dividing by 0.85 typically for larger columns and by 0.75 for smaller (<2 m in diameter) columns.

5.2.4.1 ***Tray Columns*** In the case of trayed columns, the criterion used in determining the "flooding diameter" is entrainment flooding which occurs due to carry over of liquid droplets by the vapor (for foaming liquids where this type of flooding becomes more severe, a larger increase in the column diameter over that given by dividing the "flooding area" by 0.85 may be required). The vapor flow rate at flooding corresponds to that required to suspend a liquid droplet and is related to the fluid densities (Souders and Brown, 1934) by

$$\frac{\dot{V}}{A} = C_F \sqrt{\frac{\rho_L - \rho_V}{\rho_V}}$$

where $\dot{V}$ is the vapor volumetric flow rate, A is the column net cross-sectional area available for vapor flow, that is, after subtracting the downcomer(s) cross-sectional area, ρ_L and ρ_V are the liquid and vapor densities, and C_F is an empirically determined factor for a given type of tray and is related to the ratio of the liquid to vapor mass flow rates, the ratio of the liquid to vapor densities, surface tension of the liquid and geometry of the tray including the tray spacing, and plots are presented in Treybal (1980). The downcomer area for a given tray spacing is based on an empirically determined clear liquid velocity through the downcomer and is kept low for liquids with high foaming tendency. It can be as low as 0.2 ft/s (0.06 m/s) for a high foaming liquid when tray spacing is as low as 18 in. (46 cm) to as high as 0.6 ft/s (0.18 m/s) for a low foaming liquid when tray spacing is as high as 30 in. (76 cm). Correlations presented by Kozioł and Maćkowiak (1990) or those in *Perry's Chemical Engineers' Handbook* (Green, 2007) may then be used to estimate the fraction of the liquid entrained by the vapor. Typically, the entrainment measured on a mass basis is limited to 10% of the liquid flow and if excessive, the column diameter or tray spacing is increased.

The vapor side pressure drop may be considered to be the sum of three components: (1) due to the flow of the vapor through a dry plate, (2) due to the depth of liquid on the tray, and (3) that required to overcome surface tension when bubbles are formed at the tray holes. Correlations have been developed for these pressure drop components for specific tray types and presented by Madigan (1964) for sieve trays and by Bennett et al. (1983) for valve trays. *Perry's Chemical Engineers' Handbook* (Green, 2007) also provides a methodology for calculating the pressure drop.

5.2.4.2 Packed Columns For random packing, both the flooding conditions and pressure drop can be obtained from plots of the dimensionless grouping, $(\dot{m}_L/\dot{m}_V)\left(\sqrt{\rho_V/\rho_L}\right)$ (plotted on the x-axis) versus the dimensionless grouping, $\left((\dot{m}_V/A)^2 C_F (\mu_L)^{0.2}/\rho_V\rho_L\right)(\rho_V/\rho_L)$ (plotted on the y-axis) where $\dot{m}_V$ and $\dot{m}_L$ are the vapor and liquid mass flow rates, μ_L is the liquid viscosity, and C_F is the "packing factor" that takes into account the unique geometry of a given packing (Treybal, 1980). Curves are presented for different values of pressure drop per unit packing height (as the parameter) as well as for the flooding condition. The flooding condition curve may be used in determining the cross-sectional area A corresponding to flooding, and then by increasing A by the appropriate factor mentioned previously to stay safely below flooding, the y-axis may then be recalculated to read off the pressure drop corresponding to the increased cross-sectional area using the pressure drop curves.

The column diameter should be based on the controlling tray or section of packing and requires evaluating the conditions within the column in different trays or at different packing heights where the liquid and/or vapor flow rates or other operating conditions such as temperature have changed significantly. If the diameter calculated for the top section is significantly larger than that required for the bottom section or vice versa, then it is not uncommon to design the column with the top and bottom sections with two different diameters as long as the length of each section having a different diameter is significant, typically greater than 6 m.

5.3 FLUID–SOLID SYSTEMS

5.3.1 Adsorbers

Adsorption is typically carried out in a fixed bed which consists of a packed bed of adsorbent particles. The fluid (either a gas or a liquid) flows downward through the bed in nearly plug flow. The method for estimating the pressure drop through the bed was described in Chapter 3 and is dependent on the particle size and shape. Larger particles result in lower pressure drop but smaller particles have greater affectivity. Since most of the surface area resides within the pores of the adsorbent particles, the area available for pore diffusion is reduced as the particle size is increased for a given bulk volume of the particles. The adsorbent either has to be regenerated or replaced before breakthrough can occur due to adsorbent saturation, that is, when the undesirable species can no longer be removed and starts appearing above acceptable values in the effluent stream. Typically, for drying operations where periodically the adsorbed water is stripped off using steam, two beds are provided each sized with capacity for the full flow with one operating in the adsorption mode while the other is regenerated. In the case of an activated carbon bed for trace component removal that requires only a few hours for regeneration and may be regenerated less frequently, say every 5 days, the two beds do not have to be sized for the full flow and instead about half of the total flow. Since, during regeneration period of one bed, the other bed would be fed with the entire flow, the pressure drop through the bed would be higher and has to be accounted for in the system design. The amount of adsorbent and size of the vessel are then determined based on the frequency of replacing or regenerating the adsorbent. When regeneration is performed with a fluid (say steam in the case of a drier) flowing upward through the bed, the velocity of the fluid should be limited to avoid lifting the bed unless a retaining grate is provided above the bed to hold the particles.

The adsorption process may be modeled from first principles (transport model) as outlined below but tends to be rather complex. A bulk equilibrium model may be used instead as an approximation as outlined later in this section.

5.3.1.1 Transport Model For a highly porous particle, since most of the adsorption occurs within the pores rather than on the outside surface, the mass transfer process for the species being adsorbed may be considered to include: diffusion through the film around the solid particle to the pore mouth, diffusion within the macropores to the micropore mouths as well as to the macropore wall surface and adsorption, and diffusion within the micropores to the micropore wall surface and adsorption. Diffusion within the pores can be by molecular diffusion and/or by Knudsen diffusion in the case of a gas at low density when the diameter of the pore is comparable to or smaller than the mean free path of the molecules, and by surface migration.

In Knudsen diffusion, the molecules bounce off the pore walls during their migration rather than bouncing off each other. Modeling of the overall mass transfer process is simplified significantly if just one of the above steps is the controlling step which may be the case depending on the system and its operating conditions.

Mass transfer through the film may be modeled empirically by Equation 5.17 which is similar to the convective heat transfer Equation 4.2 using concentration difference as the driving force:

$$N_A = k_C\,(C_{bulk} - C_{in}) \tag{5.17}$$

where k_C is a mass transfer coefficient, C_{bulk} is the bulk gas concentration of the adsorbate, and C_{in} is the concentration at the pore mouth (inlet). Correlations for the mass transfer coefficient, k_C for packed beds such as Equation 5.18 (Wakao and Funazkri, 1978) have been developed relating the Sherwood number, $\mathrm{Sh} = (k_C D_P/\mathcal{D}_m)$, Schmidt number, $\mathrm{Sc} = \nu/\mathcal{D}_m$ and Reynolds number, $\mathrm{Re} = \bar{v} D_P/\nu$ for $3 < \mathrm{Re} < 10{,}000$

$$\mathrm{Sh} = 2 + 1.1\,\mathrm{Re}^{0.6}\,\mathrm{Sc}^{1/3} \tag{5.18}$$

where $\bar{v}$ is the superficial fluid velocity, D_P is the particle diameter, $\mathcal{D}_m$ is the molecular diffusivity, and ν is the kinematic viscosity of the fluid.

A simplified geometric model for the pores consists of considering them as cylindrical in shape with a mean pore diameter and applying an experimentally determined "tortuosity factor," (defined as the ratio of the actual distance a molecule travels between two points to the shortest distance between those two points) to account for the pores not being straight and cylindrical but consisting of a series of tortuous and interconnecting pathways of varying cross-sectional areas. The combined mass transfer rate through the fluid diffusing through the pores and adsorption at the pore internal solid surface may be derived by considering the ideal pore and performing a mass balance on a differential element in the axial direction of the pore and then integrating across the entire length of the pore while simultaneously solving the adsorption by the solid surface. Heat effects and temperature variations should also be accounted for when the heat of adsorption is significant enough. The effective diffusivity that takes into account the molecular, the Knudsen and the surface diffusivities and the tortuosity is utilized and this methodology is similar to that used in modeling heterogeneous catalytic reactions (Carberry, 1976; Levenspiel, 1972) except that for adsorption within a particle, the analysis entails unsteady state since the particle surface gets saturated with the adsorbate over a period of time. The process of surface adsorption in general fits experimental data satisfactorily by first-order rate expressions utilizing the driving force as the displacement from equilibrium (Levenspiel, 1972). The adsorption rate J_A may then be written as

$$J_A = k_a\left(C^*_{Fpore} - C_{Fpore}\right) \tag{5.19}$$

where k_a typically shows an Arrhenius type of dependence on temperature, C_{Fpore} is the fluid phase concentrations of adsorbate species within the pore, and

C^*_{Fpore} corresponds to the fluid phase concentrations of adsorbate species within the pore in equilibrium with the solid surface. Next, a relationship of the type expressed by Equation 5.6 is used in relating C^*_{Fpore} and the concentration of the adsorbate on the solid surface.

5.3.1.2 Equilibrium Model To estimate the mass of adsorbent m_{Ad} required to treat a fluid at a rate $\dot{V}_{\text{F}}$ such that the fractional loading of the adsorbent is Ψ_{Ad} of its value in equilibrium with the specified concentration C_{out} (mass/unit volume) in the effluent fluid which is reached after time t (at which time the adsorbent bed is replaced or recharged), an overall adsorbate mass balance for the system shown in Figure 5.7 yields

$$C_{\text{in}}\dot{V}_{\text{F}}t + \mathfrak{X}_{\text{Ad}}m_{\text{Ad}} = C_{\text{out}}\dot{V}_{\text{F}}t + \Psi_{\text{Ad}}\mathfrak{X}^*_{\text{Ad}}m_{\text{Ad}} \tag{5.20}$$

where $\mathfrak{X}_{\text{Ad}}$ is the initial concentration (mass fraction) in the adsorbate, $\mathfrak{X}^*_{\text{Ad}}$ the equilibrium value corresponding to C_{out}, and C_{in} is the concentration in the fluid to be treated. Next, a relationship between C_{out} and $\mathfrak{X}^*_{\text{Ad}}$ is required. If the system can be described by the Langmuir isotherm, Equation 5.6 may be written in terms of concentrations as

$$\frac{\mathfrak{X}^*_{\text{Ad}}}{\mathfrak{X}_{\text{Ad,sat}}} = \frac{KC_{\text{out}}}{1+KC_{\text{out}}} \tag{5.21}$$

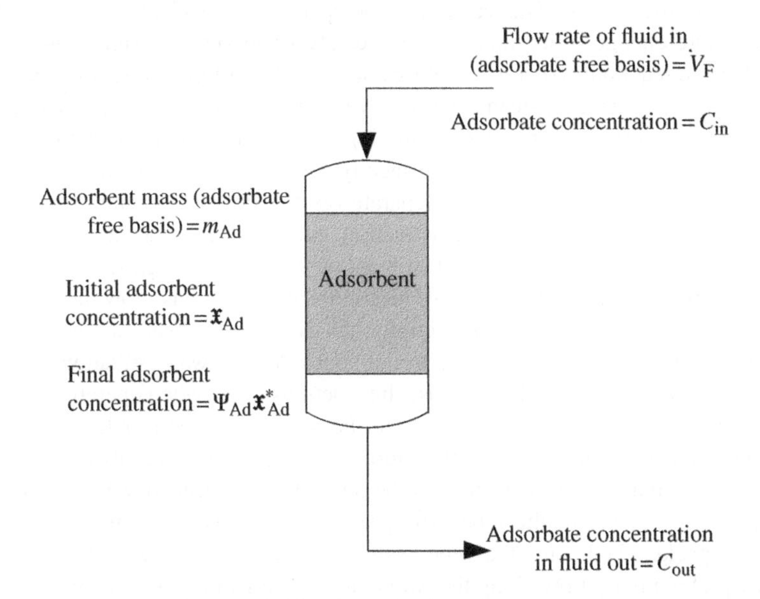

FIGURE 5.7 A batch adsorption process

where $\mathfrak{X}_{\text{Ad,sat}}$ is the maximum or saturated concentraion of the adsorbate in the adsorbent, a constant for a given system, and has to be determined from experimtal data along with K. Simultaneous solution of the Equations 5.20 and 5.21 provides the value of m_{Ad}. Note that this method cannot be used when the heat of adsorption is significant since the heat starts accumulating raising the bed temperature which effects the relationship between C_{out} and $\mathfrak{X}^*_{\text{Ad}}$.

5.3.2 Catalytic Reactors

Catalytic reactions may be carried out in packed beds (fixed beds reactors, tubular reactors, and trickle bed reactors), fluidized beds, and slurry bubble beds depending on the type of reaction system and scale of operation. A brief description of these major types of reactors follows along with an outline of the mass transfer process and reactor modeling.

5.3.2.1 Packed Bed Reactors A fixed bed reactor which consists of a packed bed of catalyst particles is most suitable for large scale, relatively slow reactions that do not have large heat of reaction. The fluid consisting of a gas or a liquid flows downward through the bed contained in a vessel in nearly plug flow. The method for estimating the pressure drop through the bed was described in Chapter 3 and is dependent on the particle size and shape. Larger particles result in lower pressure drop but smaller particles have greater surface area per unit bed volume for the catalytic affectivity. Most of the surface area resides within the pores of the catalyst as in adsorbent particles and the area available for catalytic activity is reduced for a given bulk volume of the catalyst. The reactor may be operated as an adiabatic reactor when the temperature change is acceptable from a catalyst and vessel design standpoints. For exothermic reactions, (i) either cooling coils may be installed to remove the heat by, for example, generating steam, (ii) or have multiple beds within the vessel and cooling the effluent from the first bed by mixing fresh feed before it enters the second bed, and so on, (iii) or by installing a surface type of heat exchanger with a cooling medium between the beds contained in separate vessels. Poor temperature control can cause hot spots and lead to catalyst deactivation, and in the extreme case, melting of the reactor wall. It is not only the heat of reaction but also the high rate of reaction (high rates can generate large amounts of heat in localized regions), low thermal conductivity within the bed (higher conductivity can disperse the heat) and insufficient cooling (when provided), that can lead to hot spots. A "runaway" situation can also occur as higher temperatures also increase the reactions rates (when operating not too close to the equilibrium point), which can further exacerbate the problem.

A tubular reactor is used for highly endothermic reactions, and also consists of a "packed bed" but contained inside a tube such that the heat may be transferred efficiently from outside of the tube. The pressure drop has two components, one due to the packed bed with the pipe wall effect taken into account (since the fluid may preferentially flow along the walls) and the other due to the bare pipe wall friction itself.

In a trickle bed reactor, the reacting gas and liquid phases flow downward cocurrently through a packed bed of solid (catalyst) particles. When the heat of reaction is excessive, the reactor may be designed such that a fraction of the liquid vaporizes to limit the temperature.

5.3.2.2 ***Fluidized Bed Reactors*** The catalytic particles are suspended by an upflow of fluid. An advantage with this type of reactor is that loading and removal of catalyst is easier, a feature especially useful when the catalyst has to be removed, replaced, or regenerated often. A high conversion rate with a large throughput can typically be obtained but the particles should be attrition-resistant and should not agglomerate. Cyclones are typically installed to recycle the fines to the bed. The method for estimating the pressure drop through the bed as well as the minimum fluidization velocity has been described in Chapter 3 and is dependent on the particle size and shape. A narrow particle size distribution is desirable with spherical shape being the most attrition resistant. Large heat of reaction can be easily controlled in a fluidized bed while maintaining uniform temperature distribution.

5.3.2.3 ***Slurry Bed Reactors*** Slurry reactors may be used for reacting the three phases: solid, liquid, and gas but in the synthesis of chemical coproducts from syngas, the solid is a catalyst, the liquid serves as a heat sink while the gas contains the reacting species. The (catalyst) particles are suspended in a circulating liquid while the gas bubbles up through the liquid phase. Slurry reactors are most frequently used when the heat of reaction is high. Particles must be attrition resistant while small catalyst particle sizes are desirable for greater mass and heat transfer rates but if too small can become entrained in the gas stream leaving the reactor and can also cause formation of particle clusters resulting in reduced mass transfer rates. This type of reactor has some of the same advantages of the fluidized bed with respect to loading and removal of catalyst, a feature especially useful when the catalyst has to be removed, replaced, or regenerated often, and high conversion rate with a large throughput. The gas side pressure drop can be lower which saves compression power and cost.

5.3.2.4 ***Advanced Reactors*** The use of microchannel technology is a developing field to enhance reaction rates and reduce size of the reactor system by increasing mass and heat transfer rates. These compact reactors contain thousands of channels with characteristic dimensions in the millimeter range. Channels with diameters in the range 0.1–10 mm filled with a catalyst are interleaved with coolant filled channels for exothermic reactions. The small dimensions of the channels allow large heat transfer rates and as a result, more active catalysts can be used without developing excessive local temperatures or hot spots.

Membrane reactors discussed in Chapter 11 are another promising technology for use in energy plants. These reactors utilize high temperature membranes to separate out a reaction product as it is formed forcing the equilibrium toward higher conversion whereas in conventional reactors, the conversion is constrained by equilibrium.

*5.3.2.5 **Reactor Models*** A transport model for the reactor is similar to that outlined for the adsorption process except that it entails transfer of typically two species (reactants) to the catalyst surface within the pores and the transfer of the reaction products in the opposite direction. A kinetic expression for the rate of reaction is required specific to a given system. As an example, the expression for the overall kinetics for the water gas shift reaction, $CO + H_2O = H_2 + CO_2$ which is encountered in today's IGCCs (that include decarbonization of the syngas before it is combusted in the gas turbine) is the Langmuir–Hinshelwood type expression derived with the assumption that both the reacting molecules are adsorbed and the adsorbed molecules undergo a bimolecular reaction. The rate of conversion of CO specifically over a Fe-based catalyst (Podolski and Kim, 1974) is given by:

$$r_{CO} = k K_1 K_2 \frac{p_{H_2O}\, p_{CO} - p_{H_2}\, p_{CO_2}/K_p}{\left(1 + K_1 p_{CO} + K_2 p_{H_2O} + K_3 p_{CO_2} + K_4 p_{H_2}\right)^2}$$

where k and K_p are the rate and the overall equilibrium constants for the water gas shift reaction while K_1, K_2, K_3, and K_4 are the adsorption equilibrium constants for CO, H_2O, CO_2, and H_2.

In general, a model has to take into account the rates of adsorption, followed by surface reaction and subsequent desorption, which could play a key role in this rate controlled process. Simplification of the overall reaction rate expression is of course made possible by determining which one of these three steps (adsorption, surface reaction, or desorption) is controlling.

A bulk equilibrium model may be used instead as an approximation. As an example, the calculation methods for determining the effluent composition, and outlet temperature in the case of an adiabatic reactor, or the required heat transfer rate in the case of a nonadiabatic reactor for a desired outlet temperature, are described next. A steam-methane reformer is chosen for this illustration in which H_2O is reacted with CH_4 to produce a syngas (primarily a mixture of H_2 and CO but the reactor effluent in addition will contain CO_2 which is also formed and the unreacted H_2O and CH_4) with a given set of design parameters such as the feed composition, pressure, and temperature, and pressure drop through the reactor.

These calculations are performed by solving a set of simultaneous equations consisting of elemental balances, energy balance, and the reaction quotient for each of the thermodynamically independent reactions that quantifies the approach to thermodynamic equilibrium. Such a thermodynamic model is often used in system studies where the reactor effluent composition and temperature (or the total heat transfer duty) are to be quantified and when reactor sizing can be done empirically. Determining the size of the reactor (and the amount of catalyst required for a catalytic reactor) nonempirically requires a transport model described earlier that takes into account not only the reaction kinetics but also the heat and mass transfer processes depending on which of these rate processes is or are controlling. Note that the model for the reaction kinetics would be specific to the characteristics of a given catalyst.

The thermodynamically independent reactions are:

$$CO + H_2O = H_2 + CO_2$$
$$CH_4 + H_2O = 3H_2 + CO$$

The following reaction also needs to be considered when H_2S and/or COS are present in which case a sulfur tolerant catalyst is utilized that may hydrolyze COS:

$$COS + H_2O = H_2S + CO_2$$

Each of the elemental balances gives the following set of equations with the nonreacting elements being all lumped together as "inerts":

H balance:

$$4\dot{M}_{\text{out},CH_4} + 2\dot{M}_{\text{out},H_2} + 2\dot{M}_{\text{out},H_2O} = \dot{M}_H \tag{5.22}$$

C balance:

$$\dot{M}_{\text{out},CH_4} + \dot{M}_{\text{out},CO} + \dot{M}_{\text{out},CO_2} = \dot{M}_C \tag{5.23}$$

O balance:

$$\dot{M}_{\text{out},CO} + 2\dot{M}_{\text{out},CO_2} + \dot{M}_{\text{out},H_2O} = \dot{M}_O \tag{5.24}$$

S balance (when the sulfur compounds are present):

$$\dot{M}_{\text{out},H_2S} + \dot{M}_{\text{out},COS} = \dot{M}_S \tag{5.25}$$

Inerts balance:

$$\sum \dot{M}_{\text{out},I} = \sum \dot{M}_I \tag{5.26}$$

where $\dot{M}_{\text{out},CH_4}$, $\dot{M}_{\text{out},H_2}$, $\dot{M}_{\text{out},H_2O}$, $\dot{M}_{\text{out},CO}$, $\dot{M}_{\text{out},CO_2}$,, and $\dot{M}_{\text{out},I}$ are the molar rates of each of the species in the reactor effluent to be determined and $\dot{M}_H$, $\dot{M}_C$, $\dot{M}_O$, $\dot{M}_S$, and $\dot{M}_I$ are the total moles of each of the atomic elements or inert species in the feed to the reactor. Next from equilibrium considerations:

Reforming reaction:

$$\frac{\dot{M}^3_{\text{out},H_2}\,\dot{M}_{\text{out},CO}}{\dot{M}_{\text{out},CH_4}\dot{M}_{\text{out},H_2O}\,\dot{M}^2_{\text{out},TOT}} = K_R \tag{5.27}$$

Shift reaction:

$$\frac{\dot{M}_{\text{out},H_2}\,\dot{M}_{\text{out},CO_2}}{\dot{M}_{\text{out},CO}\,\dot{M}_{\text{out},H_2O}} = K_S \tag{5.28}$$

COS hydrolysis reaction:

$$\frac{\dot{M}_{\text{out},H_2S}\,\dot{M}_{\text{out},CO_2}}{\dot{M}_{\text{out},COS}\,\dot{M}_{\text{out},H_2O}} = K_H \tag{5.29}$$

where $\dot{M}_{\text{out,TOT}}$ is the total molar rate of the reactor effluent, K_R, K_S, and K_H are the reaction quotients which correspond to the equilibrium constants for the three reactions at specified empirically determined temperatures in order to account for incomplete reactions (below the actual effluent temperature for the reforming reaction since this reaction is endothermic and above the actual effluent temperature for the shift reaction since this reaction is exothermic).

The eight equations (5.22)–(5.29) above together with the enthalpy balance on the reactor allow the determination of the effluent composition and temperature in the case of an adiabatic reactor or rate of heat absorbed (or released) in the case of a nonadiabatic reactor. Note that K_y (K_R, K_S, and K_H in this example) based on mole fractions is related to K_P discussed in Chapter 2 for a reaction $aA + bB \leftrightarrow cC + dD$ as follows:

$$K_y = K_P P^{(a+b-c-d)}$$

Example 5.3. The feed to a fixed bed shift reactor has the following composition (on a mole basis): 20% CO, 10% CO_2, 22% H_2, and 40% H_2O with the rest being N_2 and leaves the reactor at a temperature of 391°C. Calculate the effluent gas composition after the catalyst has aged some such that the effluent composition corresponds to the equilibrium composition at a temperature 14°C above the actual reactor effluent temperature (14°C "approach temperature").

Solution

In the case of a single thermodynamically independent reaction as in the case of a "sweet" shift reactor with feed gas free of any sulfur or HCN which may undergo hydrogenation to form CH_4 and NH_3, the solution is simplified by solving for molar conversion X by the reaction. From the stoichiometry of the shift reaction, the mole fractions in the effluent will be $y_{CO}-X$, $y_{CO_2}+X$, $y_{H_2}+X$, $y_{H_2O}-X$. Then

$$\frac{(y_{H_2}+X)(y_{CO_2}+X)}{(y_{CO}-X)(y_{H_2O}-X)} = K_S$$

Or

$$(1-K_S)X^2 + [y_{CO_2} + y_{H_2} + K_S(y_{CO} + y_{H_2O})]X + (y_{CO_2}\,y_{H_2} - K_S\,y_{CO}\,y_{H_2O}) = 0$$

Solving this quadratic equation with $K_S = 15.26$ which corresponds to a temperature of $391 + 14 = 405$°C (which is higher than the actual outlet temperature giving less conversion than the equilibrium conversion for an exothermic reaction) yields $X = 0.17$ resulting in an effluent composition of 3% CO, 27% CO_2, 39% H_2, and 23% H_2O by mole.

REFERENCES

Bennett DL, Agrawal R, Cook PJ. New pressure drop correlation for sieve tray distillation columns. AIChE J 1983;29(3):434–442.

Bird RB, Stewart WE, Lightfoot EN. *Transport Phenomena*. Hoboken, NJ: John Wiley & Sons, Inc.; 2007.

Brunauer S, Emmett PH, Teller E. Adsorption of Gases in Multimolecular Layers. J Am Chem Soc 1938;60:309–319.

Burgess JM, Calderbank PH. The measurement of bubble parameters in two-phase dispersions-II: The structure of sieve tray froths. Chem Eng Sci 1975;30(7):1107–1121.

Carberry JJ. *Chemical and Catalytic Reaction Engineering*. New York: McGraw-Hill; 1976.

Chilton TH, Colburn AP. Mass transfer (absorption) coefficients prediction from data on heat transfer and fluid friction. Ind Eng Chem 1934;26(11):1183–1187.

Fenske MR. Fractionation of straight run Pennsylvania gasoline. Ind Eng Chem 1932;24:482.

Garcia JA, Fair JR. A fundamental model for the prediction of distillation sieve tray efficiency. 1. Database development. Ind Eng Chem Res 2000;39(6):1809–1817.

Geddes RL. Local efficiencies of bubble plate fractionators. Trans Am Inst Chem Eng 1946;42:79.

Green DW, editor. *Perry's Chemical Engineers' Handbook*. 8th ed. New York: McGraw-Hill; 2007.

Hughmark GA. Models for vapor-phase and liquid-phase mass transfer on distillation trays. AIChE J 1971;17:1295.

King CJ. *Separation Processes*. Mineola, NY: Courier Dover Publications; 2013.

Kozioł A, Maćkowiak J. Liquid entrainment in tray columns with downcomers. Chem Eng Process Process Intensif 1990;27(3):145–153.

Krishna R, Standart GL. Mass and energy transfer in multicomponent systems. Chem Eng Commun 1979;3(4–5):201–275.

Levenspiel O. *Chemical Reaction Engineering*. 2nd ed. New York: John Wiley & Sons, Inc.; 1972.

Madigan CM. Pressure drop and stability characteristics of Sieve Tray columns [doctoral dissertation]. New York: New York University; 1964.

Neuberg HH, Chuang KT. Mass transfer modeling for GS heavy water plants. Can J Chem Eng 1982;60:504–510.

Palla R. Meeting Staged CO_2 Capture Requirements with the UOP SELEXOL™ Process. Gasification Technologies Conference; October 2009; Colorado Springs, CO.

Parente JO, Traverso A, Massardo AF. Saturator analysis for an evaporative gas turbine cycle. Appl Therm Eng 2003;23(10):1275–1293.

Piché S, Lévesque S, Grandjean B, Larachi F. Prediction of HETP for randomly packed towers operation: integration of aqueous and non-aqueous mass transfer characteristics into one consistent correlation. Sep Purif Technol 2003;33(2):145–162.

Podolski WF, Kim YG. Modeling the water-gas shift reaction. Ind Eng Chem Process Des Dev 1974;13(4):415–421.

Poling BE, Prausnitz JM, O'Connell JP. *The Properties of Gases and Liquids*. 4th ed. New York: McGraw Hill; 2001.

Prado M, Fair JR. Fundamental model for the prediction of sieve tray efficiency. Ind Eng Chem Res 1990;29(6):1031–1042.

Rachford HH, Rice JD. Procedure for use of electrical digital computers in calculating flash vaporization hydrocarbon equilibrium. J Petrol Tech 1952;4(10):19.

Rao AD, inventor; Fluor Corporation, assignee. Process for producing power. US patent 4,829,763. Washington, DC: U.S. Patent and Trademark Office; 1989, May 16.

Reddy S, Johnson D, Gilmartin J. Fluor's Econamine FG Plus^{SM} Technology for CO_2 Capture at Coal-fired Power Plants. Power Plant Air Pollutant Control "Mega" Symposium; August 2008; Baltimore, MD.

Souders M, Brown GG. Design of fractionating columns I. Entrainment and capacity. Ind Eng Chem 1934;26(1):98–103.

Treybal RE. *Mass Transfer Operations*. New York: McGraw-Hill; 1980.

Underwood AJV. Fractional distillation of multicomponent mixtures. Chem Eng Prog 1948;44(8):603–614.

Wakao N, Funazkri T. Effect of fluid dispersion coefficients on particle-to-fluid mass transfer coefficients in packed beds: correlation of Sherwood numbers. Chem Eng Sci 1978;33(10):1375–1384.

Wang GQ, Yuan XG, Yu KT. Review of mass-transfer correlations for packed columns. Ind Eng Chem Res 2005;44(23):8715–8729.

Zuiderweg F. Sieve trays—A review on the State of the Art. Chem Eng Sci 1982;37:1441.

6

PRIME MOVERS

A prime mover is a machine that converts energy (non-mechanical) into work (mechanical) and examples of such machines are the gas turbine, steam turbine, reciprocating internal combustion engine, and hydraulic turbine. These machines may be used for turning a generator to produce electricity or may be used to drive another machine such as a compressor or a pump. Another example of a prime mover, which will be discussed later in Chapter 12, is the wind turbine. The gas turbine, steam turbine, and reciprocating internal combustion engine are all heat engines, that is, convert thermal energy into work while the hydraulic and the wind turbines convert the available kinetic energy in the fluid into work. In the case of hydraulic turbines, the fluid is a liquid under pressure or with stored potential energy which is then converted to kinetic energy in the fluid as the fluid flows into the turbine.

Both the gas turbine and the reciprocating internal combustion engine are "air breathing" machines and if a flammable vapor is accidently released in their vicinity, possibility exists that the vapors may be drawn into the air intake of the engine and act as a fuel requiring these engines to be shutdown during such episodes. On the other hand, some gas turbines have been used to "treat" air contaminated with vapors that can be decomposed or oxidized within the combustor of the gas turbine, although at much lower concentrations in the air.

Sustainable Energy Conversion for Electricity and Coproducts: Principles, Technologies, and Equipment, First Edition. Ashok Rao.

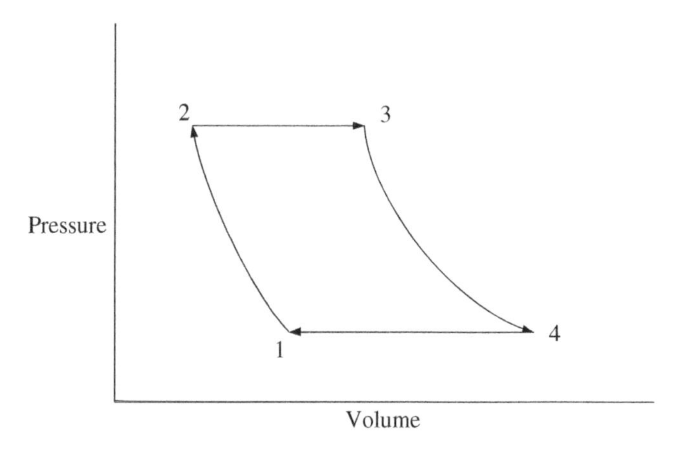

FIGURE 6.1 Air-standard Brayton cycle

6.1 GAS TURBINES

The Brayton cycle in the form of the "air-standard cycle[1]" is depicted in Figure 6.1 (Cengel and Boles, 1998; Smith et al., 2005) and consists of adiabatic compression of a gas (from statepoint 1 to 2) followed by heat addition to the pressurized gas at constant pressure (from statepoint 2 to 3), and then adiabatic expansion of the hot pressurized gas (from statepoint 3 to 4). The cycle is named after George Brayton, an American engineer. A machine incorporating this cycle is the gas turbine which is also sometimes called the combustion turbine in which the heat addition occurs by combusting a suitable fuel in the pressurized air. Unlike the ideal cycle, the combustor has pressure drop associated with it, typically about 4%. The net work produced by the cycle is the difference between that produced by the expansion step (in a turbine) and that consumed by the compression step (in a compressor). Typically, the work required by the compressor is as much as half of the power developed by the turbine in modern day gas turbines which have high turbine inlet temperatures, as high as 1600°C for a large (utility size) state-of-the-art gas turbine. Both closed and open forms of this cycle are employed, the open cycle being the more common one.

In the closed cycle, the working fluid (a gas) operates in a closed loop with heat exchange between the working fluid and the heat source or heat sink occurring through surface type heat exchangers. Such a cycle is also called an "indirectly fired cycle" in contrast to a "directly fired cycle" as in a combustion turbine. The heat source in an indirectly fired gas turbine cycle may consist of the hot combustion products generated by burning a "dirty" fuel such as coal or biomass as discussed in Chapter 9, or it may be heat generated by a nuclear reaction or solar heat concentrated

[1] An idealized cycle approximated as a closed cycle with the combustion process replaced by heat addition through an exchanger while assuming that the working fluid is entirely air behaving as an ideal gas, and that all processes are internally reversible.

by mirrors as discussed in Chapter 12. The gas leaving the expansion step after rejecting heat to the heat sink via an exchanger is provided to the compression step to complete the cycle resulting in an indirectly fired closed cycle.

6.1.1 Principles of Operation

As illustrated in Figure 6.2, in a "simple cycle" gas turbine in the open cycle form, fresh (filtered) air enters the compression step and the heat addition to the cycle occurs by directly firing a "clean" fuel (e.g., natural gas, distillate fuel oil, and syngas derived by gasification of dirty fuels such as coal or biomass) into the pressurized air. The combustion products enter the expansion step while the unconverted thermal energy is carried away by the gas after completing the expansion step. In some instances, heavy oils derived from petroleum that are "dirty" because of their metal and sulfur content are also fired directly in the gas turbine. The gas turbine in such cases is operated at lower turbine inlet temperatures to minimize the adverse effects of the metals and sulfur that are present in this type of fuel, resulting in reduced output and efficiency. Note that Equation 3.29 shows that the power developed during expansion is directly proportion to the absolute inlet temperature of the working fluid.

There are various forms of this cycle, and the most common one is the simple cycle as described above. Intercooling may be introduced in the compression step by inserting the intercooler between a low pressure compressor and a high pressure compressor. Intercooling is justified from a first law basis for very high overall compression

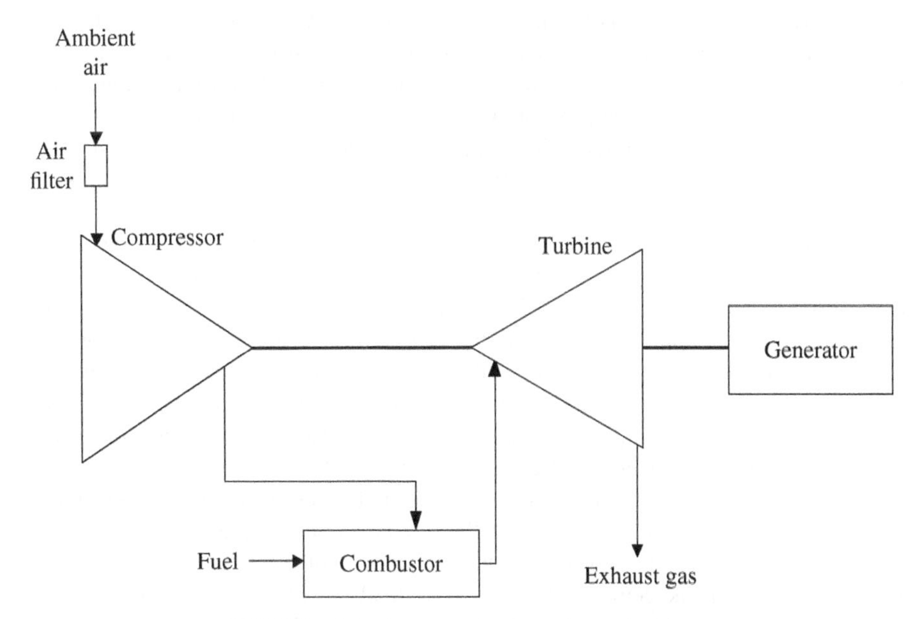

FIGURE 6.2 An open simple cycle gas turbine

ratios and the optimum location for the intercooler is such that the low pressure compressor pressure ratio is much lower than that of the high pressure compressor. The compressor discharge air before entering the heat addition step may be preheated against the turbine exhaust resulting in the "recuperated cycle." When both intercooling and recuperation are incorporated in the cycle, the optimum location for the intercooler is such that the low pressure and the high pressure compressor ratios are similar, depending on their inlet temperatures.

The useful work developed by the gas turbine, instead of being converted into electricity by turning a generator, may be used directly as mechanical energy for driving other rotating equipment such as a compressor. An aircraft jet engine is a gas turbine except that the net output of the engine is the thrust from the turbine exhaust, the turbine producing enough work to operate the compressor.

At the larger scales, there are two types of land-based gas turbines, the heavy frame engines and the aeroderivative engines. Heavy frame engines are characterized by compression ratios that are lower and tend to be physically large. Most of the frame engines had compression ratios below 15 but the compression ratio has been steadily increasing to take full advantage of the increasing turbine inlet temperatures, and some of the current state-of the-art gas turbines have pressure ratios near 25. Aeroderivative engines, as implied by the name, are derived from jet engines and have compression ratios that are much higher and tend to be quite compact. The specific power output, which is defined as the net power produced per unit of air flow rate entering the gas turbine compressor, steadily increases with pressure ratio whereas the efficiency shows a peak value at a certain pressure ratio, and this pressure ratio increases with the turbine inlet temperature. Figure 6.3 shows quantitatively the effect of compression or pressure ratio on the efficiency of an air standard simple cycle gas turbine with turbine inlet temperature of 980°C for two cases, one with compressor and turbine isentropic efficiency at 100% and the other with more realistic values for a small gas turbine (83% for the compressor and 86% for the turbine).The plots show that one can come to a totally erroneous conclusion when using unrealistic component efficiencies in the analysis of cycles, that is, in this case, the efficiency keeps monotonically increasing with pressure ratio.

The optimum pressure ratio of a gas turbine of the simple cycle configuration (i.e., without intercooling or recuperation) is much lower when used in combined cycle applications (where the exhaust from the gas turbine is used in a steam Rankine cycle) than in simple cycle applications (where the exhaust heat is lost to the atmosphere). The entire combined cycle efficiency is utilized in optimizing the pressure ratio of the gas turbine while at the same time taking into account the impact on overall system cost. Large heavy frame engines are designed for combined cycle applications and they also tend to have lower specific cost ($ on a per kW basis).

Demand for electricity changes daily and seasonally. For example, a manufacturing plant may start and stop throughout the day and week, while air conditioners are turned on and off seasonally. The largest amount of electricity is needed by consumers during peak demand above a "base load" of electricity that is needed year-round. When gas turbines were first applied in the electric power generation

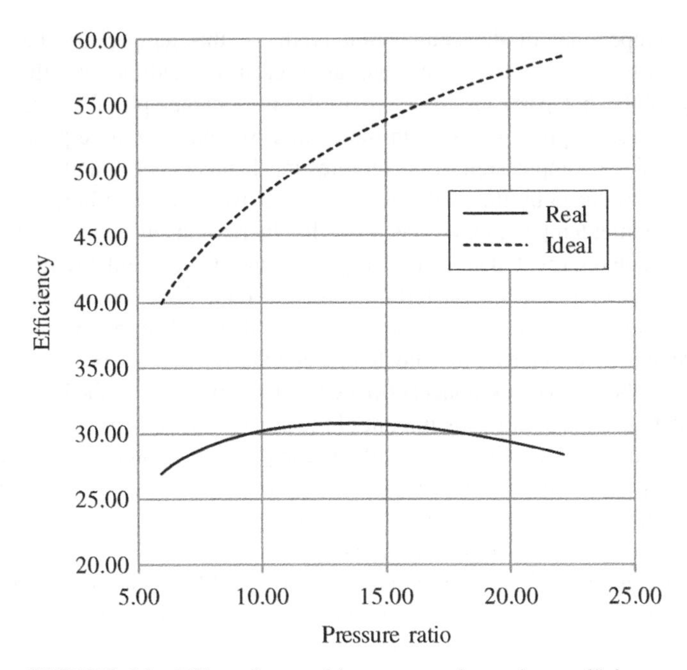

FIGURE 6.3 Effect of gas turbine compression ratio on efficiency

industry, majority of the power generated by gas turbines was for "peak load" service.[2] Since then however, with increases in efficiency and reliability, the gas turbine is being utilized more and more in base load generation. The exhaust heat from a land-based gas turbine when recovered to generate steam to produce additional electric power in a steam turbine resulting in a combined cycle, as discussed in more detail in Chapter 9, is excellent for base load power generation, or to provide process or district heating resulting in a CHP or cogeneration plant as explained in Chapter 10.

The efficiency of a gas turbine is at a maximum when temperature of the working fluid entering the expansion step is also at a maximum since the expansion work is maximized and this occurs when the fuel is burned in the presence of the pressurized air under stoichiometric conditions. When natural gas is combusted with air under stoichiometric conditions, however, the resulting temperature of the combustion products is greater than 1940°C (3500°F) (depending on the temperature of the combustion air) making it too high from materials standpoint. Therefore, it is necessary to use a large amount of excess air in the combustion step as a thermal diluent to

[2] "Exotic" materials of construction are required in modern gas turbines to withstand the extreme operating temperatures necessary to achieve high efficiency and these materials typically tend to have relatively low tolerance for thermal cycling. So gas turbine manufacturers severely limit the number of starts per year when warrantying performance of modern gas turbines for such peaking service.

reduce the temperature of the combustion products, this temperature being constrained by factors such the type of materials used in the turbine, and the cooling technology of the hot parts as discussed in the next paragraph. The necessity to use a large excess of pressurized air in the combustor creates a large parasitic load on the cycle, since compression of air requires mechanical energy and this reduces the net power produced by the cycle, as well as the overall cycle efficiency.

The turbine inlet temperature has been steadily increasing since gas turbines were first introduced and current state-of-the-art gas turbines have an inlet temperature to the first set of rotating blades in the turbine (also called the firing temperature or rotor inlet temperature) of about 1300–1400°C instead of less than 900°C in the mid-twentieth century. This increase in firing temperature has been made possible by being able to operate the turbine components (that come into contact with the hot gasses) at higher temperatures. A significant amount of the pressurized air from the gas turbine compressor is used for cooling these turbine components in order to maintain the metal temperatures within their design limits, which could range from 800 to 950°C depending on the alloys utilized, while the gas flowing through the turbine may be as high as 1300–1400°C. In the case of turbine blade cooling, the cooling air passes through internal passages built into the turbine blades and with open circuit air cooling (which is currently employed), the air then leaves through holes placed in strategic locations in the blade to combine with the working fluid flowing outside the blades. In “film cooling,” the air leaves the blade in such a manner to form an insulating layer on the outside surface of the blade (an advantage of open circuit cooling over closed circuit cooling). The need for cooling the turbine hot parts for a given turbine inlet temperature limits the overall efficiency of the gas turbine since the cooling air reduces the temperature of the hot gas (working fluid) flowing through the turbine upon mixing. Technologies are being developed in areas of materials including use of ceramics and in turbine cooling to minimize the cooling air requirement as discussed later in Section 6.1.6. Conversely, with more advanced materials and cooling technologies, increases in turbine inlet temperature are made possible without having to increase the cooling air requirement significantly, again to affect an increase in the overall cycle efficiency. The specific power output of the engine is also increased which typically translates into lower specific cost for the equipment (unless the materials used in the engine to achieve the higher efficiency and specific power output are too exotic).

Other approaches to increasing the efficiency are to incorporate closed loop steam cooling which can be possible when the gas turbine is used in combined cycle applications. In the current state-of-the-art air cooled gas turbines with firing temperature in the range of 1300–1400°C, as much as 25% of the compressed air is used for cooling. Closed circuit steam cooling in the gas turbine provides an efficient way of increasing the firing temperature without having to use a large amount of cooling air. Furthermore, steam with its very large heat capacity is an excellent coolant. Closed circuit cooling also minimizes momentum and dilution losses in the turbine (which can occur when the cooling air exiting the blade in open circuit cooling mixes with the working fluid flowing outside the blade) while the turbine operates as a partial reheater for the steam bottoming cycle. Another major

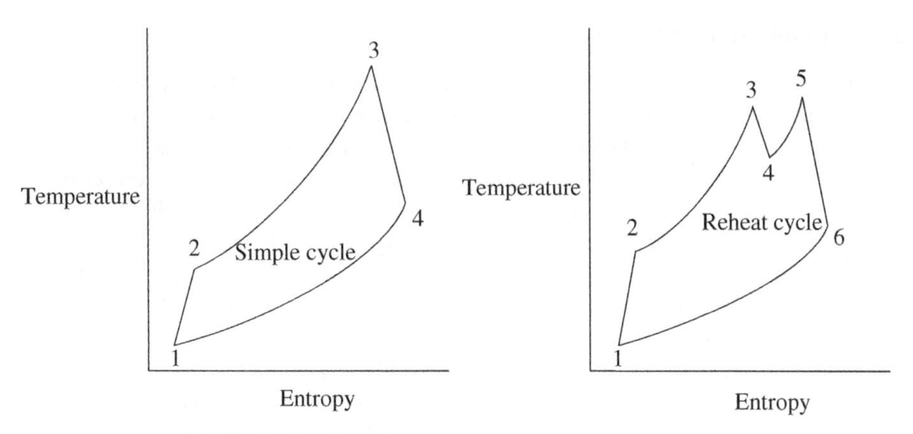

FIGURE 6.4 Simple cycle and reheat gas turbine cycles

advantage with closed circuit cooling is that the temperature drop between the combustor exit gas and the turbine rotor inlet gas is reduced since the coolant used in the first stage nozzles of the turbine does not mix with the gasses flowing over the stationary blades (vanes). Thus, the required combustor exit temperature is reduced for a given firing temperature resulting in lower NO_x emissions; control of NO_x emissions at such high firing temperatures being a major challenge. The General Electric H series gas turbines incorporate steam cooling of both the stationary and rotating blades of the high pressure stages of the turbine, while Siemens and Mitsubishi G series gas turbines incorporate steam cooling of the stationary blades of the high pressure stages.

Reheating during the expansion step of the basic Brayton cycle can be used to reduce firing temperature for a given cycle efficiency as depicted in Figure 6.4. The statepoints shown for the simple cycle case corresponds to those shown in Figure 6.1. For the reheat cycle, the gas turbine has a high pressure combustor (State points 2 to 3 represent the heat addition in this high pressure combustor) and a low pressure combustor (State points 4 to 5 represent the heat addition in this low pressure combustor). The high pressure combustor exhaust enters a high pressure turbine which expands the gases only partially (State points 3 to 4 represent the expansion in this high pressure turbine). These gases which are at a pressure significantly above atmospheric pressure enter the low pressure combustor where additional fuel is combusted and the exhaust form this low pressure combustor then enters the low pressure turbine to complete the expansion step (State points 5 to 6 represent the expansion in this low pressure turbine). The optimum pressure ratio required for such a cycle tends to be higher than that required for a simple cycle gas turbine even with a higher firing temperature to achieve the same thermal efficiency. Reheat gas turbines have been in commercial operation as offered by Alstom and this type of cycle can play a role in achieving higher efficiencies in the future, but as firing temperatures are increased to realize even higher efficiencies, compressor pressure ratio may become the limiting technology.

6.1.2 Combustor and Air Emissions

There are three main types of combustors used in modern gas turbines: can, annular, and can-annular combustors. A can combustor consists of a tubular design and has both a liner and a casing. The primary combustion air from the compressor is guided into the burner directly while the secondary combustion air also from the compressor is supplied to outside of the liner and enters the combustion zone through slits in the liner while cooling the liner. Several cans are arranged around the central shaft while the combustion products from each can flow into the turbine through a "transition piece." Major advantage of this type of combustor is due to the ease of design and testing since a single can may be tested without having to test the entire system, while servicing and/or replacing just a single can is possible. An annular combustor consists of a continuous casing with the liner sitting inside it, which provides the advantage of having more uniform combustion, shorter size, and less surface area. The amount of cooling air required can be reduced due to the less surface area, and this type of combustor would be attractive in applications where the firing temperature is higher or when the fuel is syngas since in such applications less total (primary + secondary) air is used by the combustor and thus less secondary air is available for cooling. In the hybrid can-annular design, the casing is annular while the liner is can shaped and has some of the same advantages of purely can combustors such as ease of design, testing, and servicing.

The principal pollutant associated with gas turbines is NO_x and control strategies include water or steam injection and premixed burners to reduce the local flame temperature and thus formation of NO_x. Postcombustion control may be used to further reduce the NO_x by installing a catalytic reduction unit as discussed in Chapter 9. It consists of reacting the NO (the nitrogen oxide formed within the combustor is mostly NO) with injected NH_3 in the presence of a catalyst. Dry low NO_x combustors currently offered for natural gas applications consist of premixing fuel with air and burning it under lean conditions to reduce the flame temperature. Values as low as 9 ppm by volume on a dry basis and "corrected" to 15% (by mole) O_2 content in the flue gas are guaranteed for some of the engines. NO_x control strategies incorporated within the combustion process often result in reduced combustion efficiency and thus increased emissions of CO and unburned hydrocarbons, and in order to increase the combustion efficiency while reducing the pollutants formed including NO_x, computer-based models and laser diagnostic techniques are being applied.

Within a typical can combustor, swirling action imparted to the incoming air imparts a centrifugal force and creates a low pressure in a region of the primary zone causing recirculation of the hot combustion products to the primary combustion zone and imparting stability to the combustion process. The hot combustion products contain free radicals which serve to initiate the combustion much like a spark plug in an auto-engine. The CO oxidizes to CO_2 in the secondary zone where additional air is added. The hydrocarbons in the fuel form CO in the "dome" region around which the circulation occurs into the primary zone and mixing. Pollutants are also formed in this region but fuel rich conditions maintained in the region can minimize NO_x formation. However, this results in loss of stability of the combustion process, especially when considering different modes of operation for the gas turbine such as start-up, idling, and full load. Variable geometry combustors are under development and may solve some of these issues.

6.1.3 Start-Up and Load Control

A gas turbine has to be accelerated from zero to a sufficient speed for ignition (after purging to avoid a fire) and to self-sustain its speed. When a separate electric motor or an internal combustion engine is utilized to provide the torque, then the starting system disengages via a clutch when the turbine reaches the desired speed to operate independently. For large gas turbines in electric power generation applications, it is more economical to use the generator connected to the gas turbine to operate as a synchronous motor for start-up (in which case a clutch is not required). In such power generation applications using a synchronous generator, once the engine has been accelerated to synchronize the generator with the grid, the engine speed is then controlled by the grid.

Reduction in the power output of a gas turbine is typically accomplished initially by reducing the suction air flow rate by modulating the inlet guide vanes on the gas turbine compressor. In this manner, the gas turbine firing temperature may be kept close to the design point temperature to maximize the overall efficiency of the engine. The engine pressure ratio decreases however, causing a decrease in the overall cycle efficiency compared to its full load efficiency. As a consequence, the temperature of the working fluid in the later stages of the turbine as well as the turbine exhaust temperature all increase and a corresponding reduction in the firing temperature is implemented which further decreases the cycle efficiency. More significant reductions in firing temperature are required to reduce the output (typically to below ~40% of full load) when the inlet guide vanes are "fully closed" and no further reduction in the inlet air flow is possible which is typically about 60% of the full load air flow. At this point the heat rate may be as much as 40% higher than at full load although when exhaust heat recovery is implemented for CHP applications or for generating additional power via a Rankine steam cycle as in a combined cycle, the increase in the combined system heat rate is not as significant since most of the unconverted energy exhausting the gas turbine is utilized for steam generation.

6.1.4 Performance Characteristics

Both the power output and the efficiency of a gas turbine are affected by ambient conditions of temperature, barometric pressure, and, to a lesser extent, humidity. As ambient dry bulb temperature increases say from 15 to 35°C, the power output decreases by more than 10% mainly due to less mass of air being sucked in by the compressor (air density being lower at the higher temperature). Gas turbine efficiency is also reduced as ambient temperature increases since its compressor power requirement is increased, the heat rate increasing by about 3%. For higher site elevations, say 1800 m, the power output can decrease by approximately 20% over that at sea level, again due to less mass of air being sucked in by the compressor (air density being lower at the higher elevation). Effect of air humidity on gas turbine performance depends on the gas turbine firing temperature control scheme used, that is, whether the exhaust temperature is biased by compressor pressure ratio to the approximate firing temperature. Since the performance of gas turbines is sensitive to ambient conditions, it is important to specify the set of ambient conditions to which the net power output or the efficiency corresponds to. The ISO conditions for the ambient (consisting of 15°C, sea level elevation, and 60% relative humidity) are typically specified when reporting the performance data.

Like all turbo-machinery, gas turbines experience loss in performance with time and part of this performance degradation which is typically associated with compressor fouling may be recovered at least partially by water washing or more fully by mechanical cleaning of compressor blades and vanes. Mechanical cleaning, however, requires opening the unit resulting in plant downtime, a loss in plant capacity factor. Gas turbines also undergo a nonrecoverable loss which is due mainly to increased clearances in turbine and compressor sections as well as changes in airfoil contours and surface finish of the blades. These can only be fixed by replacing the affected parts at the required inspection intervals.

Gas turbines for electric power generation are available in sizes ranging from a few kilowatts (microturbines) to several hundred megawatts. Usually, the machines for 60 cycle applications (for countries such as the United States, Canada, Mexico, South Korea, and some parts of Japan) are smaller than the corresponding 50 cycle machines (for applications in most other locations) since one machine is typically scaled from the other and the scaling criteria consists of maintaining a constant tip speed in the compressor at the corresponding stage. The large gas turbines are axial machines and predominantly of the simple cycle configuration. The first law percentage efficiency can be in the 20s or 30s for the microturbines, while for the large gas turbines can be in the 30–40s while operating on natural gas (LHV basis at ISO conditions). The microturbines consist of radial compressor and radial turbine and utilize the recuperative (also called regenerative) cycle consisting of preheating the compressor discharge air against the turbine exhaust in a heat exchanger (recuperator) before it enters the combustor to increase the overall cycle efficiency, the individual component efficiencies being quite low.

Example 6.1. The effect of ambient temperature on the power output and fuel consumption of an open cycle gas turbine is shown in Figure 6.5. Calculate the change in the first law efficiency of the gas turbine when the ambient temperature (i) rises by 20°C and (ii) falls by 20°C over the ISO temperature.

Solution

First law efficiency $\eta_{\mathrm{I}} \propto$ (Power output/Fuel consumption)

1. When the ambient temperature rises by 20°C over the ISO temperature, or at an ambient temperature of 35°C, $\eta_{\mathrm{I}} \propto (86\%/90\%) = 96\%$ of ISO efficiency.
2. When the ambient temperature drops by 20°C below the ISO temperature, or at an ambient temperature of −5°C, $\eta_{\mathrm{I}} \propto (112.5\%/110\%) = 102\%$ of ISO efficiency.

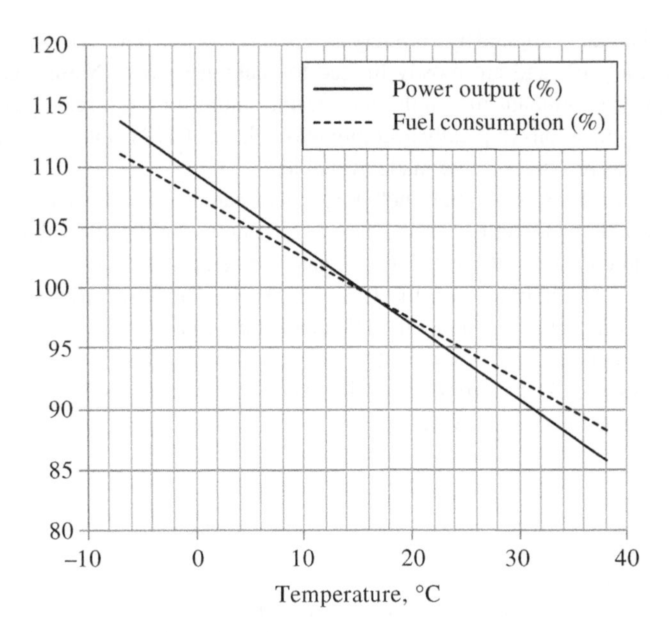

FIGURE 6.5 Effect of ambient temperature on gas turbine performance

TABLE 6.1 Characteristics of blast furnace, digester, and landfill gases

	Volume %		
	Blast furnace gas	Digester gas	Landfill gas
CH_4		57–61	20–60
H_2	1–4		
CO	25–30		
CO_2	8–16	36–38	22–60
N_2	55–60	0.6–1.1	0.6–46
O_2		0.2–1.4	0–2.5
$H_2S + CS_2 + SO_2$		40–200 ppmv	1–1700 ppmv
Cl compounds		25–200 ppbv	Present
C_2+ (paraffins)		40–175 ppmv	Trace
Aromatics		800–1800 ppbv	240–1400 ppmv
Siloxanes		Present	Present
LHV, kJ/Nm^3	3500–4300	20,000–22,000	7,900–24,000
(Btu/SCF)	(90–110)	(520–560)	(200–600)

6.1.5 Fuel Types

Suitable gaseous fuels for gas turbines include natural gas; liquefied natural gas after vaporization; gasification-derived syngas; as well as the "special opportunity" gases such as landfill gas, gas from anaerobic digester in sewage treatment plants, blast furnace gas from iron making industry, and waste gases from a refinery (Rao et al., 1996). Characteristics of some of these special opportunity gas streams are presented in Table 6.1. Liquid fuels such as distillates supplied by a refinery are also used as fuel but to a lesser extent due to the higher fuel cost and are mostly limited to peaking service. Natural gas is and will remain the predominant fuel in the foreseeable future for large-scale gas turbines in base loaded applications (typically in combined cycle and CHP plants) especially with the large shale gas resources being made available.

Composition of the gaseous fuels including that of natural gas can vary significantly and the gas turbine manufacturer establishes specifications for the allowable ranges in composition and contaminant levels for each gas turbine model to avoid damaging the gas turbine and to be able to burn the fuels efficiently. Other specifications include temperature, heating value, and a modified Wobbe Index (MWI) in the case of a gaseous fuel. This index is a measure of the relative amount of energy entering the combustor for a fixed pressure drop across a nozzle. Allowable variations in this index as calculated by Equation 6.1 are typically ±5%.

$$\mathrm{MWI} = \frac{\mathrm{LHV}}{\sqrt{(\mathrm{SG})T}} \tag{6.1}$$

where LHV, volumetric lower heating value of the fuel, SG, specific gravity of the fuel relative to air at ISO conditions, and T, absolute temperature of the fuel.

Example 6.2. A natural gas fired gas turbine is to be refueled by a digester gas supplied at the same pressure and temperature as the natural gas in order to stay within the design limits of the existing piping system. Estimate the derating in the fired duty of the gas turbine if no physical changes were made including those to the fuel nozzle, while ignoring any impacts on the turbo machinery operating parameters due to factors such as the aero-thermal characteristics of the turbine inlet gas (e.g., impact on blade wall temperatures) and the impact on the engine pressure ratio. For natural gas, use the composition and the calculated LHV as in Example 1.1, and for digester gas, use the following volumetric composition: $CH_4 = 60\%$, $CO_2 = 38\%$, $N_2 = 1.1\%$, $O_2 = 0.9\%$.

Solution

From Equation 6.1

$$\frac{\mathrm{MWI_{DG}}}{\mathrm{MWI_{NG}}} = \frac{\mathrm{LHV_{DG}}/\mathrm{LHV_{NG}}}{\sqrt{(\mathrm{SG_{DG}}/\mathrm{SG_{NG}})(T_{\mathrm{DG}}/T_{\mathrm{NG}})}}$$

The above expression may be written in terms of molecular weights of the two gases, made possible with the ideal gas behavior assumption,

$$\frac{\mathrm{MWI_{DG}}}{\mathrm{MWI_{NG}}} = \frac{\mathrm{LHV_{DG}}/\mathrm{LHV_{NG}}}{\sqrt{(M_{w\mathrm{DG}}/M_{w\mathrm{NG}})(T_{\mathrm{DG}}/T_{\mathrm{NG}})}}$$

where $M_{w\mathrm{DG}}$ and $M_{w\mathrm{NG}}$ are the molecular weights of the digester gas and the natural gas.

Substituting the calculated LHV and the M_w corresponding to the given compositions of the two gases into the above expression (for natural gas it was assumed that it contains equal amounts of N_2 and CO_2 on a volumetric basis), we have

$$\frac{\text{MWI}_{\text{DG}}}{\text{MWI}_{\text{NG}}} = \frac{20{,}439/35{,}290}{\sqrt{(26.94/17.54)\,(1)}} = 0.47$$

Thus, the derating in the fired duty of the gas turbine would be as much as 53% under the given assumptions.

In addition to the type of combustible(s) in the gas, the concentration of inerts (e.g., N_2, H_2O vapor or CO_2) that may be present in a fuel gas effect its LHV and a lower limit for the LHV exists not only for stable and efficient combustion but also from aerodynamic and aerothermal standpoints. A fuel with an LHV much lower than the design fuel can result in increasing the gas turbine compressor pressure ratio significantly due to the larger mass flow of gases flowing through the turbine caused by the larger concentration of the diluents in the fuel. This can cause compressor surge and the associated damage unless the compressor inlet guide vanes are closed to limit the amount of the engine suction air and/or to extract a fraction of the compressor discharge air if there is use for such pressurized air and if the engine can be modified for this capability. However, limits exist on how much a reduction in the air flow can be realized with the inlet guide vanes (note that the compressor can surge if there is insufficient air entering it), and on the fraction of air that may be extracted (note that the combustor and turbine cooling air may be starved if too much air is extracted). Depending on the LHV of the fuel gas, modifications to the fuel delivery system including the control valve may be required as well as to the combustor to burn the fuel efficiently and limit formation of pollutants such as CO and VOCs. Some of the gas turbines designed for natural gas operation can accommodate fuel gas with an LHV as low as approximately 4 MJ/Nm^3 after the necessary modifications are made.

Preignition and flashback can be issues if H_2 content of a fuel gas is very high when using a premixed combustor where the fuel is mixed with combustion air and then the mixture is introduced into the combustion chamber to reduce the local flame temperature and thus the NO_x formation which is highly dependent on temperature. Certain fuel gas streams such as landfill gas or gas from anaerobic sewage treatment (digester gas) may contain O_2 and upper limits exist for the O_2 content to avoid preignition and flashback. Even natural gas, as mentioned in Chapter 1, during peak demand months in certain areas of the United States (the northeast) may contain O_2 as high as 4% by volume because during such seasons of natural gas shortage, the gas company may blend in propane or butane to extend the fuel supply and air would be added as a diluent to hold the Wobbe Index within limits.

The fuel may be preheated using heat from the bottoming cycle to increase the overall combined cycle efficiency but an upper limit exists for a given gas turbine system set by design capabilities of the fuel delivery system including the materials used in the fuel control valve, as well as considerations of preignition and flashback in the case of a premixed combustion. Fuel gases can contain moisture as well as some of the heavier hydrocarbons and a lower limit is stipulated for the fuel temperature to keep the fuel gas safely above its dew point to avoid blockage of the fuel filter, carryover of any droplets that could cause degradation of fuel nozzles, or premixed flame flashback in premixed combustion with hydrocarbon droplets, or collection of liquids in piping low points, as well as to avoid formation of CH_4 and CO_2 hydrates. Most gases (except H_2 and He) have a positive Joule–Thompson coefficient above ambient temperatures and at pressures typically encountered such that the gas can undergo cooling due to expansion. This temperature drop can be significant when the pressure drop is also significant such as flow through a pressure letdown valve and this aspect should be taken into account while setting the minimum fuel gas temperature (upstream of the valve). General Electric Bulletin GEI 41040G (2002) provides the minimum superheat requirement above the hydrocarbon dew point of the gas to avoid hydrocarbon condensation, and the minimum superheat requirement above the moisture dew point of the gas to avoid moisture condensation and hydrate formation, in the form of correlations that take into account the pressure of the gas at the inlet to the gas turbine fuel flow control valve and the temperature.

Stringent limits are specified for particulate loading in the fuel gas by size to protect the gas turbine from corrosion/erosion. Typically, a filter is included in the fuel delivery line to catch the particulates even if the fuel supplied (e.g., natural gas and LNG) is free of particulates as there is a possibility of the fuel gas picking up rust or scale particles. Stringent limits are also specified for Va, total alkalis (Na and K), Pb, Ca, and Mg.

Landfill as well as digester gas can contain both inorganic and organic sulfur compounds, organic Cl compounds and siloxanes which are organic compounds containing Si atoms. The various compounds have to be removed to limits set by the gas turbine supplier, to avoid corrosion in the combustor and turbine sections and/or to satisfy the environmental emission limits in the case of the S and Cl compounds, while the siloxanes can leave Si deposits on turbine blades and on the HRSG heat transfer surfaces as well as give rise to particulate emissions. The bulk of the inorganic sulfur compounds may be removed using a "throw-away" bed of iron oxide offered for such small-scale applications. The various organic compounds along with the remaining inorganic sulfur compounds may be removed by passing the gas through a bed containing an adsorbent(s) such as activated carbon (also a throw-away process). These treatment processes located upstream of the gas turbine have to have the gas sufficiently above its dew point to avoid condensation in the equipment as well as to avoid pore condensation within the adsorbents. This may be accomplished by dehydrating the gas using a desiccant or cooling the gas sufficiently below its dew point and then preheating it. These equipment can all be located downstream of the compressor to reduce their size, a compressor being required in any case since the landfill gas is available at near ambient pressures at its source.

6.1.6 Technology Developments

Gas turbines could play a key role in the future power generation market addressing issues of producing clean, efficient, affordable, and fuel-flexible electric power. The technological advances being made or being investigated to improve the Brayton cycle include higher rotor inlet temperature of 1700°C (3100°F) or higher, blade metal temperature of approximately 1040°C (~1900°F), while limiting coolant amount that may be made possible with the use of advanced materials including advanced thermal barrier coatings and turbine cooling techniques including closed loop steam cooling. Other advances include those to the combustor liners to withstand the higher temperatures within the combustor, pressure gain (Gemmen et al., 1994) and cavity or trapped vortex (Hsu et al., 1995) combustors, high pressure ratio compressors (to take full advantage of higher firing temperature), and integration capability with high temperature ion transport membrane air separation for IGCC applications discussed in Chapter 9.

6.1.6.1 Firing Temperature Higher firing temperatures demand advanced materials in both the combustor and the turbine to withstand the severe environment created. Directional solidification blades as well as single crystal blades which have been utilized successfully in advanced turbines are replacing the conventionally cast nickel-based super alloys. Advanced thermal barrier coatings with extensive use of ceramics are being investigated to reduce metal temperatures but these coatings, however, need to withstand an environment containing water vapor at a high partial pressure, especially for application involving decarbonized syngas. Also under consideration is the development of ceramic matrix composites including single crystal oxide fibers for the hot components or sections of the turbine. Combustor materials that can withstand a combination of creep, pressure loading, high cycle, and thermal fatigue at the higher firing temperatures are being investigated.

Higher firing temperatures also require better cooling technology for gas turbine components exposed to the hot working fluid to limit the metal temperature. The open circuit cooling of the high pressure turbine blades (both stationary and rotating) in modern gas turbines is accomplished with internal convective cooling and external surface film cooling. Internal convective cooling directs the coolant into all regions of the component requiring cooling under the available pressure gradients. The heat transfer coefficients are enhanced through various surface treatments such as ribrougheners or turbulators as well employing diverters and swirl devices to force the direction of flow while film cooling consists of bleeding the internal coolant flow onto the exterior surface of the blade to provide an insulating layer. An advanced technology for open circuit cooling being investigated is transpiration cooling where the cooling gas effuses through a porous wall to create an insulating film or boundary layer on the outer surface.

6.1.6.2 Compression Ratio and Intercooling Compression ratio must be increased in order to take full advantage of higher firing temperature from an overall thermal efficiency standpoint. Higher pressure ratios are also required to limit turbine exhaust temperature and thermal stresses at the roots of the last stage turbine blades which tend to be long in gas turbines for large-scale applications and are uncooled. Direct contact intercooling by spraying water into the air discharging the low pressure compressor whereby the water evaporates into the air stream has the advantage over the shell and tube type of intercooler in terms of overall cycle efficiency by minimizing the pressure drop on the air side as well as by providing additional motive fluid (water vapor). Spray intercooling of gas turbine compressor with a high compression ratio (>30) can decrease compressor discharge temperature without decreasing overall efficiency in combined cycle applications. Reduction in compressor discharge temperature has beneficial impacts on compressor material costs as well as NO_x formation within the combustor. An intercooled gas turbine using the spray type has been commercialized by General Electric. Intercooling may also be then taken advantage of in reheat gas turbine cycles which optimize at very high overall pressure ratios.

6.1.6.3 Inlet Air Fogging Another approach to reducing the parasitic load of air compression and its exhaust temperature in a gas turbine is to introduce liquid water into the suction air (Bhargava and Meher-Homji, 2002). The water droplets need to be extremely small in size and be in the form of a fog to avoid erosion of the blades of the compressor by impingement. The air being compressed is cooled as the water evaporates within the compressor from the heat of compression. Since the compression work is directly proportional to the absolute temperature of the fluid being compressed, compression of cooler air requires less work resulting in an increase in the specific power output of the engine. Another benefit is the reduction in the NO_x due to the presence of the additional water vapor in the combustion air as well as due to the lower combustion air temperature. Such a fogging system has been installed on a number of gas turbines but care should be taken in specifying the water treatment equipment since high quality demineralized water is required and the design of the fogging system to avoid impingement of the compressor blades with water droplets.

6.1.6.4 Pressure Gain Combustor In this type of combustor, its exhaust stagnation pressure is greater than the initial state stagnation pressure, resulting in producing greater available energy in the end state than constant pressure systems. An example of such a system is the constant volume combustion in an ideal spark ignited engine. The heat rate of a simple cycle gas turbine with a pressure ratio of 10 and a turbine inlet temperature of approximately 1200°C (2200°F) can be decreased by more than 10% utilizing a constant volume combustion system (Gemmen et al., 1994). Pulse combustion which relies on the inherent unsteadiness of resonant chambers can be utilized as a pressure gain combustor. Research continues at the U.S. Department of Energy, at NASA and others for the development of pressure gain combustors. A major challenge is the interfacing of the continuous flow turbomachinery (compressor and turbine) with the combustor that may operate cyclically.

6.1.6.5 Trapped Vortex Combustor The Trapped Vortex Combustor (TVC) has the potential for numerous operational advantages over current gas turbine combustors which include lower weight, lower pollutant emissions, effective flame stabilization, high combustion efficiency, and stable operation in the lean burn modes of combustion. Swirlers used in gas turbine combustors create the toroidal motion to cause flow recirculation of a portion of the hot combustion gases to provide continuous ignition of the incoming air and fuel but swirl stabilized combustors have somewhat limited combustion stability and can blow out under certain operating conditions whereas the TVC maintains a high degree of flame stability using a trapped vortex within a cavity which provides a stable recirculation zone that is protected from the main flow in the combustor. A bluff body dome contained within the TVC next distributes and mixes the hot products from the cavity with the main air flow. Fuel and air are injected into the cavity in such a way that they reinforce the vortex that is naturally formed within it. Conceptually, the TVC is a staged combustor with two pilot zones and a single main zone, the pilot zones being formed by cavities incorporated into the liners of the combustor (Burrus et al., 2001) which operate at low power as rich pilot flame zones achieving low CO and unburned hydrocarbon emissions, as well as providing good ignition and lean blow out margins. At higher power conditions, say greater than 30% power, the required additional fuel is staged from the cavities into the main stream while the cavities are operated under fuel rich conditions. Experiments have shown an operating range 40% wider than conventional combustors with combustion efficiencies greater than 99%. In addition to use in natural gas fired applications as an alternate option for suppressing NO_x emissions, the TVC combustor holds special promise in syngas applications where premixed burners may not be employed.

6.1.6.6 Catalytic Combustor Reduction in flame temperature to reduce NO_x formation may be achieved by operating under lean conditions, but maintaining combustion stability can be a challenge. Use of a catalyst can provide stability under such conditions with a potential for simultaneous low NO_x, CO, and unburned hydrocarbon emissions (Smith, 2004). Another potential benefit is the reduction in combustion-induced pressure oscillations. The catalytic combustor could also play a special role in syngas applications for reducing NO_x emissions.

6.2 STEAM TURBINES

The working fluid, steam is generated by transfer of heat from a source such as hot combustion products generated by burning a "dirty" fuel (e.g., coal or biomass) as discussed in Chapter 8, or from the gas turbine exhaust as discussed in Chapter 9, or from a nuclear reaction or concentrated solar energy as discussed in Chapter 12.

6.2.1 Principles of Operation

Large utility steam turbines are axial machines which were described in Chapter 3 with alternating stationary blades (nozzles) which direct the steam on to rotating blades (buckets). The exhaust from the buckets of an upstream stage flows directly into the nozzles of the downstream stage. Large axial turbines must be operated such that the exhaust steam does not contain more than 10–13% of liquid since condensate droplets could seriously erode the nozzles and blades at the high velocities and proper selection of the inlet steam pressure and temperature are required. Special moisture removal stages may be incorporated in the design when the steam superheat temperature is limited, say by the temperature of the heat source.

Steam may be utilized directly in the steam turbine without any superheat as is typically done with low pressure steam as well as with high pressure steam in nuclear power plants, or superheated or also reheated to increase the cycle efficiency. Reheating as in the case of a reheat gas turbine cycle, which was discussed previously, consists of expanding the superheated steam to an intermediate pressure, heating up this partially expanded steam, and then readmitting it into the turbine to complete the expansion step. In the case of a condensing steam turbine, the steam is expanded to sub atmospheric pressure and exhausts to a condenser and the latent heat of the steam is transferred to the cooling water while transforming the steam to BFW which is then used for the generation of steam to complete the cycle. In the case of a back pressure steam turbine, the turbine is designed to exhaust at a much higher pressure as determined by the user for this exhaust steam such as for heating in a CHP application or for a process use. Steam may also be extracted at a pressure higher than its exhaust pressure for similar applications as well as for preheating the BFW by equipping the turbine with nozzles between stages, while the steam that is not extracted continues the expansion step through the turbine.

6.2.1.1 Impulse versus Reaction Blades In a turbine employing impulse blades, the steam leaves each of the nozzles as a high speed jet directed toward each bucket with the steam undergoing a pressure drop across the nozzles while its kinetic energy increases. This kinetic energy is converted into mechanical energy (shaft rotation) by the buckets as the steam jet changes its direction after impinging upon the buckets. A water wheel equipped with paddles is an example of an impulse turbine. Impulse turbines are mainly used for smaller scale applications. In a turbine employing reaction blades, steam is also directed onto rotating blades by the nozzles but the rotating blades themselves are arranged to form convergent nozzles to create a reaction force as the steam accelerates through these convergent nozzles, similar to the motion imparted to a rotating sprinkler by the water exiting the sprinkler. Reaction turbines tend to be more efficient but the number of stages required as well as the axial thrust is higher. Larger number of stages translates into higher maintenance cost.

6.2.2 Load Control

The three different methods for controlling the power output of a steam turbine attached to a synchronous generator (where the speed is controlled by the grid) are sliding pressure control, throttling control, and governing stage control. Hybrid strategies combining some of these methods are also employed. In sliding pressure control, the amount of steam admitted to the turbine is reduced without making any changes to the geometry of the system delivering the steam to the turbine such as reducing the flow using a control valve, while the steam generation pressure is allowed to fall to a value set by the physics of fluid flow. In this type of control which is typically used in combined cycle plants described in Chapter 9, the turbine isentropic efficiency remains essentially unchanged. In throttling control, all main steam control valves are simultaneous operated to regulate the pressure of the steam admitted to the turbine while the steam is generated near constant pressure. The turbine isentropic efficiency is reduced somewhat but more importantly a significant amount of irreversibility is introduced by the pressure drop across the throttling valves. Governing stage control consists of sequential operation of the control valves which vary the turbine output by increasing or decreasing the arc of admission of steam to the first (control) stage. Each control valve feeds a section of the control stage and they are operated in an order based on the allowable stresses on this stage. Since a given set of valves may be fully open or fully closed at a given load, the irreversibility introduced by throttling control is eliminated but the isentropic efficiency of the turbine itself is reduced.

6.2.3 Performance Characteristics

Steam turbines for electric power generation are available in sizes ranging from less than 100 kW to hundreds of MWs. The isentropic efficiency of steam turbines can range from 30% for the smaller machines to 80 to greater than 90% for the larger machines depending on the inlet steam conditions of flow rate, temperature and pressure, and outlet pressure. As stated in Chapter 3, for steam turbines greater than 16.5 MW, empirical correlations developed by Spencer et al. (1974) are available. The smaller turbines consist of a radial design while the large turbines are axial machines.

Example 6.3. A condensing steam turbine is designed to operate with an inlet temperature of 400°C, pressure of 42 bar, flow rate of 100,000 kg/h. Assume that cooling water to condense the exhaust steam is supplied at 30°C and is returned to the cooling tower at 40°C (i.e., a 10°C rise), while the pinch temperature in the vacuum steam condenser is 5°C (smallest temperature difference in the exchanger which is the difference in temperature of the cooling water leaving the condenser and temperature of the condensing steam) at the design point operation of the turbine. The steam flow rate to the turbine is reduced by 25% while its temperature is also reduced to 375°C. Assume that the system is designed for sliding pressure control and choked flow conditions prevail at the inlet nozzle of the turbine while its isentropic efficiency of 80% and the mechanical and generator losses amounting to 5% (of the power developed from the steam) do not change significantly at the lower steam flow. Calculate the steam turbine inlet pressure and the electric power output at the turned down point of operation.

Solution
Replacing density with pressure P and absolute temperature T using the ideal gas law, Equation 3.12 may be written in terms of mass flow rate $\dot{m}_{\text{sonic}}$ as

$$\dot{m}_{\text{sonic}}\sqrt{T}/P = K_{\text{sonic}}$$

where K_{sonic} is a constant for a given flow cross-sectioal area and for an ideal gas with the ratio of constant pressure to constant volume specific heats remaining constant.

K_{sonic} may be determined using the design point data and then P determined for the off-design point using its mass flow rate and temperature. The turbine exhaust pressure which is the surface condenser pressure at the off-design point is determined as in Example 4.1 but iteratively:

1. By assuming a condenser temperature
2. Determining the corresponding saturated steam pressure
3. Performing the steam turbine calculations
4. Using the steam turbine exhaust conditions established in step 3 to calculate the surface condenser heat transfer duty
5. Performing the surface condenser calculations as in Example 4.1
6. Checking if the steam temperature calculated in step 5 is the same as that assumed in step 1.

The condenser pressure at the design point corresponds to the saturated pressure of steam at 45°C which is calculated by adding the given pinch temperature of 5°C to the given cooling water exit temperature of 40°C.

For the steam turbine calculations, the first step is to read off the specific (on a unit mass basis) enthalpy Δh_{in}^{V} and entropy Δs_{in}^{V} of steam at the inlet conditions from steam tables (e.g., Keenan et al., 1969) at the given or calculated conditions of pressure and temperature. Since the entropy at the inlet is lower than the entropy at the outlet pressure for saturated steam, the exhaust has to be a two-phase fluid, as both vapor and liquid. So the next step is to read off the enthalpy Δh_{out}^{V} and Δh_{out}^{L}, and specific entropy Δs_{out}^{V} and Δs_{out}^{L} at the given or calculated outlet pressure for saturated vapor and liquid to determine the mass fraction of liquid $\mathcal{X}_L$ (=1 – steam quality) in the exhaust by solving the following for the isentropic path:

$$\Delta s_{in}^{V} = (1 - \mathcal{X}_L)\, \Delta s_{out}^{V} + x_L\, \Delta s_{out}^{L}$$

With $\mathcal{X}_L$ known, the corresponding enthalpy of steam at the outlet Δh_{out}^{mix} for this isentropic path is calculated from:

$$\Delta h_{out}^{mix} = (1 - \mathcal{X}_L)\, \Delta h_{out}^{V} + x_L\, \Delta h_{out}^{L}$$

The actual enthalpy of the exhaust steam Δh_{out}^{act} is then calculated using the turbine isentropic efficiency η_{turb} by:

$$\Delta h_{out}^{act} = \Delta h_{in}^{V} - \left(\Delta h_{in}^{V} - \Delta h_{out}^{mix}\right) \eta_{turb}$$

The actual moisture in the exhaust $\mathcal{X}_L^{act}$ may be calculated by solving the following:

$$\Delta h_{out}^{act} = \left(1 - \mathcal{X}_L^{act}\right) \Delta h_{out}^{V} + \mathcal{X}_L^{act}\, \Delta h_{out}^{L}$$

TABLE 6.2 Design and off-design point performance of a steam turbine

		Design point			Off-design point		
			Outlet			Outlet	
Turbine	Units	Inlet	Ideal	Actual	Inlet	Ideal	Actual
K_{sonic}		61,767			61,767		
Efficiency	fraction			0.8			0.8
Flow rate	kg/h	100,000	100,000	100,000	75,000	75,000	75,000
Temperature	C	400	45	45	375	41.5	41.5
Pressure	bar	42	0.095	0.095	30.91	0.079	0.079
Enthalpy	kJ/kg	3210.6	2130.1	2346.2	3172.2	2132.5	2,340.4
Entropy	kJ/kg-K	6.7444	6.7444	7.4239	6.8203	6.8203	7.4815
Steam quality			0.81	0.90		0.82	0.90
Power	MJ/h			86,444			62,386
Generator efficiency	fraction			0.95			0.95
Electric Power	kW			22,811			16,463
	fraction			1.00			0.72
Surface condenser duty	MJ/h			215,836			162,544
	fraction			1.00			0.75

The electric power output of the turbine $\dot{W}$ is then calculated using the mass flow rate of the steam $\dot{m}$ and the generator efficiency η_{gen} as

$$\dot{W} = \dot{m}\left(\Delta h_{in}^{V} - \Delta h_{out}^{act}\right)\eta_{gen}$$

The next step is to calculate the surface condenser heat duty $\dot{Q}$ when the exhaust steam is condensed. The specific enthalpy $\Delta h_{cond,out}^{L}$ of the condensed steam may be read off from the steam tables corresponding to the saturated liquid at the condenser pressure. Then the heat duty may be calculated by

$$\dot{Q} = \dot{m}\left(\Delta h_{out}^{act} - \Delta h_{cond,out}^{L}\right)$$

The surface condenser duties for the design and the off-design cases are utilized in the calculations as in Example 4.2.

Table 6.2 summarizes the results of this iterative procedure between the turbine and the condenser calculations. The steam turbine inlet pressure decreases from 42 to 30.91 bar, while the electric power output decreases from 22,811 to 16,463 kW or by 28% at the turned down point of operation.

6.2.4 Technology Developments

Research and development is occurring on advanced materials for steam turbines suitable for highly efficient ultra-supercritical steam cycles that operate at much higher temperatures and pressures than those in current pulverized coal power plants. High temperature super alloys are being evaluated with the ability to be shaped into a finished product that has resistance to creep, fatigue, and oxidation under the harsh steam conditions of temperature and pressure (>800°C and >350 bar). Erosion, corrosion, and oxidation resistant coatings are also being investigated. High potential candidate alloys for critical components are being identified and being evaluated for the mechanical and corrosion properties for use in the rotating as well as the stationary parts including bolting materials and casing materials.

6.3 RECIPROCATING INTERNAL COMBUSTION ENGINES

Examples of a reciprocating engine are the gasoline fired auto engine, and the diesel and gas fired engines used in both mobile as well as stationary applications. In a Diesel cycle (named after its inventor, Rudolf Diesel) engine, the combustion process is initiated by the heat of compression itself without requiring an external source of ignition and requires the compression ratio to be high enough to raise the temperature of the air where ignition can occur. On the other hand, in an Otto cycle (named after its inventor, Nikolaus Otto) engine (as in the gasoline fired auto engine), a spark plug initiates the combustion process and its compression ratio is limited to avoid autoignition. Thus, the characteristics for the fuel suitable for each of these engines are quite different. The fuel for a Diesel cycle engine needs to burn rapidly after the piston has reached the top dead center (position of the piston at the end of the compression stroke where it is closest to the cylinder head) or else it may not burn completely during the power stroke. These fuel characteristics are quantified by the cetane number for a compression ignition engine and by the octane number for a spark ignited engine as described later in this section.

6.3.1 Principles of Operation

In a reciprocating engine air or precompressed air (or air–fuel mixture in the case of certain types of engines) is compressed in a cylinder by the movement of a piston (compression stroke), then thermal energy is added to the compressed air by directly combusting the fuel with the compressed air within the cylinder, which is then followed by expansion of the hot pressurized combustion products in the cylinder against the piston to produce useful work (power stroke). The cylinder walls may be cooled by circulating a cooling medium through cylinder jackets or by incorporating fins cooled by air. The cooling of the cylinder walls provides an opportunity for heat recovery and export in CHP applications.

Both the Otto cycle and the Diesel cycle engines can be operated on gaseous fuels. Diesel cycle engines in such applications require spark ignition however, or can operate on a blend of a suitable liquid fuel (diesel) and gas, with the primary fuel being gas while the small amount of the liquid fuel serves as an ignition source. The pressure-specific volume diagrams for these two cycles in the form of the air-standard cycle are depicted in Figures 6.6 and 6.7 (Cengel and Boles, 1998; Smith et al., 2005) and consists of

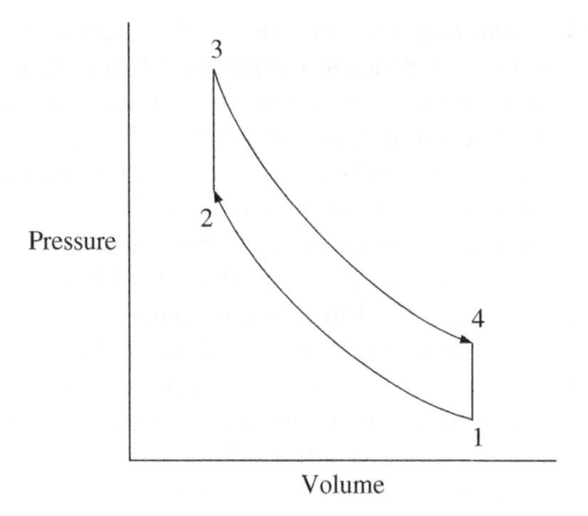

FIGURE 6.6 Air-standard Otto cycle

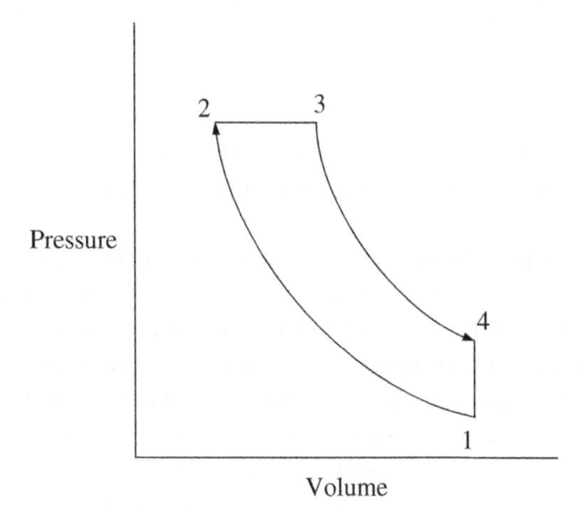

FIGURE 6.7 Air-standard Diesel cycle

adiabatic compression of a gas (from statepoint 1 to 2) followed by heat addition (from statepoint 2 to 3), and then adiabatic expansion of the hot pressurized gas (from statepoint 3 to 4). Fuel combustion occurs over a longer period of time in the case of a diesel cycle, igniting as the piston approaches the top dead center and combustion continues during the initial part of the power stroke. Thus, the heat addition step is idealized as occurring at constant pressure.

These engines may be divided into two categories determined by their method of operation, the first being engines that operate on a two stroke cycle once per revolution of the shaft, and the second being engines that operate on a four stroke cycle. The two stroke engines are used for very small power applications.

6.3.1.1 Two Stroke Cycle Engines In a two stroke engine, an up stroke and a down stroke of the piston are completed in a single revolution of the crank shaft resulting in high specific power but may be within a narrow range of rotational speeds. Combustion occurs in the region of top dead center while gas exchange (fresh air or air/fuel mixture to replace the combustion products) occurs near the bottom dead center of each stroke. Thus, the intake of fresh air or air/fuel mixture and exhausting the combustion products occur simultaneously at the beginning of the compression stroke and at the end of the expansion stroke. Thus, the different cycle stages are: (i) compression of the homogeneous air/fuel mixture in the case of a gas fueled engine by the upward stroke of the piston (for a diesel fueled engine, the air is compressed and then toward the end of the compression stroke, the fuel is injected as a spray into the cylinder to form a heterogeneous mixture), (ii) ignition of the fuel/air mixture by spark (for a gas engine) or heat of compression (for a diesel engine), (iii) expansion of the hot combustion products to perform work during the piston down stroke forcing the crank shaft to rotate, (iv) opening of an exit valve resulting in a rapid drop in pressure as combustion products leave the cylinder, (v) gas exchange during which combustion products are mostly replaced by fresh air (in diesel engines) or air/fuel mixture (in gas engines), and (vi) closing of the exit valve while recharging the cylinder until the inlet valve is closed.

6.3.1.2 Four Stroke Cycle Engine In a four stroke engine, the piston completes four separate strokes consisting of an intake stroke, a compression stroke, an expansion power stroke, and an exhaust stroke during two separate revolutions of the engine's crankshaft.

The intake stroke starts when the piston is at top dead center and ends when it is at the bottom dead center. The compression stroke starts when the piston is at bottom dead center while both intake and exit valves are closed. Expansion power stroke starts when the piston is at top dead center while both valves are closed and ends when the piston is at bottom dead center (expansion of the hot pressurized combustion products to perform work during the piston down stroke forcing the crank shaft to rotate). The exhaust stroke starts when the exhaust valve opens and ends when the piston reaches top dead center at which point the exhaust valve closes and the intake valve opens to repeat the cycle. Thus, the different stages of the cycle are: (i) compression of the homogeneous air/fuel mixture in the case of a gas fueled engine by the upward stroke of the piston (for a diesel fueled engine, a charge of air is compressed and then toward the end of the compression stroke, the fuel is injected as a spray into the cylinder to form a heterogeneous mixture), (ii) ignition of the fuel/air mixture by spark (for a gas engine) or heat of compression (for a diesel engine), (iii) expansion of the hot combustion products to perform work during the piston down stroke, (iv) opening of an exit valve resulting in a rapid drop in pressure as combustion products leave the cylinder, (v) exhausting the combustion products from the cylinder during the second upstroke, (vi) opening of the inlet valve followed by closing of the exit valve, (vii) drawing in of fresh charge of a homogeneous fuel/air mixture (gas engines) or a fresh charge of air (diesel engines) during the second down stroke of the cylinder, and (viii) begin the compression stroke during which all ports are closed.

6.3.1.3 Supercharging and Turbocharging Providing the engine with pressurized air increases the mass of air contained in a given size of cylinder allowing more fuel to be burnt thus increasing the engine power output. Pressurization is accomplished by using a compressor which may be driven directly off the engine (supercharging) or by an expansion turbine operating on the engine exhaust gas, the exhaust gas being above atmospheric pressure (turbocharging). Since the process of compression raises temperature, an intercooler is used to cool the air before it enters the engine cylinders to increase its density. In practice, the compression ratio of a turbocharged diesel engine must be reduced to avoid excessive cylinder pressure. In the case of gas engines the pressure ratio should be limited to avoid detonation.

6.3.2 Air Emissions

The amounts of pollutants emitted are dependent upon the type of fuel as well as whether it is a two stroke or a four stroke engine. In a two stroke engine, a fraction of the air and fuel being introduced into the piston leaks out through the exhaust port causing emissions. Furthermore, when a two stroke engine requires oil to be injected with the fuel (e.g., gasoline) for lubricating the crankcase, it gets burned along with the fuel increasing emissions of organic compounds and soot. Four stroke engines on the other hand have a dedicated oiling system in such cases which is kept largely separate from the combustion process.

The engine exhaust carries most of the pollutants but some organic compounds escape from cylinders past the piston rings and are emitted from the crankcase as gases and vented from the oil pan. This type of "nonexhaust" emission is much lower in diesel engines. The major pollutants from internal combustion engines are NO_x, total organic compounds (TOC), CO, and particulates mostly responsible for the visible smoke in the exhaust. NO_x formation is directly related to high pressures and temperatures during the combustion process and to the fuel bound nitrogen, while the other pollutants, TOC, CO, and smoke are primarily the result of incomplete combustion. Particulate emissions are also due to any ash and metallic additives in the fuel. SO_x can also be emitted depending upon the sulfur content of the fuel.

NO_x control methods for diesel and dual fuel engines include reducing the combustion temperature and in some cases also the pressure. One technique consists of adjusting the fuel to air ratio to change the combustion temperature. Rich-burn engines have an air to fuel ratio that is near stoichiometric while lean-burn engines have an air to fuel ratio that is significantly above stoichiometric. In the case of rich-burn, in addition to lowering the temperature, the reducing conditions (insufficient amount of O_2) cause NO_x formation to also decrease. Catalytic reduction and techniques for postcombustion NO_x control are also being applied as well as for the oxidation of the CO and the organic compounds in the exhaust.

6.3.3 Start-up

The engine is started by rotating the shaft at a speed sufficient to achieve ignition and self-sustained operation. Small engines are started by electric motors while large ones are provided with special valves such that some of the engine cylinders can be operated as air motors, utilizing high pressure air made available from outside to rotate the engine. In the case of diesel engines operating on heavy residual fuel oils, two fuel systems consisting of a heavy fuel oil system for normal operation and a light fuel oil system for starting and stopping are required.

6.3.4 Performance Characteristics

Both spark ignited and compression ignited two stroke and four stroke engines have similar operating characteristics. The speed and load carrying capacity can range from zero to full torque for all speeds within the operating range and is varied by adjusting the fuel input by using a governor that also maintains constant speed under varying load. The performance sensitivity of these engines to different site ambient conditions may be minimized by properly sizing the supercharger or turbocharger. Engines for stationary power applications are available with outputs as low as a few kilowatts to in excess of 50 MW. The first law percentage efficiency can be in the 20–40s depending on the type of engine and its operating conditions.

The performance of diesel and gas engines deteriorates as the cylinders, piston rings, and other mechanical components become worn out but performance may be recovered at least partially by a major overhaul. Engines have to be shutdown at periodic intervals for inspection and repair and properly maintained machines can result in an availability of greater than 95%. Depending on the frequency of use, these engines are more prone to breakdown than electric motors or steam turbines.

6.3.5 Fuel Types

6.3.5.1 Cetane Number Cetane number is a measure of the combustion characteristics of a fuel suitable for compression ignition engines. It measures a fuel's ignition delay which is the time period between the start of fuel injection and start of ignition; higher the cetane number, shorter the ignition delay. Generally, diesel engines run well with a cetane number ranging from 40 to 55. A substance that ignites very quickly, cetane (*n*-hexadecane) was arbitrarily given a rating of 100 while a substance that is slow to ignite, 1-methylnaphthalene was given a rating of zero (but has been replaced by isocetane with a rating of 15) and values on this scale were named cetane numbers. A specially designed engine with adjustable compression is used to determine a fuel's cetane number. The fuel being tested is injected at 13° before top dead center where the angle 13° refers to the rotation of the crankshaft that connects each of the pistons (in an engine consisting of multiple cylinders) to the load. The engine's compression ratio is then adjusted till the fuel ignites at the top center. Next, the engine is run on various blends of cetane and 1-methylnaphthalene till a blend is found for which ignition occurs at the same compression ratio as the given fuel. The corresponding percentage by volume of cetane in the mixture is the cetane number. Often without running this test, an estimate of the cetane number can be made from the fuel's specific gravity and the temperature at which 10, 50, and 90% of the sample boils away.

6.3.5.2 Octane Number In the case of spark ignited engines, the fuel should not spontaneously ignite before the spark plug is fired at the optimum ignition moment, otherwise it results in "knocking" (pinging noise can be heard during operation). The octane number measures this property; higher the octane number, the more compression the fuel can withstand before ignition. The research octane number is most common type of octane rating used worldwide and is determined by running the engine on various blends of iso-octane and n-heptane till a blend is found for which the knocking intensity is the same as the given fuel. The corresponding percentage by volume of iso-octane in the mixture is the octane number.

Fuels used in stationary industrial engines are usually petroleum derived but natural gas, syngas derived from biomass gasification, digester gas, and landfill gas are also being used. The gaseous fuels should be free of dust, noncorrosive, not detonate or preignite during the compression stroke, and release enough energy on burning to develop power. Diesel fuel of 5–6% may be required with such gaseous fuels in compression ignition engines for ignition. Liquid fuels need to be grit-free, injectable, possess sufficient lubricity to minimize the wear on the injection plungers, and of course release enough energy on burning to develop power. Preheating of the fuel may sometimes be required to lower its viscosity.

6.4 HYDRAULIC TURBINES

Hydraulic turbines transfer the kinetic energy from a flowing liquid to a rotating shaft; the liquid being typically under pressure or having stored potential energy is converted to kinetic energy in the fluid as the fluid flows into the turbine. Hydraulic turbines are used in hydroelectric power plants as well as in the process industry, although the design of these turbines differs in the two applications.

6.4.1 Process Industry Applications

In the process industry, such turbines are called hydraulic power recovery turbines and are used to recover energy from a liquid stream by reducing its pressure that would otherwise be wasted across a pressure letdown valve. A common type of a hydraulic turbine is a reverse rotating centrifugal pump. Different types of centrifugal pumps can be used as hydraulic turbines with a few design changes required due to the reverse rotation and high inlet pressure. Typically, the turbine is used as a driver for rotating another equipment in the process such as a pump. Since a process liquid may not always be available or its pressure may be variable, a hydraulic turbine is typically not used as a stand-alone driver, and is instead used along with an electric motor as primary driver to reduce the power used by the motor. The hydraulic turbine is disconnected from the motor using a clutch when the process liquid becomes unavailable or its pressure becomes too low to avoid the hydraulic turbine from becoming a drag on the system.

The performance of a hydraulic power recovery turbine is different from that of a pump. Unlike a pump which can operate from its minimum to maximum flow range, a hydraulic power recovery turbine must operate near its best efficiency point (BEP). At flow rates below its BEP, the capacity of a hydraulic turbine to recover energy is greatly reduced and it can become a drag to the system at flow rates further lower.

6.4.2 Hydroelectric Power Plant Applications

There are two main types of hydraulic turbines, the impulse and the reaction and the type of turbine selected depends on the height of standing water ("head") and flow rate. Other factors include how deep the turbine must be set in the water, efficiency, and cost. The impulse turbine uses the velocity of the water to move the "runner" (rotor) and discharges to atmospheric pressure. An impulse turbine is generally suitable for high head and low flow rate applications. The reaction turbine develops power from the combined action of water pressure and velocity. Reaction turbines are generally suitable for lower head and higher flow rates than those for the impulse turbines.

The hydraulic power developed by the turbine $\dot{W}_{Hy}$ may be calculated using an equation similar to that used for a pump, and given as follows, as long as no phase change occurs across the turbine.[3] If the height of the standing liquid (head) is known instead of the hydrostatic pressure, then the static pressure drop ΔP_{turb} across the turbine may be calculated from the head using the liquid density.

$$\dot{W}_{Hy} = \Delta P_{turb} \dot{V} \eta_{Hy}$$

where $\dot{V}$ is the volumetric flow rate, and η_{Hy} is the turbine hydraulic efficiency. The above hydraulic power should be multiplied by the mechanical and generator efficiencies to determine net power developed.

REFERENCES

Bhargava R, Meher-Homji CB. Parametric analysis of existing gas turbines with inlet evaporative and overspray fogging. ASME IGTI Turbo-Expo Conference; June 2002; Amsterdam, Holland.

Burrus DL, Johnson AW, Roquemore WM, Shouse DT. Performance assessment of a prototype trapped vortex combustor for gas turbine application. Proceedings of the ASME IGTI Turbo-Expo Conference; June 2001; New Orleans, LO.

Cengel YA, Boles MA. *Thermodynamics: An Engineering Approach*. Hightstown, NJ: McGraw-Hill; 1998.

Gemmen RS, Richards GA, Janus MC. Development of a pressure gain combustor for improved cycle efficiency. ASME Cogen Turbo Power Congress and Exposition; October 25–27, 1994; Portland, OR.

[3] In process plant applications the liquid entering the turbine may be close to its bubble point in which case a fraction of the liquid may vaporize and methodology involving entropies similar to that explained for a steam turbine may be utilized in calculating the hydraulic power.

General Electric Bulletin GEI 41040G. Specification for fuel gases for combustion in heavy-duty gas turbines. Schenectady, NY: General Electric Company; revised January 2002.

Hsu KY, Gross LP, Trump DD. Performance of a trapped vortex combustor. Paper No. 95-0810, AIAA 33rd Aerospace Sciences Meeting and Exhibit; January 1995; Reno, NV.

Keenan JH, Keyes FG, Hill PG, Moore JG. *Steam Tables: Thermodynamic Properties of Water Including Vapor, Liquid, and Solid Phases*. New York: John Wiley & Sons, Inc.; 1969.

Rao AD, Francuz DJ, West E. Refinery gas waste heat energy conversion optimization in gas turbines. ASME Joint Power Generation Conference; October 1996; Houston, TX.

Smith JM, Van Ness HC, Abbott MM. *Introduction to Chemical Engineering Thermodynamics*. 7th ed. Boston, MA: McGraw-Hill; 2005.

Smith LL. Ultra low NOx catalytic combustion for IGCC power plants. Prepared by Precision Combustion Inc., U.S. Department of Energy Topical Report; March 2004. Morgantown, WV: U.S. Department of Energy.

Spencer RC, Cotton KC, Cannon CN. A method for predicting the performance of steam turbine-generators: 16,500 kW and larger. Paper no. 62-WA-209; ASME Winter Annual Meeting; New York; Revised July 1974 version.

7

SYSTEMS ANALYSIS

Selection of a suitable process (in our case for sustainable energy conversion) or a subsystem making up this process from among various options requires systems analysis which includes preliminary systems design and cost estimation to determine the relative performance including the environmental signature and economics of each alternative. Figure 7.1 illustrates the various tasks that are required to develop such an analysis for each alternative but some of these work tasks may be left out of the analysis depending on how detailed the analysis has to be, or those work tasks may be performed on a qualitative basis. For example, for preliminary studies, off-design and dynamic performance as well as safety analysis may be addressed only on a qualitative basis, that is, by inspection of the process flow scheme to asses if the system ends up with any unusual features that may be "show stoppers." The various tasks to be performed are discussed in the following and as indicated by the diagram, the system configuration may evolve in an iterative manner as it may undergo modifications depending on the findings of some of the downstream activities. It should be noted that this chapter does not cover detailed design of a process required for construction of the plant.

7.1 DESIGN BASIS

The first step in systems analysis is to establish the design basis so that the system is suitable for the site-specific conditions of the project. Also it provides a set of design

Sustainable Energy Conversion for Electricity and Coproducts: Principles, Technologies, and Equipment, First Edition. Ashok Rao.

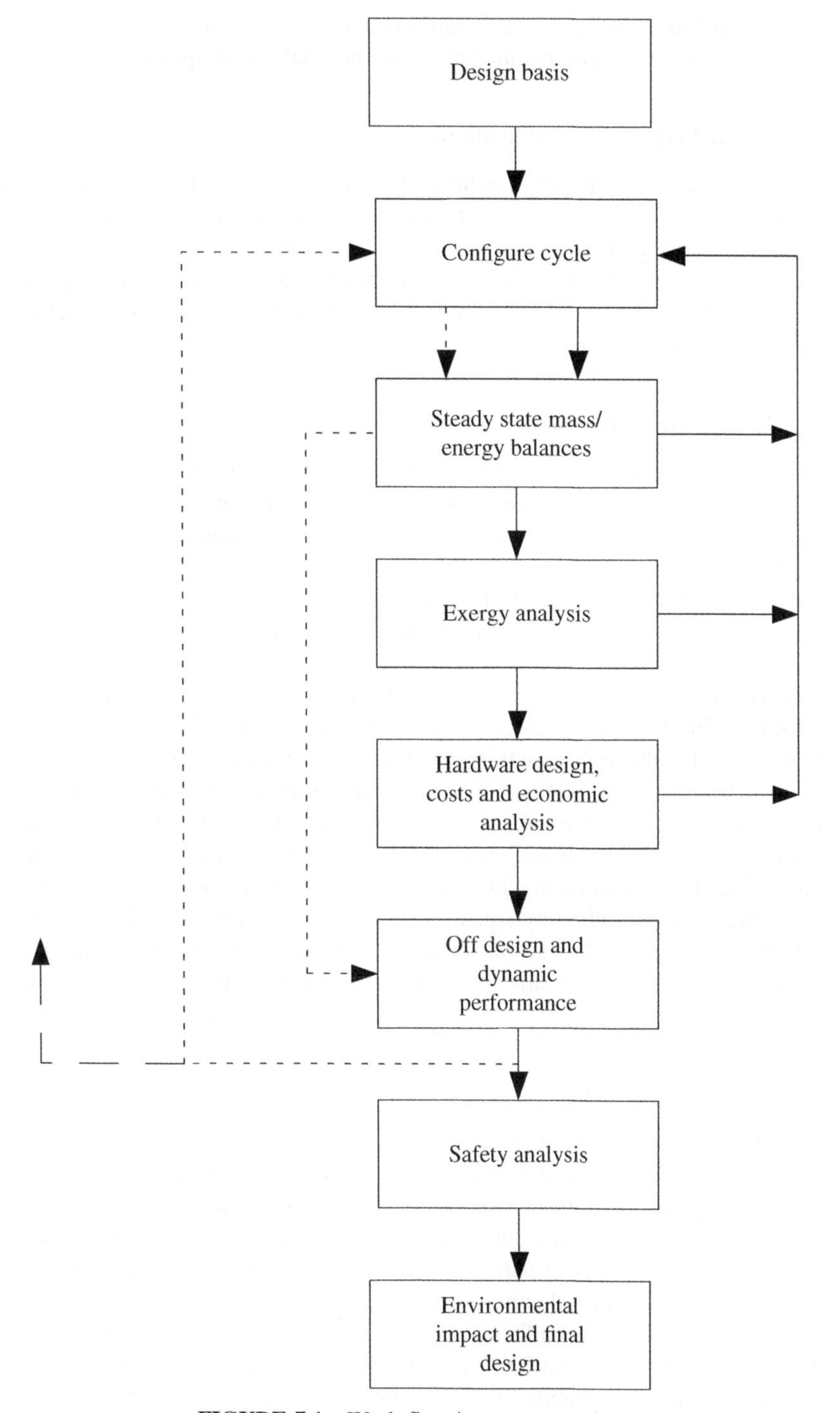

FIGURE 7.1 Work flow in systems analysis

criteria for comparing the various alternatives on a consistent basis. Listed below are some of the minimum requirements that should be established upfront.

7.1.1 Fuel or Feedstock Specifications

This includes establishing the availability and composition as well as the contaminant or impurity level present in the fuel or feedstock, its pressure and temperature where applicable. Variability in these parameters should also be specified since the system design is directly affected. As mentioned in Chapter 1, fuels such as natural gas and coal can have variability in composition and the system design should be able to accept such "off-design" fuel.

7.1.2 Mode of Heat Rejection

A significant fraction of the energy entering the system is rejected as thermal energy to the environment (unless it is a CHP plant) whenever material and energy transformations occur due to the constraints placed by the second law. There are different methods for heat rejection which are described in Chapter 8, and the most appropriate method is not only dependent on the local site conditions such as availability of water and relative humidity of the ambient air but also the environmental regulations. If the relative humidity is too high, then cooling towers may not be desirable. If a large body of water such as a lake, ocean or river can supply the cooling water on a once through basis, local environmental regulations which may pose an upper limit on temperature of water returned to the body supplying the water to avoid damaging the aquatic habitat have to be taken into account. Lowering the return temperature of the water impacts the circulation rate of the cooling water which in turn affects the circulating water pump size, its power usage, and cost. In addition, environmental regulations may also stipulate the manner in which the source point for the water intake to avoid ingesting and destroying the tiny aquatic fauna (in their larvae state) that may pass through the screens, and the manner in which it is discharged back into the body of water to avoid shocking the surrounding aquatic habitat and in which case, multiple discharge points may be required which also affects the cost of the discharge system.

7.1.3 Ambient Conditions

It is typical to select a set of ambient conditions consisting of the dry bulb temperature and wet bulb temperature (which is required to determine the humidity of the ambient air and to specify the cooling water supply temperature when wet cooling towers are utilized for plant heat rejection), and barometric pressure as the design point conditions. In addition to the performance of turbomachinery such as compressors and gas turbines as well as internal combustion engines being affected by these ambient conditions, their sizing is also affected. Performance and sizing of the heat rejection system, especially if "dry cooling" (e.g., an air cooled exchanger), are also affected by the ambient conditions. Variability in these conditions may be accounted for by specifying a minimum and a maximum dry bulb temperature and the accompanying wet bulb temperatures. As a minimum it should be qualitatively ascertained that the system can operate under these extreme conditions. For example, the plant cost may have to account for winterization and freeze protection including buildings to house equipment such as the steam turbine depending on the minimum dry bulb temperature. In a more detailed systems analysis, plant performance may have to be developed at the minimum and maximum ambient conditions as off-design analyses to estimate the changes in performance. Equipment limiting the plant performance at these extreme conditions may also be identified and an assessment may be made if increasing the size of such equipment to debottleneck the overall plant performance is justified.

7.1.4 Other Site-Specific Considerations

Some of the other site-specific conditions that should also be specified or taken into consideration especially when comparing results of one study with another include the following since these effect the plant cost.

- Site topography—typically assumed level for studies but if not level, then the additional cost for site preparation should be taken into account.
- Site access—if landlocked, then typically it may have access by rail and highway in which case an upper size limit will exist for shop fabricated equipment that needs to be transported to the site. It is usually the clearance available under bridges that sets the maximum diameter (of about 4 m) for shop fabricated vessels. For landlocked sites, larger equipment has to be field fabricated, that is, on site which requires additional cost but in some cases might be the cheaper option on a total plant cost basis since the number of trains or lines of that equipment may be reduced. For example, savings are realized in instrumentation and the associated electrical wiring and/or pneumatic piping. Savings in the process or utility piping may also be realized despite its larger diameter required for the larger capacity due to economies of scale.[1] Costs of structural support and foundations are similarly reduced on a total plant basis. If the site is located close to a waterway, however, then possibly larger equipment could be shipped on a barge.
- Supply of plant feeds—which depends on the "site access." The transportation method used for supplying the feedstocks affects their cost. Large coal-based plants may sometimes be located near the mine mouth itself to avoid such cost especially when the coal is of low rank such as lignite.
- Plant wastes disposal—whether the plant wastes such as ash from a coal-based plant can be disposed on-site in which case transportation costs may be avoided.

[1] The specific cost (cost per unit capacity) of larger equipment is also typically lower than that of smaller equipment when both are shop fabricated. In case of pressure vessels, however, the wall thickness increases with the diameter due to increase in hoop stress and should be accounted for in the trade-off between smaller diameter shop fabricated versus larger diameter field fabricated vessels.

- Carbon capture and sequestration—if CO_2 emissions are to be limited (as in coal fueled plant), then to what extent and final disposition of the captured CO_2, that is, method used for sequestration, distance from the plant site and how the CO_2 is transported including under what pressure.
- Raw water supply—whether and how much fresh water is available and its analysis which effects the treatment costs in preparing boiler feed water (BFW) and also if used as makeup to a wet cooling tower. If water is in short supply, then wet cooling towers may not be used for plant heat rejection.
- Soil conditions—any special considerations to be taken into account which affect the foundation costs such as piling and flood plain.
- Seismic zone—any special considerations to be taken into account which affect foundations and structural design.

7.1.5 Environmental Emissions Criteria

The design criteria for the environmental air emissions and liquid and solid discharges determines the need for pollution abatement measures to be included in the system configuration or selection of the type of equipment should be established. A number of factors should be considered in setting the environmental emissions criteria, including current emissions regulations, regulation trends, and the status of current best available control technology (BACT). The current emissions regulations are set by the New Source Performance Standards (NSPS). The Clean Air Act Amendments of 1970 required that the U.S. Environmental Protection Agency (U.S. EPA) develop NSPS for new and modified stationary sources. The pollutants regulated in the NSPS program are PM, SO_2, NO_x, CO, VOC, opacity, dioxins/furans, fluorides, H_2SO_4 mist, Cd, Pb, Hg, total reduced sulfur, municipal solid waste landfill emissions, and municipal waste combustor emissions of metals and acid gases (SO_2 and HCl).

Other regulations that could affect emissions limits from a new plant include the Prevention of Significant Deterioration (PSD), and the New Source Review (NSR) permitting process which requires installation of emissions control technology meeting either BACT determinations for new sources being located in areas meeting ambient air quality standards (also known as "attainment areas"), or Lowest Achievable Emission Rate (LAER) technology for sources being located in areas not meeting ambient air quality standards (also known as "nonattainment areas"). The U.S. EPA is authorized to establish regulations to prevent significant deterioration of air quality due to emissions of any pollutant for which a National Ambient Air Quality Standard (NAAQS) has been established and the environmental area designations vary by county and can be established only for specific site locations. In addition, state and local jurisdictions can impose more stringent regulations on a new facility.

CO_2 is not currently regulated in the United States but the possibility exists that limits may be imposed in the future. On September 20, 2013, the U.S. EPA announced its proposed NSPS for CO_2 emissions from future power plants but these will not apply to existing power plants. Separate national limits are proposed for natural gas and for coal fired units. Large natural gas fired power plants would need to emit less than 1000 lbs (454 kg) of CO_2 per MWh, while new small natural gas fired units would need to emit less than 1100 lbs (499 kg) of CO_2 per MWh. New coal fired units would need to emit less than 1100 lbs (499 kg) of CO_2 per MWh. Operators of these units could choose to have additional flexibility by averaging their emissions over multiple years to meet a somewhat tighter limit. Large modern natural gas fired combined cycle plants are already able to meet this standard but new (supercritical) coal fired power plants will require approximately 40% carbon capture.

7.1.6 Capacity Factor

The required plant capacity factor (CF) establishes the spare capacity in critical equipment and the need to supply a backup fuel (and store it in the case of a liquid) which affect the plant cost. The backup fuel also increases the operating cost. CF also has a more direct effect on plant economics since it affects the revenue stream over a period of time from the sale of the product and it also measures how well the capital is utilized. The North American Electric Reliability Council (NERC) defines an equivalent availability factor (EAF) as a measure of the plants CF assuming there is always a demand for the product while the EAF accounts for planned and scheduled derated hours, as well as seasonal derated hours. The average EAF for coal fired boiler plants in the 400–599 MW size range was about 85% in 2004 which is the value used in a U.S. Department of Energy (U.S. DoE) sponsored systems analysis of various fossil fired power plants (Black, 2010). The corresponding CF for a coal fed integrated gasification combined cycle (IGCC) was lower at 80% while that for a natural gas combined cycle was 85%.

Example 7.1. The environmental regulation to define the limit for NO_x emissions from a gas turbine or a gas turbine–based combined cycle in the United States is expressed in parts per million on a volume and on a dry basis while "correcting" to 15% O_2 by adding or removing air to the flue gas such that the concentration of the O_2 on a dry basis is 15% by volume ("ppmvd, 15% O_2"), rather than the NO_x emitted on a unit of power generated. Compare the NO_x emitted by the two combined cycle plants A and B on a ppmvd, 15% O_2 basis and on a unit of power generated (using $x = 2$ although NO is predominantly formed in the gas turbine combustor, the ultimate form of NO_x in the atmosphere is NO_2). Assume the dilution air consists of 20.9% O_2 on a volume dry basis, required to increase the O_2 concentration in the flue gas to 15%. Data for these two plants is presented in Table 7.1.

Solution

The calculations are summarized in Table 7.2. As can be seen, even though the two plants have the "same" NO_x emissions when expressed on a dry basis and corrected to 15% O_2 basis, plant A is more polluting since it emits 10% more NO_x than plant B when expressed on a unit net MW basis (because plant B is more efficient).

TABLE 7.1 Combined cycle plant data

Plant	A	B
Net power output (MW)	518.4	564.7
NO_x formed (kg/h)	56.2	56.0
Flue gas flow rate (kg/h)	4,179,592	3,230,636
Flue gas major components (volume %)		
O_2	14.14	12.09
$N_2 + Ar$	76.56	75.2
CO_2	3.1	4.04
H_2O	6.2	8.67

TABLE 7.2 NO_x emission calculations

Plant	A	B
NO_x molecular weight	46.01	46.01
NO_x formed (kmol/h)	1.222	1.217
Flue gas molecular weight	28.46	28.28
Flue gas molar flow rate (kmol/h)	146,872	114,242
Dry flue gas major components (volume %)		
O_2	15.07	13.24
$N_2 + Ar$	81.62	82.34
CO_2	3.30	4.42
H_2O	0	0
Total	100.00	100.00
Dry flue gas molecular weight	29.15	29.25
Dry flue gas molar flow rate (kmol/h)	137,766	104,337
Dry flue gas molar flow rate with 15% O_2, kmol/h [= dry flue gas molar flow rate × (20.9 − % O_2 in dry flue gas)/(20.9 − 15)]	136,024	135,502
NO_x emission, ppmvd (15% O_2)	9.0	9.0
NO_x emission (kg/MWh)	0.1084	0.0992

7.1.7 Off-Design Requirements

In addition to specifying the requirement for the system to operate over a certain range of ambient conditions, it is also important for a plant, especially one that produces electricity as its primary product to have definition of whether load-following capability is required (as imposed by the grid) and how low the plant output has to be turned down to. Dynamic simulation of the system is required to determine the load-following capability but the minimum turn down ratio may require developing the off-design performance of the system with equipment sized for the design point and determining if there are any operational issues with the equipment. Turn-down capability of a plant may be extended by installing multiple trains of an equipment or subsystem and as a minimum, the maximum turn-down capabilities of the major equipment in the plant should be reviewed and a decision made as to the number of individual trains of that equipment, for example, two trains with each having half of the total design capacity of the plant, or three trains with each having third of the total design capacity of the plant. Note that, as explained previously, the cost of the plant then increases since generally the specific cost (cost per unit capacity) of smaller equipment is larger, as well as costs associated with electrical, controls, and structural hardware increases as the number of trains increases.

In addition to the above, the design criteria for the equipment should also be established before the equipment can be specified and priced for the cost estimation task. The minimum driving forces such as the temperature difference between a hot and a cold stream in the case of a heat exchanger (specific to the type of fluids), or approach temperature of cooling water to the ambient wet bulb temperature in the case of a wet cooling tower should be established upfront in scoping studies since optimizing such parameters may be beyond the study scope and budget.

7.2 SYSTEM CONFIGURATION

The initial step consists of conceptualizing the system configuration at a very high level with the help of an overall block flow diagram (BFD). The BFD provides a very broad overview of the process and contains very few specific details. It shows the order and relationship between major subsections or "units" of the overall system using flow arrows. Figure 7.2 shows a BFD for a gas turbine–based combined cycle plant with the gas turbine as the topping cycle with fuel and air entering it and its exhaust entering the heat recovery steam generator (HRSG). The HRSG provides steam to run a steam turbine. Some of the other blocks included in this BFD are the heat rejection system and the water treatment/preparation system.

The next step consists of developing the process flow scheme within each of the blocks of the BFD. The development of the process flow scheme usually starts with the major power generators (such as gas turbines, fuel cells in the case of power cycles) as well as reactors (in the case of more complex systems requiring such equipment). Next columns for mass transfer operations may be added into the flow scheme. Once the inlet and outlet streams and any recycle streams are configured, compressors, blowers and pumps may be added depending on the estimated pressure at the sources and at the destination points for the streams. Heat exchangers may then be added by determining whether heating or cooling is required based on the temperature at the source and at the destination point of a stream. These temperatures may be very approximate based on a knowledge of previous work or may have been specified by the downstream and/or the upstream equipment. The fluid on the other side of the heat exchanger can be specified after generating an initial simulation of the system based on mass and energy balance calculations. Pinch analysis described in the next section of this chapter can be beneficial in developing the heat recovery and rejection schemes.

The development of the process flow scheme often has to be performed in parallel with the steady state simulations of each of the subsystems or a group of major subsystems at the design point. There are number of computer simulation tools available such as Aspen Plus® and Pro/II® suitable for complex energy plants such as an IGCC,

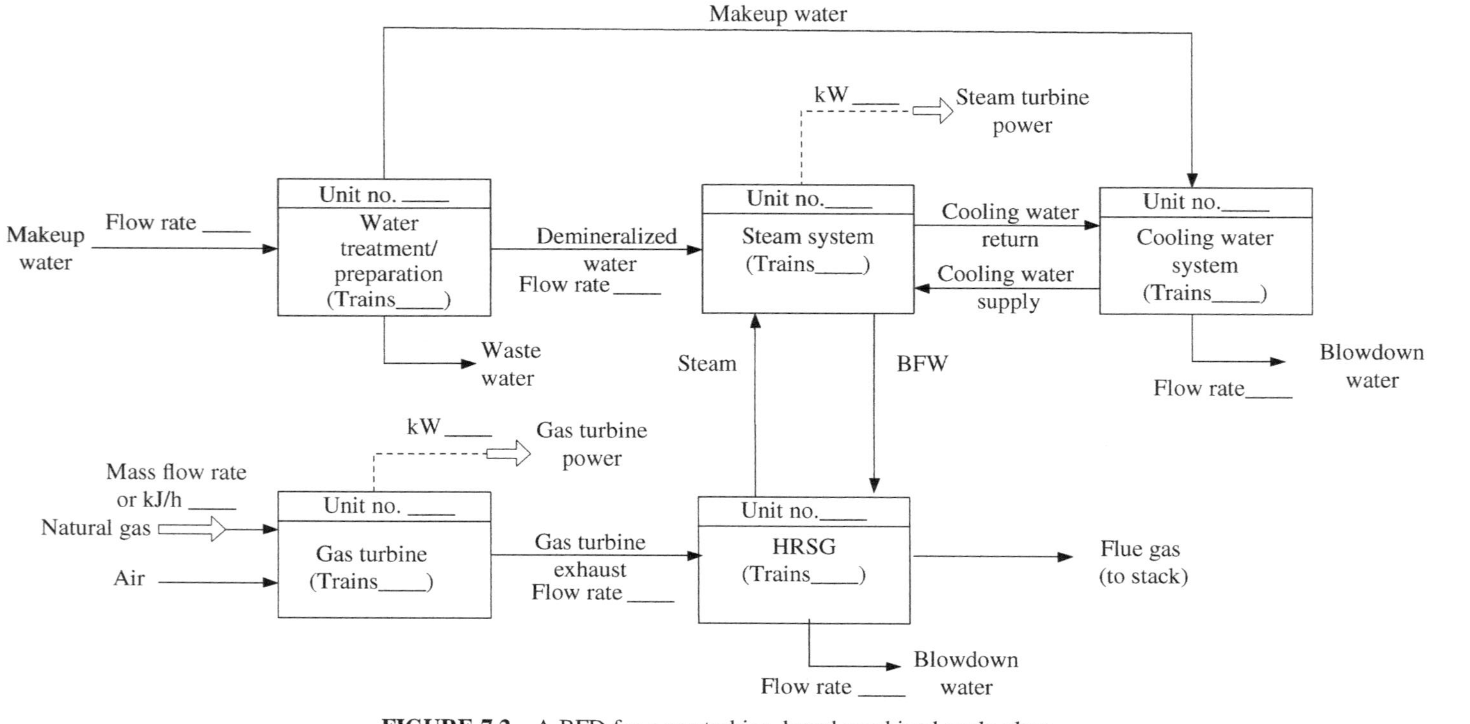

FIGURE 7.2 A BFD for a gas turbine–based combined cycle plant

GateCycle™ and Thermoflex more suitable for natural gas fired power plants. These software programs have user friendly graphical interfaces and systems engineers often develop the process integration scheme directly using this graphical interface rather than sketching it out beforehand. After providing the necessary input consistent with the design basis to run the simulation, the process integration scheme may have to be modified. Depending on the type of software program being used, the simulation consisting of performing the mass and energy balances on each of the unit operations and processes is either constrained by specified net driving force for heat or mass transfer, or by approaches to equilibrium for mass transfer and chemical reactions. Rate-based equipment models are also included in software such as Aspen Plus but it is more efficient to develop the simulations initially using the more simplified thermodynamic models, and after being satisfied with the design scheme, to replace the simpler models with the more complex rate-based models. A system or process flow sketch that shows the type of equipment in the process and how they are physically connected and the utilities used may then be developed utilizing the system configuration illustrated by the graphical interface for use in the subsequent steps.

7.3 EXERGY AND PINCH ANALYSES

After configuring a process scheme and performing the steady-state system simulation, the second law of thermodynamics may be applied to identify areas of inefficiency quantitatively and thus, areas where modifications to the configuration or design parameters will make significant improvement in the overall system efficiency. Specifically, exergy and pinch analysis provide insight into the thermodynamics of the process which enables identifying potentially advantageous process changes in configuration as well as operating conditions to reduce the energy requirement of the process.

7.3.1 Exergy Analysis

As explained in Chapter 2, exergy analysis requires calculating the exergy of each of the streams (entering, leaving and within the system) to quantify the amount of work potential destroyed across each major equipment within the system as well as quantifying the upper limit for the system efficiency. It also provides a quantitative explanation of why one configuration is better than another from an efficiency standpoint. For example, exergy analysis of a combined cycle shows that the highest amount of exergy destruction occurs across the gas turbine combustor. Thus, the greatest improvement in the combined cycle efficiency is to choose a gas turbine with a higher firing temperature. Exergy analysis also helps in identifying research and development needs for the long term. For example, improvements in firing temperature may be realized by developing advanced materials that can withstand the higher temperatures in the combustor, the transition piece (connecting the combustor and the turbine) and in the turbine. In order to make further advancements in the manner in which the fuel bound energy is converted to mechanical energy, alternatives to the near isobaric combustion process are required. One can then perform literature search or come up with new ideas (say, in a brain storming session) on how to reduce this wastage of exergy. More transformative or revolutionary improvements instead of the previous evolutionary improvement (which occurs in smaller increments of temperature) may be identified for research and development. One such technology is replacing the isobaric combustion process with a constant volume process. Such a combustor is the constant volume or pressure gain or pulse detonation combustor (Tangirala et al., 2007) discussed in Chapter 6, which shows a much lower amount of exergy destruction (Hutchins and Metghalchi, 2003). Major challenges remain at the present time to be solved associated with interfacing the unsteady flow through the combustor with the compressor and the turbine. Yet another approach is to use a high temperature fuel cell as described in Chapter 11 to convert the fuel bound energy (or at least a majority of it) electrochemically, although the major challenge with fuel cells at the present time is their high initial cost. Exergy analysis may thus have to also include the effect on cost and economics. An example of such a method is described in the context of optimizing the operating parameters of an HRSG of the combined cycle (Franco and Russo, 2002). Since decreasing exergy loss by reducing the heat transfer temperature driving forces in the HRSG increase the size and thus the cost of the HRSG, the methodology consists of minimizing an objective function that is a sum of exergy losses and a representation of the cost. The minimization of such a combined objective function can allow finding an optimum compromise between efficiency and cost. Exergy destruction has also been used as a measure of resource depletion (Lombardi, 2001).

7.3.2 Pinch Analysis

Pinch technology which is also based on thermodynamic principles provides a systematic methodology for developing an efficient heat integration scheme involving both recovery and rejection (Linnhoff et al., 1988). This technology has evolved into an analysis that has been combined with exergy (Staine and Favrat, 1996), economic analyses (Gundersen and Naess, 1990; Linnhoff, 1993), and has been applied to IGCC power plant applications (e.g., Emun et al., 2010). The following provides a brief overview of this technology and the reader is referred to the above mentioned references to acquire working knowledge for applying this technique.

The basic approach of pinch technology consists of first generating the enthalpy versus temperature data for streams requiring cooling and streams requiring heating. A single composite plot of enthalpy versus temperature is then developed representing all the hot streams (by adding up the corresponding enthalpies at a given temperature). A similar composite plot is also developed representing all the cold streams drawn on the same temperature enthalpy diagram which now represents the total heating and cooling requirements of the process. The hot composite curve can be used to heat up the cold composite curve by process stream to process stream heat exchange (using an "interchanger") only along the enthalpy axis where the two curves overlap. Any overhangs at either end indicate the need for hot and cold utility requirements such that the top of the cold composite curve needs an external heat source $\dot{Q}_{\mathrm{H,min}}$ and the bottom of the hot composite curve needs external cooling $\dot{Q}_{\mathrm{C,min}}$. The two curves come closest to touching each other at the pinch point and for a chosen minimum approach temperature ΔT_{min} ("pinch target") there corresponds a region of overlap representing the maximum possible amount of process stream to process stream heat exchange, and the corresponding $\dot{Q}_{H,min}$ and $\dot{Q}_{C,min}$ are then the minimum hot utility (e.g., steam) and cold utility (e.g., cooling water) requirements and provide the "energy targets." To achieve these energy targets, the plant heat integration should be configured such that heat must not be transferred across the pinch. Also, there must not be any outside cooling above the pinch nor any outside heating below the pinch. Note that ΔT_{min} can be increased or decreased by shifting one or both of the composite curves parallel to the enthalpy axis because the composite curves represent changes in enthalpy and not absolute values. When ΔT_{min} is increased, the overlap decreases increasing the energy targets but equipment cost decreases and the proper choice of ΔT_{min} is facilitated by a trade-off analysis between efficiency and plant cost.

From an efficiency standpoint, the entire amount of utility for supplying $\dot{Q}_{H,min}$ and the entire amount of utility for rejecting $\dot{Q}_{C,min}$ do not have to be at the highest and the lowest temperatures, respectively, as indicated by the composite curves. For example, if heat is supplied by condensing steam, then instead of using all high pressure steam, some lower pressure steam may be substituted to supply a portion of the heat. This type of assessment may be made by constructing a "grand composite curve" from the two composite curves, the procedure for which is described in Linnhoff (1993). Pinch technology in addition to configuring the heat exchanger network has also been applied in identifying opportunities for heat pumps as well as heat engines to improve overall system efficiency in process plants.

Safety considerations should be taken into account while performing heat integration, that is, in selecting compatible fluids to heat exchange with each other. Heat exchangers may spring a leak or have a tube rupture in which case the fluid on one side of the tube will come into contact with the fluid on the other side resulting in a fire or even an explosion, depending on the type of fluids and their operating conditions involved.

A very simple application of the use of composite curves to evolve the heat integration scheme for an exothermic reactor is presented next in Example 7.2 to illustrate the essence of this methodology. Exothermic reactors may be found in the synthesis of CH_3OH, CH_4, Fischer–Tropsch liquids, NH_3, and shifting of syngas to adjust the H_2 to CO ratio.

Example 7.2. The feed to an exothermic reactor (stream 1) has to be heated up from 40°C to a temperature high enough for the reaction to initiate, this temperature being dependent on the catalyst and on the synthesis process, and in this example the required temperature is 200°C (stream 2). The reactor effluent is at a temperature of 220°C (stream 3) and has to be cooled to 50°C before the synthesized product is separated. One option is to simply use steam (a utility stream) to preheat the reactor feed gas and then generate steam against the reactor effluent to an intermediate temperature (stream 4) depending on the pressure of the steam being generated followed by cooling water to cool the effluent to the final temperature of 50°C (stream 5), as illustrated in Figure 7.3. The plots of the hot and the cold curves corresponding to this

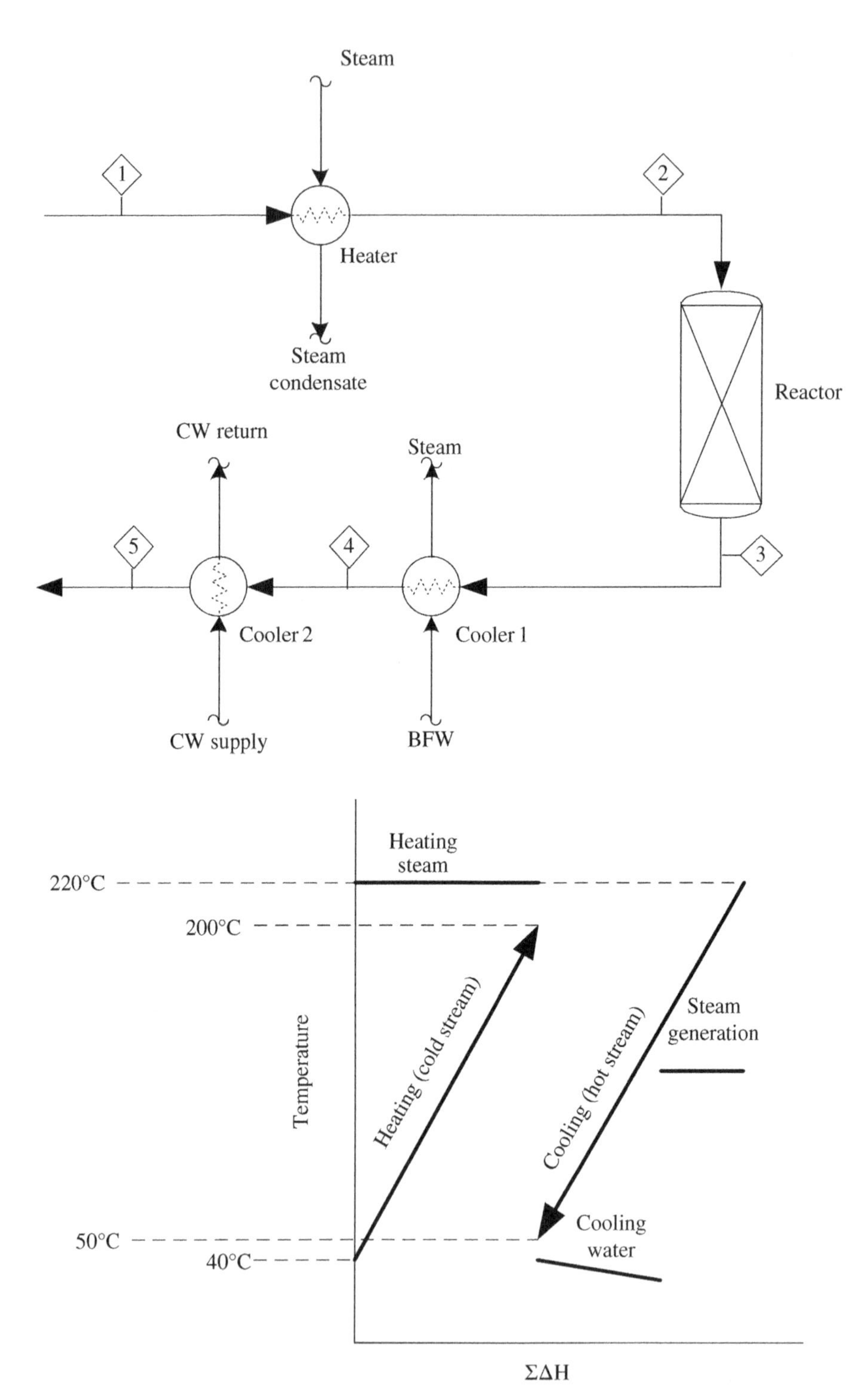

FIGURE 7.3 Configuration with high utility stream usage

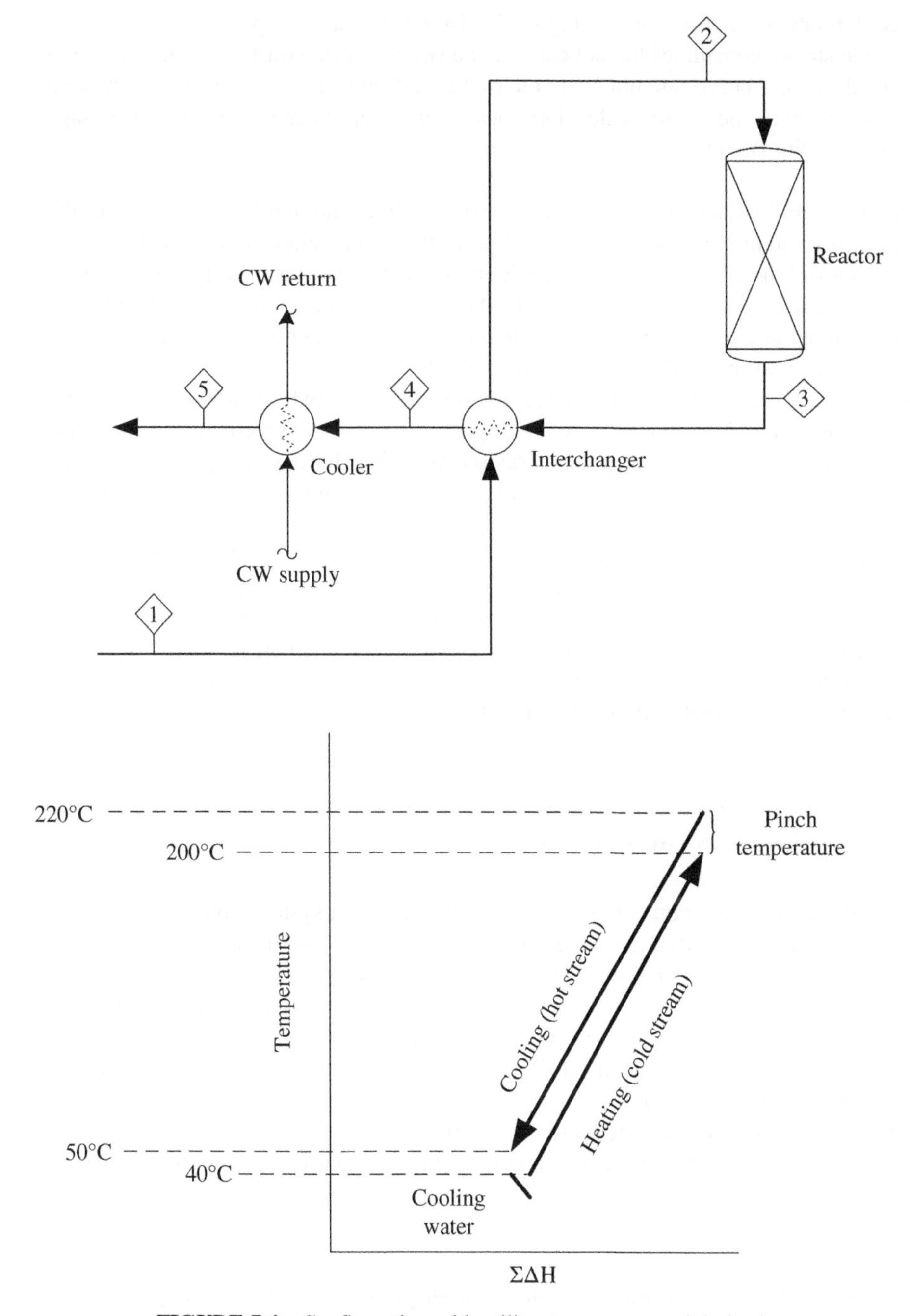

FIGURE 7.4 Configuration with utility stream usage minimized

configuration are also shown in Figure 7.3. Note that in this simple example chosen, a single stream constitutes the hot curve and a single stream constitutes the cold curve and thus these curves are not being referred to as "composite" curves. Using the data represented by these curves, develop a heat integration scheme to minimize the usage of the utility streams.

Figure 7.4 illustrates the configuration resulting from moving the cold curve to the right till a minimum temperature of 20°C (pinch temperature) is obtained between the hot and the cold curves.[2] This configuration now has a process stream to process stream exchanger or interchanger (a feed/effluent exchanger in this case) while the only utility stream used is a small amount of cooling water to cool stream 4 to the final temperature of 50°C resulting in stream 5.

When a process involves a single hot stream and a single cold stream as in the above example, it is may be easy to design an optimum heat recovery scheme intuitively but in the case of complex energy plants such as IGCCs or coproduction facilities where the number of streams is quite large, application of pinch technology can be found to be quite useful.

When a unit is designed with a feed/effluent interchanger, an alternate heat source should be made available for preheating the feed gas during startup, for example, providing an electrically or a steam-heated exchanger. Sizing of this heater does not have to be for the full design flow rate of the syngas since the unit may be started at a lower flow rate, say 25% of the design flow rate but its cost should be accounted for when studying the different options for heat integration.

7.4 PROCESS FLOW DIAGRAMS

A process flow diagram (PFD) also sometimes called a system flow diagram, may next be developed utilizing the system configuration evolved by the preceding steps. A PFD is a simplified graphical description of the basic process flow scheme, showing only the equipment, piping, and key instruments that are necessary for process clarity. Symbols to represent each type of equipment and flow line (and valve and instrument when included in the PFD) are established and documented upfront. A PFD provides important information regarding the type of equipment and how they are physically connected and the utilities used in a process, the major instruments located in each area of the plant illustrating the basic strategy for the overall

[2] In this case, the pinch occurred at the hot end but depending on the relative slopes and shapes of the two curves, the pinch may occur at the cold end, or somewhere between the hot and the cold ends when the curve or curves are not straight lines. When a stream undergoes phase change (e.g., a vapor component undergoes condensation as the stream is cooled or vice versa), the curve will not be a straight line. This is also the case when there is a large variation in the specific heat.

controllability of the process scheme evolved. In addition the PFD provides information required for:

- Preparing equipment specification forms
- Keeping other engineering disciplines informed and actively involved in the design function
- Developing and reviewing operating procedures (including start-up) and preparing operating manuals
- Developing material selection diagrams to specify the materials of construction for the equipment and piping.
- Checking overall process continuity and integrity
- Developing piping and instrumentation diagrams for detailed systems analysis when a much higher accuracy plant cost estimate is required or when the plant is actually going to be constructed.

Some major guidelines for drawing PFDs are listed below:

- Landscape page orientation should be used and generally flowing from left to right
- Cluttering should be avoided and instead multiple sheets should be used
- Equipment should be drawn in approximate relative size (e.g., columns taller than vessels such as those used for vapor-liquid separation, heat exchangers larger than pumps)
- Major columns and reactors should be located close to a single level
- Streams should be drawn with minimum direction changes
- Streams should be shown entering across the battery limits on the left while leaving on the right
- Streams that enter the next sheet should be shown leaving on the right while streams recycling to a previous sheet should be shown leaving on the left.

The following summarizes the information shown on a PFD depending on the level of detail of the systems analysis:

- All material streams entering and leaving the process should be identified by source and destination. Where there are two or more identical trains, only one representative train is shown along with a note identifying the number of parallel trains. Sometimes steams entering or leaving the battery limits of that train indicate the number of trains by branching the streams.
- Normal operating temperature and pressure for process streams, including streams entering and leaving the process battery limits should be shown. New conditions are shown when a stream changes temperature or pressure. When catalytic reactors are involved, values for start of run (SOR) and end of run (EOR) should also be shown since the catalytic activity varies with time resulting in

different temperatures, stream compositions and sometimes the pressure drop from those at normal operation.

- All process streams (including air entering an air-breathing engine or a reactor, water used for quenching or reaction, etc.) should be named (the name can appear within the box at the beginning of the line showing the source or at the end of the line showing the destination), and numbered with a table showing the stream data such as compositions, mass and molar flow rates, phase (solid, liquid or vapor), molecular weight or density on the PFD itself or instead on a separate sheet. The stream data for SOR, EOR, and any other operating cases should also be shown in a table format.
- Data for following utility streams should include:
  - Cooling water
    - Inlet and outlet temperatures
    - Flow rate (typically volumetric).
  - Steam and condensate
    - Inlet and outlet pressures
    - Temperature if condensate or superheated steam or identified as saturated steam
    - Mass flow rate.
- Each equipment should be identified by a tag number. The tag number identifies (i) the unit number that the equipment belongs to, (ii) the train number, (iii) type of equipment using a one or two letter identifier, and (iv) the equipment number in that particular category. When there are multiple equipments due to having installed spare equipment within a train, a suffix letter such as A or B is added. For example, the tag number 20-2-P-3A or (20-2-PU-3A) indicates that the equipment is a pump (identified by the letter P or PU) and is located in the second train of unit 20.
- Data for the following equipment should include:
  - Vessels
    - An outline of all important internals such as catalyst, packing, demisters
    - Operating pressure and temperature
    - For columns, top and bottom trays numbered, and those connected to process lines and/or instruments
    - At top of PFD above each vessel:
      - Equipment tag number (also shown on the vessel)
      - Description of the vessel
      - Inside diameter and T/T length
  - Fired heaters
    - Type of fuel
    - Process piping for only one pass
    - At top of PFD above each heater:

    - Equipment tag number
    - Descriptive title
    - Absorbed heat duty
- Heat exchangers including air cooled
  - Tube side flow identified by a dashed line
  - Equipment tag number while denoting multiple shells by a letter suffix (A, B, C, etc.).
  - Heat duty exchanged.
- Pumps (below each pump):
  - Equipment tag number
  - Normal operating volumetric flow rate
- Compressors
  - Each stage of a multistage compressor shown by repeating the compressor symbol
  - Equipment tag number shown below each compressor.

PFDs typically do not show pipe classes, piping line numbers, flanges, minor bypass lines, process control instrumentation such as sensors and final elements, isolation and shutoff valves, maintenance vents and drains, and relief and safety valves.

Figure 7.5 shows the PFD for the power generation subsystem of the combined cycle plant consisting of the gas turbine and the steam cycle but with blanks for the equipment data that is typically included in such diagrams. A stream data table would accompany the diagram and hence the major streams are tagged (numbered) to cross reference the table. Each diagram is actually assigned a drawing number and streams originating from another unit or leaving this combined cycle unit while entering another unit in the plant are identified by the drawing number.

7.5 DYNAMIC SIMULATION AND PROCESS CONTROL

7.5.1 Dynamic Simulation

Many of the operations in an energy plant are dynamic such as the startup and shutdown of a unit, responding to changes in the power load demand, the changeover from one fuel to another, or sudden failure of a system component such as the gas turbine or the steam turbine. The effects of moving from one operating mode to another can be unanticipated and sometimes unsafe. In addition to safety, environmental and economic factors highlight the importance of an understanding of the operation of the plant. In addition, testing of control schemes, development of operating procedures and training of plant operators in advance of actually operating the real (physical) plant for smooth and safe operation may be accomplished using dynamic simulation of the plant or a subsection of the plant performed on a computer, providing an economically effective solution to these needs. The steady-state simulations developed

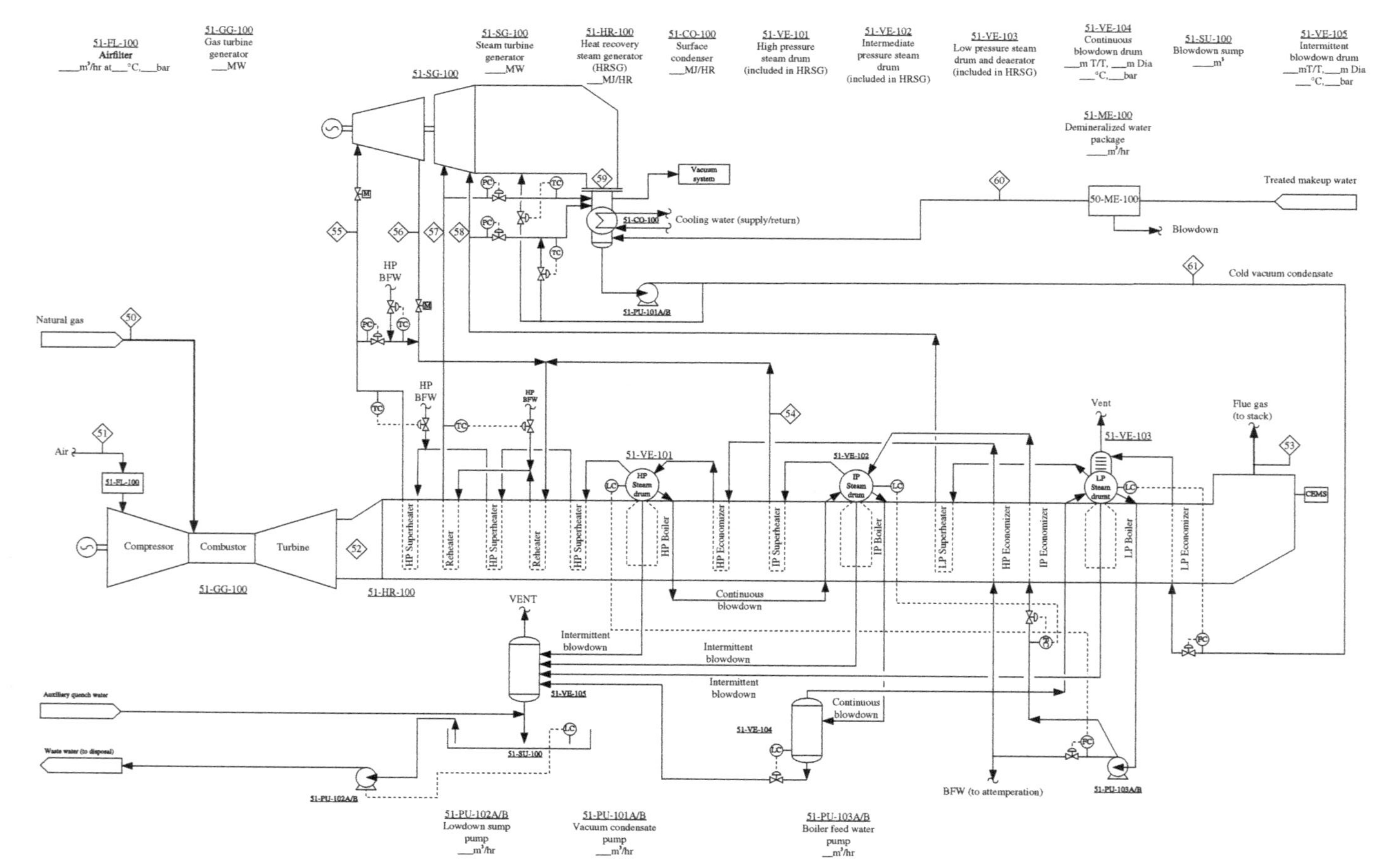

FIGURE 7.5 PFD for the power generation subsystem of the combined cycle plant

say on Aspen Plus or Pro/II can be converted to dynamic simulations using the corresponding software such as Aspen Plus Dynamic and Dynsim, respectively, for developing the dynamic simulations.

The major difference between steady-state and dynamic simulation is that in the case of steady-state simulation, many of the unit operations and processes occurring within the plant are modeled on a thermodynamic basis, that is, assuming equilibrium or a specified approach to equilibrium whereas in the case of dynamic simulation, rate-based models are required including a definition of the geometry of the equipment and piping, metal masses, desired holdups, valve characteristics, and the process control scheme. Thus, dynamic simulation requires significantly more definition of the plant for accurately capturing its unsteady state behavior than steady-state simulation. The steady-state simulation is utilized however, to define the "geometry" of the components within the system. The following simple example may provide the essence of dynamic simulation.

Consider a well-agitated vessel as illustrated in Figure 7.6 equipped with a jacket for condensing steam at a temperature T_S. A liquid stream of constant heat capacity C_p, at temperature T_{in}, flowing at a rate $\dot{m}_{in}$ is heated by the condensing steam to a temperature T_{out}, while flow rate of the heated liquid leaving the vessel is $\dot{m}_{out}$. It is assumed that the vessel is at a uniform temperature due to the agitation.

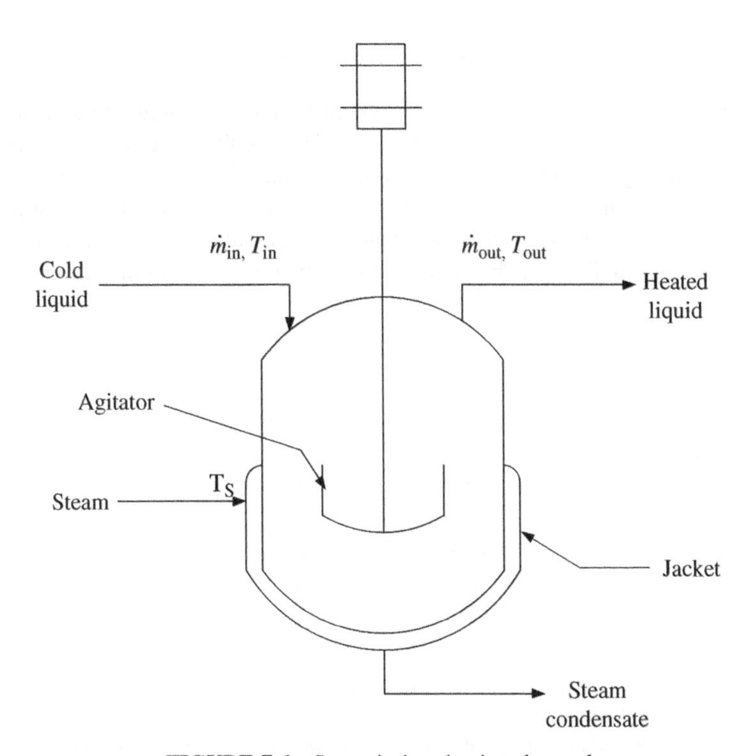

FIGURE 7.6 Steam jacketed agitated vessel

The mass balance around the system with m being the mass of liquid contained within the vessel at any time t may be written as:

$$\dot{m}_{in} - \dot{m}_{out} = \frac{dm}{dt} \tag{7.1}$$

If the system is at steady state with respect to the liquid mass flow rate, then the accumulation term $dm/dt = 0$, and $\dot{m}_{in} = \dot{m}_{out}$. Next, the energy balance around the system for this special case of a well-agitated vessel in which the temperature of liquid contained within the vessel at any time t is at a uniform temperature and equal to the outlet temperature T_{out}, may be written as:

$$\dot{m}_{in}\, C_p\, (T_{in} - T_{out}) + \mathcal{U}_o\, A_o\, (T_s - T_{out}) = m\, C_p\, \frac{dT_{out}}{dt} \tag{7.2}$$

where $\mathcal{U}_o$ is the overall heat transfer coefficient, a calculated value (Chilton et al., 1944; Kern, 1950) based on the jacketed surface area A_o. The amount of steam condensed within the jacket will vary depending on the amount of heat transferred, that is, will vary as T_{out} changes for constant $\mathcal{U}_o$ and T_s. The jacketed surface area A_o as part of defining the "geometry" of this component may be determined from Equation 7.3,[3] the steady-state energy balance obtained from Equation 7.2 with $dT_{out}/dt = 0$ and $T_{out} = T_{out,R}$, the steady-state value:

$$\dot{m}_{in} C_p (T_{in} - T_{out,R}) + \mathcal{U}_o\, A_o (T_s - T_{out,R}) = 0 \tag{7.3}$$

Assume there is a step change in the inlet temperature of the liquid to temperature $T_{in,f}$. Dynamic simulation now allows us to predict how the outlet temperature T_{out} varies with time t by solving Equation 7.4 below obtained from Equation 7.2 by replacing T_{in} with $T_{in,f}$ (since $T_{in} = T_{in,f}$ for $t > 0$), and replacing m with known variables, density of the liquid ρ (assumed independent of temperature), and volume of vessel occupied by the liquid V (which is assumed to remain constant in this case).

$$\dot{m}_{in}\, C_p\, (T_{in,f} - T_{out}) + \mathcal{U}_o\, A_o\, (T_s - T_{out}) = \rho\, V\, C_p \frac{dT_{out}}{dt} \tag{7.4}$$

If all other variables appearing in the above equation remain constant (except of course for T_{out} and t), then the solution to this equation below shows that T_{out} will not also undergo a step change but an exponential change with t.

$$T_{out} = T_{out,R} + \frac{(\mathcal{U}_o A_o / \dot{m}_{in} C_p) T_s + T_{in,f}}{1 + (\mathcal{U}_o A_o / \dot{m}_{in} C_p)} \left(1 - e^{-t/T}\right)$$

[3] Note that the conditions used in this equation should correspond to the minimum value for T_{in} and maximum value for $\dot{m}_{in}$ that the process may see in order to provide enough heat transfer surface area A at these extreme conditions if temperature of the stream leaving the vessel T_{out} is to be maintained at the desired value at all conditions including these extremes depending on the design basis chosen.

where $\mathcal{T} = \rho V / \dot{m}_{in}$ is called the time constant and is a measure of how quickly the system responds to change, and is faster when the volume of the vessel (capacitance) is lower or when the mass flow rate of the liquid is higher. Dynamic modeling of an actual process system, however, is complicated by delays caused by capacitance of the piping system, by the instrumentation and control strategy used.

7.5.2 Automatic Process Control

Controlling the temperature of the heated liquid T_{out} (the process variable or PV in the above case) by comparing it against the required value $T_{out,R}$ (the set point) and varying the steam supplied to the jacket (which is the manipulated variable in the above case) constitutes a closed control loop. The supply of steam is controlled by opening or closing a control valve which also changes its pressure downstream of the valve and consequently its condensation temperature T_s. A change in the amount of heat transferred is affected since in addition to T_{out}, T_s has also changed, while $\mathcal{U}_o$ not changing significantly. Note that in the region where the steam saturation pressure is $< \sim 34$ bar, an adiabatic reduction in its pressure as in a control valve imparts superheat to the steam but the heat transferred by the superheated steam is quite small compared to the heat transferred by condensation.

In feedback control, empirical approaches in the form of algorithms are used which define a relationship between the deviation (or "error") in the PV from its desired set point and the controller output to vary the manipulated variable. The basic approach consists of making a change in the controller output $\mathcal{Y}$ over $\mathcal{Y}_o$, its output when error $e = T_{out,R} - T_{out} = 0$ (also called controller bias), in proportion to the error. This type of control is called proportional control while the constant of proportionality is called the proportional gain $\mathcal{K}_p$, that is, $\mathcal{Y} = \mathcal{K}_p e + \mathcal{Y}_o$. This proportioning function of the controller is active in a defined range called the proportional band over which is covered the full operating range of the final control element. For a given change in input, the change in the controller output is increased as the proportional band is decreased. The drawback with this type of control is that $\mathcal{Y}_o$ corresponds to a particular set of process conditions (a certain flow rate of the liquid being heated in the above example), and when that flow rate changes, the error cannot become zero and the process variable is offset from its set point known as "droop." The problem is further exacerbated when there is a continuous disturbance (the inlet temperature of the liquid in the above case). This method of control then leads to both the process variable and the manipulated variables to continuously fluctuate in addition to the off-set or droop occurring in the process variable from the set point. Proportional control may be improved by adding integral control which provides a corrective action based on summing up the instantaneous errors over time. This accumulated error is multiplied by a constant, the integral gain $\mathcal{K}_i$ and may be added to the output obtained from proportional only control resulting in proportional-integral (PI) control, that is, $\mathcal{Y} = \mathcal{K}_p e + \mathcal{K}_i \int_0^t e\,dt + \mathcal{Y}_o$. A shortcoming of PI control, however, depending on the values chosen for $\mathcal{K}_p$ and $\mathcal{K}_i$

is that process variable may oscillate around the set point with either a constant amplitude or one that may grow instead of decaying, leading to instability. Addition of derivative control which provides an output by multiplying the time derivative of the error by a constant, the derivative gain $\mathcal{K}_d$ can speed up the process and its stability, resulting in proportional-integral-derivative (PID) control, that is, $\mathcal{Y} = \mathcal{K}_p e + \mathcal{K}_i \int_0^t e\,dt + \mathcal{K}_d(de/dt) + \mathcal{Y}_o$. Arriving at the optimum values for the three gains $\mathcal{K}_p$, $\mathcal{K}_i$, and $\mathcal{K}_d$, especially with respect to achieving a stable control system is called tuning. Instead of using human experts to tune these parameters (based on their knowledge and experience), algorithms are being or have been developed that use approaches such as neural networks, fuzzy logic, and genetic algorithms (Siddique, 2013).

In cases where a load disturbance occurs often (in the above case the inlet temperature of the liquid T_{in}), feedback control can be improved by adding feedforward control. In pure feedforward control, the disturbance itself is measured and used to predict an impact on process variable or PV to be controlled (T_{out} in this case), and then compute the required controller output to vary the manipulated variable (flow of steam supplied to the jacket in the above case).

Cascade control utilizes two or more controllers in which the output from one controller provides the input to another in defining its set point. For example, if the level in a vessel is to be controlled, a level controller (as the primary controller) may be used to provide the input to a flow controller (as the secondary controller) which measures the flow rate and compares it against the set point provided by the primary controller, to adjust the control valve on the outlet of the vessel. The added complexity of cascade control may be justified when the dynamics of the inner loop (with the secondary controller) are fast (at least three times as fast) compared to those of the outer loop (with the primary controller). Cascade control is also used when the control valve has a nonlinear flow characteristic (i.e., the relationship between flow rate and valve stem position is nonlinear) or useful when valve can be sticky (which tends to oscillate between sticking and slipping).

7.6 COST ESTIMATION AND ECONOMICS

7.6.1 Total Plant Cost

The total plant cost is the cost required for designing, procuring all the physical hardware such as equipment, piping, instruments, electrical wiring, structural steel, concrete, and then constructing the plant. The total plant cost estimate for systems analysis is typically developed either by capacity factored or by equipment modeled techniques or sometimes supplier quotes for subsystems. Total plant cost is subdivided into direct field material costs, direct field labor costs, subcontractor costs, indirect field costs, home office costs, and other miscellaneous costs such as contractor's risk and profit fee. These various costs are described in the following.

7.6.1.1 Direct Field Material Costs Direct field material costs are the costs of all materials used in the construction of the permanent physical plant facilities and include the following elements along with freight to jobsite:

- Equipment—all new machinery and equipment installed on a permanent basis in the plant which includes equipment such as tanks, vessels, heat engines, compressors, heat exchangers/HRSGs, fired heaters, pumps, material processing equipment, material handling equipment, and miscellaneous equipment such as filters and strainers.
- Material—bulk materials used in the construction of the permanent plant such as cement, aggregate, sand, steel, building materials, piping and fittings, valves, wire and conduit, instruments, insulation, and paint.

7.6.1.2 Direct Field Labor Costs The direct field labor costs are typically obtained by multiplying estimates of field labor hours by an effective "wage rate." This wage rate is an average wage rate for an appropriate mix of construction crafts and includes salaries or wages, incidental overtime, payroll burdens, and travel/living allowance. Other costs included in this effective wage rate account for the construction equipment, small tools and consumables, temporary construction facilities, construction services, field (foreman) supervision, and engineering firm's overhead costs and profit.

7.6.1.3 Subcontractor Costs These costs include the supply and erection costs and include equipment and materials furnished by the local subcontractors such as buildings, field fabricated tanks and vessels, cooling towers, solids storage and handling systems, and water treating systems. These costs also include all installation labor, indirect costs, and overhead and profit of the subcontractors. Lump sum turnkey costs are also included where a single firm provides all of the engineering, material, and construction services required to build a certain area of the plant.

7.6.1.4 Indirect Field Costs These costs include construction management which is the total cost of the construction supervisory staff provided by the management contractor. Included are salaries, subsistence allowances, burdens, benefits, overhead and any international expenses. Costs for other related managing contractor's expenses such as temporary field offices, and all other applicable office services and supplies are included. Also included are costs for specialized heavy lift and heavy haul equipment and cranes to move heavy machinery and equipment.

7.6.1.5 Home Office Costs These costs include engineering, design, and procurement costs; office expenses such as computer costs; reproduction; communication and travel; and office burdens, benefits, overhead costs, and fee.

7.6.1.6 Other Miscellaneous Costs In addition to the aforementioned costs, contractor's fee, costs for removal of obstructions, above or below ground, removal of contaminated soil or toxic materials or site remediation, soil compaction/dewatering, construction permitting, land use, environmental permitting, property taxes during construction, project development such as legal and financial consultants, office and laboratory equipment, and plant mobile operating equipment.

As summarized in Table 7.3, The Association for the Advancement of Cost Engineering (AACE) International identifies five classes of cost estimates designated as class 1, 2, 3, 4, and 5 (Dysert, 2001). A class 1 estimate is associated with the highest level of project definition or maturity while a class 5 estimate is associated with the lowest. The characteristics used to distinguish one class of estimate from another are the degree of project definition, end use of the estimate, estimating methodology, estimating accuracy, and effort to prepare the estimate.

7.6.1.7 Capacity-Factored Estimate This type of estimate (class 5 estimate) is developed during the feasibility stages to decide between alternative designs or plant sizes. This early screening method may be used to estimate the cost of the entire system or subsystem or individual equipment items. Capacity factored estimates are based on multiplying the cost of a system, subsystem, or equipment for which the costs are known from previous work by the ratio of the new unit's capacity to the capacity of the known unit. Capacity ratios are adjusted by an exponent chosen on the basis of the unit type. For example, if cost Cost_1 of a unit with a capacity Capacity_1 (characterized by flow rate of a major stream, or power produced, or volume, or heat transfer rate depending on the type of unit) is known, then the cost Cost_2 of a similar but larger or smaller unit with capacity Capacity_2 is estimated by Equation 7.5:

$$\frac{\text{Cost}_1}{\text{Cost}_2} = \left(\frac{\text{Capacity}_1}{\text{Capacity}_2}\right)^n \tag{7.5}$$

The exponent n is determined by correlating cost with capacity data, and ranges typically from 0.6 to 0.85 but for some types of units, values as low as 0.37 and values as high as 1.2 have been reported (Peters and Timmerhaus, 1968). This method loses considerable accuracy when the ratio of the two capacities is less than 0.5 or greater than 2. Any costs that are not applicable to the new unit should be deducted from the known reference unit cost. Similarly, any additional costs that are required for the new unit but were not included in the known reference unit cost should be added. The costs may then be adjusted for differences in location by multiplying with a factor that accounts for differences such as the labor efficiency or shipping cost, market conditions, and time frame by applying an escalation factor derived from published cost indices such as the Engineering News-Record Construction Index or the Chemical Engineering's Plant Cost Index. A cost index shows the cost at a given time relative to a certain base time.

7.6.1.8 Parametric Cost Estimation A parametric cost model (may also be used in class 5 estimates) is useful in preparing early conceptual estimates when there are limited technical data to provide a basis for using more detailed estimating methods. It is

TABLE 7.3 Types of cost estimates

Estimate class	Project definition (% of complete definition)	Purpose of estimate	Estimating method	Accuracy range	Preparation effort (index relative to project cost)
5	0–2	Screening	Capacity factored, parametric models	Low = −20 to −50%; High = 30 to 100%	1
4	1–15	Feasibility	Equipment factored, parametric models	Low = −15 to −30%; High = 20 to 50%	2 to 4
3	10–40	Budget authorization or cost control	Semi-detailed unit cost estimation with estimated line items	Low = −10 to −20%; High = 10 to 30%	3 to 10
2	30–70	Control of bid or tender	Detailed unit cost estimation with assumed material takeoff	Low = −5 to −15%; High = 5 to 20%	4 to 20
1	50–100	Check estimate, bid or tender	Semi-detailed unit cost estimation with detailed material takeoff	Low = −3 to −10%; High = 3 to 15%	5 to 100

an empirical cost relationship involving several independent variables that correlate the capacity of a system and its cost obtained by a regression analysis. Erratic or outlier data points are removed and the best-fit mathematical relationship is tested to ensure that it properly explains the data, as well as it makes physical sense. The algorithm may be a linear relationship such as

$$\text{Cost} = C_0 + C_1\,(\text{Capacity}_1) + C_2(\text{Capacity}_2) + \cdots$$

Or a nonlinear relationship such as

$$\text{Cost} = C_0 + C_1\,(\text{Capacity}_1)^{n1} + C_2(\text{Capacity}_2)^{n2} + \cdots$$

where Capacity_1 and Capacity_2 are input variables (physical and functional characteristics) defining the capacity of a unit, while the constants C_0, C_1, and C_2, and the exponents $n1$ and $n2$ being derived from a regression analysis.

7.6.1.9 Equipment-Factored Estimates This type of estimate (class 4 estimate) is also developed during the feasibility stages when engineering is approximately 1–15% complete but unlike the previous method, is based upon the specific process design resulting in much higher accuracy. It requires the development of process flow diagrams and then equipment lists. Equipment modeled estimates for each process equipment (e.g., heat engine, fuel cell, reactor, heat exchanger, compressor, pump, column, vapor liquid separator, and storage tank) in a subsystem are developed based on equipment type, size, design temperature and pressure, and metallurgy. These "bare" equipment costs may be developed by designing the equipment or supplied by the manufacturer in which case specifications have to be provided, or estimated by the capacity factored method using Equation 7.5. Adjustment to the cost has to be made if the metallurgy and operating conditions are different.

The equipment-factored estimation method relies on the existence of a ratio between the cost or capacity of an equipment item and costs for the associated nonequipment items such as piping and insulation, instrumentation and control system, electrical (e.g., wiring), civil and structural (e.g., foundations, equipment structures), as well as the engineering, procurement and construction required when building a plant. Systems analysis computer programs such as Aspen Plus and Pro/II can generate the plant costs using this method. Since the material cost of equipment can represent 20–40% of the total project cost for process plants, it is extremely important to estimate the equipment costs as accurately as possible. The equipment installation factors may have to be adjusted for specific project or process conditions. For example, if the plot layout requires placement of equipment much closer to each other, adjustments for the shorter piping, wiring, and conduit may have to be made. Adjustment to seismic zone may also be required for cost of the foundations and structural steel. After developing equipment-factored costs, costs that are not covered by the equipment installation factors should be added. Costs such as the indirect field costs and home-office cost are also sometimes accounted for by applying factors to the equipment costs using data derived from other similar installations.

7.6.1.10 Detailed-Cost Estimation Detailed estimates are typically prepared to support final budget authorization, contractor bid tenders, cost control during project execution and change orders, and can be quite accurate (class 3 through class 1). However, a significant detail on the project design is required. At the very least, (1) PFDs and utility flow drawings (utility flow diagrams provide similar information as PFDs for utility systems such as cooling, process, potable and fire water, plant and instrument air, inert gases, steam, and BFW), (2) piping and instrumentation diagrams or P&IDs (showing interconnection of process equipment and controls in more details than the PFD including items such as pipe sizes and fittings, block valves, safety relief valves, control loops, equipment design conditions, and support equipment), (3) equipment specifications (which define the performance requirements of the equipment including stream data along with the required physical and thermodynamic properties, design conditions, capacity, materials of construction, fabrication methods and procedures, test and inspection requirements), (4) motor lists, (5) electrical diagrams, (6) piping isometrics (a three-dimensional representation of the piping and fittings shown on a preprinted plane sheet with lines of equilateral triangles forming 60°, with symbols modified to adapt to the isometric grid), (7) equipment and piping layout drawings, and (8) overall plot plan are required. In the development of the layout of the overall plot plan, the process units should be located logically in terms of piping networks, safety, and allowance for any future expansion.

7.6.2 Economic Analysis

In developing the process economics, various other costs in addition to the plant, land and feedstock costs have to be accounted for as described in the following while taking credit for any co- or by-product.

7.6.2.1 Organization and Start-Up Costs Organization and start-up costs are intended to cover pre-project administrative costs and operator training, equipment checkout, changes in plant equipment, unplanned maintenance, and inefficient use of feedstock and other materials during plant commissioning and start-up. These are typically taken as the sum of 1 month of fixed operating and maintenance costs, 2 months of consumable costs excluding fuel cost (calculated at full capacity), 1 month of fuel inefficiency (25% excess fuel at full capacity) and 2% of the total plant cost to account for changes that may be required.

7.6.2.2 Working Capital Working capital is typically taken as the sum of 2 months of consumable costs excluding fuel cost (calculated at full capacity), 2 months supply of fuel (in the case it is stored onsite such as coal or biomass) at full capacity, 3 months of operating and maintenance labor costs, spare parts inventory at 0.5% of the total plant cost, and a contingency of 25% of the total of the above four items.

7.6.2.3 Operating and Maintenance Costs The operating and maintenance costs are divided into fixed and variable cost components.

The fixed operating costs are essentially independent of the plant capacity factor and are composed of operating labor, maintenance costs, and overhead charges. Operating labor accounts for the operating personnel per shift required for the plant including payroll burdens. Fixed maintenance costs are typically estimated in systems studies as a percentage of the installed plant section cost since it varies with the type of unit (about 2–5% of the installed plant section cost). The unit by unit annual maintenance cost factors are divided into fixed and variable maintenance costs (typically 65 and 35%, respectively), and the fixed maintenance cost is further divided into labor and materials (typically 40 and 60%, respectively). Overhead charge is a charge for administrative and support labor, which is typically taken as 30% of the operating and maintenance labor.

The variable operating costs vary with the plant load and include raw water acquisition, its treating and pumping costs, catalyst and chemicals and other consumables, disposal of plant wastes, and variable maintenance labor and materials (typically taken as 35% of total maintenance cost of the plant and is divided into labor and materials typically as 40 and 60%, respectively). Feedstock costs also vary with plant load but are typically reported separately.

7.6.2.4 ***Cost of Product*** When the product is electricity, a levelized cost of electricity is often used in systems analysis to compare alternative designs and assess their relative economic sustainability. The levelized cost of electricity is defined as the constant price per unit of electricity (i.e., per kWh) that causes the investment to just break even, that is, earn a present discounted value equal to zero. It accounts for installation costs, financing costs, taxes, operating and maintenance costs, feedstock costs, revenue from sale of co- or by-products, salvage value, any tax incentives, and net quantity of electricity the system generates over its lifetime. The methodology for calculating the levelized cost of electricity is described in the Technical Assessment Guide (TAG) published by the Electric Power Research Institute (Ramachandran, 1993).

In cases where the value of the coproduct is to be assessed, one approach would be to develop a base case on a consistent basis to produce electricity without that coproduct and establish the levelized cost of electricity. The cost of producing the coproduct may then be determined (say, by trial and error) such that the resulting levelized cost of electricity for this coproduction case is the same as that given by the base case.

7.6.2.5 ***Simple Payback Period Calculation*** A simple payback period calculation is often used to assist in the evolution of a system configuration. It may be used in selection of a type of equipment or subsystem from among competing alternatives, or deciding whether to include a certain unit operation or process within a system design. The payback period is the time it takes to recoup the initial investment and in the case where the annual return on investment is uniform, may be calculated by

$$\text{Payback time period, } n_{\text{years}} = \frac{\text{Initial investment}}{\text{Annual return on investment}}$$

If the annual returns on investment are not uniform, then their cumulative amount over that period of time that equals the initial investment is the payback period. An improvement in the payback period calculation would be to account for the time value of money in which case it is called discounted payback period. If the investment is depreciable, then the annual depreciation is taken credit for by adding it in the denominator of the above expression for the payback period.

Lower the payback period, the more attractive the alternate is. Payback period may also be used to assist in deciding whether to install additional equipment. When deciding whether to include a certain unit operation or process within a system design, an absolute payback period is required and typically $n_{\text{years}} = 3\text{–}5$ years for projects in the United States with the annual return on investment on a pretax basis. The payback period may also be calculated on a differential basis as illustrated in Example 7.3.

Example 7.3. A process stream is to be cooled from a temperature of 110–60°C. Option A consists of using a water cooled exchanger while option B consists of using an air cooled exchanger. The installed cost (including both direct and indirect costs) for the exchanger of option A is $250,000 and that of option B is $450,000. The annualized costs for the supply of the cooling water for option A are $100,000/year while those for the electricity to operate the air fans for option B are $15,000/year. Using a payback period calculation, recommend which of the two alternates should be selected.

Solution

Additional initial investing for option B over option A = $450,000 – $250,000 = $200,000.

Savings in annual costs for option B over option A = $100,000/year – $20,000/year = $80,000/year.

$$n_{\text{years}} = \frac{\$200,000}{\$80,000/\text{year}} = 2.5 \text{ years.}$$

The above calculated payback period of only 2.5 years indicates that option B of using an air cooled exchanger is advantageous.

7.7 LIFE CYCLE ASSESSMENT

As discussed in Chapter 1, system analysis that is conducted on purely an economic sustainability basis, that is, without assigning a monetary value to the environmental impact is not complete. Life cycle assessment (LCA) quantifies how much energy and raw materials are required, and how much waste is generated at each stage of the product's life, from the mining of the raw materials used in its production, product's distribution, use, and possible reuse or recycling or its eventual disposal in the case of a material product. Thus, it is a cradle to grave analysis.

The boundary of the studied system should be clearly specified to identify the various streams entering and leaving the system, and the three phases consisting of construction, operation, and dismantling of the system should be considered. Construction includes assembly of the equipment required in the construction of the plant while taking into account raw materials, production processes, transportation, and on-site work. Data on the physical components of the plant such as weights, type of materials, recyclability, and the processes involved in the manufacturing of each item is also required. System operation should include the delivery of materials to the plant for operation and maintenance. Eventual dismantling of the plant at the end of its life should consider transportation and recycling or disposal of materials. For example, significant portions of materials such as steel, cast iron, copper, and aluminum may be recycled while concrete and asphalt may be crushed and reused as low quality landfill. Rubber and steel in reinforced concrete might be landfilled while plastic could possibly be recycled or used for its fuel value. The commercially available software program SimaPro includes a significant database required for performing an LCA.

The analysis of the compiled data sometimes requires judgment in addition to scientific interpretation. For example, if one process produces more SO_x but less NO_x than another, the two types of pollutants will have to be weighted such that they may be compared on "an apples to apples" basis. One possible approach is to aggregate the various impacts into categories; for example, the possible impact on the O_3 layer, or on acid rain. Thus, a standardization of the method for interpreting the collected data would be useful.

REFERENCES

Black J. Cost and performance baseline for fossil energy plants, volume 1: Bituminous coal and natural gas to electricity. U.S. Department of Energy Report; 2010. Report nr. DOE/NETL-2010/1397. Morgantown, WV: U. S. DoE.

Chilton TH, Drew TB, Jebens RH. Heat transfer coefficients in agitated vessels. Ind Eng Chem 1944;36(6):510–516.

Dysert LR. Sharpen your cost estimating skills. Chem Eng 2001;108(11):70–81.

Emun F, Gadalla M, Majozi T, Boer D. Integrated gasification combined cycle (IGCC) process simulation and optimization. Comput Chem Eng 2010;34(3):331–338.

Franco A, Russo A. Combined cycle plant efficiency increase based on the optimization of the heat recovery steam generator operating parameters. Int J Therm Sci 2002;41(9):843–859.

Gundersen T, Naess L. The synthesis of cost optimal heat exchanger networks: An industrial review of the state of the art. Heat Recovery Syst CHP 1990;10(4):301–328.

Hutchins TE, Metghalchi M. Energy and exergy analyses of the pulse detonation engine. J Eng Gas Turbines Power 2003;125(4):1075–1080.

Kern DQ. *Process Heat Transfer*. New York: McGraw-Hill; 1950.

Linnhoff B. Pinch analysis: A state-of-the-art overview: Techno-economic analysis. Chem Eng Res Des 1993;71(5):503–522.

Linnhoff B, Polley GT, Sahdev V. General process improvements through pinch technology. Chem Eng Prog 1988;84(6):51–58.

Lombardi L. Life cycle assessment (LCA) and exergetic life cycle assessment (ELCA) of a semi-closed gas turbine cycle with CO_2 chemical absorption. Energy Convers Manag 2001;42(1):101–114.

Peters MS, Timmerhaus KD. *Plant Design and Economics for Chemical Engineers*. 4th ed. New York: McGraw-Hill; 1968.

Ramachandran G. EPRI TAG—Technical Assessment Guide (electric supply). Electric Power Research Institute Report; June 1993. Report nr. TR-102275, Volume 1, Revision 7. Palo Alto, CA: Electric Power Research Institute.

Siddique N. *Intelligent Control: A Hybrid Approach Based on Fuzzy Logic, Neural Networks and Genetic Algorithms (Studies in Computational Intelligence)*. Cham: Springer; 2013.

Staine F, Favrat D. Energy integration of industrial processes based on the pinch analysis method extended to include exergy factors. Appl Therm Eng 1996;16(6):497–507.

Tangirala VE, Rasheed A, Dean AJ. Performance of a pulse detonation combustor-based hybrid engine. GT2007-28056. Proceedings of the ASME IGTI Turbo-Expo Conference; May 2007; Montreal, Canada.

8

RANKINE CYCLE SYSTEMS

The Rankine cycle using steam as the working fluid generates about 90% of electric power on a worldwide basis. By far, the majority of such power plants have been fueled by coal. Present-day nuclear power plants also utilize steam-based Rankine cycles; so do most geothermal and solar thermal power plants. The cycle is named after Professor William John Macquorn Rankine of Glasgow University. Many of the impracticalities of the Carnot cycle, which was described in Chapter 2, may be overcome with a Rankine cycle. For example, it is difficult to build an engine utilizing a single phase working fluid with isothermal expansion (before the adiabatic expansion step). Heat transfer required during the isothermal expansion would require a piston-type engine with present-day technology and the frictional losses in such machines are quite high limiting the net work produced. The basic Rankine cycle, however, most closely approaches the Carnot cycle with respect to the isothermal heat addition and rejection steps. This is made possible by utilizing an appropriate single-component vapor as the working fluid that changes phase during these steps, evaporating during the heat addition step and condensing during the heat rejection step. The working fluid temperatures for the heat addition and rejection steps are set by the source and sink temperatures while utilizing reasonable temperature differences between the source or sink and the working fluid to keep the size of the heat transfer equipment reasonable. The appropriateness of choosing a given working fluid then depends on the corresponding evaporation and condensation pressures being reasonable. Other factors that should be considered in choosing the working fluid are the

Sustainable Energy Conversion for Electricity and Coproducts: Principles, Technologies, and Equipment, First Edition. Ashok Rao.

latent heats of evaporation and condensation, which dictate the amount of fluid being circulated within the cycle as well as those based on safety considerations such as flammability and toxicity. By far, steam has been the working fluid of choice for the Rankine cycle in moderate to high temperature applications (>150°C) due to the aforementioned attributes while organic fluids such as propane and butane have been used in lower temperature applications.

8.1 BASIC RANKINE CYCLE

Figure 8.1 shows the idealized Rankine cycle, while the temperature-entropy (T-S) diagram of this cycle is presented in Figure 8.2. The cycle consists of the following steps:

1. Isobaric Heat Addition: Subcooled liquid pressurized by the feed pump is preheated to saturation temperature and then completely evaporated to form saturated vapor.
2. Adiabatic Expansion: The pressurized vapor formed in the preceding step is expanded adiabatically in an ideal turbine to produce work. Condensation of some of the vapor may occur within the turbine as work is extracted from the fluid.
3. Isobaric Heat Rejection: The vapor–liquid mixture leaving the expansion turbine is then condensed while rejecting heat to the heat sink. Note that in order to cause this required phase change, the pressure of the fluid should correspond to the saturated vapor pressure at a temperature set by the heat sink temperature (reasonably above the heat sink temperature to limit the size of the heat exchanger).

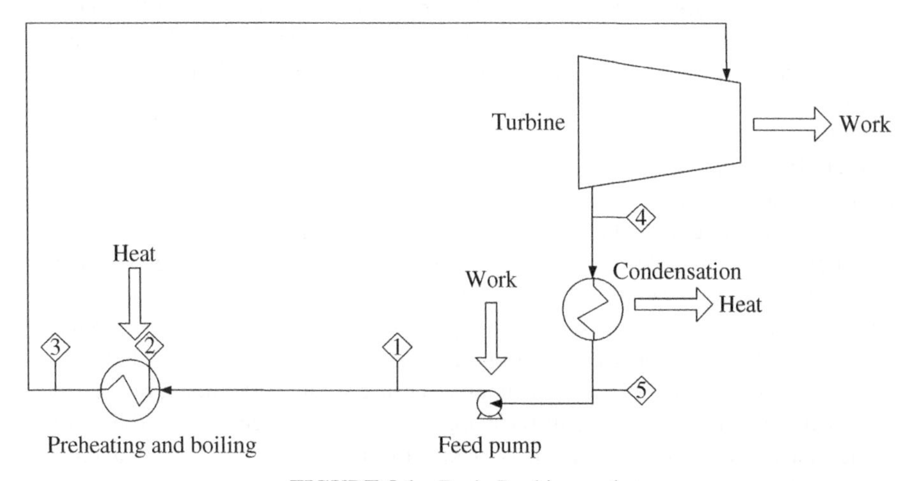

FIGURE 8.1 Basic Rankine cycle

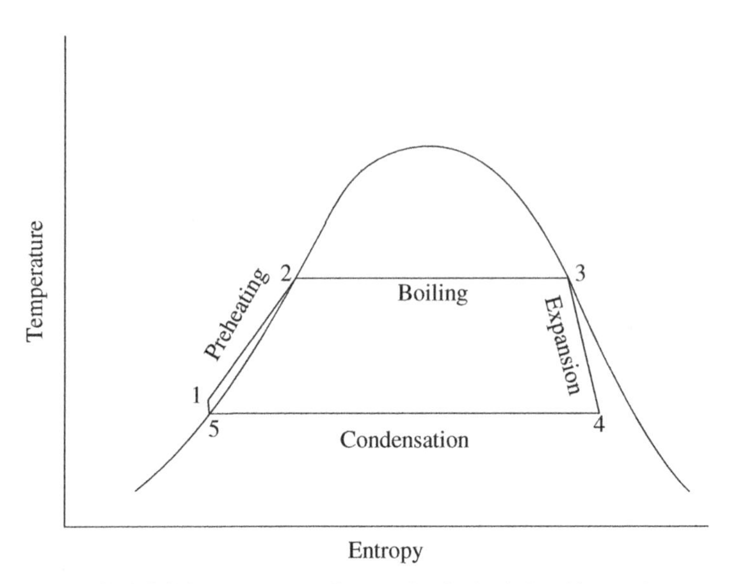

FIGURE 8.2 *S* versus *T* diagram for the basic Rankine cycle

Thus from a thermodynamic standpoint, the outlet pressure of the expansion turbine of the previous step is limited by the temperature of the heat sink.

4. Isentropic Pressurization: The pressure of the liquid formed in the preceding step ("condensate") is then raised in the feed pump (ideal machine) and supplied to the evaporator to complete the cycle.

Note that in the analysis of a real-world (nonidealized) Rankine cycle, the pressure drop through the equipment, piping, valves, and so on., and the inefficiencies of the turbine and the pump should be taken into account, although the pump power usage is quite small as compared to that produced by the turbine and has a small effect on the overall first law efficiency of the cycle. Another constraint that should be taken into consideration is the fraction of vapor condensed within the turbine. As pointed out in a previous chapter, large axial turbines must be operated such that the exhaust steam does not contain more than 10–13% of liquid since condensate droplets could seriously erode the nozzles and blades at the high velocities and proper selection of the inlet steam pressure and temperature are required, for example, by superheating the saturated vapor leaving the evaporator and before entering the expansion turbine. Special moisture removal stages may have to be incorporated in the design when the steam superheat temperature is limited, say by the temperature of the heat source. These issues are especially of concern in applications involving large steam turbines where the isentropic efficiency of the turbine is high (in excess of 80%) such that the work extracted from the fluid is high enough to cause significant amount of condensation within the turbine. For small-scale applications with much lower turbine efficiencies, the degree of superheating may be kept to a minimum (if not totally avoided to simplify the required heat exchange equipment) and reduce the design temperatures of the superheater, the piping, the valves as well as the turbine. Modification to the basic Rankine cycle by the addition of the superheater is discussed next.

8.2 ADDITION OF SUPERHEATING

A Rankine cycle incorporating superheating of the working fluid before it enters the expansion turbine is depicted in Figure 8.3, while the entropy versus temperature diagram of this cycle is presented in Figure 8.4. In addition to avoiding problems described earlier, superheating of the saturated vapor can also increase the cycle efficiency when the heat source temperature is much higher than the saturation temperature of the working fluid corresponding to maintaining a reasonable evaporator pressure. The irreversibility associated with heat transfer between the heat source and the working fluid is reduced for a given amount of heat input to the cycle and this reduction manifests itself as a lower flow rate for the working fluid within the cycle and thus reducing the amount of heat rejected in the condenser, and an increase in power developed by the turbine. Furthermore, when a material stream such as the products of combustion from burning a fuel or a hot process stream not undergoing phase change is the heat source, its temperature does not remain constant as heat is transferred to the working fluid of the cycle while the heat absorption in the evaporator occurs essentially at constant temperature (the pressure drop on the working fluid side within the evaporator being quite small during the phase change). This causes a divergence in the heat release and heat absorption lines during the evaporation step as seen in Figure 8.5 where the cumulative heat exchanged between the two fluids ($\Sigma\dot{Q}$) is plotted against temperature (T) of the two fluids for the basic Rankine cycle. The superheated steam cycle is also presented in this figure and it may be seen that by incorporating superheating, this heat transfer irreversibility is reduced. Economic

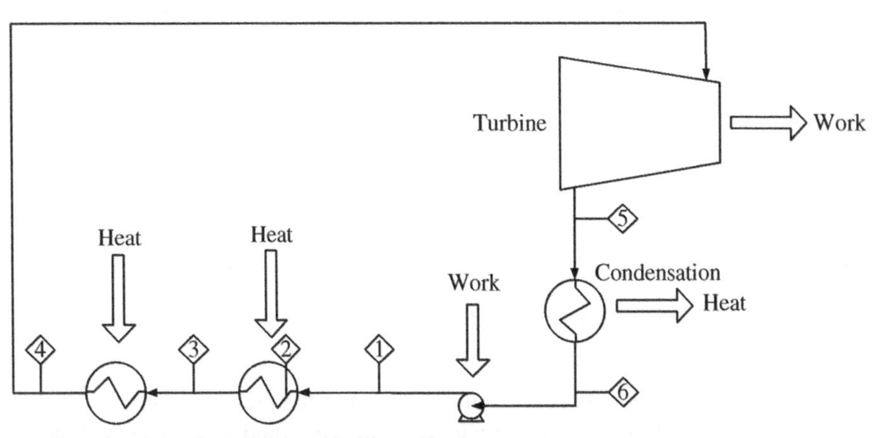

FIGURE 8.3 Rankine cycle with superheating

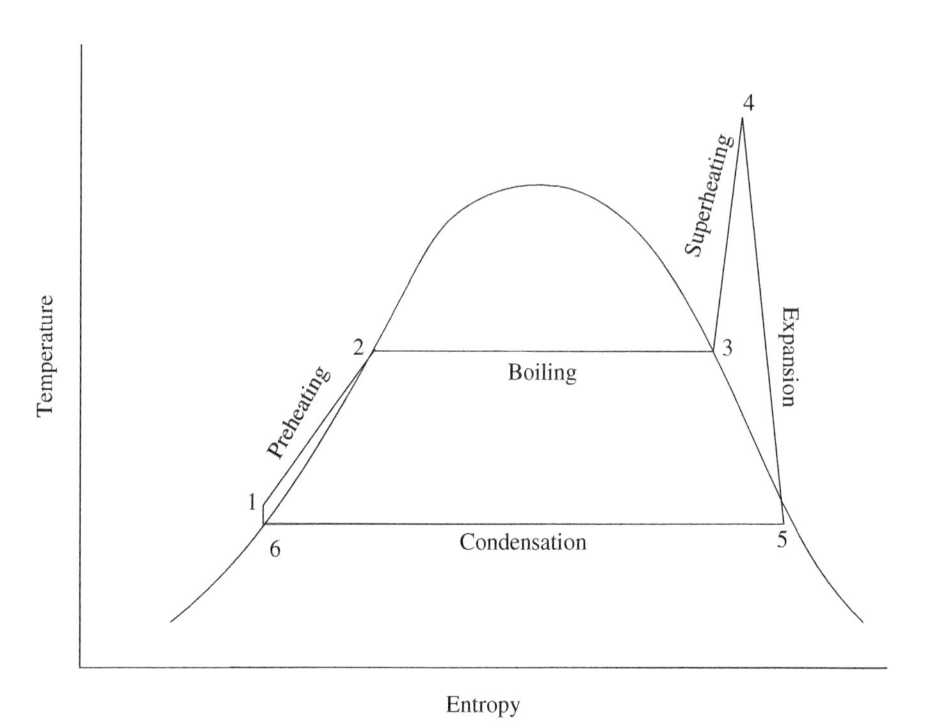

FIGURE 8.4 *S* versus *T* diagram for Rankine cycle with superheating

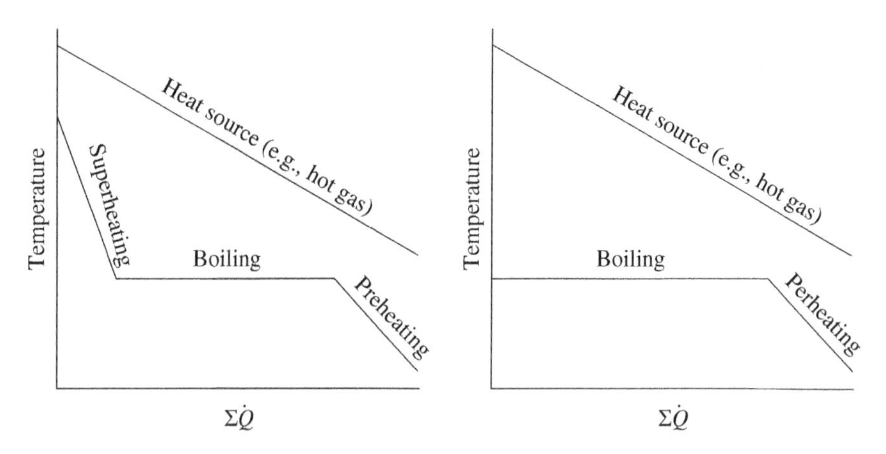

FIGURE 8.5 $\Sigma\dot{Q}$ versus T diagram for Rankine cycle with and without superheating

justification of including superheat requires increasing the pressure over the previously described basic Rankine cycle at which the working fluid evaporates such that a significant fraction of the thermal energy added in the superheater is converted to work in the turbine and does not end up leaving in the turbine exhaust.

An expression for the Carnot cycle efficiency when the heat source and the heat sink temperatures do not remain constant but change linearly as heat is being transferred may be derived as follows and should be used in place of the classical expression given by Equation 2.30. Consider a heat source such as a flue gas being cooled from a temperature T_{Gin} to a temperate T_{Gout} without phase change and a heat sink consisting of cooling water being supplied at a temperature T_{Win} and returned at a temperature T_{Wout}. Since entropy is conserved for a reversible process, that is, entropy in = entropy out, we have:

$$\frac{d\dot{Q}_{in}}{T_G} = \frac{d\dot{Q}_{out}}{T_W}$$

where $d\dot{Q}_{in}$ is the differential rate of heat transferred in from the flue gas at a temperature T_G, and $d\dot{Q}_{out}$ is the differential rate of heat transferred out to the cooling water at a temperature T_W. If $\dot{m}_G$ and $\dot{m}_W$ are the mass flow rates of the flue gas and the cooling water, and when the specific heats for the flue gas and cooling water ($C_{p,G}$ and $C_{p,W}$ on a mass basis) remain constant within these temperature ranges, we have:

$$\dot{m}_G\, C_{p,G}\, dT_G\,/T_G = \dot{m}_W\, C_{p,W}\, dT_W/T_W$$

Integration yields:

$$\dot{m}_G\, C_{p,G} \ln(T_{G,in} - T_{G,out}) = \dot{m}_W\, C_{p,W} \ln(T_{W,out} - T_{W,in}) \tag{8.1}$$

From the first law, we have:

$$\dot{m}_G\, C_{p,G}\,(T_{G,in} - T_{G,out}) = \dot{m}_W\, C_{p,W}\,(T_{W,out} - T_{W,in}) + \dot{W}$$

The Carnot cycle efficiency is then:

$$\eta_{Carnot} = \frac{\dot{W}}{\dot{Q}_{in}} = \frac{[\dot{m}_G\, C_{p,G}\,(T_{G,in} - T_{G,out}) - \dot{m}_W\, C_{p,W}\,(T_{W,out} - T_{W,in})]}{[\dot{m}_G\, C_{p,G}\,(T_{G,in} - T_{G,out})]}$$

$$= 1 - \frac{[\dot{m}_W\, C_{p,W}\,(T_{W,out} - T_{W,in})]}{[\dot{m}_G\, C_{p,G}\,(T_{G,in} - T_{G,out})]} \tag{8.2}$$

Substituting Equation 8.1 into Equation 8.2:

$$\eta_{Carnot} = 1 - \left[\frac{(T_{W,ut} - T_{W,in})}{\ln(T_{W,out} - T_{W,in})}\right] \left[\frac{(T_{G,in} - T_{G,out})}{\ln(T_{G,in} - T_{G,out})}\right]$$

$$= 1 - \frac{(\Delta T_W)_{ln}}{(\Delta T_G)_{ln}} \tag{8.3}$$

This expression given by Equation 8.3 shows that it is the logarithmic mean temperatures that should be used when the heat source and the heat sink temperatures do not remain constant but change linearly as heat is being transferred.

8.3 ADDITION OF REHEAT

Reheating is a further step in the same direction as the addition of superheating and is typically justified in large central station type power plant applications. A Rankine cycle incorporating reheating in addition to superheating of the working fluid is depicted in Figure 8.6, while the entropy versus temperature diagram of this cycle is presented in Figure 8.7. As depicted, the steam after expansion to an intermediate pressure is reheated before it is fed back to the turbine to complete the expansion process. The corresponding $\Sigma\dot{Q}$ versus T diagram of this cycle is presented in Figure 8.8. Once again, economic justification of including reheat may require increasing the pressure over the previously described superheated Rankine cycle at which the working fluid evaporates such that a significant fraction of the thermal energy added in the reheater is converted to work in the turbine. A more efficient turbine converts a greater fraction of the thermal energy to work and the reheat cycle may also be the economic choice in applications consisting of large turbines which tend to be more efficient. Geometrical constraints of a small steam turbine may also preclude the incorporation of the reheat cycle.

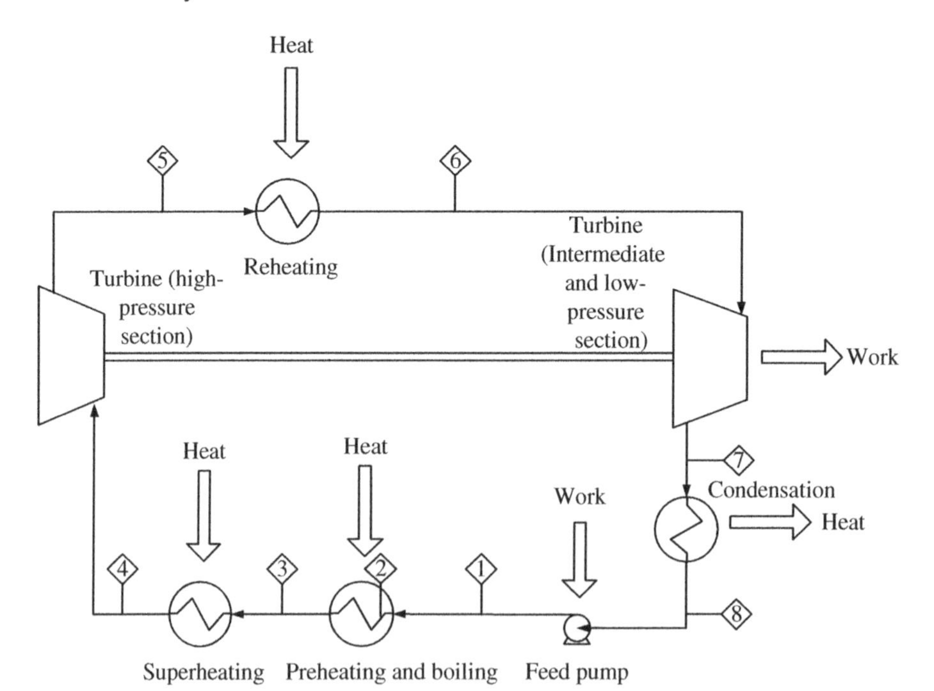

FIGURE 8.6 Rankine cycle with superheating and reheating

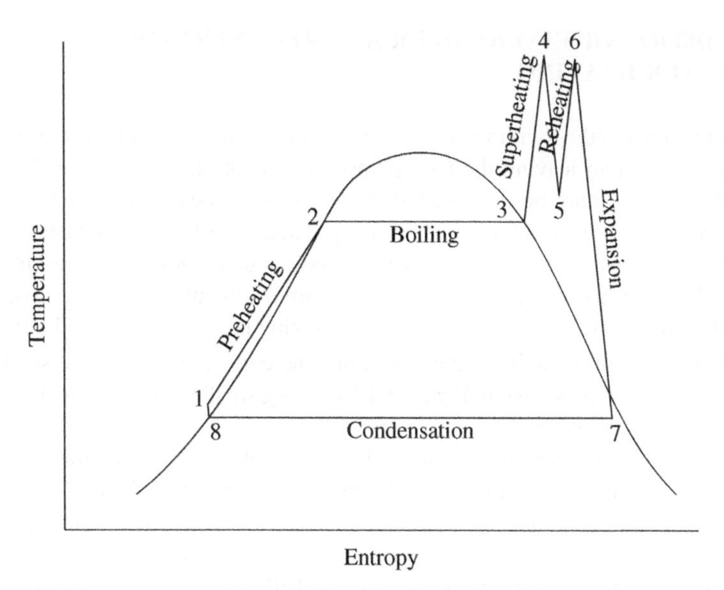

FIGURE 8.7 *S* versus *T* diagram for Rankine cycle with superheating and reheating

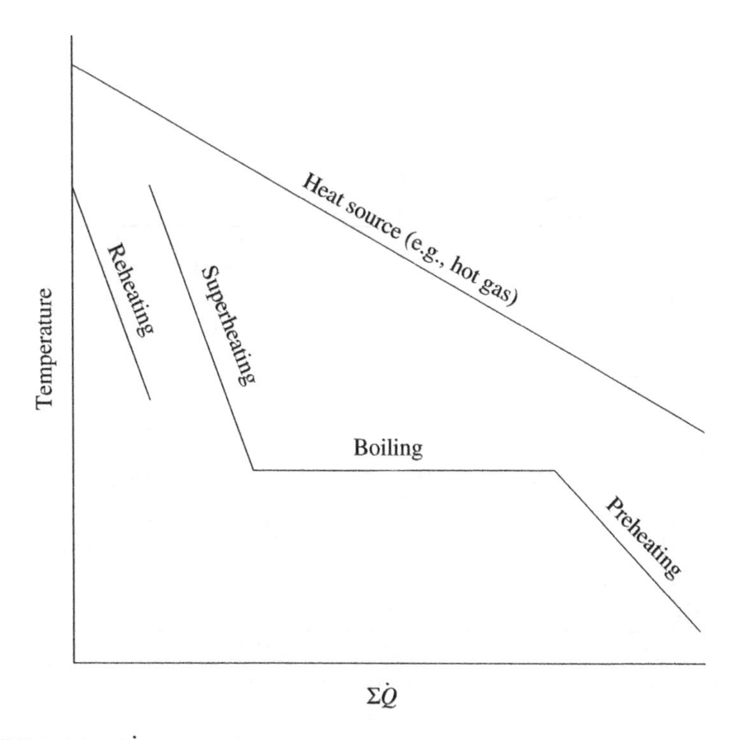

FIGURE 8.8 $\Sigma\dot{Q}$ versus *T* diagram for Rankine cycle with superheating and reheating

8.4 ADDITION OF ECONOMIZER AND REGENERATIVE FEEDWATER HEATING

The various forms of the Rankine cycle discussed in the previous sections consisted of introducing the liquid leaving the feed pump directly into the evaporator. The efficiency of the cycle may be increased if the liquid is preheated and then provided to the evaporator since more liquid may be evaporated without using that heat for preheating the liquid water. In the case where the heat source consists of a hot material stream such as flue gas (or a process stream), the preheating of the liquid may be done in an "economizer" depicted in Figure 8.9 by utilizing heat remaining in the flue gas (or the process stream) leaving the evaporator. The corresponding $\Sigma\dot{Q}$ versus T diagram for this cycle is presented in Figure 8.10 showing smaller temperature difference between heat source and boiling.

In some cases, however, the amount of heat available for the economizer may be limited depending on constraints placed by the temperature limit for the stream providing the heat and leaving the economizer such as its acid dew point, water dew point, and the associated plume or buoyancy requirements if entering a flue gas stack. To avoid using expensive tube materials such as Teflon®-coated tubes in the economizer, the tube surface temperatures should be kept above the acid dew point of the gas. The limiting acid dew point is typically set by sulfuric acid (H_2SO_4) formed by

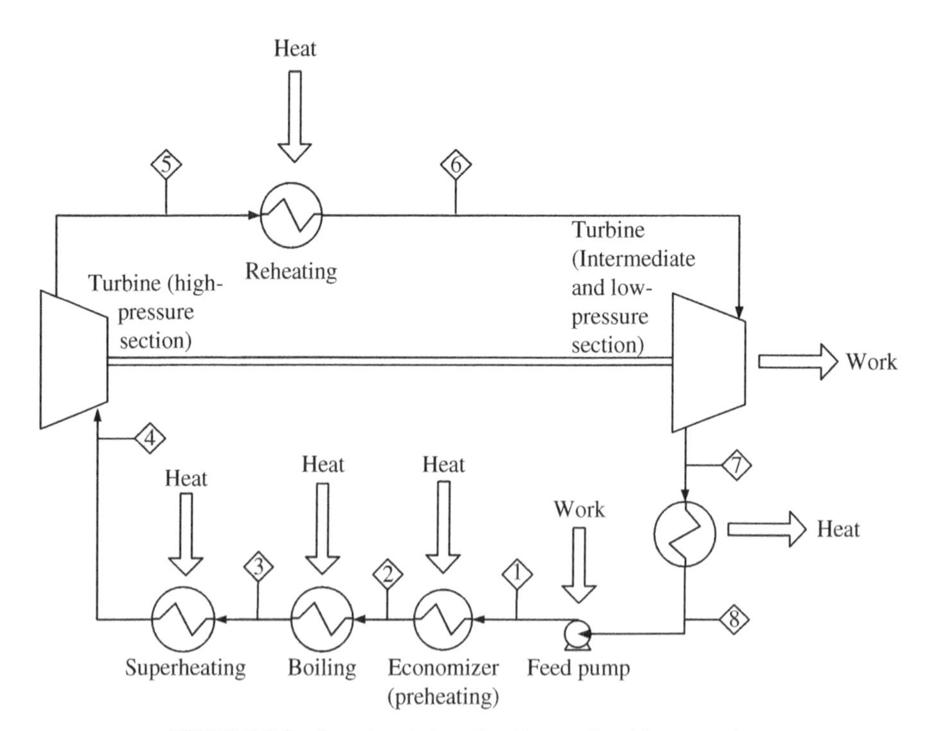

FIGURE 8.9 Superheat/reheat Rankine cycle with economizer

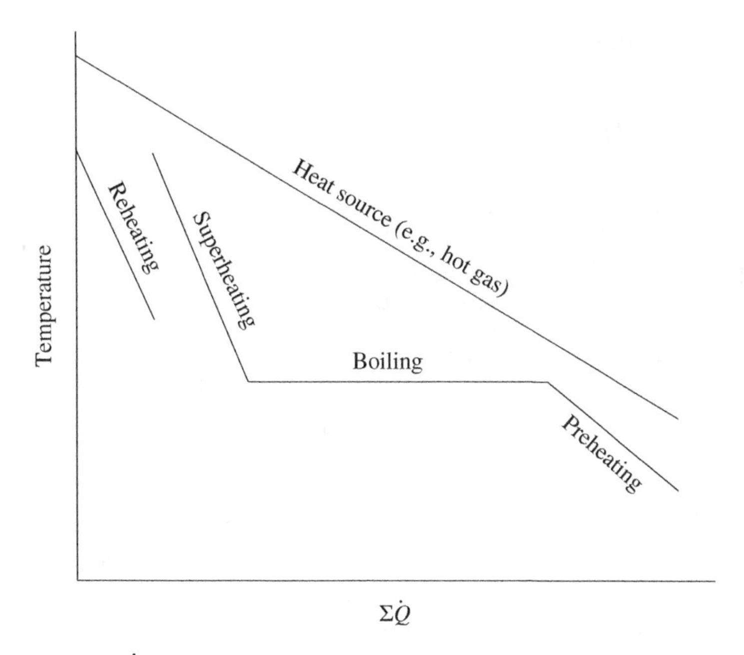

FIGURE 8.10 $\Sigma\dot{Q}$ versus T diagram for superheat/reheat Rankine cycle with economizer

the oxidation of sulfur present in the fuel to SO_3[1] during combustion of the fuel and subsequently the SO_3 combining with water vapor to form the acid. An estimation of the dew point temperature T_{DP} (in K) may be made from the partial pressure P_i (in atmospheres) of each of the species i (H_2O vapor and SO_3) using Equation 8.1 (Pierce, 1977):

$$\frac{1000}{T_{DP}} = 1.7842 + 0.0269 \log P_{H_2O} - 0.1029 \log P_{SO_3} + 0.0329 \log P_{H_2O} \log P_{SO_3} \quad (8.1)$$

Furthermore, in fired boilers, the combustion air is typically preheated against the flue gas to increase the overall system efficiency, which limits the quantity of heat remaining, if any, for the economizer. In all such cases, other means of preheating the liquid before it enters the evaporator may be required. Figure 8.11 shows a thermodynamically efficient way of accomplishing this, by extracting vapor from the turbine at appropriate pressures to provide the heat for preheating the liquid. In utility scale steam boiler plants, a series of regenerative heaters (as many as seven including the deaerator) are typically employed, the steam supplied to each of these heaters being extracted at the appropriate pressure from the turbine. The heat is transferred

[1] Formation of SO_3 varies greatly with boiler design, fuel properties, and operating conditions. Some of it is formed in the convective pass of the boiler and additional amounts may be generated in the selective catalytic reduction (SCR) unit (depending on the type of catalyst employed) when used for reducing NO_x emissions.

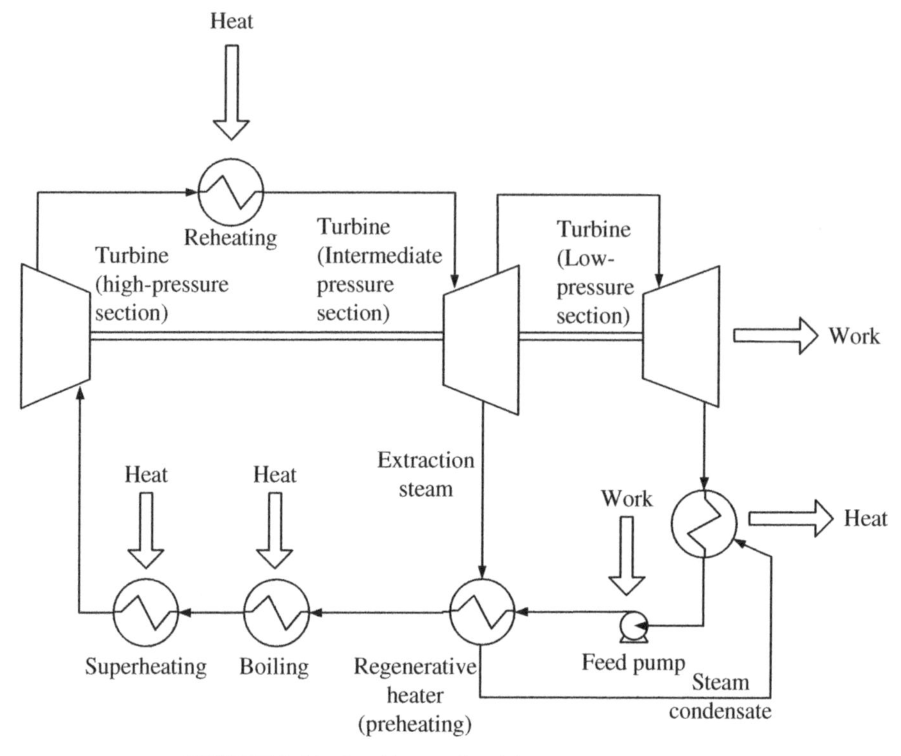

FIGURE 8.11 Rankine cycle with regenerative heating

primarily by the condensation of the extracted vapor and thus the pressure at which it is extracted from the turbine is set by the requirement that its saturation temperature be higher than the temperature of the liquid being preheated (not too high from an efficiency standpoint but not too low from the standpoint of reducing the cost of the regenerative heat exchangers).

In the case of a steam boiler operating at a pressure below the critical pressure of the steam, the liquid (BFW) leaving the economizer or the regenerative heaters enters a steam drum from where the liquid is circulated by either natural convection or by a circulating pump into the evaporator tubes which are exposed to the hot combustion products. The temperature of the water entering the steam drum is kept below its saturation temperature to avoid sudden change in phase of the BFW to steam that may occur due to the pressure drop across the control valve located on the BFW line between the steam drum and the economizer (with regenerative heaters, however, the BFW temperature is always below the saturation temperature). This also avoids boiling within the economizer especially during part-load operation as well as during start-up since this subcool temperature (difference between the saturation temperature and the actual fluid temperature) decreases. The BFW entering the evaporator tubes from the steam drum, however, may be considered to be at the saturation temperature since the subcooled BFW entering the drum heats up to this temperature rapidly by mixing, a portion of the steam exiting the evaporator tubes providing the required heat by condensation.

8.5 SUPERCRITICAL RANKINE CYCLE

Addition of superheating and reheating does reduce the irreversibility associated with heat transfer between the heat source and the working fluid especially in cases where the heat source temperature does not remain constant but superheating and reheating do not minimize this irreversibility since bulk of the heat addition still occurs in the evaporator, essentially isothermally. By operating the aforementioned cycle, above the supercritical pressure of the working fluid, the isothermal phase change step is eliminated and thus the corresponding irreversibility associated with the evaporator of the subcritical cycles. This can be seen in the corresponding $\Sigma\dot{Q}$ versus T diagram for a supercritical Rankine cycle as presented in Figure 8.12.

8.6 THE STEAM CYCLE

The present trend is to select the supercritical cycle for large coal-fired central station power plants (nominally >500 MW net power output). A typical cycle may consist of the following conditions: 24.1 MPa/593°C/593°C (3500 psig/1100°F/1100°F), where

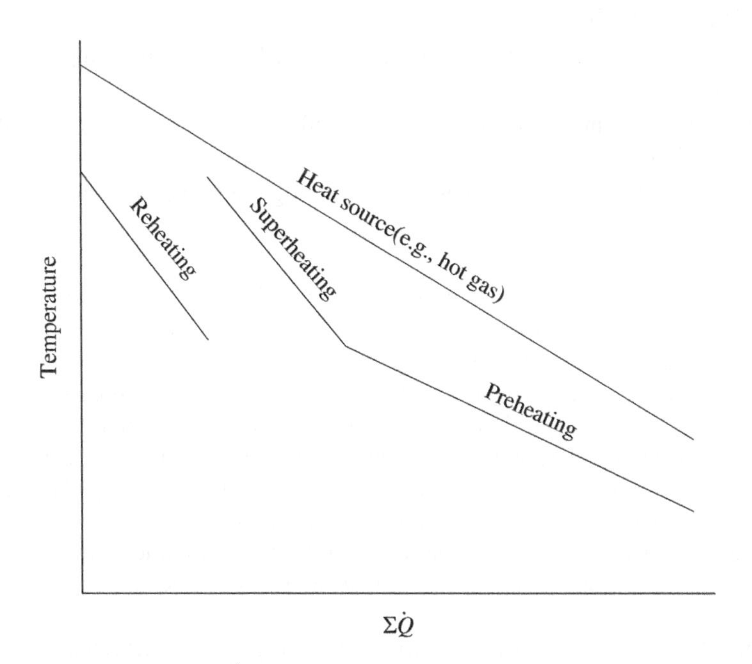

FIGURE 8.12 $\Sigma\dot{Q}$ versus T diagram for supercritical Rankine cycle

the superheated steam admitted to the high-pressure steam turbine is at a pressure of 24.1 MPa and temperature of 593°C and after expansion in the high-pressure turbine is reheated also to 593°C before admitting into the intermediate turbine, while typical conditions for a subcritical steam cycle may be 16.5 MPa/566°C/566°C (2400 psig/1050°F/1050°F). Materials for ultra-supercritical steam cycle conditions such as 34.5 MPa/732°C/760°C (5000 psig/1350°F/1400°F) are being developed for introduction in the near future. With these severe operating conditions, the corrosiveness of the fuel can be an issue especially due to its sulfur content and such applications may be limited to low sulfur content coals such as the Powder River Basin sub-bituminous coal, which is found in the southeast Montana and northeast Wyoming regions of the United States. High sulfur coals such as the bituminous Illinois No. 6 could cause an exponential increase in tube material wastage rates in the higher temperature portions of the superheater and reheater due to corrosion, requiring replacement of pressure parts in approximately 10–15 years and this additional cost may offset the fuel savings and emissions reduction (due to the higher efficiency) of the ultra-supercritical cycle.

Unlike the large multistage turbines used in the earlier applications, single-stage machines are used for very small applications (<2 MW) where the steam inlet conditions are quite modest.

Nonutility (industrial)-scale steam turbines are divided into classes based on the maximum operating steam temperature and pressure at the inlet:

- Class I—maximum steam pressure of 250 psig (17.2 bar) and temperature of 500°F (260°C)
- Class II—maximum steam pressure of 600 psig (41.4 bar) and temperature of 600°F (316°C)
- Class III—maximum steam pressure of 250 psig (17.2 bar) and temperature of 750°F (399°C)
- Class IV—maximum steam pressure of 850 psig (58.6 bar) and temperature of 750°F (399°C)

The cost of the turbine which depends on the metallurgy required varies with the steam cycle conditions and thus the class. More significant breakpoints in cost occur at 750°F (399°C), and especially at 950°F (510°C), in which case, X-ray quality castings are required for the blades which make the turbine cost substantially higher. Systems analysis is required to evaluate the trade-off between efficiency and capital cost to select the suitable steam cycle conditions. It should be borne in mind, however, that higher efficiency means typically lower emissions per net power generated (especially those of CO_2 emissions) and this factor should be accounted for in the systems analysis by assigning a monetary value to emissions. Steam turbines are also offered for admitting saturated steam at pressures as low as 35 psig (3.4 bar) for the smaller-scale applications.

Example 8.1. A turbine is supplied with 100,000 kg/h of superheated steam at 42 bar. The exhaust is at 0.06 bar. The superheat temperature in one case is at

400°C and another case at 300°C. Assume the turbine has an isentropic efficiency of 80% for both cases and the directly connected generator has an efficiency of 95%. Ignore mechanical losses.

Determine the amount of moisture in the turbine exhaust and the electric power generated for the two superheat temperature cases.

Solution

The first step is to read-off the specific enthalpy $\Delta h_{\mathrm{in}}^{\mathrm{V}}$ and entropy $\Delta s_{\mathrm{in}}^{\mathrm{V}}$ of steam at the inlet conditions from steam tables (e.g., Keenan et al., 1969) at the given conditions of pressure and temperature. This entropy corresponding to the inlet conditions is found to be lower than the entropy $\Delta s_{\mathrm{out}}^{\mathrm{V}}$ corresponding to saturated steam at the outlet pressure read-off from the steam tables. This indicates that the exhaust has to be a two-phase fluid consisting of both vapor and liquid. So the next step is to read-off the specific enthalpy of saturated steam and water $\Delta h_{\mathrm{out}}^{\mathrm{V}}$ and $\Delta h_{\mathrm{out}}^{\mathrm{L}}$, and in addition to entropy $\Delta s_{\mathrm{out}}^{\mathrm{V}}$, to read-off entropy of saturated water $\Delta s_{\mathrm{out}}^{\mathrm{L}}$ at the outlet pressure from steam tables to determine the fraction liquid $\mathscr{X}_{\mathrm{L}}$ by solving the following for the isentropic path:

$$\Delta s_{\mathrm{in}}^{\mathrm{V}} = (1-\mathscr{X}_{\mathrm{L}})\,\Delta s_{\mathrm{out}}^{\mathrm{V}} + \mathfrak{x}_{\mathrm{L}}\,\Delta s_{\mathrm{out}}^{\mathrm{L}}$$

With $\mathfrak{x}_{\mathrm{l}}$ known, the corresponding enthalpy of steam at the outlet $\Delta h_{\mathrm{out}}^{\mathrm{mix}}$ for this isentropic path is calculated from:

$$\Delta h_{\mathrm{out}}^{\mathrm{mix}} = (1-\mathscr{X}_{\mathrm{L}})\,\Delta h_{\mathrm{out}}^{\mathrm{V}} + \mathfrak{x}_{\mathrm{L}}\,\Delta h_{\mathrm{out}}^{\mathrm{L}}$$

The actual enthalpy of the exhaust steam $\Delta h_{\mathrm{out}}^{\mathrm{act}}$ is then calculated using the turbine isentropic efficiency η_{turb} by:

$$\Delta h_{\mathrm{out}}^{\mathrm{act}} = \Delta h_{\mathrm{in}}^{\mathrm{V}} - \left(\Delta h_{\mathrm{in}}^{\mathrm{V}} - \Delta h_{\mathrm{out}}^{\mathrm{mix}}\right)\eta_{\mathrm{turb}}$$

The actual moisture in the exhaust $x_{\mathrm{l}}^{\mathrm{actual}}$ is then calculated by solving the following:

$$\Delta h_{\mathrm{out}}^{\mathrm{act}} = \left(1-\mathscr{X}_{\mathrm{L}}^{\mathrm{act}}\right)\Delta h_{\mathrm{out}}^{\mathrm{V}} + \mathscr{X}_{\mathrm{L}}^{\mathrm{act}}\,\Delta h_{\mathrm{out}}^{\mathrm{L}}$$

The electric power output of the turbine $\dot{W}$ is then calculated using the mass flow rate of the steam $\dot{m}$ and the generator efficiency η_{gen} as:

$$\dot{W} = \dot{m}\left(\Delta h_{\mathrm{in}}{}^{\mathrm{V}} - \Delta h_{\mathrm{out}}^{\mathrm{act}}\right)\eta_{\mathrm{gen}}.$$

The results of these calculations are summarized in Table 8.1 for the two superheat temperatures. As can be seen, the higher superheat temperature results in a higher power output, while the fraction of moisture in the turbine exhaust is not excessive as it is in the 300°C superheat temperature case.

TABLE 8.1 Power output from steam turbine at two different superheat temperatures

		Case 1 (400°C superheat)			Case 2 (300°C superheat)		
			Outlet			Outlet	
	Units	Inlet	Ideal	Actual	Inlet	Ideal	Actual
Flow rate	kg/h	100,000	100,000	100,000	100,000	100,000	100,000
Pressure	Bar	42	0.06		42	0.06	
Temperature	°C	400	36		300	36	
Vapor enthalpy	kJ/kg		2,566.6			2,566.6	
Liquid enthalpy	kJ/kg		151.5			151.5	
Enthalpy	kJ/kg	3,210.6	2,076.5	2,303.3	2,954.4	1,948.7	2,149.8
Vapor entropy	kJ/kg-K		8.3290			8.3290	
Liquid entropy	kJ/kg-K		0.5208			0.5208	
Entropy	kJ/kg-K	6.7444	6.7444	7.4777	6.3311	6.3311	6.9814
Moisture	Fraction		0.2030	0.1090		0.2559	0.1726
Shaft power	MJ/h			90,730			80,455
Electric power	kW			23,943			21,231

8.7 COAL-FIRED POWER GENERATION

It is expected that the use of coal for power generation will increase in most regions of the world, especially the emerging markets such as China and India, while in the United States coal is being pushed out by natural gas. Coal has the highest C-to-H ratio among all fossil fuels and it is of utmost importance that research and development of technologies aimed at increasing the overall power plant efficiency in order to reduce the CO_2 footprint (being in inverse proportion to the efficiency) be pursued. Furthermore, CO_2 capture and sequestration in coal-fired power plants is another area of great importance for the sustainability of our planet. Thus, it is imperative to understand the current state-of-the art so that improvements can be sought.

8.7.1 Coal-Fired Boilers

Coal-fired boilers providing heat for the steam Rankine cycle are identified by the heat transfer method (e.g., water-tube or fire-tube), the arrangement of the heat transfer surfaces (i.e., horizontal or vertical, straight, or bent tubes), and the firing method (i.e., suspension, stoker,[2] or fluidized bed). The most common heat transfer method for coal-fired boilers is using water-tubes which are exposed to the hot combustion products while the boiler water and steam are contained within the tubes. Coal-fired water-tube boilers include pulverized coal and cyclone units, stokers, and fluidized beds.

In pulverized coal-fired boilers, the fuel is pulverized to a fine powder with at least 70% of the particles passing through a 200-mesh sieve (<74 μm) and pneumatically injected with a portion of the combustion air through burners into the furnace.

[2] A mechanical fuel feeding system onto a grate where the fuel is primarily combusted in a fixed bed.

Combustion occurs while the coal is in suspension in the furnace. Pulverized coal-fired boilers can be either dry bottom or wet bottom depending on whether the ash is removed in a solid state or molten state (the molten ash or slag withdrawn from the furnace is solidified by quenching it with water). Coals with high fusion temperatures are combusted in dry bottom furnaces while coals with low fusion temperatures are combusted in wet bottom furnaces. There are two firing methods in pulverized coal-fired boilers: wall and tangential. In wall-fired boilers, burners mounted either in one wall of the furnace or two opposing walls fire horizontally, while in tangential (or corner-fired) boilers, burners are mounted in the corners of the furnace with the fuel and air injected tangentially. Cyclone furnaces, which are often classified as pulverized coal-fired systems, actually feed much larger coal particles, as large as approximately 4-mesh (4.76 mm) which is introduced tangentially into a horizontal cylindrical furnace with the primary air. The smaller particles are combusted in suspension while the larger particles stick to the molten slag layer on the combustion chamber wall to combust with the secondary air.

In stoker-fired systems, the fuel is primarily burned on the bottom of the furnace or on a grate. These units account for the vast majority of coal-fired boilers for the smaller-scale (industrial) applications. There are three main types of stokers: underfeed, overfeed, and spreader stokers. In the underfeed stoker, the feed coal is either introduced horizontally with ash discharging on the side or the fuel is gravity fed with ash discharging at the rear. In the overfeed stoker, a moving grate assembly is used to feed the coal from a hopper onto a continuous grate which transports the fuel into the furnace. In a spreader stoker, pneumatic or mechanical feeders spread the fuel uniformly over the surface of a moving grate. The combustion of the finer sized, dryer, and more reactive fuel particles is initiated while in suspension as it is injected into the furnace and onto the grate with a greater percentage of heat release occurring above the bed.

In a fluidized bed combustor, the coal along with a sorbent such as limestone for capturing the SO_x and chlorides or an inert material such as sand form the bed which is fluidized by the combustion air. Fluidized bed combustors have gained popularity in small to large utility scale applications. Majority of these are the atmospheric pressure units, operating at or near ambient pressure while pressurized units operate from 4 to 30 bar. The atmospheric fluidized bed combustors can either be bubbling beds or be circulating beds. The fluidization velocity is relatively low in the bubbling bed in order to minimize solids carryover or elutriation while in circulating fluidized bed combustors, the fluidization velocity is such that the carryover or circulation of the solids occurs. The solids carried over in either type are captured by high temperature cyclones and returned to the primary combustion chamber. In the circulating fluidized bed combustor, however, a high recycle rate is maintained by returning the carried-over solids which increases the residence time for the solids with a correspondingly higher combustion efficiency and better sorbent utilization.

8.7.2 Emissions and Control

As mentioned in Chapter 1, coal combustion undergoes a complex process of drying/devolatilization followed by oxidation of the released gases and the remaining char, while a significant fraction of the ash forms particles typically 10 μm or less in size called fly ash. The major pollutants from coal combustion are particulate matter (PM), SO_x, and NO_x, and some unburned combustibles including CO and various organic compounds formed during the devolatilization step and are generally emitted even under proper boiler operating conditions.

8.7.2.1 ***Particulate Matter*** Composition and amount of PM emitted depend on the configuration and operation of the boiler, control equipment and coal properties. It consists of fly ash (fraction of the ash that does not settle out in the boiler but is carried by the flue gas) and any unburned carbon. Soot blowing periodically used to dislodge ash from the heat transfer surfaces is also a source of PM emissions. PM emissions can also be caused by vapors (mostly inorganics in the case of coal) that later condense to form aerosol particles. Postcombustion control of PM emissions can be accomplished by one or more of the following:

- Cyclones
- Electrostatic precipitator (ESP)
- Fabric filter (or baghouse)
- Wet scrubber

Cyclones are often used upstream of an ESP, fabric filter, or wet scrubber for capture of coarser particulates (PM10) as in fluid bed boilers. Typical overall collection efficiency can range from 90 to 95%. ESPs come in modular designs and are thus applicable to a wide range of plant sizes. Fabric filters have been used extensively in coal-fired plants and consist of multiple filtering elements or bags and a bag cleaning system enclosed in a shell structure equipped with hoppers for the captured particulates. The efficiency of fabric filters can be as high as 99.9% and is dependent on particle cohesion characteristics and the electrostatic properties of the fabric and the particles in addition to the size distribution. Examples of wet scrubbers are venturis, trayed and spray columns and are used primarily in the control of SO_x but they also capture PM. Scrubber efficiency can range from 95 to 99% for 2 μm particles and depends on gas side pressure drop and droplet size of the water (or scrubbing slurry) in case of spray, in addition to PM size distribution. Fugitive PM emissions from fly ash handling operations in most modern plants are minimized by utilizing pneumatic systems or enclosed and hooded systems venting through small fabric filters or other dust control devices.

8.7.2.2 ***SO_x*** Sulfur present in coal is in the form of pyritic, organic, and sulfate forms and during the combustion process, forms primarily SO_2 with much smaller fraction of SO_3 and sulfates. The ash can capture some of the sulfur in the furnace to form various sulfate salts depending on its alkaline nature. Typically, subbituminous coals have higher alkalinity ash and can capture more than 5% of the sulfur. Several techniques are used to reduce the SO_x emissions including precombustion coal cleaning. Physical cleaning reduces the pyritic form of sulfur which is present in the mineral content of coal while chemical cleaning techniques are being developed to remove the organic sulfur. Postcombustion flue gas desulfurization (FGD) can capture the SO_2 by using slurry or dry powder of an alkaline reagent and these can be either the more common throwaway or regenerable processes.

Wet systems consist of contacting the flue gas in a venture scrubber, packed column, or spray column with water slurry containing the alkaline sorbent. The sorbent can be lime, CaO which forms hydrated lime, or $Ca(OH)_2$ by reacting with the slurry water, or limestone which is mostly $CaCO_3$, or dolomite which is mostly $MgCO_3$ and $CaCO_3$. Wet systems can be designed to remove 90–98% of the SO_2 from the flue gas depending on whether limestone or the more reactive lime is used, to form sulfite and bisulfite. The sulfite and bisulfite are then oxidized using air to form the more stable sulfate in a separate operation. Calcium sulfate, $CaSO_4$, also known commercially as gypsum, can be used for making drywalls and also as a soil conditioner. PM (fly ash) reduction of greater than 99% is possible with wet scrubbers but typically the PM is collected upstream in an ESP or baghouse to avoid erosion of equipment, especially the pumps and piping and possibly interference with the chemical reactions in the FGD unit. Furthermore, the amount of sludge produced for disposal is reduced. Sodium hydroxide (NaOH) or Sodium carbonate or (Na_2CO_3) also known as soda ash and occurs naturally as the mineral, Trona, is utilized in smaller applications because of the higher reagent cost but can capture more than 96% of the SO_2. The reagent cost may be reduced by using a dual alkali system where the sodium may be recovered by regeneration with the cheaper lime or limestone to produce a calcium sulfite and sulfate sludge.

Dry systems are located upstream of the PM control device and inject the lime or limestone either as a dry powder or as water slurry into the hot flue gas and can remove 90–95% of the SO_2. A baghouse as the PM control device will typically require less sorbent than an ESP to achieve similar removal of the acid gas since it provides additional contact time between the sorbent and the acid gases as the gas passes through the cake of sorbent formed on the bag surface. If limestone is injected, the temperature of the flue gas should be above its decomposition temperature to form the more reactive CaO. Lime as well as the dry hydrated lime absorbs SO_2 to form sulfite and some sulfate. These reaction products along with the unreacted hydrated lime, some CaO and $CaCO_3$, and fly ash are then captured in the downstream PM control device. A fraction of these solids may be circulated to reduce the reagent usage while the remainder requiring disposal or possibly used as construction material. The circulation of the solids may also be accomplished by using a circulating fluidized bed in which the flue gas and the dry sorbent come into intimate contact. SO_2 capture efficiencies as high as 98% have been reported with the circulating fluidized bed dry scrubber.

Another process, although having limited commercial applications, uses anhydrous or aqueous ammonia to react with SO_2 in the flue gas to form a high value crop fertilizer, $(NH_4)_2SO_4$. SO_2 removal efficiency greater than 98% has been reported.

8.7.2.3 NO_x NO_x is formed by the oxidation of the N_2 present in the combustion air (thermal NO_x), oxidation of the fuel bound nitrogen (mainly present in coal's aromatic ring structures), and by the reaction of various hydrocarbon radicals like

CH and CH_2 produced by the devolatilization of the coal (after the initial drying step) with the molecular N_2 (prompt NO_x). Some (<1%) N_2O is also formed during coal combustion but the dominant form by far (~95%) is NO with the rest being NO_2. Thermal NO_x formation is exponentially dependent on temperature and increases with the O_2 concentration and gas residence time, and cyclone boilers typically tend to generate higher emissions of thermal NO_x. Bituminous and sub-bituminous coals can contain from 0.5 to 2% by weight of nitrogen and can account for up to 80% of the total NO_x. Two types of approaches (often in combination) are used for limiting NO_x emissions, the more widely used combustion controls and postcombustion controls. Combustion controls include low excess air (LEA), overfire air (OFA), low NO_x burners (LNBs), and reburn and flue gas recirculation while postcombustion control approaches are selective catalytic reduction (SCR) and selective noncatalytic reduction (SNCR).

With OFA, a fraction of the total combustion air is injected above the burners to delay and extend the combustion process resulting in less intense combustion and lower flame temperatures, and also to reduce the concentration of O_2 in the combustion zone where fuel bound nitrogen containing compounds are evolved. This method which may be used for tangential and wall-fired as well as stoker boilers can reduce NO_x by 20–30%. LNBs control the concentration of O_2, temperature profile of the combustion process as well as reduce the residence time at peak temperature in each burner zone to limit conversion of fuel bound nitrogen to NO_x and the thermal NO_x formation. LNBs may be used in tangential as well as wall-fired boilers but not in cyclone boilers or stokers. This method when combined with OFA can reduce NO_x emissions by 40–60% over uncontrolled levels. Reburn involves supplying approximately 60% (at full load) of the total fuel energy to the main combustion zone while the remaining in the form of natural gas, oil, or pulverized coal injected with either air or flue gas above the burners to create a reburn zone to create a fuel-rich zone that reduces the NO_x formed in the main combustion zone to N_2 and H_2O vapor. This method may be used in tangential, wall-fired, and cyclone boilers, and can reduce NO_x by 50–60% over uncontrolled levels. Flue gas recirculation consists of taking a fraction of the flue gas from downstream of the PM control device and feeding it back into the furnace using a fan as part of the secondary "combustion air." Due to the reduced concentration of O_2 in the mixture of the recirculated flue gas and fresh air, the flame temperature is reduced and thus the NO_x formation.

Both SCR and SNCR involve injecting NH_3 (or urea in the case of SNCR) into the flue gas to react with the NO to form N_2 and H_2O vapor (in the presence of a catalyst in the case of SCR). Performance of these control methods depends on flue gas temperature (typically about 400°C for SCR and 800–1150°C for SNCR), fuel sulfur content, inlet NO_x concentration, reaction residence time, and the NH_3 (or urea)-to-NO_x ratio used. The catalyst activity is reduced by some of the fly ash components such as As, Ca, P, and K. NO_x reductions by as much as 75–85% have been demonstrated with SCR while much more modest levels with the more commercially proven SNCR, in the range of 25–50%. The capital cost is higher with SCR but the reagent cost is lower. A drawback with the SCR is that it can catalyze the oxidation of the SO_2 to form additional SO_3 which in turn forms sulfurous acid and is corrosive to downstream equipment such as air heater, baghouse, and ESP. SO_3 can also react with chlorides in the flue gas to form submicron aerosol particles that create an opaque plume, and with unreacted NH_3 to form sticky ammonium salts which can deposit on heat exchange surfaces. It is advantageous to locate the SCR upstream of the PM control device for the fly ash to "catch" these salts.

8.7.2.4 ***CO and Organic Compounds*** The three most important factors that affect the amount of CO and organic compounds in the combustion products are residence time, temperature, and turbulence (referred to as "the three Ts") in the combustion zone. The primary mechanism for increased emissions of CO and the organic compounds is poor mixing of the fuel and air during the combustion process such that fuel-rich regions persist and cannot find enough O_2 to burn out while mixing can be improved by turbulence. These emissions may also be caused by a very low excess air setting. Smaller boilers, heaters, and furnaces typically emit higher amounts of these pollutants because of less residence time in the high-temperature region. Design of combustion controls used to reduce NO_x emissions that are based on reducing the residence time and temperature should take into account the impact on emissions of CO and organic compounds. Organic compounds emitted include various hydrocarbons (alkanes, alkenes, and aromatics such as benzene, toluene, xylene, and ethyl benzene), oxygenates (such as aldehydes and alcohols), as well as chlorinated compounds such as dioxins and furans. Dioxins and furans can also be formed in the downstream air pollution control equipment with a maximum potential in the temperature range of 230–350°C. Formation of these compounds following combustion may be minimized by quickly cooling the flue gas and by limiting the amount of certain metals known to promote or catalyze the formation of these compounds in the downstream air pollution control equipment.

8.7.2.5 ***Trace Metals*** Mn, Be, Co, and Cr are distributed roughly equally in the fly ash and in the bottom ash, and their emissions depend on the total PM emitted. As, Cd, Pb, and Sb partition themselves predominantly into the fly ash (as opposed to bottom ash) while showing higher concentrations with decreasing particle size. Thus, the emissions of these volatile metals are dependent on the collection efficiency of the fine PM. Hg and in some cases Se remain in the flue gas and other means specifically designed to capture these elements is required. The amount of Cl in the coal affects the fractions of Hg present in the flue gas in the elemental form and as a chloride. $HgCl_2$ is soluble in water and can be removed in a wet scrubbing process such as the wet FGD. A relatively simple and proven approach for capturing the Hg from the flue gas consists of injecting activated carbon into the flue gas to adsorb the Hg upstream of the PM control device, and separating the activated carbon along with the adsorbed Hg from the flue gas in the PM control device. Concentrations of SO_x, NO_x, and H_2O in the flue gas can affect the adsorption of Hg. Regenerable sorbents, catalysts, scrubbing liquors, combustion modification, and ultraviolet radiation are all being investigated to replace this "throwaway" process.

8.7.2.6 ***Halogens*** Cl and F present in the coal are emitted primarily in the form of HCl and HF with lesser amounts emitted as elemental Cl and F gas, while some of these emissions may be absorbed by the fly ash or bottom ash. Both HCl and HF are highly water-soluble and can be removed in a scrubbing process such as the wet FGD. A side benefit of using limestone for limiting SO_x emissions in a fluidized bed boiler (as well as in other type of boilers using lime of limestone-based dry FGD) is that it also captures significant amounts of these acid gases.

8.7.2.7 ***Greenhouse Gases*** In addition to CO_2, other greenhouse gases produced during coal combustion are N_2O and CH_4. N_2O emissions are significant for fluidized bed boilers, typically two orders of magnitude higher than that for all other types of coal boilers. Formation of N_2O is minimized when combustion temperatures are kept high (>850°C) and the excess air is minimized (<1%). CH_4 emissions are highest during start-up or shutdown of a coal-fired boiler when the combustion process is incomplete but it is the CO_2 which is emitted continuously during the boiler operation that requires the major attention. The following describes a flue gas scrubbing process (postcombustion capture) using an aqueous amine solution ("amine wash") for capturing CO_2 but its impact on plant cost and performance is quite significant, resulting in almost doubling the cost of electricity when the CO_2 leaves the plant at the pressure (15.3 MPa or 2215 psia) required for transportation by pipeline to the sequestration site. Thus, technologies are being investigated that involve using other types of solvents, membranes, and adsorbents. Use of high-purity oxygen instead of air for the combustion process ("oxy-combustion") is another approach being developed. It produces a flue gas with much higher concentration of CO_2 which may be further purified. However, the capital cost and the energy requirements of the air separation unit (to provide the high-purity oxygen stream) and the CO_2 purification unit can also be quite significant. Separating the CO_2 is just one step in controlling this emission and its capture requires finding a "home" for it and as explained in Chapter 1, geologic sequestration as well as using it for coal bed methane recovery is being investigated, while commercial use in enhanced oil recovery has been practiced for a number of years.

An amine wash process for separating the CO_2 from the flue gas (Chapel et al., 1999) uses an aqueous MEA solution with proprietary additives. The process consists of an absorber/stripper system with the solvent circulating between the two columns as briefly described in Chapter 5 and depicted in Figure 5.2. The cooled flue gas after PM removal and desulfurization is supplied to the absorber where it comes into contact with the MEA solution and about 85–90% of the CO_2 can be absorbed by the solution. The rich solvent (loaded with the CO_2), is regenerated in the stripper using steam to produce a high-purity CO_2 stream at near atmospheric pressure. A small slipstream of the lean solvent is fed to a reclaimer for the removal of high-boiling nonvolatile impurities including heat-stable salts that are formed, volatile acids, and iron products from the circulating solvent solution. Reclaiming is accomplished primarily by an ion-exchange process. The CO_2 released in the stripper near atmospheric pressure is compressed to pipeline pressure by a multistage intercooled centrifugal compressor. The CO_2 has to be also dehydrated, which is accomplished by treating the CO_2 stream at an appropriate pressure before completing

TABLE 8.2 HCl emissions from a bituminous coal and a lignite

Coal type	Bituminous	Lignite
Cl, fraction by weight	0.0029	0.000012
Heating value		
HHV (kJ/kg)	27,135	15,391
HHV (Btu/lb)	11,666	6,617
Molecular weight of Cl	35.45	
Cl content in coal (lb mol/10^6 Btu)	0.007012	0.000051
HCl emitted (lb mol/10^6 Btu)	0.007012	0.000051
Molecular weight of HCl	36.46	
HCl emitted (lb/10^6 Btu)	0.2557	0.001865
HCl emission allowed (lb/10^6 Btu)	0.002	
HCl removal required (%)	99.22	None required

the compression to the final pressure to avoid formation of crystalline hydrates which can form deposits and corrosion by wet CO_2. A dew point of −40°C (−40°F) is typically required and may be accomplished using triethylene glycol to scrub the wet CO_2.

Example 8.2. A utility Power Plant burns a bituminous coal with the analysis shown in Table 1.4. The plant just became aware of the proposed new Air Toxics rule passed by the U.S. EPA, which would impose the limit on HCl emission at 0.002 lb/10^6 Btu HHV. If they continue to utilize this coal, what level of HCl removal will be required? If instead, they switch the coal to a lignite with the analysis shown in Table 1.4, what level of HCl removal will be required, if any? Assume all of the Cl in the coal shows up as HCl in the flue gas.

Solution

The calculations for the two coals are summarized in Table 8.2. As can be seen, switching to lignite with its low Cl content will not require adding the equipment for HCl removal from the flue gas. However, the impact of switching to the lignite has to be carefully considered while taking into account the changes required to the boiler as well as the impact on the thermal performance and power output of the plant. Transportation costs for the lignite, which has a much lower energy density, also has to be taken into consideration.

8.7.3 Description of a Large Supercritical Steam Rankine Cycle

The following describes the major sections of a state-of-the-art large pulverized coal-fired boiler plant with a supercritical steam Rankine cycle and postcombustion CO_2 capture for a central station (typically >500 MW net power) application (Black, 2010). A simplified overall plant block flow diagram is presented in Figure 8.13.

8.7.3.1 Boiler There are different arrangements for the BFW and steam circuitry through a boiler depending on the boiler design and one such arrangement is presented

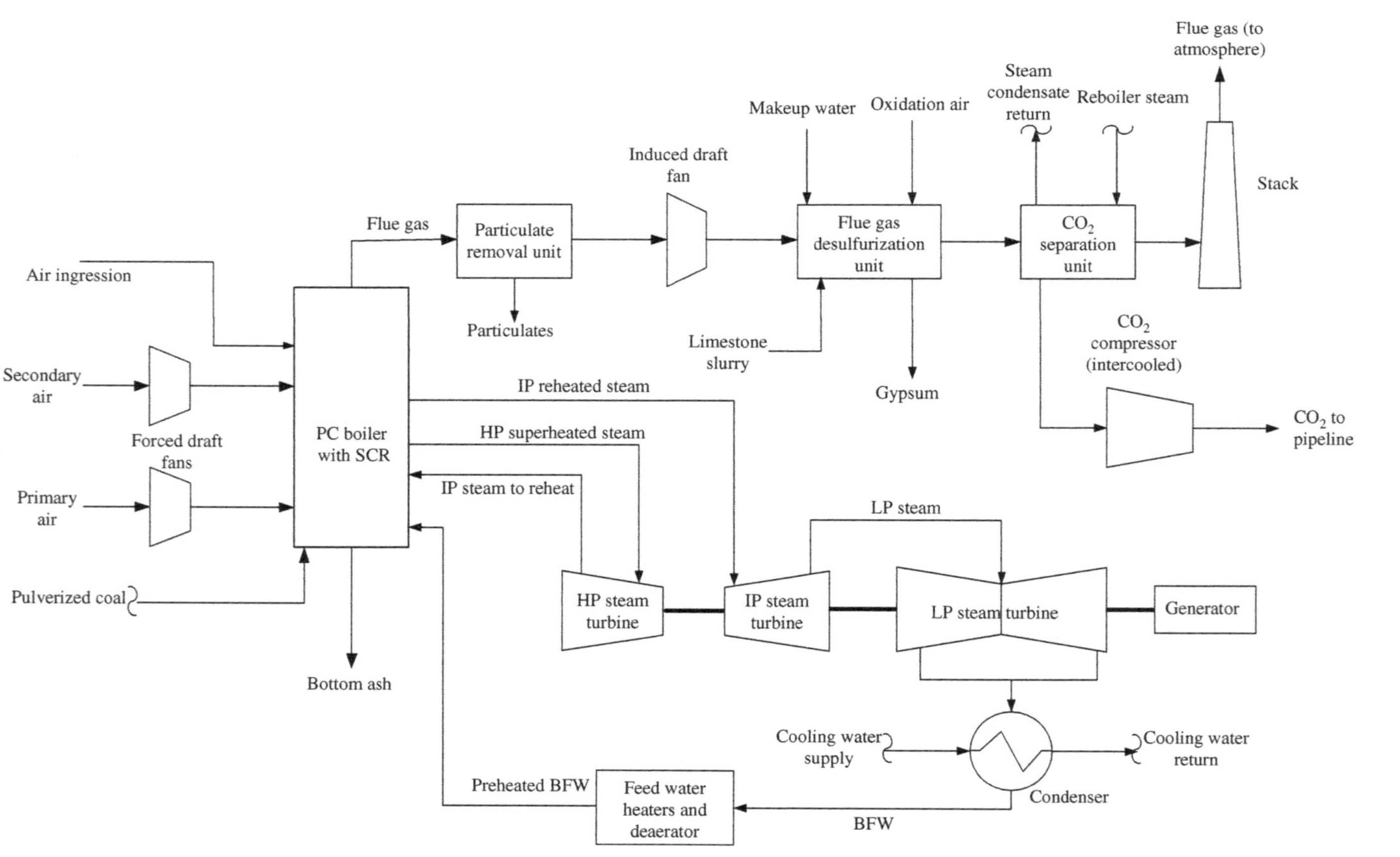

FIGURE 8.13 Central station power plant with supercritical steam Rankine cycle

here. The BFW after deaeration and preheating in regenerative feedwater heaters is further heated in an economizer located within the boiler. It then enters the furnace wall tubes forming the supercritical steam. Note that in a supercritical boiler, a steam–water separator (steam drum) is not required and the BFW flows through the boiler on a once-through basis (whereas in the case of a subcritical boiler, the BFW enters a steam drum from where it is distributed into the boiler tubes in which a fraction of the BFW changes phase and the steam along with the unevaporated BFW is returned to the steam drum for separation of the two phases). The steam then flows to a primary superheater located in the convection pass, followed by a first stage of water attemperation, and then to the platen superheater, which consists of flat panels of tubes located in the upper section of the furnace exposed to high-temperature gases such that heat transfer is predominantly by radiation. The steam next flows through a second stage of attemperation, then to an intermediate temperature superheater, followed by a final superheater. Steam returning from the high-pressure section of the turbine passes through the primary reheater, followed by interstage attemperation in crossover piping which feeds the steam to the final reheater banks, and then out to the intermediate pressure section of the turbine. Interstage attemperation during superheating and reheating is used to provide temperature control during load changes. The flue gas leaving the economizer enters the combustion air heater, which is typically a regenerative type heater, in which heat is transferred from the hot flue gas to a heat storage medium and then to the combustion air; a common type being the rotary Ljungstorm preheater. Recuperative air heaters are also utilized, which consist of transferring heat directly from the hot flue gas to the combustion air across heat exchanger tubes. The flue gas then flows to a SCR unit for NO_x control which is followed by a baghouse for PM control. The flue gas then enters an induced draft fan to provide the pressure required to flow through the downstream FGD. The flue gas is next cooled by direct contact with water to reduce the power consumption of a downstream blower which provides the pressure required to flow through the CO_2 absorber and wash column before the flue gas enters the stack.

Metallurgy for each of the heat transfer services would be different based on the design temperature/pressure and corrosive nature of the environment each service is exposed to, both on the flue gas/ash as well as on the BFW/steam sides.

8.7.3.2 Steam Turbine The steam turbine consists of a high-pressure section, an intermediate-pressure section, and two low-pressure condensing sections, each being double flow (i.e., the steam is divided between two turbine sections built "back to back" with the two steam streams flowing away from each other and exhausting on opposite sides) to limit the length of the low-pressure stage blades, the specific volume, and total flow rate of steam being high. Each section (high pressure, intermediate pressure and low pressure) enclosed in a separate pressure casing, is mounted on a common shaft driving a hydrogen cooled generator. The generator stator windings are cooled with a closed-loop water system that includes water circulating pumps and heat exchangers (either shell and tube, or plate type). The turbine bearings are lubricated by a closed-loop water-cooled pressurized oil system. Method of heat rejection from the closed loop systems to the environment depends on the particular heat rejection system employed which is discussed later in this chapter. Turbine shafts are sealed against air in-leakage or steam out-leakage ("blowout") using a labyrinth gland arrangement connected to a low-pressure steam seal system. The high-pressure superheated steam from the boiler enters the high-pressure section of the turbine at 24.1 MPa/593°C (3500 psig/1100°F) while the reheated steam from the boiler enters the intermediate-pressure section of the turbine at 593°C (1100°F). After passing through the intermediate-pressure section, the steam enters a crossover pipe which provides the steam to the two low-pressure sections which exhaust into the surface condenser at a vacuum, this negative pressure being set by the temperature of the cooling medium used in the condenser.

There are numbers of ways of rejecting heat from the surface condenser and other sections of the plant depending on site conditions and economic parameters as discussed in Chapter 7. Cooling towers (introduced in Chapter 5 from a mass and heat transfer standpoint) may be utilized where freshwater is available as makeup to the cooling towers. Airflow through the cooling tower may either be forced by mechanical means (using fans) or induced by natural circulation. The mechanical draft towers have lower capital cost but higher electrical power requirement (to operate the fan) when compared to natural draft cooling towers. When the plant is located close to a large body of water, once-through cooling is often utilized but if the water is brackish or seawater, then appropriate materials for the heat exchangers should be chosen such as titanium. Air-cooled heat exchangers are employed in dry locations but the first law efficiency of the plant is compromised since the surface condenser operating pressure has to be increased due to the higher cooling medium (air in this case) temperature and due to the need for larger temperature approach within the condenser to compensate for the poor heat transfer coefficient of the air (as compared to cooling water).

When cooling water is supplied by wet cooling towers operating under ISO site ambient conditions of 15°C dry bulb temperature and 60% relative humidity, the corresponding cooling water supply temperature can be as low as 15°C (with ~4°C approach temperature to the ISO wet bulb temperature, i.e., ~4°C above the wet bulb temperature). The corresponding operating pressure for the surface condenser can then be as low as 4.8 kPa while maintaining a reasonable temperature rise for the cooling water of about 11°C and a reasonable pinch temperature of 5–6°C in the surface condenser.

The makeup BFW to the steam cycle consists of demineralized water since steam is generated at high pressure. The demineralizer consists of an initial cartridge filter, followed by a reverse osmosis (RO) membrane assembly, and finally an electrodeionization unit which utilizes an electrolytic cell to separate out remaining dissolved ions. The partially purified water leaving the RO units is passed between an anode (positive electrode) and a cathode (negative electrode) and ion-selective membranes which allow the positive ions to migrate toward the negative electrode while the negative ions migrate toward the positive electrode resulting in high-purity deionized water. This makeup BFW is required since a small fraction of water/steam leaks out from the steam turbine and also a small fraction of steam is vented from the deaerator used to remove the dissolved gases such as O_2 and CO_2 that can make the water corrosive to the downstream equipment. Additionally, an O_2 scavenging chemical which can also be a metal passivator is injected into the BFW. Examples of scavenging chemicals are hydrazine (which however, is going out of favor because of its carcinogenic character), hydroquinone, diethyhydroxylamine, methylethylketoxime, carbohydrazide, and erythorbic acid. The deaerator also functions as a regenerative heater using the steam directly to preheat the BFW in addition to getting the stripping action required to remove the dissolved gases.

8.8 PLANT-DERIVED BIOMASS-FIRED POWER GENERATION

The main incentive driving plant-derived biomass use for energy is the near CO_2 neutrality of biomass cultivated sustainably or recovered as residues and wastes. These resources can be large enough to make a significant impact on reducing our dependence on high carbon content fossil fuels such as coal. Combustion followed by a steam Rankine cycle is the most mature technology currently available for biomass-based energy plants and improvements in efficiency, emissions footprint, and cost will further the use of this resource.

8.8.1 Feedstock Characteristics

Plant-derived biomass differs from conventional fossil fuels in physical structure, chemical structure, and since it is "alive" when harvested, has high moisture content since water is needed for life and is present in cell structures and on its surface. The moisture content varies with species, location, season, and handling practices with natural levels ranging from 30% to greater than 65%. Thus, and as seen from data presented in Table 1.6, biomass can have significantly higher moisture content than the high rank coals. Although biomass can be combusted with moisture content as high as 60%, it is advantageous especially from an efficiency standpoint to predry the fuel using lower temperature thermal energy from the system instead of its chemical energy as the case is when wet biomass is directly combusted without predrying. Vapor compression drying which has been proven on lignites can be applied to biomass. The drying operation consists of compressing the evaporated water vapor and then using its latent heat to provide heat for the drying (Schippers, 2010). The compression does increase electric power usage but the overall system net efficiency can be higher. The quality of latent heat of the compressed vapor is increased and made useful. This compressed vapor is supplied to the heat exchange tubes located within the fluidized bed dryer to provide the heat for drying. This type of drying technology consisting of a "heat pump" increases the quality of the thermal energy contained in the evaporated water vapor by using the much higher quality mechanical energy in the vapor compressor. Note that this is consistent with the second law of thermodynamics discussed in Chapter 2 as long as the overall quality of energy leaving the system is less than that entering the system.

Plant-derived biomass has high "volatile matter" content; as much as 70–80% in dry wood which is released from the wood structure upon heating at relatively low temperatures of approximately 260°C or 500°F (balance is mostly "fixed carbon" or “char” with little ash). Because of the high volatile matter content, spontaneous combustion of the dry biomass can occur during storage and an inerting gas system may be required in the storage silos or bins. Another characteristic of biomass as compared to fossil fuels is its low energy density (even worse with undried biomass) which makes transportation over long distances prohibitive and requires the plant utilizing this feedstock to be located close by (<50 miles) to the resource.

The ash constituents of plant-derived biomass include alkalis such as K which lower the ash fusion temperature leading to slagging and fouling of heat transfer tubes in boilers. Some plant-derived biomasses contain much higher concentration of chlorides than many coals which can form corrosive deposits on heat transfer tubes in boiler. Solid biomasses are less friable than coals but drying of the biomass increases its friability and reduces the grinding power requirement. An approach being investigated to increase both the friability and energy density is torrefacation (Tumuluru et al., 2011) involving mild pyrolysis where the biomass is heated in absence of O_2 to decomposition temperatures such that some of the volatile matter is driven off leaving a char that is quite friable. A drawback with this approach is that a gas containing tars and oils is produced and requires removal of these before the gas can be compressed and utilized, in addition to doing something with the tars and oils such as processing them separately unless they are also combusted.

Most biomasses are lower in S and Hg content than coal on a unit energy basis and can avoid disposal charges of the wastes generated from throwaway processes used for capturing SO_x or Hg. Agricultural biomass has higher contents of K which can lead to higher emissions of particulates, increased ash corrosion and deposits. Boiler plants burning municipal solids waste, urban waste wood, and demolition wood should be designed with efficient flue gas cleaning for the abatement of toxic pollutants such as heavy metals and chlorine compounds including dioxins and furans.

Cost components to be considered for biomass as a fuel are the collection method which is one of the largest costs for forest residue while for agricultural residues; it can be 20–25% of the delivered cost. Furthermore, costs associated with transportation and storage, whether stored where collected or at point of use, should be taken into account.

8.8.2 Biomass-Fired Boilers

Due to the immense diversity in biomass characteristics, a number of designs for boilers are offered and the most frequently used ones are the grate and the fluidized bed boilers similar to the ones described for coal firing. Underfeed stoker systems are generally suitable for small-scale systems and for low ash fuels such as wood chips and sawdust due to ash removal problems. Bubbling fluidized beds are typically suitable for applications greater than10 MWth, while circulating fluidized beds are more suitable for larger applications (>30 MWth).

8.8.3 Cofiring Biomass in Coal-Fired Boilers

Due to its low energy density (and being a distributed resource), transportation of biomass collected from various locations to a central plant over long distances makes it economically prohibitive. Thus, it is desirable that the biomass energy plant be located in close proximity to the feedstock but this may limit the size of the plant, and economies of scale of a large plant may not be taken advantage of to reduce the specific capital cost. One option is to cofeed the biomass with coal in a central station type energy plant located close to the biomass source. An added benefit with cofiring agricultural biomass with coal is that the seasonal availability of this biomass resource will not be a major impediment for its use; by building a plant designed for firing varying proportions of coal and the biomass, the plant capacity utilization may be maximized by increasing the coal feed rate to the plant during times when the supply of the biomass resource is low (as long as the feed handling systems for the two feedstocks are properly sized).

Cofiring of various types of biomass such as herbaceous as well as woody types (residues and energy crops) with various types of coals (lignite, subbituminous, bituminous) as well as petroleum coke has been demonstrated at commercial scales in different type of boilers such as tangentially fired, wall fired, and cyclone boilers. The technical challenges associated with cofiring biomass are due to differences in the energy density, ash characteristics, and friability as compared to those of coal and should be thoroughly addressed in the design of equipment for commercial operation. Since both the heating value and the bulk density of biomass are significantly lower than that of coal, plants cofiring biomass require shipping, storage, and on-site fuel handling systems that are disproportionately higher when compared to coal on a basis of the relative heat contribution. Communition (particle size reduction) of biomass results in a nonfriable and fibrous product and generally it is not possible to reduce its size or shape similar to that of a coal particle. The biomass particles tend to be larger and with larger aspect ratios creating challenges for pneumatically transporting the fuel to the boiler. Large and aspherical particles may also pose challenges for combustion efficiency. Many forms of wood waste produce relatively minor deposits but herbaceous materials produce higher ash deposition rates, which can also be quite corrosive. Leaching out alkali (potassium salts) and chlorides with water has been shown to mitigate this ash-related issue but then the water has to be treated for recovery of these salts to minimize the makeup and discharge water flow rates and any blowdown discharge water from the system has to be treated to meet the environmental requirements. Furthermore, the treated biomass from this leaching process would be saturated with water increasing the drying load.

8.8.4 Emissions

SO_x is usually not an issue due to the low content of sulfur in most biomasses. Since biomasses commonly contain more moisture than coal even after drying, the peak flame temperatures can be lower resulting in lower thermal NO_x formation. Woody biomass also has lower nitrogen content than many coals which can also reduce the fuel NO_x.

Example 8.3. A lignite-fired boiler plant (without FGD) with fuel characteristics shown in Table 1.4 is being converted to burn a woody biomass with characteristics

TABLE 8.3 SO_x emissions from a biomass and a lignite

Feedstock	Woody biomass	Lignite
Sulfur, fraction by weight	0.0002	0.0063
Heating value, HHV (kJ/kg)	9813	15,391
Molecular weight of S	32.07	
S content in feedstock (k mol/10^3 MJ)	0.000636	0.012766
SO_x emitted (k mol/10^3 MJ)	0.000636	0.012766
SO_x reduction (%)	95.02	

shown in Table 1.6. Calculate the reduction in SO_x emissions on a fired HHV basis due to the fuel switch, assuming none of the sulfur remains in the ash.

Solution
The calculations for the two feedstocks are summarized in Table 8.3. As can be seen, switching to the biomass with its low sulfur content will reduce the SO_x emissions significantly. However, the impact of switching to the biomass has to be carefully considered while taking into account the changes required to the boiler as well as the impact on the thermal performance and power output of the plant. Costs associated with the collection and transportation of the biomass, which may be a distributed resource and has a low energy density to a central facility has to be also taken into consideration.

8.9 MUNICIPAL SOLID WASTE FIRED POWER GENERATION

Municipal solid waste (MSW) can be directly combusted as a fuel with minimal processing, known as mass burn, or it can be preprocessed before being directly combusted as refuse-derived fuel (RDF). Both these technologies present the opportunity for electricity generation while providing an alternative to landfilling or composting the MSW. In contrast to many other fuels, the MSW has a negative price ("tipping fee"), which is comparable to the fee charged for disposal of the garbage at a landfill.

Due to the very heterogeneous and odorous nature of MSW, the feed receiving and handling systems in an MSW energy recovery plant are quite different than say in a coal-based energy plant. The incoming trucks bringing in the MSW deposit the refuse into pits where cranes mix the refuse and remove any bulky or large noncombustible items such as large appliances. The refuse storage area is maintained under a vacuum in order to prevent odors from escaping.

8.9.1 MSW-Fired Boilers

The refuse is loaded into boiler feed hoppers by cranes and the boilers typically used consist of grate systems. Other type of combustors such as rotary kilns, multiple hearth furnaces, and fluidized beds have been considered or applied but to a limited extent. One form of a stoker grate used in this application has the grate sloping downward and is composed of alternating rows of fixed and moving grate bars. The grate bars push upward against the natural downward movement of the waste. This ensures that the burning waste is continually agitated and pushed back to serve as underfire for freshly fed waste. A forced draft fan supplies primary combustion air underneath the grate, and overfire air is injected through the front and rear walls of the furnace. While the combustion gases move through the boiler, bottom ash slowly makes its way to the end of the grate where it falls into a water quench trough.

8.9.2 Emissions Control

MSW combustion systems include air pollution control equipment similar to a coal-fired plant for control of SO_x, NO_x, PM, and Hg but modified or additional equipment are required focusing on chlorides, furans, and dioxins removal.

8.9.2.1 Dry Scrubbing Process The dry FGD process consisting of injecting lime slurry into the flue gas (upstream of the PM control device) in addition to capturing the SO_x, also captures HCl by forming the respective salt. The resulting PM may then be captured by the downstream baghouse; 95% of the evolved Hg may be captured by injecting activated carbon which can also capture some of the dioxins and furans. A baghouse is the preferred PM control device in these cases as the cake consisting of the FGD sorbent and the activated carbon building up on the surface of the bags provides additional residence time for the removal of the pollutants. Controlled high-temperature combustion at over 1100°C can destroy a significant fraction of the dioxins within the furnace. NO_x formation may be minimized with combustion controls and the NO_x emissions may be further reduced by an SNCR. An SCR may instead be employed for the reduction of NO_x which when combined with an oxidation catalyst can simultaneously destroy the dioxins and furans by catalytically reacting with the available O_2 in the flue gas (Finocchio et al., 2006). Dedicated catalysts for the destruction of the dioxins are also being developed (Liljelind et al., 2001), which may be used when an SCR is not utilized. One proposed technology concept consists of adding honeycomb catalyst modules for dioxin destruction above each compartment of the fabric filter.

The presence of CO, CH_4, and other organic compounds which are products of incomplete combustion can be minimized by proper design of the combustion system.

8.9.2.2 Wet Scrubbing Process Wet scrubbing of the flue gas may be employed for removal of acid components, metals, and dioxins. The flue gas after heat recovery enters a baghouse to first remove the particulates and then the HCl is removed by water wash and the HCl laden water may be further processed for the recovery of HCl or its neutralization. SO_x is removed by washing with a lime slurry. Both the HCl and gypsum (formed by oxidizing the sulfites and bisulfites from the FGD) may be marketable by-products depending on the quantities produced. The ADIOX® technology (Andersson et al., 2003) consists of fabricating packing material used in wet scrubbers such as the wet FGD and the HCl scrubbing processes out of a polymer material consisting of polypropylene doped with carbon particles. The carbon particles which have affinity for the dioxins capture it irreversibly by absorption/adsorption.

Solid residues from the combustion process are bottom ash collected at the bottom of the combustion chamber, and fly ash which exits with the flue gas. The combined ash and the additional solids generated by the FGD process are typically 20–25% by weight of the original refuse fed to the plant and much denser resulting in a significant reduction in the landfill requirement for its disposal. Depending on the makeup of the MSW, however, this ash residue may or may not be considered a hazardous material and if hazardous, will have to be disposed at landfills designed and classified for handling such wastes. Cement solidification is applied for coal-derived fly ash disposal but cannot be used for MSW-derived fly ash due to the high concentration of chlorine compounds present in the form of dioxins and/or alkali chlorides. A more prudent approach would be to prevent the sources in the MSW that create the hazardous residue from entering the system. Treatment of the ash to make it nonleachable (e.g., by vitrification) and nonhazardous should be investigated. If the residue is nonhazardous, it could be mixed with soils for use as landfill cover or possibly sold as pavement aggregate.

8.9.2.3 Refuse-Derived Fuel RDF typically consists of pelletized or shredded ("fluff") MSW and is a by-product of a resource recovery operation consisting of separating ferrous materials, glass, grit, and other noncombustible materials from the MSW. The remaining material is then sold as RDF and typically both the RDF processing facility and the RDF combustion facility are located in close proximity to each other to minimize transportation of this low energy density fuel.

8.10 LOW-TEMPERATURE CYCLES

Many older chemical process plants have low-temperature "waste" heat that gets simply rejected to the environment. Depending on the temperature and the quantity of this thermal energy being rejected, it may actually be economical with today's energy prices and environmental issues to generate power utilizing this resource. Even in new plants, again depending on the temperature and the quantity of the thermal energy, its recovery for the generation of power may be economical since rejection of heat also requires equipment and is not free. From a sustainability standpoint, waste heat recovery provides the opportunity to generate pollution free electricity (with respect to air emissions including those of greenhouse gases and thermal pollution while conserving natural resources).

8.10.1 Organic Rankine Cycle (ORC)

An organic fluid with a low boiling point as compared to water is used as the working fluid in Rankine cycles used to generate power from low-temperature heat sources. With such fluids, the phase change from liquid to vapor occurs at pressures much higher than that for water for a given temperature allowing the use of more compact hardware (turbine, heat exchangers, piping, and valving) while also providing room for pressure drops associated with these hardware. The turbine design is also simplified without requiring many expansion stages for obtaining a reasonable isentropic efficiency when a higher molecular weight fluid (compared to H_2O) is used which is typically the case with organic compounds (hydrocarbons and refrigerants being commonly used), an important consideration when dealing with small-scale applications. Thus, the Organic Rankine Cycle (ORC) is useful not just in low-temperature waste heat applications but also in small biomass, geothermal, and solar pond applications.

8.10.1.1 Selection of the Working Fluid The working fluid in addition to having optimal characteristics from a thermodynamic standpoint should be safe to handle (from a toxicity and flammable standpoint) and have low environmental impact, for example, from ozone depletion and global warming potentials if leakage occurs (turbine seals should be designed to avoid or minimize leakage from around the shaft). Other characteristics include low freezing point, high chemical stability temperature, noncorrosivity to ordinary carbon steel such that expensive metallurgy is not required, and high latent heat of vaporization so that the required flow rate, size of equipment, and pump power usage are all reduced.

Typically, ORCs are designed with minimal superheating, if any, of the working fluid entering the turbine since either the heat sources do not have a wide temperature range to justify it from a thermodynamic standpoint, or are too small to warrant the addition of another heat exchanger (the superheater) from a cost and simplicity standpoint. However, it is advantageous to avoid entering the two-phase region during expansion in the turbine to avoid erosion of the turbine blades. Oftentimes an upper limit may be set for the turbine inlet pressure based on criteria such as maximum operating pressure for an unmanned facility (for small plants, having an unmanned operation may be necessary to make it economically feasible). In such cases, for a given turbine pressure ratio, the exhaust from the turbine even without superheating the inlet stream may be at a temperature much higher than the heat rejection temperature, and a regenerative heater to recover the heat from the exhaust may be justified. Such a cycle is depicted in Figure 8.14, where the liquid organic fluid leaving the condenser is preheated against the turbine exhaust before entering the evaporator.

It is advantageous to select cycle conditions (turbine inlet conditions and pressure ratio) such that the exhaust from the turbine is above ambient pressure in order to reduce size of equipment downstream of the turbine.

In cases where the heat source temperature range is large, then a supercritical Rankine cycle, or a binary mixture where the bubble point varies with composition discussed under geothermal energy in Chapter 12 may be used.

Finally, if thermal stability is a concern for the working fluid in cases where the temperature of the heat source is high enough to cause excessive tube wall temperatures resulting in decomposition of the organic fluid, then a heat transfer or thermic fluid may be utilized such that heat from the source is transferred to the heat transfer

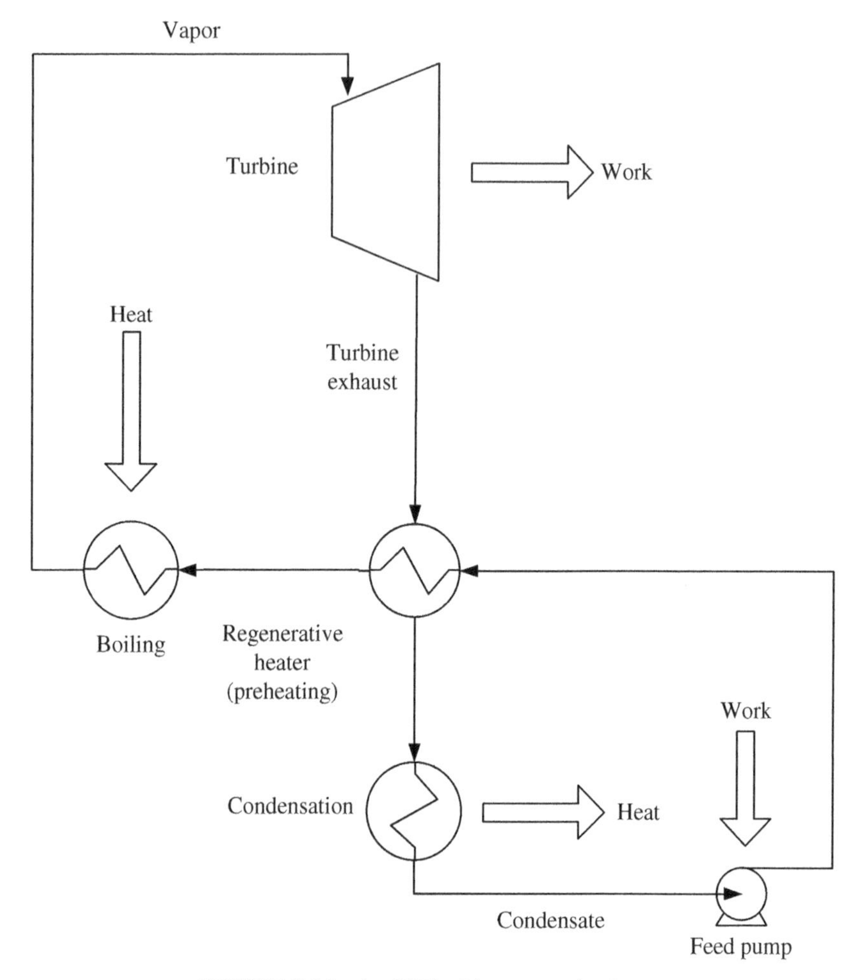

FIGURE 8.14 An ORC with regenerative heater

fluid which in turn transfers the heat at a lower temperature to the working fluid in a second exchanger. Various such fluids are commercially available consisting of petroleum-derived oils as well as synthetic silicon-based fluids.

REFERENCES

Andersson S, Kreisz S, Hunsinger H. Innovative material technology removes dioxins from flue gases. Filt Sep 2003;40(10):22–25.

Black J. Cost and performance baseline for fossil energy plants, volume 1: Bituminous coal and natural gas to electricity. U.S. Department of Energy Report; 2010. Report nr. DOE/NETL-2010/1397. Morgantown, WV: U.S. Department of Energy.

Chapel DG, Mariz CL, Ernest J. Recovery of CO_2 from flue gases: Commercial trends. Canadian Society of Chemical Engineers Annual Meeting; October 4–6, 1999; Saskatoon, Saskatchewan, Canada.

Finocchio E, Busca G, Notaro M. A review of catalytic processes for the destruction of PCDD and PCDF from waste gases. Appl Catal B Environ 2006;62(1):12–20.

Keenan JH, Keyes FG, Hill PG, Moore JG. *Steam Tables: Thermodynamic Properties of Water Including Vapor, Liquid, and Solid Phases*. New York: John Wiley & Sons, Inc.; 1969.

Liljelind P, Unsworth J, Maaskant O, Marklund S. Removal of dioxins and related aromatic hydrocarbons from flue gas streams by adsorption and catalytic destruction. Chemosphere 2001;42(5):615–623.

Pierce RR. Estimating acid dewpoints in stack gases. Chem Eng 1977;11:125–128.

Schippers F. High efficiency with lignite—Experience with lignite drying at Niederaussem. Prepared by RWE Power, IEA Workshop; October 2010; Moscow, Russian Federation. p 25–27.

Tumuluru JS, Sokhansanj S, Hess JR, Wright CT, Boardman RD. A review on biomass torrefaction process and product properties for energy applications. Ind Biotechnol 2011;7(5):384–401.

9

BRAYTON–RANKINE COMBINED CYCLE SYSTEMS

9.1 COMBINED CYCLE

A combined cycle combines two power cycles in series to obtain an overall efficiency significantly higher than the individual efficiencies of the two cycles making up the combined cycle. Figure 9.1 depicts simplified sketches showing the energy flows in systems employing a single cycle (e.g., the basic gas turbine cycle described in Chapter 6 without intercooling, reheat, or recuperation, i.e., the simple cycle form) and in a combined cycle. As illustrated in this diagram, for the simple cycle with a first law efficiency of 40%, 40 units of electrical energy are produced when 100 units of fuel energy are supplied while 60 units of energy not converted to electricity are rejected (primarily through its exhaust gas). The combined cycle consists of installing a second or "bottoming" cycle with a first law efficiency of 30% in series with the previous cycle of 40% efficiency. This results in an additional 18 units of electrical energy being developed from the energy rejected by the "topping" cycle resulting in an overall first law efficiency of as high as 58% for this combined cycle. Note that the generator, heat, and mechanical losses as well as the small change in efficiency of the topping cycle when its exhaust pressure is increased to accommodate the heat recovery for the bottoming cycle have been neglected in this analysis. A Brayton cycle or gas turbine is utilized for the topping cycle and a steam Rankine cycle for the bottoming cycle in the combined cycles discussed in this chapter although the topping

---

Sustainable Energy Conversion for Electricity and Coproducts: Principles, Technologies, and Equipment, First Edition. Ashok Rao.

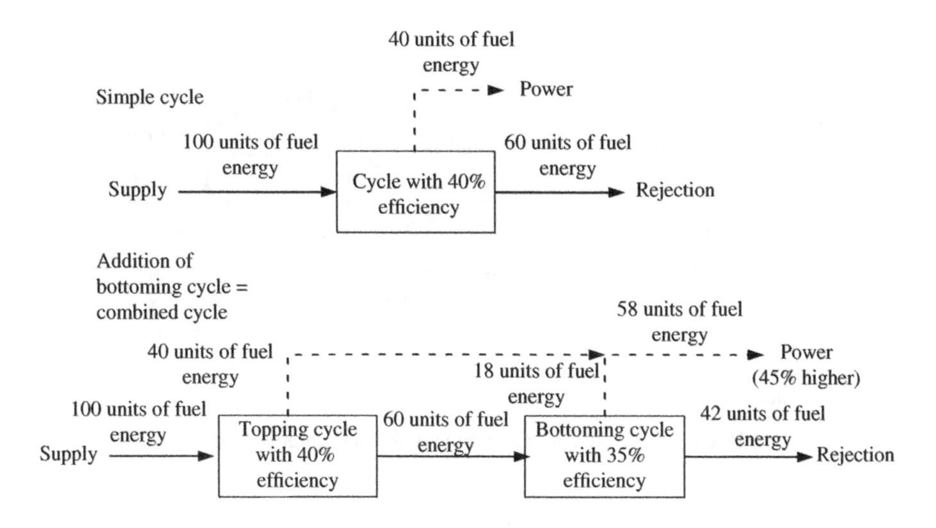

FIGURE 9.1 Energy flows in simple cycle and combined cycle

cycle can consist of a reciprocating engine or a fuel cell as described in Chapter 12 while the bottoming cycle can consist of a Rankine cycle using a working fluid other than steam or even a gas turbine in the case of the topping cycle being a fuel cell.

Combined cycles come in a variety of sizes depending on the size and number of gas turbines utilized; combined cycle sizes may range from less than 10 MW to in excess of 500 MW, say when using a single General Electric H class 50 cycle gas turbine.

9.1.1 Gas Turbine Cycles for Combined Cycles

Most combined cycle applications employ the basic Brayton cycle configuration or the simple cycle consisting of adiabatic compression and expansion, and near constant pressure heat addition in the combustor. The optimum pressure ratio for the gas turbine in this application is much lower than that required for peak thermal efficiency of a simple cycle gas turbine. A number of variations are possible to the basic Brayton cycle configuration, and among the variations, the one most useful in combined cycle applications incorporates reheat during the expansion step by the addition of a second combustor as described in Chapter 6, while intercooling may also be then taken advantage of in such cases since the reheat cycle optimizes at very high overall gas turbine pressure ratios. Reheat can either increase the cycle efficiency for a given turbine inlet temperature or lower the required turbine inlet temperature for a given thermal efficiency. A reheat gas turbine with spray intercooling is illustrated in Figure 9.2.

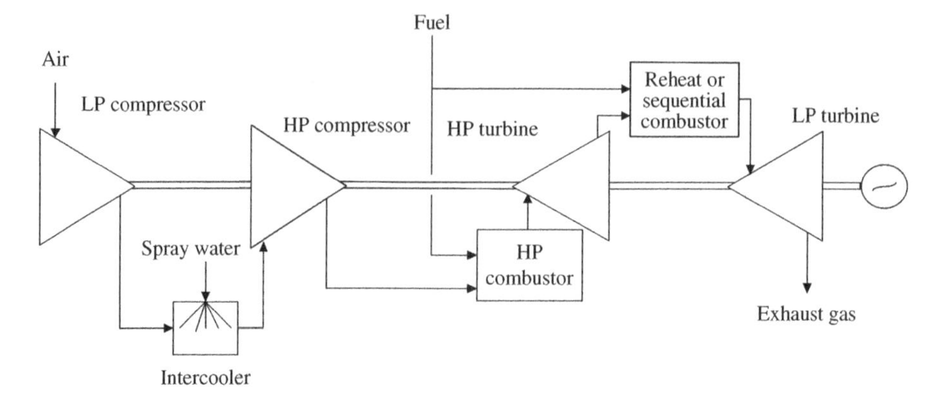

FIGURE 9.2 Reheat gas turbine with spray intercooling

9.1.2 Steam Cycles for Combined Cycles

Both saturated steam generation and its superheating (to increase the efficiency while reducing the moisture content in the steam turbine exhaust and the associated wear of the turbine blades and exhaust piping) are performed in the gas turbine exhaust before it is supplied to a steam turbine for expansion. Reheating of the steam can be advantageously used when the gas turbine exhaust temperature is high, typically in excess of 550°C and the steam turbine is large to be efficient enough to convert a significant fraction of the thermal energy added in the reheater to work. The amount of heat recovered from the gas turbine exhaust and consequently the work produced is limited by the "pinch" temperature, which is the minimum temperature between the heat source, the gas turbine exhaust in this case, and the heat sink, the evaporating boiler feed water (BFW) in this case. The design pinch temperature is typically 5–10°C depending on the value of energy recovered. By lowering the pressure of steam, more steam may be generated but the thermodynamic efficiency of converting the recovered heat to work is reduced and hence an optimum pressure exists for a given gas turbine exhaust temperature that maximizes the efficiency. Instead, by adding a second steam generator operating at a lower pressure allows more heat to be recovered and correspondingly more work to be produced. Both the high pressure superheated and the low-pressure steam (that may also be superheated to avoid droplet carryover) are admitted into a single steam turbine, each at the appropriate stage. Thus, in order to maximize the amount of heat recovered from the gas turbine exhaust and its cycle efficiency, steam at one or more lower pressures may also be produced but the economic justification of using a dual or triple pressure steam cycle depends on the gas turbine exhaust temperature and size of the plant. Today's large state-of-the-art combined cycles consist of generating steam at three different pressures. The plots of cumulative heat exchanged between the two fluids $\Sigma\dot{Q}$ plotted against temperature T of the two working fluids for single and dual pressure superheat-reheat steam cycles are depicted in Figure 9.3. It can be seen that additional heat from the lower temperature region is recovered by adding the low-pressure steam generation.

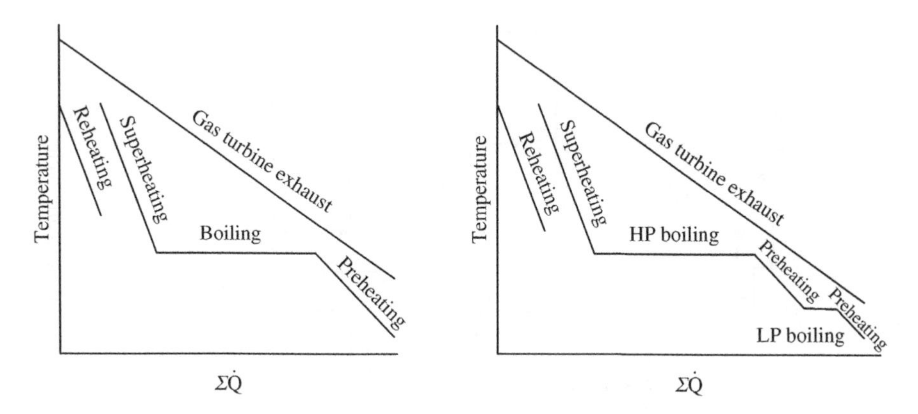

FIGURE 9.3 $\Sigma\dot{Q}$ versus T diagrams for single and dual pressure steam cycles

Figure 9.3 shows water entering the evaporator at its saturation temperature, which is an idealization. In practice, the economizer is designed to heat up water to a temperature typically below its saturation temperature by 5–10°C at the design point to avoid sudden change in phase of the liquid to vapor that may occur due to the pressure drop across the control valve located on the BFW line between the steam drum and the economizer as explained in Chapter 8. This also avoids boiling within the economizer especially during part-load operation as well as during start-up since this "subcool temperature" (difference between the saturation temperature and the actual fluid temperature) decreases. The BFW entering the evaporator coil from the steam drum, however, may be considered to be at the saturation temperature since the subcooled BFW entering the drum heats up to this temperature by mixing a portion of the steam exiting the evaporator coil providing this heat by condensation.

9.2 NATURAL GAS-FUELED PLANTS

9.2.1 Description of a Large Combined Cycle

A closed-circuit steam cooled gas turbine combined with a triple pressure reheat steam cycle (Smith, 2004) is depicted in Figure 9.4 in the form of a process flow diagram but without the equipment data that are typically included with process flow diagrams. Some of the modern gas turbines employ this type of steam cooling for cooling the turbine hot parts; the steam being provided by the bottoming cycle (instead of air flowing in an open circuit provided by the gas turbine compressor). Ambient air, after passing through a filter to remove air-borne particulates (especially those that are larger than 10 μm), enters the gas turbine air compressor. This gas turbine uses the premixed type of burners to reduce NO_x formation. Hot exhaust from the gas turbine flows through an heat recovery steam generator (HRSG). The design steam cycle conditions depend on the gas turbine exhaust temperature and flow rate and the steam

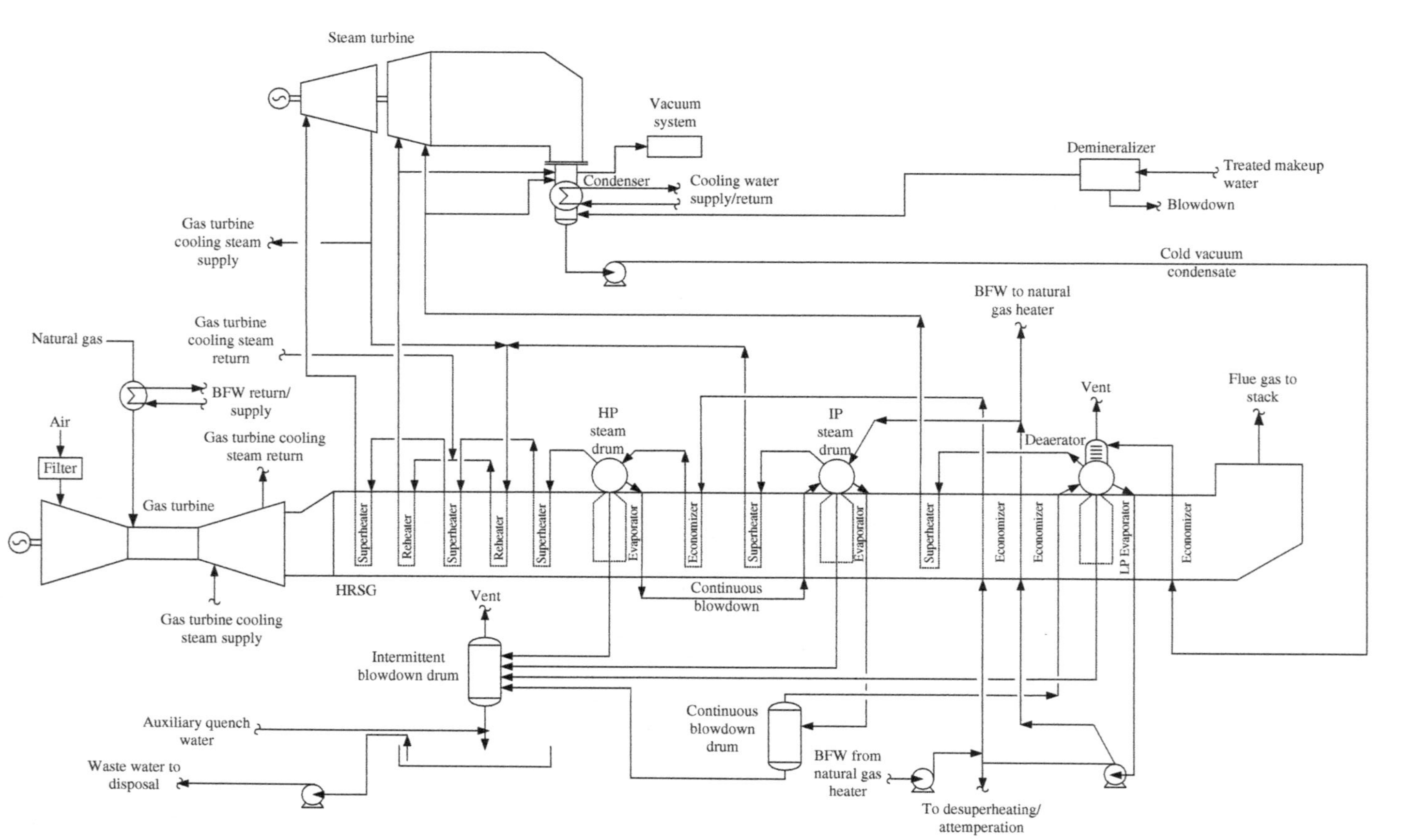

FIGURE 9.4 Closed-circuit steam cooled gas turbine combined cycle with triple pressure reheat subctritical steam cycle

cycle conditions presented in the following are typical of a steam-cooled H class gas turbine-based combined cycle.

The makeup BFW to the steam cycle consists of demineralized water since in larger combined cycle plants steam is generated at high pressures. The demineralizer consists of mixed-bed ion exchangers filled with cation and anion resins with internal regeneration in a single vessel. The system typically consists of two units, one unit in operating mode while another undergoing regeneration. The system also includes the required chemical storage and a neutralization basin. The resins are regenerated by using a strong acid such as HCl or H_2SO_4 as a source of hydronium ions for the cations, and liquid caustic NaOH as a source of hydroxyl ions for the anions.

The makeup BFW is typically sprayed directly into the surface condenser. This makeup is required since a small fraction of water from the steam drums is blown down to limit the buildup of solids in the steam drums. Also a small fraction of steam is vented from the deaerator whose function is described later. The surface condenser condenses steam leaving the low-pressure section of the steam turbine under a vacuum, this negative pressure being set by the temperature of the cooling medium used in the surface condenser. There are a number of ways of rejecting heat from the surface condenser depending on site conditions and economic parameters as described in Chapter 8. Cooling towers introduced in Chapter 5 are utilized where the supply of fresh water is not constrained, while for plants located close to a large body of water, once-through cooling may be considered. Air-cooled surface condensers have to be employed in dry locations but the first law efficiency of the overall plant is compromised due to the higher surface condenser operating pressure.

Again as mentioned in Chapter 8, when wet cooling towers are employed operating under ISO ambient site conditions, the operating pressure for the surface condenser can be as low as 4.8 kPa resulting from maintaining a reasonable temperature rise for the cooling water of about 11°C and a reasonable pinch temperature of 5–6°C in the surface condenser. The combined stream of condensed steam ("vacuum condensate") and makeup BFW is drawn from the surface condenser well by the vacuum condensate pump and preheated in an economizer located within the HRSG. It is then supplied to a dual functionality "integral" deaerator, which is part of the HRSG and removes dissolved gases such as O_2 and CO_2 in the BFW that can cause corrosion, as well as functions as a low pressure steam generator in the HRSG, generating steam at about 460 kPa. Additionally, an O_2 scavenging chemical that can also be a metal passivator is injected into the BFW. Examples of scavenging chemicals as mentioned in Chapter 8 are hydrazine (which is going out of favor because of its carcinogenic character), hydroquinone, diethylhydroxylamine, methylethylketoxime, carbohydrazide, and erythorbic acid.[1] The temperature of the BFW entering the deaerator is typically held at about 10–15°C below the deaerator temperature to ensure that enough steam is utilized for stripping out the dissolved gases. The excess steam generated within the integral deaerator after superheating, which in addition to increasing cycle

[1] Sodium sulfite (Na_2SO_3) is less expensive and is used in lower steam pressure boilers (up to 42 bar) but the sulfate formed cannot be tolerated in higher pressure boilers as it increases the dissolved solids content of the BFW.

efficiency also avoids condensation of steam into droplets, is provided to the low-pressure section of the steam turbine. A multi-stage pump withdraws the BFW from the deaerator to supply both intermediate-pressure BFW (extracted at an intermediate stage of the pump) and high-pressure BFW. The intermediate-pressure BFW flows through the intermediate pressure economizer in the HRSG and then enters the intermediate pressure steam drum of the HRSG to produce saturated steam at about 2850 kPa. The intermediate pressure steam thus generated is superheated to an intermediate temperature in the intermediate temperature section of the HRSG and then combined with steam exiting the high-pressure section of the steam turbine and reheated to about 570°C depending on the gas turbine exhaust temperature. This reheated steam is fed back to the intermediate pressure section of the steam turbine. Saturated high-pressure steam generated at about 17,400 kPa in the HRSG is superheated to about 570°C, again depending on the gas turbine exhaust temperature and supplied to the high-pressure section of the steam turbine. Typical design pinch temperature for the high-pressure evaporator coil is 6–9°C.

Control of temperature of the superheated and reheated steam entering the steam turbine is accomplished by attemperators in which the steam comes into direct contact with BFW supplied by the main pump, whereby the steam is cooled through the evaporation of the water. A fraction of the high-pressure steam turbine exhaust is supplied to the gas turbine for its closed-circuit cooling requirements. Steam returned from the gas turbine is also combined with the intermediate pressure steam before it is reheated within the HRSG. Typical design pinch temperature for the intermediate pressure evaporator coil is also 6–9°C. The reheater coils are generally interspersed among the superheater coils since the temperature of both the superheated and the reheated steam entering the steam turbine is typically held the same. As explained previously, the steam drums are continuously purged to limit the amount of build-up of dissolved solids and the continuous blowdown is cascaded from the high-pressure steam drum to the intermediate-pressure steam drum while blowdown from the intermediate pressure steam drum is routed to a drum where low-pressure steam is recovered. Water discharging from this last drum is depressurized in a second low-pressure drum and steam produced by this pressure reduction (flashing) is vented to atmosphere.

High gas velocities can be maintained in the HRSG to enhance heat transfer since the gas turbine exhaust is essentially free of particulates, but at the expense of a higher pressure drop across the HRSG. As the pressure drop increases, both the gas turbine output and efficiency are decreased and this inefficiency manifests itself as higher gas turbine exhaust temperature. A trade-off exists between the overall combined cycle efficiency versus size of the HRSG and consequently its cost; only a portion of this additional heat leaving with the gas turbine exhaust is converted to work by the steam cycle. Triple pressure reheat HRSGs are typically designed for a pressure drop of 28 mm Hg while those for a cycle without reheat are typically designed for a slightly lower pressure drop of 24 mm Hg. These values are inclusive of stack losses.

Steam from the bottoming cycle may be exported in CHP applications. Duct or supplemental firing may be utilized to increase steam production. It consists of combusting fuel gas in duct burners utilizing O_2 contained in the gas turbine exhaust flowing through the HRSG. Attention should be given to decrease in overall system efficiency and increase in emissions as well as impact on tube metallurgy due to the higher gas temperatures. Duct burners may be installed either upstream of superheater/reheater coils or more downstream within the HRSG, that is, after the gas has been cooled down somewhat in order to limit temperature rise when a significant degree of duct firing is required. The increase in the gas turbine exhaust to accommodate the additional pressure drop across the duct burner is quite low, on the order of 1 mm Hg.

To avoid using expensive tube materials such as Teflon®-coated tubes in the lower temperature sections of the HRSG, the tube surface temperatures should be kept above the acid dew point of the gas. The limiting acid dew point is typically set by sulfuric acid (H_2SO_4) formed by the oxidation of sulfur present in the fuel to SO_3 (typically 1–5%) within the gas turbine combustor (Ganapathy, 1989) and subsequently the SO_3 combining with water vapor to form the acid. As described in Chapter 8, an estimation of this acid dew point temperature T_{DP} (in K) may be made from the partial pressure P_i (in atmospheres) of each of the specie i (H_2O vapor and SO_3) using Equation 9.1 (Pierce, 1977) reproduced here for convenience:

$$\frac{1000}{T_{DP}} = 1.7842 + 0.0269 \log P_{H_2O} - 0.1029 \log P_{SO_3} + 0.0329 \log P_{H_2O} \log P_{SO_3} \tag{9.1}$$

The tube surface temperature of the vacuum condensate heater coil can be lower than the acid dew point when burning natural gas that contains the typical amounts of sulfur compounds in which case heated condensate is recirculated to raise the temperature of the fluid entering the coil and thus the tube surface temperature to avoid acid condensation.

If operation of the gas turbine is required when the steam cycle is not operating, diversion dampers to bypass the gas turbine exhaust directly to the stack can be provided upstream of the HRSG. Such an arrangement is typically avoided, however, since leakage across the diversion valve is a concern and during normal operation this leakage flow is not available for heat recovery. Furthermore, continuing operation of the gas turbine while performing maintenance or repair on the steam cycle may require an additional damper in series while maintaining a buffer gas in between the two dampers to avoid leakage of hot gas toward the HRSG, which would be a safety concern.

A single HRSG is paired with a gas turbine but multiple such trains (sometimes as many as four gas turbine/HRSG units) may be combined with a single steam turbine. Each of the turbomachineries in this arrangement has its own electrical generator while in a "single-shaft" arrangement offered for some gas turbine models, the gas and the steam turbines are both mounted on a common shaft connected to a single generator. In such cases, however, as many steam turbines as gas turbine trains are required.

Example 9.1. Calculate the H_2SO_4 dew point for the flue gas at a total pressure of 1 atm for Example 7.1, Plant B, assuming the fuel to the combined cycle was natural gas with 1 grain/100 standard ft^3 (0.0242 g/Nm^3) of sulfur and that 5% of the sulfur in the fuel gets converted to SO_3. The flue gas flow rate and composition correspond to a net power output of 555 MW and a heat rate of 6,466 kJ/kWh (LHV). Molecular weights of the flue gas and the natural gas fuel are 28.38 and 17.33 while the LHV of the natural gas is 47,454 kJ/kg.

Solution

Flue gas flow rate from Example 7.1 = 3,230,636/28.38 = 113,823 kmol/h

Natural gas flow rate = 555 MW × 1000 kW/MW × 6,466 kJ/kWh / (47,454 kJ/kg × 17.33 kg/kmol) × 22.14 Nm^3/kmol = 96,613 Nm^3/h

SO_3 in flue gas = 0.05 × 0.0242 g/Nm^3 × (kg/1000 g) / (32.06 kg/kmol) × 96,613 Nm^3/h = 0.00365 kmol/h

$$P_{SO_3} = \frac{0.00365 \text{ kmol/h}}{113{,}823 \text{ kmol/h}} \times 1 \text{ atm} = 0.03203 \times 10^{-6} \text{ atm}$$

$P_{H_2O} = 0.0867 \times 1 \text{ atm} = 0.0867 \text{ atm}$ (without having to make any adjustments to the mole percentage for H_2O given in Example 7.1 because of the very low concentration of the sulfur compounds in the flue gas).

Next, substituting the preceding two partial pressures into Equation 9.1, $T_{DP} = 359$ K or 85°C.

9.2.2 NO_x Control

With dry low NO_x combustors currently offered for natural gas applications consisting of premixing fuel with air and burning it under lean conditions to reduce flame temperature and thus formation of NO_x, values as low as 9 ppm by volume on a dry basis and "corrected" to 15% (by mole) O_2 content in the flue gas are guaranteed for some of the engines. Values as low as 2 ppm are being required, however, in a number of locations within the United States due to more stringent environmental emission standards. A selective catalytic reduction (SCR) unit is essential at the current time to approach such stringent emission requirements. NH_3, typically in aqueous form and which is easier to store, is injected upstream of the SCR located within the HRSG to react with the NO_x to form elemental N_2 and H_2O. Gas turbine back pressure is increased in order to accommodate pressure drop across the SCR but the impact on overall combined cycle performance is quite small since pressure drops as low as 4–5 mm Hg are typical. Optimum location within an HRSG for a vanadium pentoxide catalyst (3% V_2O_5 as the active material) based SCR is typically in the 300–400°C temperature zone. Any unreacted NH_3 slipping through the SCR can be an environmental concern and catalysts for NH_3 oxidation are under development for installation within the HRSG downstream of the SCR to oxidize the NH_3 to elemental N_2 and H_2O. The additional pressure drop required for such units is expected to be similar to that of a SCR.

9.2.3 CO and Volatile Organic Compounds Control

Oxidation catalysts typically formulated with platinum group metals can provide greater than 90% destruction of CO, volatile organic compounds (VOCs), formaldehyde, and other toxic compounds, and can be housed within an HRSG like the SCR. Conversion rates increase with temperature and thus it is advantageous to place this catalyst near the HRSG inlet. Catalyst life can typically be more than 10 years with continuous operation for this noble metal-based catalyst and occasional washing may be required to maintain its performance. Spent catalyst is typically recycled for the precious metal value. Typical pressure drop is as low as 3 mm Hg such that the impact on overall combined cycle performance is quite small.

9.2.4 CO_2 Emissions Control

Two basic options are available for CO_2 capture in gas turbine combined cycle plants: postcombustion and precombustion capture. In postcombustion capture, CO_2 is captured from flue gas before it enters the atmosphere while in precombustion capture CO_2 is captured from the fuel before combustion in the gas turbine. Separating the CO_2 is just one step in controlling this emission and its capture requires finding a "home" for it, and as explained in Chapter 1 geologic sequestration as well as using it in coal bed methane recovery is being investigated while commercial use in enhanced oil recovery has been practiced for a number of years. Depending on the sequestration method employed and distance between the sequestration site and the power plant, compression of the captured CO_2 to a pressure in the range of 11–15 MPa is typically required at the plant site.

9.2.4.1 Precombustion Control Fossil fuels such as natural gas are reformed catalytically (in the presence of a Ni-based catalyst) by reaction with steam, $CH_4 + H_2O = 3H_2 + CO$, or partially oxidized using air by reaction with O_2, $CH_4 + ½O_2 = 2H_2 + CO$ (in the presence of water vapor to minimize soot formation), to form a syngas with combustibles being primarily H_2 and CO. Reformers are discussed in more detail in Chapter 10. The syngas after some cooling/heat recovery and steam addition if required is then catalytically shifted to convert the CO to CO_2 while forming H_2 by the water gas shift reaction: $CO + H_2O = H_2 + CO_2$. A set of two adiabatic reactors in series with intercooling between the two reactors may be utilized to increase the conversion of this exothermic reaction. The first reactor may be operated at a higher temperature (as there is more heat released due to the larger amount of CO being converted) using a catalyst such as a Cu-added Fe-Cr formulation. The second reactor may be operated at a lower temperature (which favors higher conversion of the CO from a thermodynamic equilibrium standpoint) using a more active Cu-based catalyst. Shift reactors are also discussed in more detail in Chapter 10. After cooling the reactor effluent while recovering the heat, and after a final cooling step to near ambient temperature using cooling water, the syngas is contacted with a suitable solvent in an absorber column for the separation of the CO_2, and the solvent loaded with the CO_2 is regenerated in a stripper column using steam to release a high purity CO_2 stream, while the solvent circulates between the two columns. Typically, 85–90% of the CO_2 is absorbed into the solution in the absorber and the pressure at which the CO_2 is released depends on the type of solvent used. Physical solvents such as mixtures of the dimethyl ethers of polyethylene glycol are suitable for higher syngas pressures, typically in excess of 4 MPa, while chemical solvents such as amine solutions are suitable for lower syngas pressures. The syngas leaving the absorber, which is now mostly H_2 (or a mixture of H_2 and N_2 in the case of partial oxidation of the fuel using air) with residual amounts of CO_2 and CO, is available as a decarbonized fuel and may be combusted in gas turbines with reduced CO_2 emissions to the atmosphere (Rao et al., 2000). Compared to postcombustion capture, the partial pressure of the CO_2 present in the gas (syngas before decarbonization) is much higher than that in flue gases with the potential for lowering the energy penalty associated with carbon capture. The high H_2 content of the decarbonized syngas, especially with the fuel reforming option, precludes use in gas turbines with current design premixed combustion to limit NO_x formation since auto-ignition and flash-back are major challenges. Addition of a thermal diluent such as water vapor is required in such cases to limit NO_x generation when utilizing "diffusion" type combustion. Water vapor may be introduced into the fuel gas stream either as steam or generated *in situ* by direct contact of the fuel gas with hot water in a countercurrent column while recovering low-temperature waste heat. This second method is more attractive when there is a significant amount of low-temperature waste heat available in the plant. The upper limit on how much diluent may be added is not only constrained by the combustor design with respect to achieving stable combustion while limiting CO emissions but also by the surge margin of the gas turbine compressor. The pressure ratio of a gas turbine designed for operation on natural gas increases when firing a fuel gas that has a much lower calorific value than natural gas since the flow rate of gas entering the turbine section is increased[2] while inputting similar amount of fuel-bound energy. This increase in pressure ratio is dependent upon the amount and nature of diluent added and degree to which the gas turbine compressor inlet guide vanes are closed. Surge margin available in the compressor could thus constrain the amount of diluent that may be added and the resulting reduction in NO_x emissions. Air extraction from the compressor may be utilized in case of the partial oxidation scheme in order to limit the increase in engine pressure ratio, since the extracted air after cooldown, heat recovery, and booster compression can be efficiently used for the partial oxidation. The quantity of air that may be extracted is constrained, however, by the minimum air required to flow through the combustor liner for its cooling. Due to these constraints, use of diluent alone cannot reduce NO_x emissions sufficiently to meet the stringent requirement of 2 ppm (by volume on a dry basis, 15% O_2) and a SCR is still required with today's diffusion type burners.

Water vapor content of the working fluid flowing through the turbine section, especially with the reforming option, will be significantly higher when firing syngas as compared to that when firing natural gas in the gas turbine. Reduced turbine firing temperature may be required to limit hot gas path temperatures due to different aerothermal characteristics of the fluid and deterioration of thermal barrier coatings as well as any ceramics that may be utilized in future advanced gas turbines (ceramics are susceptible to high water vapor partial pressure environments at the high

[2] "Choked flow" (Mach number = 1) conditions prevail within the first stage nozzle of the turbine.

temperatures). Higher cooling air temperatures resulting from increase in the engine pressure ratio should also be taken into account in determining the reduced firing temperature. In steam-cooled gas turbines, however, the required reduction in firing temperature may be less significant since the temperature of the steam used for turbine cooling may be maintained independently of the gas turbine pressure ratio, as long as the lower pressure air-cooled stages of the gas turbine do not become limiting.

Both plant efficiency and cost are significantly compromised with precombustion capture of CO_2. When compared to a plant without CO_2 capture where natural gas is directly fired in gas turbines, a more than 30% increase in heat rate and a more than doubling of the specific plant cost ($/kW basis) may be expected (Rhudy, 2005). High-temperature H_2 separation membranes described in Chapter 11 are under development (Roark et al., 2003) and should provide some improvement in performance and possibly cost for the reforming option.

9.2.4.2 Postcombustion Control In postcombustion capture, the fossil fuel such as natural gas is first combusted in the gas turbine and CO_2 formed during the combustion process is separated from the HRSG exhaust gas for sequestration using commercially proven amine solvent processes as described in Chapter 8 (Chapel et al., 1999) such as aqueous MEA solution with proprietary additives. A major difference as compared to capture from a boiler plant flue gas is the much lower CO_2 concentration for a gas turbine derived flue gas, approximately 3% versus greater than 10% by volume while the O_2 concentration for a gas turbine derived flue gas is significantly higher (a large excess air being used in a gas turbine to reduce the temperature of the working fluid entering the turbine). Additionally, flue gas from a "dirty fuel" such as coal can contain many trace components that can form unwanted compounds by reacting with the CO_2 capture solvent, their concentrations depending upon the degree of pollution control measures taken upstream. The CO_2 separation process consists of an absorber/stripper system with the solvent circulating between the two columns (illustrated previously in Fig. 5.2). The HRSG exhaust is first cooled by direct contact with water rather than in a surface-type heat exchanger to minimize pressure drop, then pressurized in a blower to overcome the various pressure losses and then supplied to the absorber where it comes into contact with the MEA solution and about 85–90% of the CO_2 can be absorbed by the solution. The rich solvent, loaded with the CO_2, is regenerated in the stripper using steam to produce a high-purity CO_2 stream at near atmospheric pressure. An advantage of this postcombustion capture approach is that it does not impact the main combined cycle plant design except for equipment added downstream of the HRSG. Again, both plant efficiency and cost are significantly compromised with precombustion capture of CO_2, although the efficiency penalty is less severe when comparing current pre- versus postcombustion capture technologies. When compared to a plant without CO_2 capture where natural gas is directly fired in gas turbines, a more than 20% increase in heat rate and again more than doubling of the specific plant cost ($/kW basis) may be expected (Rhudy, 2005). A fraction of the flue gas prior to entering the absorber may be recycled to the gas turbine to increase the concentration of the CO_2 in the flue gas to reduce both the efficiency and the cost penalties. A reduction in NO_x formation in the combustor may also be expected. Effect on the gas turbine compressor from an aerodynamic standpoint should be taken into account if flue gas recycle is incorporated since the molecular weight and specific heat of CO_2 is much higher.

Another approach to postcombustion capture that is under investigation, which is analogous to oxy-combustion of coal mentioned in Chapter 8, consists of using O_2 provided by an air separation unit for combusting the gas turbine fuel such that the cooled flue gas consists of essentially all CO_2, which can simply be pressurized for sequestration after some treatment, thereby eliminating the CO_2 separation process (the amine unit). However, this approach requires the addition of an air separation unit that at the present time uses cryogenic technology[3] and requires significant amount of electrical energy for its operation and is capital intensive. Also, a special type of combustor is required for this application such as the one being developed by Clean Energy Systems (Martinez-Frias et al., 2002), which uses a combustor derived from rocket engine technology operating at a pressure in excess of 10 MPa and a temperature of 540–760°C. Recycled water and in some cases steam is injected into the combustor to control the combustor exhaust temperature (which is the turbine inlet temperature). In the Clean Energy Systems process, the combustion products consist of approximately 90% H_2O vapor and 10% CO_2 by volume (and a small amount of residual O_2) when natural gas is the fuel. This gas stream is introduced into a high-pressure turbine that exhausts at a pressure of approximately 1 MPa to be reheated to a temperature in excess of 1240°C by combusting additional natural gas with additional O_2. The reheated gas then enters an intermediate pressure turbine followed by a low-pressure turbine that exhausts the gas into a condenser at atmospheric or sub-atmospheric pressure to condense out the H_2O vapor and leaving a high-purity CO_2 stream. The majority of the condensate is recirculated to the high-pressure combustor after preheating in the turbine exhaust. Depending on the specifications for the O_2 content of the CO_2 stream, the humid CO_2 exiting the condenser after some compression may be treated in a catalytic combustor where additional natural gas can be added to use up the residual O_2. A variant of this oxy-combustion cycle is the Graz cycle developed by Graz University of Technology (Sanz et al., 2005). It utilizes a single combustor, that is, a non-reheat cycle and operating at a much lower pressure of about 4 MPa as compared to Clean Energy System's high-pressure combustor that operates around 10 MPa while the combustor exhaust temperature is controlled by recycled steam and a compressed mixture of CO_2 and H_2O vapor.

9.2.5 Characteristics of Combined Cycles

Some of the advantages of the combined cycle when compared to a Rankine only cycle (i.e., boiler-based power plant) are listed in the following. It should be borne in mind, however, that the combined cycle requires a cleaner fuel such as natural gas, syngas (derived from coal or biomass), or distillate, which are significantly more

[3] Cryogenic air separation consists of liquefying air and then distilling out the N_2 and to a lessser extent the Ar to produce a high-purity O_2 stream.

TABLE 9.1 J class gas turbine combined cycle features[a]

Turbine inlet temperature	1600°C
Generator frequency	60 Hz
Gas turbine pressure ratio	23:1
Gas turbine power output	327 MW
Combined cycle output	470 MW
Combined cycle efficiency	61.5% (natural gas LHV)

[a] Data from Hada et al. (2012).

expensive than dirty fuels such as coal and biomass, which the boiler power plant can utilize directly in the combustion process.

- High thermal efficiencies, greater than 60% for natural gas on a LHV basis with a current state-of-the-art combined cycle utilizing a "J class" gas turbine at an ambient temperature of 15°C (see Table 9.1).
- Overall cycle efficiencies utilizing advanced technology gas turbines approaching 65% on natural gas on a LHV basis may be expected in the 2020–2025 time-frame (Dennis, 2008).
- Outstanding environmental performance including lower cooling water requirement.
- Easy (and can be designed for quick) startup and shutdown that complement intermittent renewables such as wind and solar where the combined cycle is required to make up the difference in power generation when power output from such renewables goes down. The number of starts per year does affect the life of the gas turbine as well as that of the tubes in the high-temperature sections of the HRSG due to the thermal stresses, however.
- Higher operating reliability.
- Significantly lower staffing, capital cost, and construction time requirements; a combined cycle takes approximately one-third the time it takes to build a pulverized coal plant.
- Smaller plot space requirement and capability for phased construction, that is, the gas turbine can be installed without the HRSG or the steam cycle during the initial phase (when utilizing a non-steam cooled gas turbine) for peaking service, and may be converted to a combined cycle later for base-loaded power generation.

Combined cycle plants are often designed for normal operation on natural gas with distillate fuel oil for backup to handle any interruption in the natural gas supply. However, because of additional emissions of SO_x formed from the sulfur present in fuel oil as well as possible deactivation of the CO oxidation catalyst that is included in the turbine exhaust in many installations, this practice has become quite uncommon in recent years. Similar to a coal-fired boiler discussed in Chapter 8, the presence of

SO_3 in the gas turbine exhaust leads to the undesirable formation of ammonium salts (ammonium bisulfate, sulfate, and bisulfite) by reaction between NH_3 slipping through the SCR (which is also typically installed in the turbine exhaust) with SO_3. These salts deposit in the lower temperature sections of the HRSG increasing the resistance to heat transfer in HRSG tubes requiring more frequent washes, while the fractions that do not deposit give rise to particulate emissions.

Both the power output and the efficiency of a gas turbine are affected by ambient conditions of temperature, barometric pressure, and to a lesser extent humidity as described in Chapter 6 but the steam bottoming cycle tends to dampen the effects on the overall combined cycle efficiency. The effect of barometric pressure on the combined cycle performance is similar to that on a simple cycle gas turbine discussed in Chapter 6 while the effect of ambient temperature on heat rate of a combined cycle is not as significant as it is on power output and the exact magnitude of the sensitivity depends on the gas turbine exhaust temperature and flow rate (corresponding to a certain set of ambient conditions) selected for optimizing the bottoming steam cycle design. The combined cycle heat rate may actually show a minimum at the ambient temperature used in the design of the bottoming cycle and may increase at lower temperatures. When wet cooling towers are utilized for plant heat rejection, the cooling water temperature is affected by the relative humidity, higher the humidity (higher is the wet bulb temperature), higher is the cooling water supply temperature. Increase in cooling water temperature increases the steam surface condenser operating pressure that then reduces the output of the steam turbine and of the combined cycle. Plant capital cost is directly affected by power output while decrease in efficiency affects plant operating cost, both on a dollar per kilowatts basis.

Power output of a combined cycle can be reduced by closing the gas turbine compressor inlet guide vanes discussed in Chapter 6 to reduce the air flow rate without having to reduce the gas turbine firing temperature. With this mode of control, the overall combined cycle heat rate is not increased significantly but further reductions in power output require the gas turbine firing temperature to be reduced, which has a significant effect on the combined cycle heat rate. For a 50% decrease in power demand from its rating point, the heat rate can increase by as much as 20%. The bottoming cycle of a combined cycle is typically designed as a floating pressure steam system that is more efficient at part load. During part load operation, however, steam pipes should be properly sized for the higher fluid velocities corresponding to the lower steam pressures causing the reduction in fluid density despite the lower mass flow rates. The heat transfer surfaces within the HRSG will be "over-sized" at part load operation with a potential for steaming in the intermediate pressure and low-pressure economizers unless proper design measures are taken upfront such as designing these economizers and piping for the two-phase flow.

Example 9.2. The part load performance of a natural gas fired combined cycle consisting of two gas turbines with an efficiency of 60% at full load is shown in Figure 9.5. Estimate the efficiency of the plant at 50% power output for the following two scenarios: (1) both gas turbines operating and (2) a single gas turbine operating while the other is shut down.

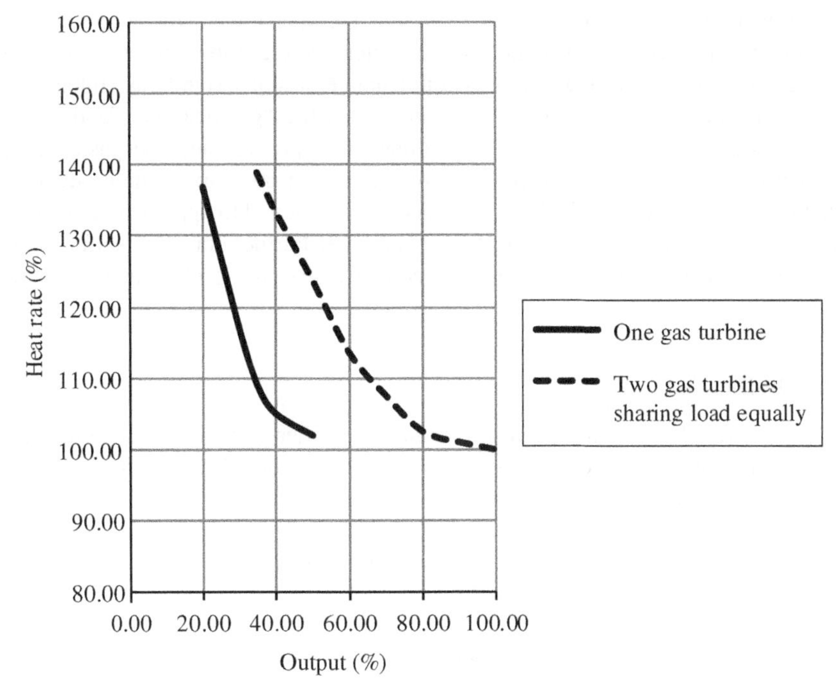

FIGURE 9.5 Part load performance of combined cycles

Solution

1. Both gas turbines operating

At 50% load, heat rate as % of base rating = 124%
Corresponding efficiency = 60% × (100%/124%) = 48%

2. Single gas turbine operating

At 50% load, heat rate as % of base rating = 102%

Corresponding efficiency = 60% × (100%/102%) = 59% showing the advantage in having multiple (smaller) trains, which provides the option of shutting down some of the trains to maintain higher overall efficiency, but the plant cost tends to increase with multiple trains and a system study is required to select the best configuration from an economic and emissions standpoint.

9.3 COAL AND BIOMASS FUELED PLANTS

Fuels such as coal and biomass that are too "dirty" to be fired directly in a gas turbine require conversion into a clean fuel gas. This conversion process involves gasification that consists of partially oxidizing the fuel under sub-stoichiometric conditions to produce a syngas and requires extensive cleanup to meet the engine fuel specifications as well as the environmental emissions standards. Integrated gasification combined cycle (IGCC) that integrates the "gasification island" with the combined cycle "power island" is commercially proven, and the gasification process itself has been in commercial use for more than 60 years in the production of chemicals or synthetic fuels. IGCC plants can be designed to approach zero emissions using the precombustion CO_2 capture process. Furthermore, the sulfur compounds removed from the syngas may be converted to salable elemental sulfur or sulfuric acid.

9.3.1 Gasification

The major overall reactions occurring within a gasifier (without considering the decomposition of the coal or biomass by a pyrolysis or devolatilization process of the organic constituents) are as follows:

$$\text{Partial oxidation: } 2C + O_2 = 2CO;\ \Delta H < 0 \tag{9.2}$$

$$\text{Steam gasification: } C + H_2O = CO + H_2;\ \Delta H > 0 \tag{9.3}$$

$$\text{Shift: } CO + H_2O = CO_2 + H_2;\ \Delta H < 0 \tag{9.4}$$

$$\text{COS hydrolysis: } H_2O + COS = H_2S + CO_2;\ \Delta H < 0 \tag{9.5}$$

$$\text{Methanation: } 3H_2 + CO = CH_4 + H_2O;\ \Delta H < 0 \tag{9.6}$$

All the preceding reactions are exothermic except for reaction (9.3), which is endothermic and shows that H_2O in the form of steam or liquid water may be utilized to control the gasifier temperature. The highly exothermic hydrogasification reaction $C + 2H_2 = CH_4$ occurs only to a limited extent[4] and the majority of the CH_4 produced by most gasifiers is due to the decomposition of the volatile matter in the feed. The sulfur present in the feed is evolved primarily as H_2S with some as COS while other sulfur compounds such as CS_2 and mercaptans are produced in insignificant amounts when the gasifier operating temperature is high (>1100°C). The nitrogen present in the feed is evolved primarily as elemental N_2, some NH_3, and much smaller amount of HCN when the gasifier operating temperature is high. The Cl present in the feed is evolved primarily as HCl.

There are three major types of gasifiers classified based on the way the solids flow through the gasifier, the moving bed gasifier, the fluidized bed gasifier, and the entrained bed gasifier:

- Moving Bed Gasifier: This type of gasifier is exemplified by the Lurgi as well as the British Gas Lurgi (BGL) gasifiers that consist of feeding coarse solids at the

[4] Unless the operating pressure of the gasifier is very high, the operating temperature not very high, and a suitable catalyst such as a potassium salt is present.

top of the gasifier while the oxidant and steam are introduced into the gasifier at the bottom, the solids flowing through the gasifier by gravity. The gasifier can be divided into four distinct zones: the drying/preheating zone at the top followed by the devolatilization zone, then the gasification zone, and at the very bottom is the combustion zone. The gases flow countercurrently to the solids through the four zones and exit the gasifier at the top containing primarily H_2, CO, CO_2, H_2O, CH_4, and other hydrocarbonaceous components such as oils, tars, and oxygenates. The hot gases produced by the combustion reactions in the bottom zone undergo gasification reactions in the zone above it, these gases in turn heat up and devolatilize the feed, and finally the gases flowing upward perform the drying of the feed. The concentrations of COS, CS_2, mercaptans, NH_3, and HCN in the syngas are more significant than the fluidized bed and the entrained bed gasifiers. Some of the sulfur and nitrogen is also distributed in the tars. The efficiency of the gasifier is high but since it produces the tars and oils, the gas cleanup is more complex, and the gasifier can handle limited amount of fines in the feed. A variant of the original Lurgi gasifier, the BGL gasifier recycles the tars and oils separated from the syngas to the gasifier by introducing these components into the bottom section of the gasifier along with fines. The much higher CH_4 content of the syngas limits the degree to which the syngas may be decarbonized to control CO_2 emissions in an IGCC (with precombustion capture) unless a reformer is utilized to convert the CH_4 into H_2 and CO but this would add to the plant cost and complexity.

- Fluidized Bed Gasifier: This type of gasifier is exemplified by the high temperature Winkler and the TRIG™ gasifiers and consists of introducing dried feed that is typically less than ¼ in. (0.63 cm) in size at the bottom of the gasifier, while the fluidization medium is a mixture of the oxidant and the moderator steam introduced at the bottom of the gasifier. Recycle syngas may be required to maintain the required gas velocity for fluidization of the bed material especially when the oxidant is not air but O_2 (typically 95% on a mole basis). In the bottom of the gasifier where oxidizing conditions prevail, the temperature may be high enough to possibly fuse the ash particle together to form agglomerates. The gasification reactions predominate in the top section of the gasifier and the temperatures are lower in the neighborhood of 900–1050°C. The raw syngas exiting the gasifier at the top in this temperature range is higher than that of a moving bed gasifier and is essentially free of hydrocarbons heavier than CH_4 although the CH_4 content may be high enough to limit the degree of syngas decarbonization required for control of CO_2 emissions. In biomass applications, however, such a gasifier may still be desirable as CO_2 capture may not be required. Fluidized bed gasifiers are very suitable for biomass feedstocks since biomass tends to be very reactive, while it is much easier to produce a feed with the suitable physical characteristics such as particle size for a fluidized bed. Furthermore, some of the biomass feedstocks such as municipal solid waste tend to be quite heterogeneous and again in this respect, a fluidized bed gasifier with its large capacitance due to the large inventory of solids contained within the gasifier has large enough response time for making any adjustments required to the flow rates of the oxidant and the steam.

- Entrained Bed Gasifier: The single stage downflow entrained bed gasifier is exemplified by the Siemens and General Electric gasifiers and consists of introducing finely ground feed as a dry power or as water slurry in the case of solids into the gasifier along with the oxidant. Steam is added as the moderator when the feed is in the form of a dry powder as in the Shell entrained bed gasifier. The General Electric gasifier introduces the feed in the form of slurry although a dry feed system is under development. Siemens offers gasifiers with either type of feed system. With the dry feed system, feed drying is required that may be combined with the feed pulverization operation in an "air swept" mill. The drying operation does represent a thermal penalty on the IGCC system but on the other hand, the wet slurry feed systems require additional oxidant for the gasifier to generate the heat required for the *in situ* drying that also represents a thermal penalty on the IGCC system. These single stage entrained bed gasifiers have highest operating temperatures among the three types, the gasification occurring at temperatures typically in excess of 1200°C with the ash forming a slag. Because of the very high operating temperature, the gases leaving the gasifier have very little CH_4 content and are essentially free of any of the heavier hydrocarbons. The E-Gas (slurry feed) gasifier and the Mitsubishi (dry feed) gasifier are also entrained beds but incorporate two stages of feed introduction. The bottom stage operates at a much higher temperature where the oxidant is also introduced and hot gases leaving this stage provide the heat required to dry, devolatilize, and gasify the feed introduced into the second stage. The result is that the specific O_2 consumption is reduced while the cold gas efficiency (measured by the fraction of the chemically bound energy of the feed that is conserved as the chemically bound energy of the syngas) is increased. On the other hand, the CH_4 content is higher but not as high as the moving bed gasifiers.

9.3.2 Gasifier Feedstocks

In addition to coal and biomass, the various other feedstocks that may be gasified include paper mill black liquor, refinery residues including petroleum coke, asphaltenes, visbreaker bottoms, refinery and petrochemical wastes, and oil emulsions (e.g., Orimulsion™). The following lists the important feedstock characteristics to be taken into account for gasifier selection:

- Reactivity: Lower reactivity feedstocks (typically higher rank coals and petroleum coke) have limited carbon conversion in fluidized bed gasifiers.
- Moisture content: High moisture content is especially a thermal penalty for both dry and slurry feed gasifiers (in the case for slurry feed gasifier, when it is the bound moisture, while in dry feed gasifiers, due to the energy required for drying). Typically lower rank coals and biomasses have higher moisture contents.
- Ash content: High ash content is a thermal penalty for slurry feed gasifiers while low ash content is a challenge for gasifiers that utilize a membrane wall such as the Shell gasifier and may require ash recycling.

- Ash fusion temperature under reducing conditions:[5] High ash fusion temperature is a thermal penalty for slagging gasifiers. It can shorten refractory life in refractory-lined gasifiers and may require a fluxing agent. Low ash fusion temperature is a challenge for non-slagging gasifiers and may increase gasifier steam requirement.
- Particle size: Can limit the fraction of feed that may be utilized in a moving bed gasifier due to the lower limit on fines and may require briquetting.
- Free swelling index: Can cause plugging in a moving bed gasifier and may require a stirrer to break up agglomerates (bituminous coals typically have higher free-swelling indices).

9.3.3 Key Technologies in IGCC Systems

In addition to the gasifier, other key technologies in an IGCC system are the supply of oxidant to the gasifier, the systems associated with treating the raw syngas produced by the gasifier before it may be utilized in a gas turbine, and the gas turbine itself.

9.3.3.1 Air Separation Technology The cold gas efficiency of a high-temperature gasifier is especially increased while the size of the downstream equipment reduced when the O_2 concentration in the oxidant is increased. Thus, high-temperature gasifiers utilize a high purity O_2 stream provided by an air separation unit (typically 95% purity on a volume basis) instead of directly using air that is about 21% O_2 on a volume basis. With air as the oxidant, a significant portion of the feed-bound chemical energy is degraded to thermal energy to heat up the N_2 that enters the gasifier (as part of the oxidant air stream) to the gasifier effluent temperature. When CO_2 capture is required or when the syngas is to be utilized for producing or coproducing products such as H_2 or C_2H_5OH, it is advantageous to have an "O_2-blown" gasifier instead of an "air-blown" gasifier. The use of air as an oxidant in a gasifier can be attractive in some situations such as the following:

- In biomass applications where the plant scale tends to be smaller and the cost of producing O_2 at smaller scales becomes unattractive.
- In lower temperature gasifiers such as fluidized bed gasifiers where the cold gas efficiency is affected to a lesser extent.
- Especially when limestone is injected into such low-temperature gasifiers for capture of the sulfur compounds evolved so that an acid gas removal (AGR) system downstream of the gasifier is not required. Note that the volume of syngas to be treated is increased (due to the N_2) while there is a corresponding reduction in the partial pressure of the sulfur compounds present in the syngas.

[5] Typically lower under reducing conditions than under oxidizing conditions for coals while higher for petroleum cokes.

Conventional air separation technology for large-scale applications consists of cryogenically separating the air to produce the high purity O_2 stream. In an IGCC using an O_2-blown gasifier, both the O_2 used in the gasifier and the N_2 used in the gas turbine for NO_x control are required at elevated pressures. In such applications, an air separation unit (ASU) designed with the distillation of the air (to produce the high purity O_2 stream) occurring at an elevated pressure is preferred over a low-pressure design where distillation of the air occurs near atmospheric pressure. A reduction of about 2% is typically realized both in plant heat rate and cost. Distillation pressure within the ASU "cold-box" affects the liquid bubble point and as this pressure is increased, the cold-box temperatures increase, that is, less severe cold temperatures are required. There is a corresponding reduction in the pressure ratio across the cold-box (ratio of incoming air pressure to outgoing pressures of the O_2 and N_2 streams) resulting in a net increase in the overall IGCC plant efficiency. As this distillation pressure is increased, the relative volatility between N_2 and O_2 approaches unity requiring more stages in the distillation operation and an optimum upper limit exists for the pressure, which is much lower than the gasifier or the gas turbine combustor operating pressure. Thus, both the O_2 and the N_2 streams exiting the cold-box require compression.

Optimum O_2 purity in IGCC applications with either the elevated pressure or the low-pressure ASU is 95% on a volume basis with the number of distillation stages decreasing steeply as purity is reduced from 99.5 to 95%. It remains quite insensitive when purity is further reduced but the size of most equipment downstream of the cold-box increase (slightly) while the efficiency of the gasification unit decreases and the O_2 compression power increases. Both the Demkolec (Hannemann and Kanaar, 2002) and Polk County IGCC plants utilize an elevated pressure ASU with 95% purity O_2 but in coproduction plants such as those coproducing high purity H_2 (utilizing pressure swing adsorption), higher purity O_2 may be required to limit its Ar content.

Air supply pressure to the cold-box of a low-pressure ASU is in the range of 350–600 kPa while that of an elevated pressure ASU is typically set by the gas turbine pressure ratio when air is extracted from gas turbine to provide at least a fraction of the total air required by the ASU. An advantage of extracting air from the gas turbine compressor discharge is that it makes room for the injection of the large amount of thermal diluents required in the gas turbine combustor for NO_x control while keeping its pressure ratio safely below the surge line of the compressor, gas turbines currently being designed for fuels such as natural gas. When the gas turbine pressure ratio is very high, partial expansion of the extracted air may be required in order to limit the cold-box operating pressure such that the relative volatility between O_2 and N_2 does not get too close to unity.

The power usage of the cryogenic ASU is typically more than half of the total power consumed by an IGCC or 5–10% of the total power produced. In order to reduce this severe penalty, other technologies are being developed such as air separation membranes utilizing semi-conductor materials that operate at temperatures around 800–900°C with a goal of reducing power usage and capital cost by as much as 30% (Armstrong, 2005) and discussed in more detail in Chapter 11. Integration of this membrane unit with a gas turbine is required to make it economical with the feed air to the membrane being extracted from the compressor discharge (followed by preheating the extraction air by directly firing syngas), while returning the hot depleted air (at ~800°C) from the membrane unit to the gas turbine combustor. As much as 50% of the compressor discharge air is required, but this very large amount of air extraction or the hot depleted air return (by which combustor liner design and materials would be impacted) is not possible with current large-scale gas turbines.

9.3.3.2 Syngas Treatment The following paragraphs describe commercially available "cold-gas cleanup" technologies for the treatment of syngas before it can be fired in gas turbines. "Warm gas cleanup" technologies that are under development are operated at higher temperatures (300–400°C) such that the syngas is kept above its water dew point eliminating extensive waste-water treatment typically required by gasification using the cold-gas cleanup, as well as minimizing the irreversibilities associated with cooling the syngas down to near ambient temperatures for the cold-gas cleanup. Some of the warm gas-cleanup processes are in the demonstration phases (Butz et al., 2003; Gangwal et al., 2004; Krishnan et al., 1996).

In the case of cold-gas cleanup, raw gas leaving a gasifier is either cooled by heat exchangers while generating steam and/or quenched with water by direct contact. The cooled gas is further cleaned by scrubbing with water to remove contaminants such as particulates, and soluble compounds of alkalis, halogens, and nitrogen. The exact nature and concentration of contaminants depends on the type of feedstock and type of gasifier and in Table 9.2 are listed the various contaminants that can be present in raw syngas derived by the gasification of coal. In the case of biomass consisting of agricultural products, potassium salts can be in significant concentrations while in the case of biomass consisting of municipal solid waste, chlorides can be in significant concentrations in the syngas.

9.3.3.3 Acid Gas Removal The sulfur compounds in syngas as well as the CO_2 are acidic in nature and the processes used for the removal of such components are called acid gas removal (AGR) processes. The available processes for capturing sulfur compounds from syngas include the following:

- Selective amine scrubbing using an aqueous methyl diethanol amine solvent with additives
- Rectisol that uses methanol as the solvent

TABLE 9.2 Contaminants in coal-derived raw syngas

Sulfur compounds	H_2S, COS, CS_2, mercaptans
Chlorides	NaCl, HCl
Nitrogen compounds	NH_3, HCN
Particulates	Ash (Si, Al, Ca, Mg, Na, K, and sulfate ions), soot
Volatile metals	Hg, As, Cd, Se
Carbonyls	$Ni(CO)_4$, $Fe(CO)_5$

- Selexol™ that uses a mixture of dimethyl ethers of polyethylene glycol as the solvent (flow diagram was illustrated in Fig. 5.1).

All the preceding processes use an absorber column and a stripper column with the solvent circulating between the two columns. The acid gases are scrubbed out from the syngas by the solvent within the absorber while the stripper releases the absorbed acid gases from the solvent. Typically, steam is used for the stripping operation that regenerates the solvent for recirculating it back to the absorber.

The amine AGR process requires a significant amount of steam to release the absorbed acid gases since amines are "chemical solvents," that is, chemical bonds, although weak, are formed between the amine and the acid gas component. Additives can improve selectivity between H_2S and CO_2 absorption rates so that the fraction of CO_2 co-absorbed by the solvent is minimized. This type of AGR process is suitable in applications where precombustion CO_2 capture is not required and the loss of CO_2 from syngas has the detrimental effect of reducing the amount of motive fluid available at pressure for expansion in the gas turbine while increasing the amount of thermal diluent required by the combustor for NO_x control. In addition, the solvent circulation rate and utility usages including the stripping steam demand are increased. The efficiency and cost of the downstream sulfur recovery unit (SRU) and the tail gas treating unit are negatively affected due to the presence of the large concentration of CO_2 in the acid gas released in the AGR stripper. The SRU converts the captured H_2S and any co-absorbed COS in the acid gas to elemental sulfur. The tail gas treating unit hydrogenates trace amounts of elemental sulfur as well as its oxygen containing compounds in the gas leaving the SRU and then recovers the additional H_2S thus formed in a second amine unit before venting the tail gas to the atmosphere to minimize the overall system sulfur emissions. The amine process is used in the two U.S. commercial IGCC plants: Polk County, Florida (McDaniel and Shelnut, 1999) and Wabash, Indiana (Keeler, 1999). In the Wabash plant, tail gas is recycled to the gasifier after hydrogenation, instead of treating it in an additional amine unit to recover the H_2S for recycling to the SRU.

Both the amine and the Selexol™ processes require hydrolysis of COS to convert it into H_2S upstream of these units since only a small fraction of the COS is absorbed by these solvents while the Rectisol process does not require it because the COS is readily absorbed by its solvent. The hydrolysis reaction $COS + H_2O = H_2S + CO_2$ is accomplished in a fixed bed reactor containing an activated alumina or titanium oxide catalyst that has been shown to be more robust.

Both Rectisol and Selexol™ processes utilize physical solvents and work best when the acid gas partial pressures are high. Both desulfurization and decarbonization of the syngas may be combined in these processes while producing a separate acid gas stream rich in the sulfur compounds and another that is essentially all CO_2 released from the solvent at pressures significantly higher than atmospheric to reduce the power usage for compressing the CO_2 stream to pressures required for sequestration. With typical acid gas concentrations in the syngas to be treated, the syngas pressure required is typically greater than 2800 kPa for Rectisol and greater than 4100 kPa for Selexol™. The solvent is chilled by refrigeration to enhance its solubility, especially in the Rectisol process. Rectisol is especially suitable for application where a very low specification for syngas sulfur content is required (down to 0.1 ppmv) such as in synthesis processes that contain catalysts that are susceptible to sulfur poisoning without requiring a separate trace sulfur removal step. Selexol™ may be used in IGCC plants where CO_2 emissions are to be controlled while the sulfur specification is not as stringent as 0.1 ppmv (Rao et al., 2005).

9.3.3.4 Trace Component Removal Some capture of the metal carbonyls (such as those of Ni and Fe) can occur by the adsorption on the carbon particles accompanying the raw syngas leaving the gasifier and collected in barrier filters located upstream of the raw syngas scrubber. The Rectisol solvent captures carbonyls by converting them into their respective sulfides. A bed containing activated carbon may also be provided upstream of the AGR to remove the carbonyls. Hg, As, Cd, and Se that enter the gasifier as constituents of the feed are typically volatilized within the gasifier in metallic form or as compounds and leave with the raw syngas. A sulfided activated carbon bed has been utilized to remove As from syngas at the coal to chemicals Tennessee Eastman gasification plant (Trapp et al., 2002). It was found that Hg was also being removed by this bed. It is captured predominantly as a sulfide and some as elemental. The spent carbon has to be treated as a hazardous waste, although attempts are being made by technology developers to recover elemental Hg. Calgon offers a sulfided-activated carbon to remove Hg down to 0.01–0.1 μg/Nm^3 concentration levels in syngas at near ambient operating temperature and low moisture content; its capture being reduced as the temperature and the relative humidity are increased. Their experience has shown that As is also captured by this activated carbon. SudChemie also offers activated carbons for removal of As and its compounds while a Cu-impregnated carbon is offered when As is present as an organic compound in the syngas. Simultaneously, capture of Cd and Se compounds by the activated carbon is also expected to occur.

9.3.3.5 Gas Turbine The constraints imposed on gas turbines in natural gas-fired combined cycles with precombustion CO_2 capture discussed earlier also apply in IGCC applications.

9.3.4 Description of an IGCC

An IGCC plant design utilizing commercially available technologies for near zero emissions (i.e., including CO_2 capture) is depicted in Figure 9.6 in the form of a simplified block flow diagram using slurry-fed high-pressure entrained bed gasifiers. A detailed description along with the process flow diagrams may be found in Rao et al. (2008). Ninety-five percent purity O_2 is supplied by a cryogenic elevated pressure ASU for the gasifiers and the SRU. The ASU also produces intermediate pressure N_2 for injection into gas turbines. The coal slurry is prepared in wet rod mills and introduced into the gasifiers operating at a nominal pressure of 7350 kPa and generate hot gas (raw syngas), slag, and char particulates. The raw syngas leaving the gasification zone is cooled by direct contact with water ("total quench design").

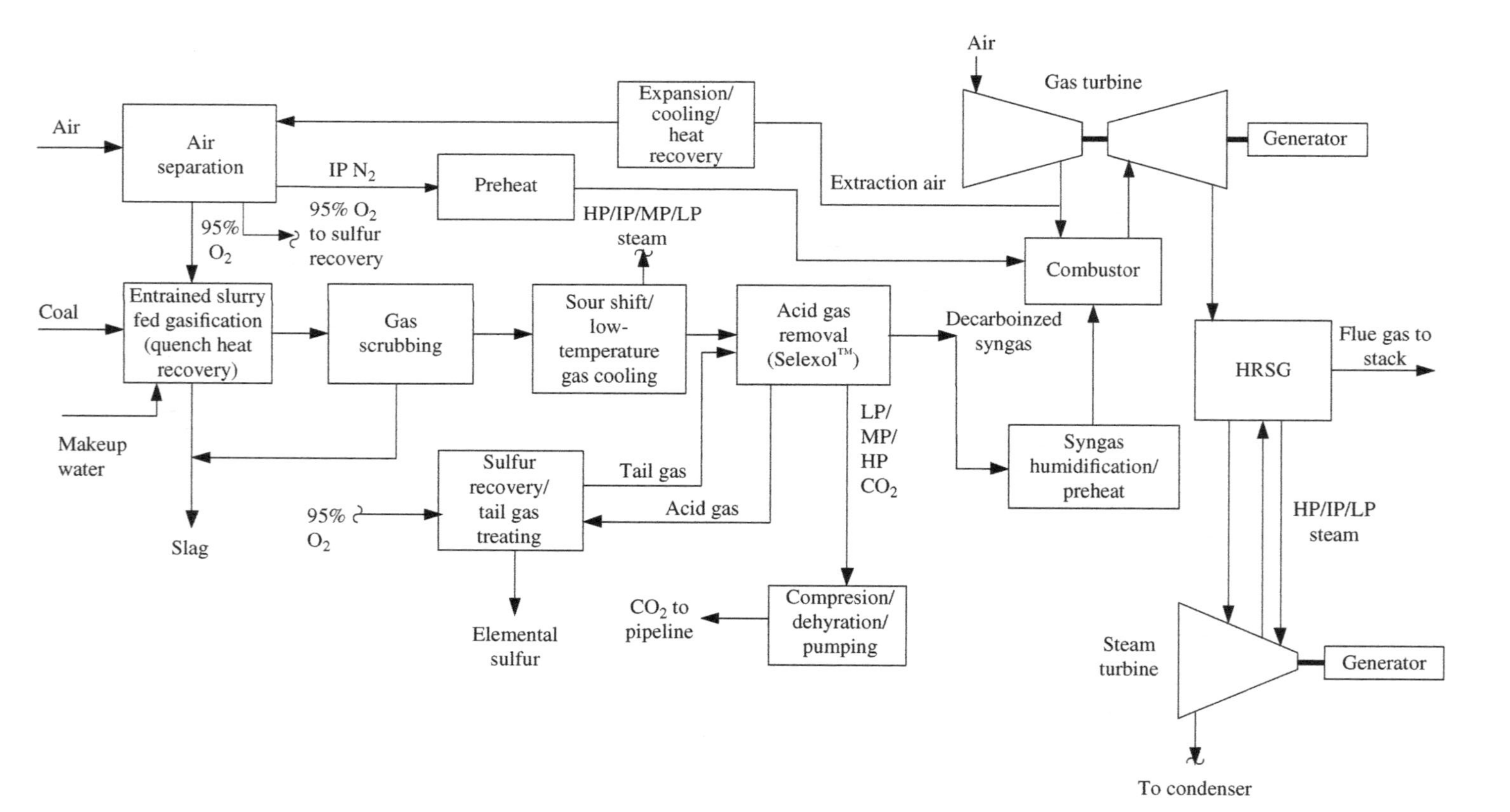

FIGURE 9.6 A near zero emission IGCC

This method of cooling the gas eliminates the need for syngas coolers that tend to be expensive; the quenching operation introduces water vapor into the syngas that is required for reacting with the CO present in the syngas to form H_2 and CO_2 within the shift reactors. The gas leaving the quench section of the gasifier is wet scrubbed to remove any entrained solids. The other contaminants such as soluble alkali salts, hydrogen halides, and a small fraction of the NH_3 are also removed. The contaminated water ("black water") is treated to remove the fine slag while the dissolved gases are removed by depressurization and sent to the SRU. A fraction of water ("gray water") is recycled to the scrubber while the remaining is treated before discharge.

The scrubbed gas is preheated to a temperature high enough for the catalytic shift reaction to occur and then fed into the first of a series of two fixed bed adiabatic reactors with intercooling. A suitable catalyst that is not poisoned by the sulfur compounds present in the syngas is the sour shift catalyst consisting of oxides of Co and Mo. Shift reactors are discussed in more detail in Chapter 10. Most of the CO present in the gas is reacted with water vapor contained in the gas introduced by the gas quenching operation without typically requiring any steam addition to form H_2 and CO_2. Heat generated by the exothermic shift reaction is recovered by the generation of high-pressure and medium-pressure steam. The shifted gas leaving the shift unit is further cooled in a series of shell and tube heat exchangers consisting of low-pressure steam generation, syngas humidifier circulating water heater, vacuum condensate heater, and a trim cooler against cooling water. The cooled syngas is superheated by about 11°C (20°F) to avoid pore condensation and then fed to a sulfided-activated carbon bed that removes the Hg and other trace metals. The effluent gas is then cooled against cooling water and fed into a Selexol™ AGR unit. The high-temperature condensate separated from the gas is recycled to the scrubber while the colder condensate is fed to a sour water stripper. The sour gases stripped off from the water are routed to the Claus SRU. The Selexol™ process scheme consists of a multi-column design and utilizes solvent refrigeration to produce the clean syngas, acid gas that is supplied to the SRU and CO_2 streams at three different pressures (illustrated previously in Fig. 5.1). About 50% of the CO_2 is produced at a pressure of 1034 kPa (150 psia), about 40% is at a pressure of 345 kPa (50 psia), and the remainder at a pressure of 115 kPa (16.7 psia) and provided to the CO_2 compression system. The CO_2 is compressed to the mixture supercritical pressure, dehydrated, and is then pumped to the pipeline pressure of 13,800 kPa (2,015 psia). The tail gas leaving the SRU that contains CO_2 and residual sulfur compounds as well as elemental sulfur vapor is hydrogenated to form H_2S from the sulfur species. The hydrogenated tail gas is then recycled back to the AGR unit. This recycle scheme not only eliminates having an amine unit to desulfurize the hydrogenated tail gas but also an additional emissions source in the plant. The clean decarbonized syngas consisting mostly of H_2 with some CO and inerts is humidified in a countercurrent column by direct contact with hot water to introduce water vapor as a thermal diluent for NO_x control (columns for such operations were described in Chapter 5). Typically, enough water vapor is introduced such that the humidified gas has a heating value of 7870 kJ/Nm^3 (200 Btu/SCF) on a LHV basis. The heat required by the humidifier is recovered from the low-temperature heat contained in the shifted gas (downstream of the shift unit in the low-temperature gas cooling area). The humid gas is preheated to safely remain above the dew point prior to firing in the gas turbines. Preheating also increases the overall plant efficiency when heat from the bottoming cycle is utilized. The preheated syngas is then fed to the gas turbines consisting of typically two units each with output greater than 200 MW such that the size of the gasification island is large enough to take advantage of economies of scale. Intermediate pressure N_2 from the ASU after preheating is also fed to the gas turbines, a significant fraction through separate nozzles within the combustors so that it does not have to be compressed to the same pressure as the syngas that has to overcome the higher pressure drop associated with the fuel control valve. The introduction of the thermal diluents consisting of N_2 and the moisture (via the syngas) in addition to reducing the formation of NO_x within the combustor of the gas turbine (by reducing the flame temperature) also increases the net power output of the gas turbine since these diluents increase the amount of motive fluid entering the turbine. The humidification of the syngas allows for efficient recovery of the low temperature heat from the plant to provide the additional motive fluid at pressure. Note that the specific heat of the triatomic H_2O molecule is significantly higher than that of the diatomic N_2 molecule on a mole basis and thus the relative amounts of diluents required, that is, water vapor versus N_2 on a volumetric or mole basis by a given amount of syngas are quite different for a given NO_x reduction target. Another factor that should be taken into consideration, however, is the effect on the gas turbine firing temperature. A gas turbine designed for a certain firing temperature on natural gas would see derating of the firing temperature due to the increased concentration of H_2O vapor in the working fluid when humidification of the syngas is employed due to the different aerothermal properties of the working fluid. This derating in firing temperature would be in addition to that caused by the increase in the pressure ratio, the temperature of the cooling air being increased as the pressure ratio is increased.

As pointed out previously, in the case of a steam-cooled gas turbine however, derating of the firing temperature due to the increase in pressure ratio may be less significant (since the cooling steam temperature may be maintained independently of the gas turbine pressure ratio), unless the temperature in the low-pressure air-cooled stages of the gas turbine become limiting. The choice of the diluent to be utilized, that is, H_2O vapor versus N_2 or their relative amounts, should be included in a trade-off/optimization systems study by taking into account not only the gasification island design such as the heat recovery options but also the impact on the gas turbine firing temperature.

Exhaust gas from the gas turbines enters triple pressure HRSGs (each HRSG connected to a gas turbine) that provide the superheated steam to a single condensing steam turbine. A SCR unit may be included in each of the HRSGs to further reduce the NO_x emissions. Note that the ultra-low sulfur content of the decarbonized syngas to the gas turbines makes it feasible to utilize the SCR without being constrained by the formation of ammonium salts. In addition to these "on-site units," "off-site units" also called general facilities such as a cooling water system, instrument air, flare, waste water treating, and so on, have to be provided.

9.3.5 Advantages of an IGCC

Air emissions from an IGCC are significantly below U.S. Clean Air Act standards and emissions of SO_x, NO_x, CO, volatile metals, and particulates can be significantly better than those obtained by scrubber-equipped pulverized and circulating fluidized bed coal combustion plants. The economic benefits of an IGCC can become more significant as more stringent air emission standards are passed because this technology can achieve greater emission reductions at lower incremental costs. For example, the cost of removing greater than 90% of the volatile Hg from a coal-fueled IGCC is about one-tenth of that in a pulverized coal-fired boiler plant (Stiegel et al., 2002). Note that Hg removal in gasification-based systems occurs prior to combustion, from a much smaller volume (of pressurized) syngas as compared to removal from the boiler exhaust that is at near atmospheric pressure and diluted with N_2.

IGCC can take advantage of the advanced gas turbines being developed that will affect not only its efficiency but also its cost. The amount of syngas required to generate a unit of electric power is reduced in direct proportion to the increase in efficiency of the power island that reduces the gasification island size and thus its cost. Technological advancements aimed at increasing thermal efficiency of boiler plants are also taking place, however.

9.3.6 Economies of Scale and Biomass Gasification

Plant size has a significant effect on IGCC economics and typical size to take full advantage of economies of scale is typically in excess of 400 MW. As stated in Chapter 8, biomass resources tend to be distributed and due to its low energy density, transportation of biomass collected from various locations over long distances to a large central plant is economically prohibitive. Consideration should be given to building a large IGCC to take advantage of the economies of scale, fed by both biomass and coal (or other fuel such as petroleum coke) and located in close proximity to a biomass resource. An added advantage with this approach is that during seasons of low biomass feedstock availability, the plant could increase its coal usage to off-set a decrease in biomass-derived syngas. In the case of co-feeding with a high rank coal (or petroleum coke) with biomass, two types of gasifiers may be required, one optimized for the coal (or petroleum coke) such as an entrained bed gasifier and the other suitable for the significantly more reactive biomass feedstock such as a fluidized bed gasifier. Note that a dry feed gasifier is required for biomass gasification.

9.4 INDIRECTLY FIRED CYCLE

A potential application for the indirect cycle discussed in Chapter 6 is for smaller scale power plants (<250 MW) where technologies such as the IGCC may not be cost competitive. Interest in coal as a fuel continues, but considerable effort is now being directed toward biomass-fired CHP applications. The heat source for this indirectly fired cycle can also be solar thermal energy concentrated by mirrors discussed in

Chapter 12. The externally fired combined cycle technology has been pursued for more than five decades but modern commercial scale power systems have yet to be realized. The critical component of this system is the heat exchanger that replaces the combustor of a gas turbine to transfer the heat released by combustion of the "dirty" fuel into the gas turbine working fluid, which is simply air.

Subscale tests with a small gas turbine using this concept have been conducted by a number of organizations. Tests of the coal-fired high-temperature heat exchanger, the key enabling technology, have also been carried out. Both ceramic and metallic heat exchangers have been designed and tested at laboratory scales with a varying degree of success. Research continues on heat exchanger test rigs and on materials, mainly metallic, in government, university, and industrial laboratories. Two versions of the indirectly fired cycle were pursued as early as 1939, the open cycle gas turbine by Brown Boveri AG, and the closed cycle[6] gas turbine by Escher Wyss AG. The Brown Boveri machine consisted of a conventional Brayton cycle with a turbine inlet temperature of 540°C (1000°F) and a resulting efficiency of slightly over 17% in simple cycle mode. The Escher Wyss closed cycle gas turbine allowed a more sophisticated thermodynamic cycle than the Brown Boveri machine resulting in a higher efficiency at essentially the same level of turbine aerodynamic technology. The turbine inlet temperature was approximately 650°C (1200°F). With this higher temperature and with a more complex cycle, the efficiency was over 31% (HHV), rivaling then current boiler steam power systems. The turbine inlet temperatures are constrained by the limitations of the metallurgy used in the coal-fired heater. The U.S. Department of Energy's High Performance Power Plant program in the 1990s aimed at achieving overall efficiency greater than 47% (HHV), while meeting emission standards that were one-tenth of the U.S. EPA NSPS for then current coal-fired boiler power plants.

Example 9.3. Derive an expression for the first law efficiency η_{I} of a closed cycle gas turbine as a function of the compressor isentropic efficiency η_{comp}, turbine isentropic efficiency η_{turb}, generator efficiency (including mechanical losses) η_{gen}, compressor and turbine pressure ratios π_{comp} and π_{turb} (ratio of the discharge pressure to its inlet pressure in the case of the compressor, and vice versa in the case of the turbine, these ratios being different when the pressure drop across the exchangers to add and reject the heat are accounted for), compressor inlet absolute temperature $T_{\text{comp,in}}$, turbine inlet absolute temperature $T_{\text{turb,in}}$, and κ the ratio of constant pressure specific heat C_{p} to constant volume specific heat C_{v} for an ideal gas working fluid with κ remaining constant and negligible heat losses. Using the derived expression, compare η_{I} for a monoatomic (e.g., He), a diatomic (e.g., N_2), and a triatomic (e.g., CO_2) gas as working fluid with the following parameters: $\eta_{\text{comp}} = 0.88$, $\eta_{\text{turb}} = 0.9$, $\eta_{\text{gen}} = 0.97$, $\pi_{\text{comp}} = 10$, $\pi_{\text{turb}} = 8$, $T_1 = 310$ K, and $T_2 = 1100$ K.

[6] The closed cycle contains the working fluid within the cycle by recirculating the turbine exhaust after heat recovery and/or heat rejection to the compressor inlet. The working fluid can be air or a gas such as He for the high temperature gas-cooled nuclear reactor discussed in Chapter 12.

Solution

The net power developed by the gas turbine is the difference between the turbine power produced as given by Equation (3.28) and the compressor power consumed as given by Equation (3.25), multiplied by η_{gen}. The volumetric flow rate appearing in these equations may be written for an ideal gas as

$$\dot{V}_{in} = \frac{\dot{m}R_u T_{in}}{M_w P_{in}}$$

where $\dot{m}$ is the mass flow rate of the gas, R_u the universal gas constant, M_w is the molecular weight, and T_{in} and P_{in} are the inlet temperature and pressure. Using this relationship, the net power developed by the gas turbine may be written as

$$\dot{W}_{net} = \frac{\kappa \dot{m} R_u \eta_{gen}}{(\kappa-1)M_w}\left[T_{turb,in}\,\eta_{turb}\left\{1-\left(\frac{1}{\pi_{turb}}\right)^{(\kappa-1)/\kappa}\right\} - \frac{T_{comp,in}}{\eta_{comp}}\left\{\left(\pi_{comp}\right)^{(\kappa-1)/\kappa}-1\right\}\right] \tag{9.7}$$

The heat added to the cycle is given by

$$\dot{Q}_{in} = \left(\dot{m}\,C_p\right)\left(T_{turb,in} - T_{comp,out}\right) \tag{9.8}$$

The compressor outlet temperature $T_{comp,out}$ may be related to its inlet temperature by solving the following relationships for the power required by the compressor

$$\dot{W}_c = \left(\dot{m}\,C_p\right)\left(T_{comp,out} - T_{comp,in}\right) = \frac{\kappa \dot{m} R_u\, T_{comp,in}}{(\kappa-1)\,M_w\,\eta_{comp}}\left[\left(\pi_{comp}\right)^{(\kappa-1)/\kappa}-1\right]$$

resulting in Equation 9.9

$$T_{comp,out} = T_{comp,in} + \frac{\kappa R_u\, T_{comp,in}}{(\kappa-1)\,C_p M_w\,\eta_{comp}}\left[\left(\pi_{comp}\right)^{(\kappa-1)/\kappa}-1\right] \tag{9.9}$$

Combining Equations 9.7–9.9, the first law efficiency may be written as while noting that $\left(R_u/C_p M_w\right) = (\kappa-1)/\kappa$

$$\eta_I = \eta_{gen} \times \frac{T_{turb,in}\,\eta_{turb}\left[1-\left(1/\pi_{turb}\right)^{(\kappa-1)/\kappa}\right] - \left(T_{comp,in}/\eta_{comp}\right)\left[\left(\pi_{comp}\right)^{(\kappa-1)/\kappa}-1\right]}{T_{turb,in} - T_{comp,in}\left\{1+\left(1/\eta_{comp}\right)\left[\left(\pi_{comp}\right)^{(\kappa-1)/\kappa}-1\right]\right\}} \tag{9.10}$$

The ratio of specific heats for the three gases as well as the first law efficiencies calculated by substituting the given numerical values into the derived

TABLE 9.3 First law efficiencies for the closed cycle gas turbine

Type gas (ideal)	Monoatomic	Diatomic	Triatomic
Ratio of specific heats κ	1.67	1.40	1.29
First law efficiency η_I	9.98%	24.27%	22.91%

Equation 9.10 earlier are presented in Table 9.3. These efficiencies do not represent the optimum for each of the gases since the pressure ratio was held constant. The pressure ratio corresponding to the optimum efficiency may be easily calculated using Equation 9.10 and will be lowest for the monoatomic gas while highest for the triatomic gas among the three types of gases.

REFERENCES

Armstrong PA. ITM oxygen: The new oxygen supply for the new IGCC market. Gasification Technologies Conference; October 2005; San Francisco, CA.

Butz JR, Lovell JS, Broderick TE, Sidwell RW, Turchi CS, Kuhn AK. Evaluation of amended silicates™ sorbents for mercury control. Proceedings of the Mega Symposium; May 2003; Washington, DC.

Chapel DG, Mariz CL, Ernest J. Recovery of CO_2 from Flue Gases: Commercial Trends. Canadian Society of Chemical Engineers Annual Meeting; October 4–6, 1999; Saskatoon, Saskatchewan, Canada.

Dennis RA. DOE advanced turbine program ceramic material needs for advanced hydrogen turbines. U.S. Advanced Ceramics Association Spring Meeting; May 2008; Arlington, VA.

Ganapathy V. Cold end corrosion: Causes and cures. Hydrocarb Process January 1989;57–59.

Gangwal SK, Turk BS, Coker D, Howe G, Gupta R, Kamarthi R, Leininger T, Jain S. Warm-gas desulfurization process for Chevron Texaco Quench gasifier syngas. 21st Annual Pittsburgh Coal Conference; 2004; Pittsburgh, PA.

Hada S, Masada J, Ito E, Tsukagoshi K. Evolution and future trend of large frame gas turbine for power generation—A new 1600 degree C, J class gas turbine. GT2012-68574. Proceedings of ASME Turbo Expo; June 2012; Copenhagen, Denmark.

Hannemann F, Kanaar M. Operational experience of the buggenum IGCC plant and enhanced V94.2 gas turbine concepts for high syngas reliability and flexibility. Gasification Technologies Conference; October 2002; San Francisco, CA.

Keeler CG. Wabash river in its fourth year of commercial operation. Gasification Technologies Conference; October 1999; San Francisco, CA.

Krishnan GN, Gupta R, Ayala RE. Development of disposable sorbents for chloride removal from high-temperature coal-derived gases. U.S. Department of Energy Report; 1996. Report nr. DOE/MC/30005-97/C0748. Morgantown, WV: U.S. Department of Energy.

Martinez-Frias J, Aceves S, Smith JR, Brandt H. Thermodynamic analysis of zero-atmospheric emissions power plant. ASME International Conference; November 2002; New Orleans, LA.

McDaniel JE, Shelnut CA. Tampa electric company polk power station IGCC project—project status. Gasification Technologies Conference; October 1999; San Francisco, CA.

Pierce RR. Estimating acid dewpoints in stack gases. Chem Eng 1977;11:125–128.

Rao AD, Francuz DJ, Scherffius J, West E. Electricity production and CO_2 capture via partial oxidation of natural gas. Prepared by Fluor Daniel Inc., CRE Group Ltd Report; April 2000. Report nr. PH3/21. IEA Greenhouse Gas R&D Programme, Cheltenham, U.K.

Rao AD, Verma A, Samuelsen GS. Acid gas removal options for the FutureGen plant. Clearwater Coal Conference; April 2005; Clearwater, FL.

Rao AD, Francuz DJ, Maclay JD, Brouwer J, Verma A, Li M, Samuelsen GS. Systems analyses of advanced Brayton cycles for high efficiency zero emission plants. Prepared by University of California, Irvine, U.S. Department of Energy Report. December 2008. Available at http://www.osti.gov/scitech/servlets/purl/951990. Accessed December 4, 2014.

Rhudy R. Retrofit of CO_2 capture to natural gas combined cycle power plants. Prepared by The International Energy Agency Greenhouse Gas Program for EPRI Report; 2005. EPRI, Palo Alto, CA.

Roark SE, Machay R, Sammells AF. Hydrogen separation membranes for vision 21 energy plants. 28th International Technical Conference on Coal Utilization and Fuel Systems; March 10–14, 2003; Clearwater, FL.

Sanz W, Jericha H, Luckel F, Göttlich E, Heitmeir F. A further step towards a Graz cycle power plant for CO_2 capture. ASME Turbo Expo; June 6–9, 2005; Reno-Tahoe, NV.

Smith D. *H System Steams On*. Modern Power Systems; February 2004, 17–20.

Stiegel GS, Longanbach JR, Klett MG, Maxwell RC, Rutkowski MD. The cost of mercury removal in an IGCC plant. U.S. Department of Energy Final Report; September 2002. Morgantown, WV: U.S. Department of Energy.

Trapp W, Denton D, Moock N. Gasification plant design and operation considerations. Gasification Technologies Conference; October 2002; San Francisco, CA.

10

COPRODUCTION AND COGENERATION

Coproduction of synthetic fuels and chemicals in coal and biomass-fueled power plants is gaining significant attention due to the synergy between electricity generation and coproduction, especially with intermittent renewables supplying a larger fraction of power to the grid. Chapter 12 discusses the importance for the non-intermittent power sources to be able to quickly ramp up or down their power output to meet the demand created when the output of intermittent renewables goes down and vice versa. Changing the split of syngas going to the power block as fuel versus that going to the coproduction unit allows this capability in integrated gasification combined cycle (IGCC) type of plants. Coproduction may also be accomplished in a fuel cell (discussed in Chapter 11) based plant with this capability by varying the fraction of fuel-bound energy converted to electricity in the fuel cell.

The incremental cost of coproducing H_2 in an IGCC plant that requires decarbonization of the syngas before it is combusted in a gas turbine for carbon emission control is especially attractive. The various other coproducts that may be synthesized are illustrated in Figure 10.1. Natural gas-fueled power plants may also be designed for coproduction. There are other benefits of coproduction that will be discussed along with the synthesis processes of some of the key coproducts, and finally some of the coproduction plant integration concepts in both natural gas and IGCCs.

Sustainable Energy Conversion for Electricity and Coproducts: Principles, Technologies, and Equipment, First Edition. Ashok Rao.

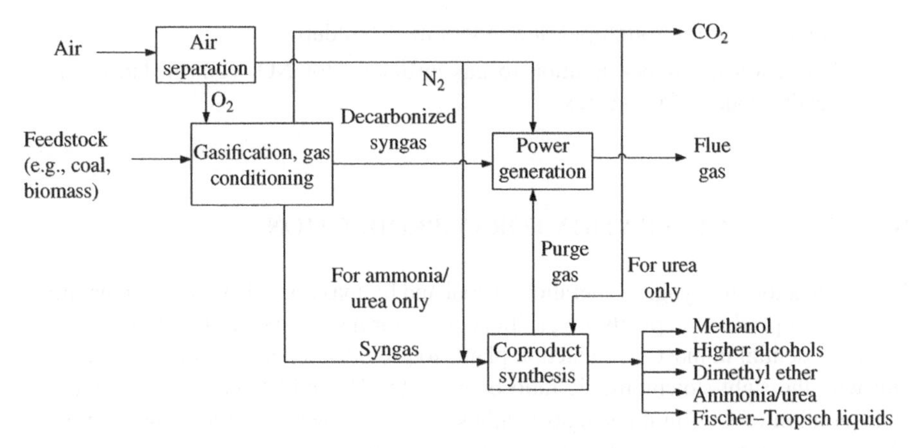

FIGURE 10.1 Coproduction in an IGCC

10.1 TYPES OF COPRODUCTS AND SYNERGY IN COPRODUCTION

In addition to hydrogen, there are a number of other coproducts that may be produced from syngas such as methanol, ethanol, and other higher alcohols, dimethyl ether (by direct synthesis instead of converting methanol), Fischer–Tropsch liquids (separated to yield naphtha, jet fuel, diesel, and waxes), and the fertilizers, ammonia, and urea. Note that some of these coproducts may then be used for synthesizing other chemicals, methanol being one such example. In addition to being useful by itself (as an industrial solvent for inks, resins, adhesives to wood items and dyes, as a pharmaceutical solvent in the manufacture of cholesterol, streptomycin, vitamins, and hormones, as an antifreeze for automotive radiators, as an ingredient in paint and varnish removers, in wastewater denitrification, and in biodiesel transesterification, as a fuel in fuel cells as well as in transportation), it is used to make hundreds of other chemicals including formaldehyde, acetic acid, and dimethyl ether.

Synergies in coproduction include:

- Realizing economies of scale of building a larger plant to produce both electricity and the coproduct, and sharing of common facilities such as utilities including plant heat rejection system, flares, instrument air, and waste water treating.
- Efficient use of steam generated by units associated with coproduction in a single large steam turbine.
- Opportunity for the plant to follow load imposed by the grid by changing the split of the syngas going to the power block versus the coproduction unit.
- In case of coproducts such as methanol, ethanol, and Fischer–Tropsch liquids (as explained later in this chapter):
  - Reduced synthesis loop recycle rate with unconverted purge gas fired in gas turbine(s).

- Higher reactor through-put due to less inert buildup.
- Reduction in diluent addition to gas turbine(s) for NO_x control due to low heating value of purge gas.

10.2 SYNGAS GENERATION FOR COPRODUCTION

The key reactors for syngas generation in coal and biomass-based as well as in natural gas-based coproduction plants are gasifiers and reformers, respectfully. Shift reactors are also essential in most coproduction plants to provide the feed gas to the synthesis unit with the right stoichiometric ratio between the H_2 and CO (while the required concentration of CO_2 in the syngas is adjusted in the downstream acid gas removal (AGR)). These reactors are discussed in the following.

10.2.1 Gasifiers

As described in Chapter 9, the partial oxidation reaction for carbon contained in the fuel (feed) is the exothermic reaction $C + ½O_2 = CO$ and H_2O in the form of steam or liquid water is introduced into the gasifier to moderate the temperature of the reaction by the endothermic reaction $C + H_2O = H_2 + CO$. Other reactions include the exothermic water gas shift reaction $CO + H_2O = H_2 + CO_2$ and the highly exothermic hydrogasification reaction $C + 2H_2 = CH_4$, although only to a limited extent and majority of the CH_4 produced by most gasifiers is due to the decomposition of the volatile matter in the feed. The sulfur is evolved primarily as H_2S with some as COS while other sulfur compounds such as CS_2 and mercaptans are produced in insignificant amounts when the gasifier operating temperature is high (>1100°C). The nitrogen is evolved primarily as elemental N_2, some NH_3, and much smaller amount of HCN when the gasifier operating temperature is high. The Cl present in the feed is evolved primarily as HCl.

There are three major types of gasifiers classified based on the way the solids flow through the gasifier, the moving bed gasifier, the fluidized bed gasifier, and the entrained bed gasifier again as described in more detail in Chapter 9. Higher CH_4 content of the syngas limits its use for synthesizing most coproducts (except substitute natural gas or SNG) for a number of reasons. In coproduction plants, the degree of decarbonization (for controlling CO_2 emissions) of the syngas used as fuel for the gas turbine is limited by the amount of CH_4 present. It is the CO and the H_2 that are the building blocks for most coproducts and CH_4 "ties up" the C and H atoms that otherwise could be used in making the CO and H_2 and hence the coproduct. Furthermore, the CH_4 content decreases the partial pressure of the reactants (H_2, CO, and CO_2) requiring a larger purge stream from the synthesis unit to limit the inert buildup as most of the processes recycle the unconverted gas back to the synthesis reactor to increase the overall conversion. A reformer could be utilized to convert the CH_4 into H_2 and CO after syngas cleanup but this would add to the plant cost and complexity.

10.2.2 Reformers

Catalytic reformers used for generating syngas from natural gas are the steam methane reformer (SMR), the auto-thermal reformer (ATR), and a combination of the two. A Ni-based catalyst is utilized and the major overall reactions occurring within a SMR are given by reactions (10.1) and (10.2) while all three of the following reactions occur in an ATR:

$$\text{Reforming: } CH_4 + H_2O = 3H_2 + CO; \Delta H > 0 \tag{10.1}$$

$$\text{Shift: } CO + H_2O = CO_2 + H_2; \Delta H < 0 \tag{10.2}$$

$$\text{Partial oxidation: } 2CH_4 + O_2 = 4H_2 + 2CO; \Delta H < 0 \tag{10.3}$$

10.2.2.1 SMR This type of reformer is a fired reactor to provide the heat required by the endothermic reaction (10.1) and consists of multiple catalyst-filled tubes located in a firebox. The feed consisting of a mixture of preheated (to about 500–650°C) desulfurized natural gas (to avoid poisoning the Ni catalyst) and steam is fed to the tubes and the reforming reaction occurs primarily in the temperature range from 700 to 1100°C. Steam is added to avoid soot formation and carbon deposition on the catalyst surface by the reversal of the Boudouard reaction $2CO = C + CO_2$, as well as to avoid metal dusting in the downstream equipment and piping occurring in the critical temperature range of about 400–750°C. Metal dusting can be a severe corrosion problem. The firebox is equipped with burners to combust the fuel (natural gas) and the heat transfer to the tubes occurs by radiation and convection. Downstream of the firebox is located the convective section where the natural gas to be reformed is preheated, steam is generated and superheated, BFW is preheated, and prereformer feed (when a prereformer is included especially when the natural gas contains significant amounts of higher hydrocarbons) is preheated. A prereformer is an adiabatic low-temperature steam reformer (a fixed bed reactor) and preconverting the higher hydrocarbons results in stable and less severe operating conditions for the SMR, thus ensuring its reliable operation and increasing the reformer tube life. The amount of steam required in the feed to the SMR (expressed usually as the molar ratio of H_2O to C in the feed mixture) is also reduced and thereby a reduction in the overall fuel consumption.

In some designs suitable for smaller scale applications, the radiant section has been eliminated and the entire heat required for the reforming reactions is provided by convection in a "heat exchange reformer." It consists of a tube bundle with each tube assembly consisting of three concentric tubes. The flue gas from the combustion chamber passes through the annular space between the outermost tube and the second inner tube providing heat for preheating and reforming reactions while the gas to be reformed passes through catalyst-filled annular space between the second inner tube and the inner-most tube. The reformed gas then passes through the inner-most tube while transferring heat to the gases being preheated and reformed. The advantage with this design is that significantly higher fraction of the heat released by combusting the fuel is used for the reforming process than in the previous SMR design where a significant fraction of the heat also enters the steam system.

10.2.2.2 ATR This type of reformer contains a combustion zone at the top and a catalyst-filled fixed bed at the bottom. Natural gas mixed with a sub-stoichiometric amount of O_2 (or air in the case of NH_3 synthesis) is combusted in the combustion zone and passes through an intermediate recirculation section where the hot gases continue to react. The gases next enter the bottom section containing the catalyst and exit the bed at temperatures typically from 850 to 1000°C with a close approach to chemical equilibrium. ATRs are suited for making large volumes of synthesis gas, especially with relatively low H_2 to CO ratio (1.5–3 on a mole basis), which is required for the synthesis of Fischer–Tropsch liquids. CO_2 separated from the syngas downstream of the shift unit may also be recycled to achieve a low H_2 to CO ratio.

10.2.2.3 Combined SMR and ATR The two types of reformers may be combined in certain applications depending on the desired H_2 to CO ratio, the heat required by the SMR being supplied by the hot effluent gas from the ATR without requiring combustion of fuel in the SMR. This configuration has been used for production of H_2 or syngas for methanol synthesis. The gas-heated SMR in such applications is primarily a heat exchanger. Proper amount of steam addition as well as limiting the operating pressure are required to avoid carbon deposition and metal dusting issues.

10.2.3 Shift Reactors

An adiabatic shift reactor consists of a vertical vessel containing a fixed bed of catalyst in which the water gas shift reaction $CO + H_2O = CO_2 + H_2$ occurs to change the ratio of H_2 to CO in the syngas generated by a gasifier or by a reformer. Typically two adiabatic reactors in series with intercooling between the two to maximize the conversion are utilized since the reaction is exothermic and the equilibrium conversion decreases with increase in temperature. The first reactor operates at higher temperatures (than the second) where bulk CO conversion occurs and the temperature rise across the reactor is much larger. The outlet temperature from the second reactor is much lower, which is conducive to higher conversion from a chemical equilibria standpoint. The process is depicted by the temperature versus CO concentration plot in Figure 10.2. The equilibrium CO concentration is plotted against the temperature, and as can be seen the conversion in the first reactor is limited by its outlet temperature. So by staging the overall conversion process in two reactors while cooling the effluent from the first reactor before it enters the second reactor, much higher conversions can be achieved, the second reactor outlet temperature being much lower and a correspondingly much lower equilibrium CO concentration as its upper limit.

Non-adiabatic reactors such as the "isothermal reactor" with internal cooling coils are also utilized when the inlet syngas CO concentration is very high but the syngas gas has to be ultra clean. Such reactors are limited to lower temperature sweet shift applications, which is discussed in the following.

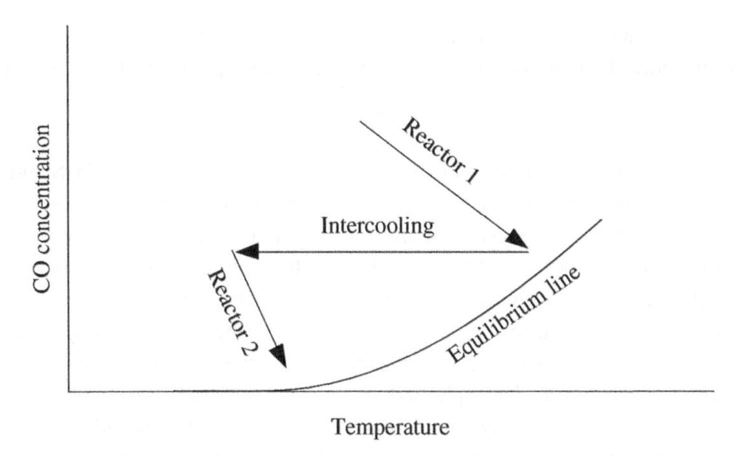

FIGURE 10.2 Temperature versus CO conversion

Shift reactors are typically designed with feed/effluent interchangers to preheat the feed gas while recovering heat from the effluent (see Example 7.2) and a heat source should be made available for preheating the feed gas during startup by providing an electric heater or a steam-heated exchanger depending on the heat transfer duty required. Sizing of this heater does not have to be for the full design flow rate of the syngas since the unit may be started at a lower flow rate, say 25% of the design flow rate.

10.2.3.1 Sweet Shift When the syngas is essentially free of the sulfur species ("sweet gas") as is the case when it is generated by a conventional Ni catalyst-based reformer or their concentrations very low (some sweet shift catalysts tolerate close to 100 ppmv), then catalysts containing Fe_2O_3 and Cr_2O_3 are utilized in higher temperature service (operating range of ~350–480°C). Some formulations contain CuO, which allows reducing the steam to gas ratio without the associated problem of making hydrocarbons by the Fischer–Tropsch reactions. For medium (operating range of ~200–320°C) or low (operating range of ~200–260°C) temperature services, catalysts containing CuO, ZnO, and Al_2O_3 are utilized. The system design should take into account the increase in the minimum inlet temperature required to initiate the reaction and the decrease in CO conversion as the catalyst ages (progressively deactivates). The reactor(s) and heat exchanger(s) may be designed with a fraction of the syngas bypassing the reactor(s) such that as the catalyst ages, the fraction of bypass may be reduced to control the overall CO conversion, that is, to hold a certain CO to H_2 ratio in the syngas leaving the shift unit. The reactor feed gas should be preheated sufficiently above its dewpoint to avoid condensation of the H_2O vapor in the catalyst pores, a requirement typical of most catalysts and solid adsorbents. A superheat of 20–25°C may suffice but the catalyst supplier should be consulted for guidelines specific to a given catalyst.

The nonadiabatic reactors or (nearly) isothermal reactors, that is, with internal tubes to transfer the heat of reaction to generate steam, can eliminate the need for multiple reactors (with intercooling) but the cost of such a reactor tends to be higher and a careful tradeoff study is required to select the right type of reactor(s) for a given application.

10.2.3.2 Sour Shift When the syngas contains sulfur species such as H_2S and COS ("sour gas") as is the case when syngas is generated by the gasification of coal or other sulfur bearing feedstocks, catalysts that are not poisoned by the sulfur species are utilized formulated from CoO, Mo_2O_3, and Al_2O_3 with proprietary promoters and activated by sulfidation. The operating temperature range for such catalysts are specific to the manufacturer; the minimum inlet temperature can be as low as approximately 200°C while the outlet temperature can be as high as approximately 430°C depending on the specific formulation. As explained under "Sweet Shift," the variation in the required inlet temperature and CO conversion as the catalyst ages should be accounted for in the system design. Again, the reactor(s) and heat exchanger(s) may be designed with a fraction of the syngas bypassing the reactor(s) to control the CO conversion (to hold a certain CO to H_2 ratio in the effluent) as the catalyst ages. The sulfided catalysts require a minimum concentration of the sulfur species in the syngas to maintain their activity, some as much as 100 ppmv although higher concentrations may be preferred to keep the catalyst in the optimum sulfided state. Any COS present in the syngas is hydrogenated or hydrolyzed to H_2S under most operating conditions while the HCN is hydrogenated to CH_4 and NH_3. A minimum H_2O concentration in the syngas is required to avoid carbon formation and deposition on the catalyst surface by the reversal of the Boudouard reaction, and also to avoid Fischer–Tropsch reactions forming hydrocarbons. This minimum H_2O concentration defined by the molar ratio between H_2O and CO may be as high as 2.0 (twice the stoichiometric requirement for CO conversion). Again, the reactor feed gas is preheated to a temperature high enough to avoid condensation of the H_2O in the catalyst pores, 20–25°C above the H_2O dew point.

10.3 SYNGAS CONVERSION TO SOME KEY COPRODUCTS

10.3.1 Methanol

The overall reactions to be considered for the synthesis of methanol are as follows (Tijn et al., 2001):

$$2H_2 + CO = CH_3OH; \Delta H < 0 \tag{10.4}$$

$$CO + H_2O = CO_2 + H_2; \Delta H < 0 \tag{10.5}$$

$$CO_2 + 3H_2 = CH_3OH + H_2O; \Delta H < 0 \tag{10.6}$$

If only reaction (10.4) is considered, then the stoichiometric ratio or number defined by moles H_2/moles CO would be 2.0. When the water gas shift reaction (10.5) also takes place, the stoichiometric number should be corrected since a mole of H_2 is consumed by each mole of CO_2 to produce a mole of CO. The appropriate stoichiometric ratio or number to be considered then becomes (moles H_2 – moles CO_2)/(moles CO + moles CO_2) and a slight excess of the stoichiometric requirement of 2.0 is used, typically 2.05 in most of the conventional gas phase synthesis reactors. Note that only two of the preceding reactions are thermodynamically independent; reaction (10.6) may be obtained by combining reactions (10.4) and (10.5). Reaction (10.4) indicates that equilibrium conversion is favored by lower temperatures and higher pressures. Excessively high pressures require more expensive equipment (the geometrical size of equipment does decrease with increase in pressure and typically a trade-off exists between size and say the wall thickness of a vessel or pipe). Over the years, the operating pressure for methanol synthesis reactors has been brought down by the use of catalysts comprised primarily of Cu and Zn, which are more active. The fresh catalyst consists of an oxide of Cu that is then reduced by CO to its active metallic form. When the catalytic methanol synthesis reactor is fed with a CO-rich syngas, overreduction may occur, which may cause enhanced sintering of the metallic particles (Kurtz et al., 2003).

For illustrative purposes, the liquid phase methanol synthesis process as developed by Air Products, the LPMEOH™ technology, is depicted in Figure 10.3 (Heydorn et al., 2003). A demonstration plant built at Eastman Chemical's complex at Kingsport, Tennessee, was operated successfully at a production rate of 80,000 gal (303 m^3) per day of methanol while converting coal-derived syngas. The process was shown to be robust enough to operate either in a continuous, base load manner or intermittently, and was easily ramped up and down in capacity, which is required for a coproduction plant to complement the intermittent renewables. The LPMEOH™ process uses commercially available methanol catalyst in fine powder form, suspended in an inert mineral oil. Another feature of this process is that clean syngas can be fed directly to the synthesis unit without any adjustments required to the ratio between H_2, CO, and CO_2, which can reduce costs associated with the water gas shift and the CO_2 removal units. Data collected from the demonstration plant have indicated that no unusual deactivation rates were observed even when the molar H_2 to CO ratio was lower than 1. It should be noted that higher the CO_2 content in the syngas for a given H_2 to CO ratio, higher will be the moisture content of the raw methanol produced, which increases the operating and capital costs of the downstream distillation unit required to purify the crude methanol. A careful system analysis is required to determine the optimum CO_2 content in the syngas based on this trade-off, especially in coproduction plants incorporating carbon capture and sequestration since a syngas decarbonization unit is included anyway, although its size may have to be increased. When the syngas is produced from natural gas in an SMR, it is advantageous to recover CO_2 from the flue gas leaving the SMR (using an amine process such as that described in Chapter 9) and add it to the synthesis unit feed gas in order to optimize the ratio between H_2, CO, and CO_2, natural gas being a hydrogen-rich feedstock, that is by utilizing CO_2 capture technologies available or being developed for post combustion capture.

Desulfurized syngas with the required ratio between the H_2, CO, and CO_2 is supplied to the synthesis unit where it is first passed through a guard bed to remove any carbonyls present (pointed out in Chapter 9) when the syngas is generated from feedstocks such as coal and petroleum coke and then compressed to the required synthesis

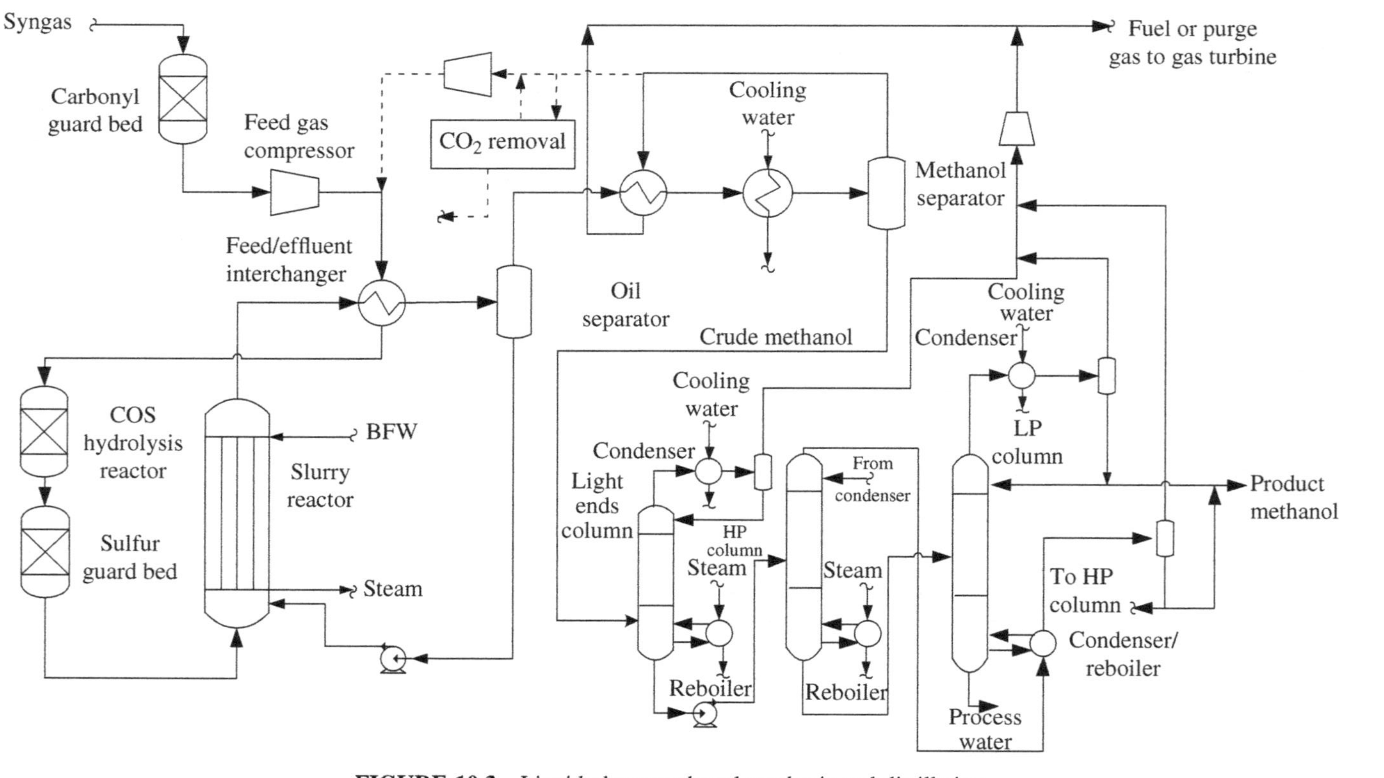

FIGURE 10.3 Liquid phase methanol synthesis and distillation

loop pressure (about 56 bar). It is then combined with the synthesis loop recycle syngas; the fraction of syngas recycled versus that purged and provided to the gas turbine in a coproduction facility may be small or none depending on results developed by an overall systems analysis study. The combined stream is then preheated in a feed/effluent interchanger and then passed through a COS hydrolysis reactor to convert any COS remaining in the syngas to H_2S followed by a sulfur guard bed to remove the traces of H_2S, when the syngas is generated from sulfur bearing feedstocks such as coal and petroleum coke. It is then fed to a slurry reactor containing fine catalyst particles suspended in an inert hydrocarbon liquid (a mineral oil). The mineral oil acts as a temperature moderator and a heat removal medium, transferring the heat of reaction from the catalyst surface via the liquid slurry to boiling water in an internal tubular heat exchanger. Intermediate pressure steam is generated from the heat. The amount of heat generated in the reactor is higher when the CO content in the feed gas is higher but the more efficient heat transfer in the liquid phase reactor enables a uniform temperature to be maintained throughout the reactor. This also allows a much higher conversion rate while the number of recycle passes required is reduced from the typical 5:1 with conventional synthesis using gas phase reactors to as low as 1:1 to 2:1. The reactor effluent at 260°C is cooled in a series of heat exchangers including the feed/effluent interchanger. The condensate collected is then fed to the methanol purification unit. In addition to methanol and water, a number of other compounds are formed although in much smaller fractions such as hydrocarbons (up to C_{12}), higher alcohols (C_2–C_5), esters, ketones, ethers, and amines. The purification unit consists of a light ends column to remove the dissolved light components (mostly gases), and a set of energy-saving heat-integrated distillation columns (high-pressure and low-pressure columns with the condenser of the high-pressure column providing heat for the reboiler of the low-pressure column) to produce the methanol meeting the required specifications. Table 10.1 presents those required for automobile fuel when used by itself (M-100), that is, without blending it with gasoline, and Table 10.2 summarizes those for chemical grade methanol.

10.3.2 Urea

The process step proceeding the synthesis of urea is the synthesis of NH_3 by reacting N_2 with H_2 in the presence of an Fe catalyst, the H_2 being derived from syngas by the water gas shift reaction followed by desulfurization and CO_2 separation to generate a high purity H_2 stream. The overall reaction for synthesizing NH_3 is:

$$N_2 + 3H_2 = 2NH_3; \Delta H < 0 \tag{10.7}$$

In the case of O_2 blown gasification-based coproduction facility with CCS, a fraction of the decarbonized syngas is fed to a pressure swing adsorption (PSA) to produce a high purity H_2 that is then combined with the stoichiometric amount of N_2 provided by the air separation unit. Deoxygenation of the mixture is required since the N_2 contains small amounts of O_2. The deoxygenation reaction may be accomplished by reacting the O_2 with the H_2 in the presence of a noble metal catalyst. When the starting

TABLE 10.1 Specifications for automobile fuel (M-100 as established by California Air Resources Board)

Specification	Value	Test method
Methanol	≥96 vol %	As determined by the distillation range
Distillation range	4.0°C	ASTM D 1078-86; At 95% by volume distilled; must include 64.6 ± 0.1°C
Other alcohols and ethers	≤2 wt %	ASTM D 4815-89
Hydrocarbons, gasoline, or diesel fuel derived	≤2 wt %	ASTM D 4815-89, and then subtract concentration of alcohols, ethers, and water from 100 to obtain percent hydrocarbons
Specific gravity at 20°C	0.792 ± 0.001	ASTM D 891-89
Acidity as acetic acid	≤0.01 wt %	ASTM D 1613-85
Total chlorine as chloride	≤0.0002 wt %	ASTM D 2988-86
Lead	≤2 mg/1 (no added *Pb*)	ASTM D 3229-88
Phosphorous	≤0.2 mg/1 (no added *P*)	ASTM D 3231-89
Sulfur	≤0.002 wt %	ASTM D 2622-87
Gum, heptane washed	≤5 mg/l	ASTM D 381-86
Total particulates	≤5 mg/l	ASTM D 2276-89, modified to replace cellulose acetate filter with a 0.8 μm pore size membrane filter
Water	≤0.3 wt %	ASTM E 203-75
Appearance	Free of turbidity, suspended matter and sediment	Visually determined at 25°C by Proc. A of ASTM D 4176-86
Bitterant	Must have a distinctive and noxious taste at ambient conditions for purposes of preventing purposeful or inadvertent human consumption	
Odorant	Upon vaporization at ambient conditions, must have a distinctive odor potent enough for its presence to be detected down to a concentration in air of not over one-fifth of the lower limit of flammability	

TABLE 10.2 Specifications for chemical grade (AA) methanol (U.S. Federal Specifications)

Specification	Value	Test method
Methanol	≥99.85 wt %	ASTM E-346
Water	≤1000 ppmw	ASTM E-1064
Ethanol	≤10 ppmw	ASTM E-346
Acetone	≤20 ppmw	ASTM D-1612
Acidity as acetic acid	≤30 ppmw	ASTM D-1613
Alkalinity as ammonia	≤30 ppmw	ASTM D-1614
Iron	≤0.05 ppmw	ASTM E-394
Nonvolatiles	≤10 mg/100 ml	ASTM D-1353
Permanganate (discharge of color)	≥30 min	ASTM E-1363
Color (Pt-Co scale)	≤5	ASTM D-1209
Specific gravity at 25°C	≤0.7928	ASTM D-891
Initial boiling point	64.7 ± 0.2°C	ASTM D-1078
Distillation range at 760 mm Hg	≤1.0°C	ASTM D-1078 must include 64.6 ± 0.1°C
Dry point	63.7–65.7°C	ASTM D-1078
Odor	Characteristic, nonresidual	ASTM D-1296
Appearances	Clear	ASTM E-346
Hydrocarbons	Pass	ASTM D-1722
Carbonizables (Pt-Co scale)	≤30	ASTM E-345

feedstock is natural gas, the syngas generation process includes an SMR followed by an air-blown ATR such that the feed gas to the NH_3 synthesis unit has the right stoichiometric ratio between H_2 and N_2. The Fe-based catalyst cannot tolerate carbon oxides and if present in the syngas should be methanated by reacting with the H_2 by the overall (highly exothermic) reactions:

$$3H_2 + CO = CH_4 + H_2O; \Delta H << 0$$

$$4H_2 + CO_2 = CH_4 + 2H_2O; \Delta H << 0$$

As illustrated in Figure 10.4, the syngas with a H_2 to N_2 molar ratio of 3 to 1 is compressed to a pressure greater than the synthesis loop pressure (which operates at about 120 bar), fed through the deoxygenation reactor when the N_2 stream is supplied by the air separation unit, and then combined with the synthesis loop recycle gas. The combined stream is then preheated in a feed/effluent interchanger before being fed to the NH_3 synthesis reactor containing the Fe-based catalyst. The reactor contains individual beds of catalyst with the feed gas introduced into the reactor in multiple locations just upstream of each bed in order to cool the effluent from a preceding bed by directly mixing with the cooler feed gas added and thus limit the temperature of the gas entering the subsequent bed (the "cold shot" reactor configuration). The reactor effluent at approximately 400°C is then cooled in a series of heat exchangers including the feed/effluent interchanger and finally a refrigerated exchanger. The NH_3 condensate collected is pumped to a pressure of 158 bar and then vaporized before

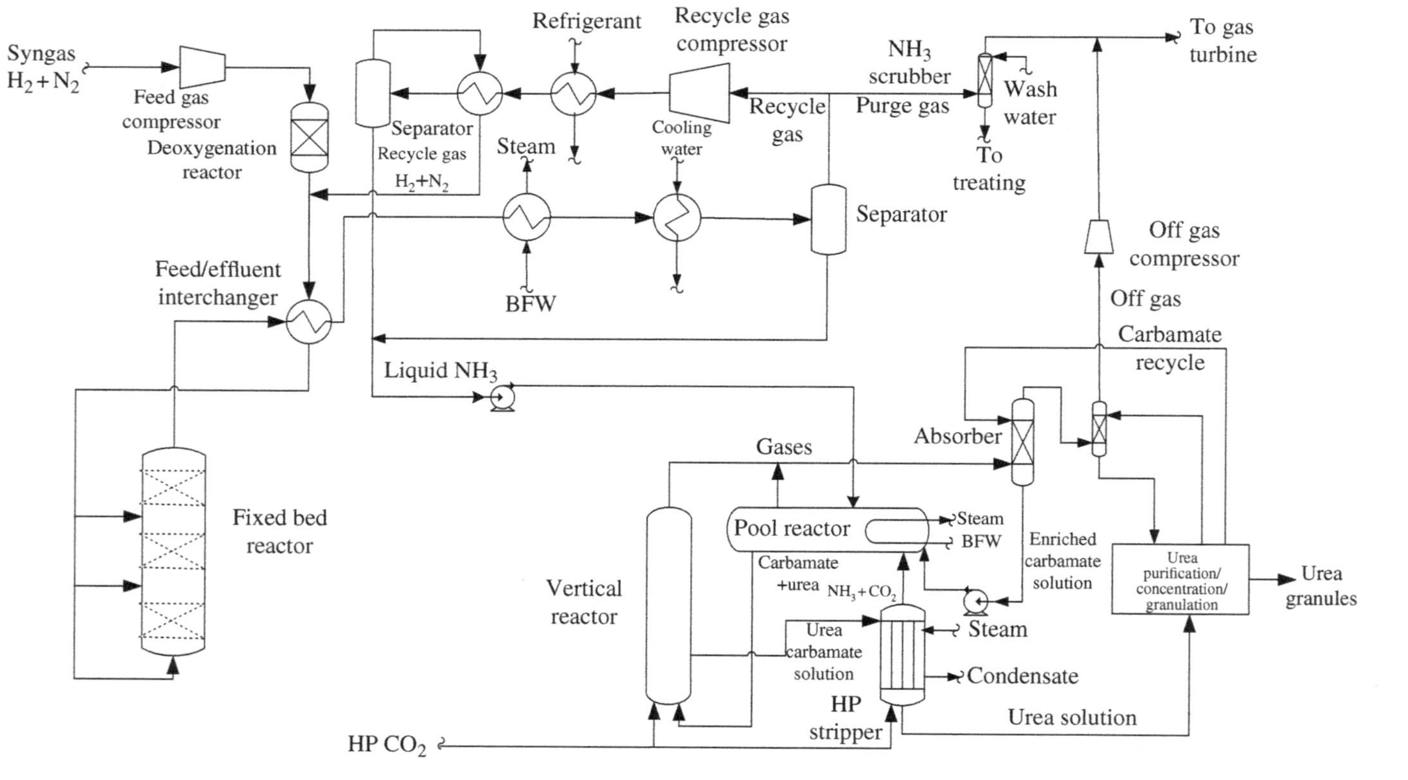

FIGURE 10.4 The Avancore urea process

feeding it to the urea synthesis unit consisting of the Avancore® process offered by Stamicarbon chosen for illustrative purposes. The NH_3 is first reacted with compressed CO_2 (also at 158 bar) supplied from the syngas decarbonization step to form ammonium carbamate (NH_2COONH_4) by the exothermic condensation reaction (10.8) and then the endothermic dehydration of the carbamate to urea and water by reaction (10.9):

$$2NH_3 + CO_2 = NH_2COONH_4; \Delta H < 0 \tag{10.8}$$

$$NH_2COONH_4 = NH_2CONH_2 + H_2O; \Delta H > 0 \tag{10.9}$$

In the case of a feedstock such as natural gas that has a low carbon to hydrogen ratio, the CO_2 recovered from the syngas decarbonization step may be insufficient and additional CO_2 may have to be recovered from the SMR flue gas using an amine process such as that described in Chapter 9.

In the Avancore urea process (Soemardji and Engineer, 2012) included in Figure 10.4, pressurized NH_3 feed is introduced into a pool reactor along with the carbamate solution coming from a downstream recirculation section. Off-gas from a downstream high-pressure stripper consisting of NH_3 and CO_2 formed by the dissociation of the carbamate is also fed into the pool reactor (gases rise as bubbles through a "pool" of liquid). The NH_3 and CO_2 react in the pool reactor to form condensed carbamate while a significant conversion to urea also occurs. The heat released by the reaction is recovered by generating low-pressure steam. Most of the pressurized CO_2 feed to the unit enters the stripper to countercurrently contact the urea-carbamate solution leaving a vertical reactor. The stripper (falling film type) consists of a shell and tube design with condensing steam on the shell side to provide the thermal energy for stripping. The remaining fraction of the fresh CO_2 supplied to the synthesis unit enters the vertical reactor along with the urea-carbamate liquid from the pool reactor producing sufficient heat for the endothermic urea reaction and the final conversion to urea occurs. The urea carbamate solution leaving the reactor enters the stripper. Gases leaving the vertical reactor along with gases leaving the pool reactor are fed into the scrubber where they are contacted with the carbamate solution from the low pressure recirculation stage. The carbamate solution thus enriched is then fed to the pool reactor. NH_3 and CO_2 are recovered from the purge gas before venting the inert gases. The urea solution leaving the stripper is then purified and concentrated to meet the requirements of the downstream granulation process.

10.3.3 Fischer–Tropsch Liquids

The overall reactions in the synthesis of hydrocarbons are as follows:

$$(2n+1)H_2 + nCO = H{-}(CH_2-)_n{-}H + nH_2O; \Delta H < 0 \tag{10.10}$$

$$2mH_2 + mCO = C_mH_{2m} + mH_2O; \Delta H < 0 \tag{10.11}$$

$$CO + H_2O = CO_2 + H_2;\ \Delta H < 0 \tag{10.12}$$

While the unwanted reaction producing oxygenates also occurs, its extent being dependent on the type of catalyst and operating conditions:

$$2pH_2 + pCO = C_{p-1}H_{2p-1}CH_2OH + (p-1)H_2O;\ \Delta H < 0 \tag{10.13}$$

The formation of methane ($n = 1$) is also unwanted. Most of the alkanes produced tend to be straight-chain, suitable as diesel fuel. In addition to alkane formation, competing reactions give small amounts of alkenes by the reaction (10.11), as well as alcohols and other oxygenated hydrocarbons by the reaction (10.13).

The catalysts commercially available today are Fe or Co based, while Ru based are being developed and have the highest activity allowing operation at lower reaction temperatures but are extremely susceptible to sulfur poisoning while the high price are limiting their use. Table 10.3 summarizes the characteristics of the Fe- and Co-based catalysts. Co-based catalysts are preferred for syngas with high H_2 to CO ratios because they have less reactivity toward the water gas shift reaction for converting CO to H_2. Co catalysts are used in low temperature Fischer–Tropsch processes (200–240°C) because at higher temperatures, excess CH_4 is produced.

A range of hydrocarbons are produced depending on the type of catalyst and operating conditions. Higher temperatures (~350°C) favor gasoline and light olefins while lower temperatures (~250°C) favor heavier hydrocarbons such as distillate and waxes (hydrocarbons that are solid at room temperature). Table 10.4 presents the product yield data from a pilot plant unit operating with a slurry reactor containing a Co-based catalyst (Bertoncini et al., 2009). If the hydrocarbon chain formation occurs by the insertion or addition of C_1 intermediates in a step-wise manner with constant probability of growth α (which is primarily a function of the catalyst type and operating conditions), then the chain length distribution is given by the Anderson–Schulz–Flory distribution (Kamali et al., 2012):

$$\mathfrak{x}_n/n = \left(1-\alpha_p\right)^2 {\alpha_p}^{n-1} \tag{10.14}$$

TABLE 10.3 Fe- versus Co-based catalyst for Fischer–Tropsch synthesis

Characteristics	Fe	Co
Cost	Low	Expensive
Life	Short	Long
Type	Precipitated/fused	Supported
Syngas H_2 to CO ratio	0.7 (higher water gas shift activity)	2.0+ (lower water gas shift activity)
Tolerance	S and NH_3 tolerated	None to S and NH_3
Toxicity/disposal	Nontoxic/disposable	Toxic/need Co reclaiming
Product types	More olefinic	More paraffinic
Products molecular weight	Higher	Lower
By-products	CO_2, steam and oxygenates	Steam and less oxygenates

TABLE 10.4 Fischer–Tropsch product yields with Co-based catalyst in slurry reactor

Carbon number	Paraffins	Olefins	Oxygenates	Total
C_1	8.17		0.14	8.32
C_2–C_4	5.18	4.68	0.49	10.35
C_5–C_9	11.62	4.42	0.61	16.64
C_{10}–C_{13}	12.10	1.72	0.61	14.43
C_{14}–C_{21}	22.73	1.09	0.68	24.49
C_{22}–C_{24}	4.58	0.07	0.12	4.78
C^{25+}	20.91	0.06	0.02	20.99
Total	85.29	12.04	2.67	100.00

$$y_n = \left(1 - \alpha_p\right) \alpha_p^{\,n-1} \tag{10.15}$$

where ξ_n and y_n are the weight and mole fractions of the hydrocarbon with n carbon atoms, and α_p is the probability of chain growth. The majority of the experimental data show that in the C_4–C_{12} region, the Anderson–Schulz–Flory distribution holds for a family of hydrocarbons and α_p may be used to characterize catalysts for Fischer–Tropsch synthesis. Typically, a high α_p gives rise to larger fraction of waxes ($>C_{20}$) that may be hydrocracked to produce diesel (C_{13}–C_{17}) or kerosene (jet fuel) (C_{11}–C_{13}). With a low α, oligomerization consisting of forming larger chain liquid hydrocarbons from the low carbon number fractions is used to produce olefinic gasoline and some diesel. Catalyst promoters and the preparation methods can alter the selectivity of a catalyst toward a high or a low α_p. Researchers have shown that the oxygenates in addition to the paraffinic and olefinic hydrocarbons also follow an Anderson–Schulz–Flory distribution (Bertoncini et al., 2009).

Fixed bed tubular reactors have been used commercially but fluidized bed and slurry reactors have certain advantages, the characteristics of each of these reactor types are summarized in the following.

10.3.3.1 Fixed Bed Tubular Reactor The catalyst (~2.5 mm precipitated) is placed inside tubes through which the syngas gas flows with operating conditions of 220–250°C and 27 bar resulting in a higher yield of waxes. Removal of the high exothermic heat of reaction is a major challenge with this type of reactor. The synthesis loop uses a large recycle to improve isothermicity within the reactor and increase overall conversion, but it increases the power consumption as well as the plant capital cost since a large recycle compressor is required. The reactor also tends to be mechanically complex with the internal tube design while catalyst replacement is more difficult compared to the other reactor types.

10.3.3.2 Fluidized Bed Reactor The catalyst particles (<70 μm fused) are suspended by the gas and the resulting space velocity is much higher than the fixed bed. The operating conditions are typically 320° and 27 bar with a recycle ratio of 1.5 to increase gas velocity through the reactor required for fluidization, improve isothermicity, increase heat transfer coefficient and overall conversion. Higher fractions of the lighter hydrocarbons are produced due to the higher operating temperature. An advantage of this type of reactor is that the catalyst replacement can occur online without having to shut down the unit. Also regeneration of the catalyst online is possible.

10.3.3.3 Slurry Reactor This type of reactor similar to the one in the LPMEOH™ process utilizes small suspended catalyst particles (<40 μm fused or 40–150 μm precipitated) within a liquid with low-vapor pressure such as high boiling waxes. Boiling water within coils installed in the slurry bed control the temperature resulting in virtually an isothermal operation allowing operation with higher average reaction temperatures (and reaction rates) without developing local hot spots. The higher reaction temperatures are conducive to less wax production. The reactor is simpler to construct while online catalyst replacement is possible and turndown ratio is good, which is important for coproduction plants to complement the intermittent renewables.

As illustrated in Figure 10.5, the synthesis loop recycle gas is fed to an autothermal reformer to convert the CH_4 and other undesirable hydrocarbons back to H_2 and CO in order to maximize the production of more valuable hydrocarbons. It is combined with fresh desulfurized syngas and is cooled in a series of heat exchangers before treatment in an amine wash unit to remove most of the CO_2 formed primarily in the autothermal reformer. The combined stream is then preheated in heat exchangers including a feed/effluent interchanger before being fed to the synthesis reactor. The reactor effluent is cooled in a series of heat exchangers including the feed/effluent interchanger while the raw product condensate collected is fed to the stabilization unit that consists of a column to remove the dissolved light ends from the Fischer–Tropsch liquids formed. The recycle gas that contains CH_4 and other hydrocarbons is compressed before being fed to the autothermal reformer.

Example 10.1. Fischer–Tropsch synthesis yield data for a syngas with a H_2 to CO ratio of 2.0 on a mole basis over a Co catalyst at two different temperatures is shown in Table 10.5 (estimated from plotted data of Johnson et al., 1991). Calculate the probability of chain growth α_p for the two sets of data.

Solution

Taking the natural logarithms on both of Equation 10.15 and rearranging

$$\ln y_n = (\ln\alpha_p)\, n - [(\ln\alpha_p) - \ln(1-\alpha_p)] \tag{10.16}$$

shows that a plot of $\ln y_n$ versus n should yield a straight line with a slope of $\ln \alpha_p$ if the Anderson–Schulz–Flory distribution holds. Figure 10.6 shows the plots for the two sets of data and the slope of the best fit straight line for each set is:

$$\text{At } 456\text{ K}, \ln\alpha_p = -0.411 \text{ or } \alpha_p = 0.66$$

$$\text{At } 484\text{ K } \ln\alpha_p = -0.661 \text{ or } \alpha_p = 0.52$$

This result indicating that α_p is lower at the higher temperature is to be expected.

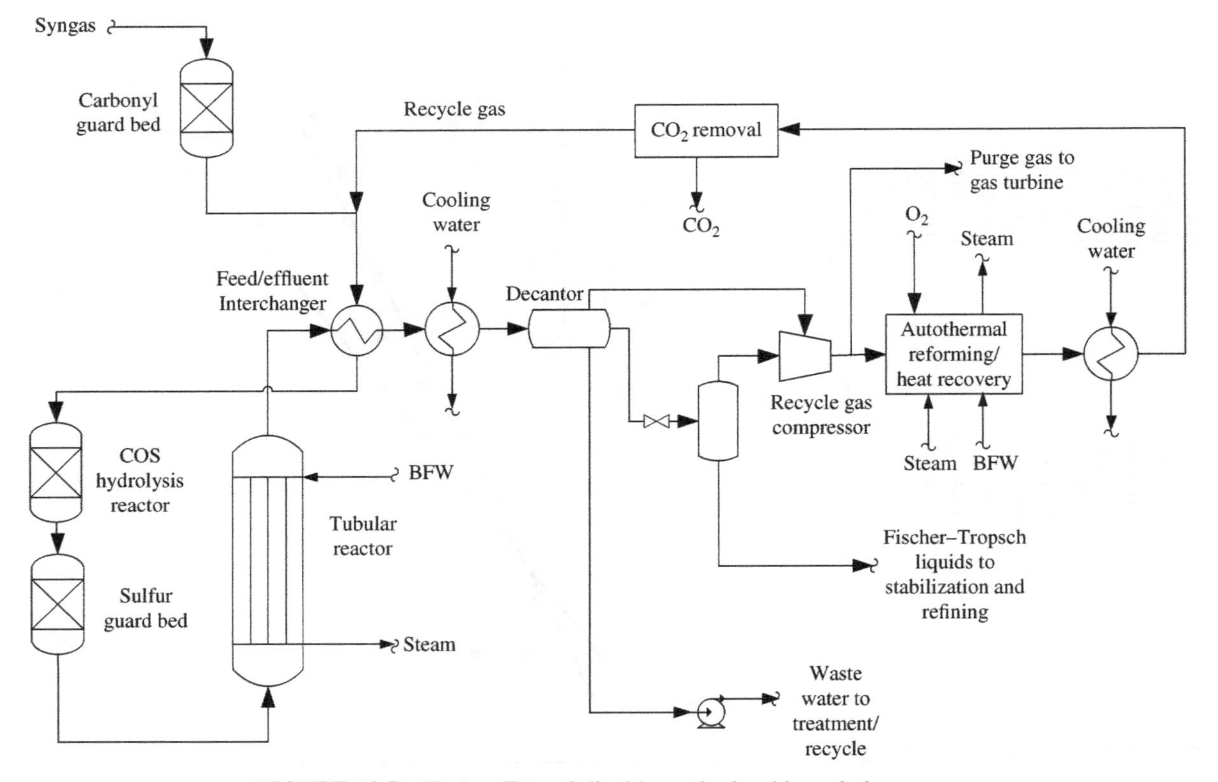

FIGURE 10.5 Fischer–Tropsch liquids synthesis with a tubular reactor

TABLE 10.5 Fischer–Tropsch synthesis yield data

Carbon no. (n)	Mole fraction (y_n) at 484 K	Mole fraction (y_n) at 456 K
1	0.6065	0.6065
2	0.04979	0.02237
3	0.06081	0.05502
4	0.04505	0.04076
5	0.02024	0.03688
6	0.01111	0.02024
7	0.005517	0.01005
8	0.003028	0.005517
9	0.002029	0.004992
10	0.000825	0.003346
11	0.000275	0.003028
12	0.000184	0.002029
13		0.001008

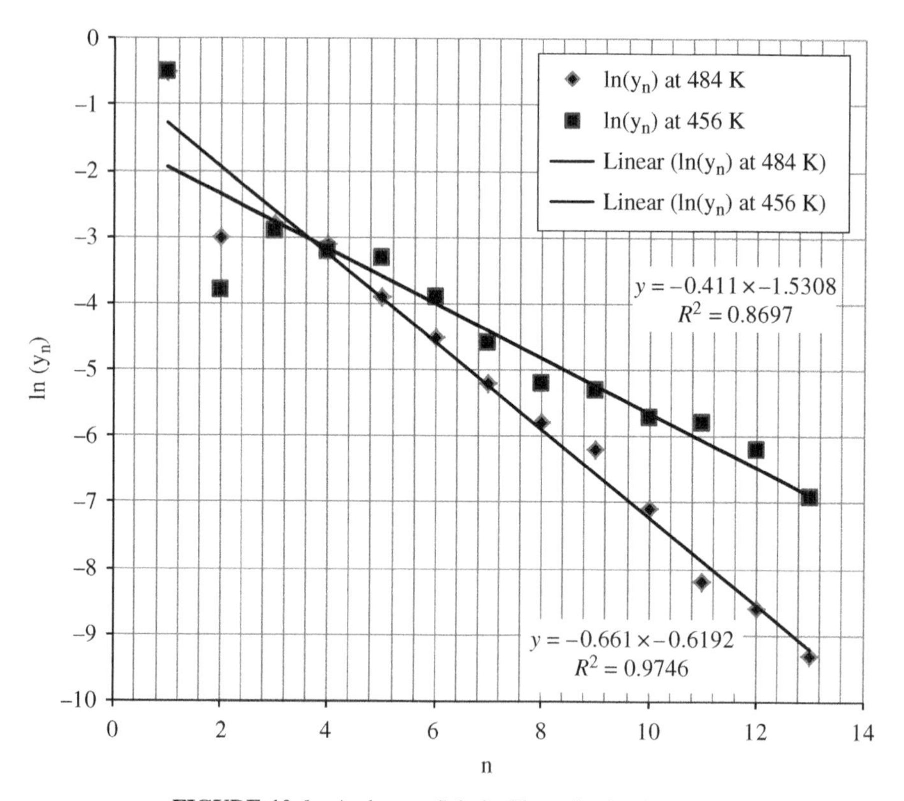

FIGURE 10.6 Anderson–Schulz–Flory distribution plots

10.4 HYDROGEN COPRODUCTION FROM COAL AND BIOMASS

The transportation sector accounts for one-fifth of global CO_2 emissions (Kırtay, 2011). Development and deployment of hybrid vehicles using proton exchange membrane fuels cells that require high purity H_2 as the fuel can help mitigate the greenhouse gas emission if the H_2 can be provided sustainably. In the following are described the plant configurations of a coal plus biomass-fed H_2 coproducing plant with CO_2 capture utilizing (i) current technology IGCC and (ii) advanced technology consisting of an integrated hydrogasifier and a solid oxide fuel cell, the integrated gasification fuel cell (IGFC). The thermal and economic performance of the IGFC case has been calculated to be significantly better than the IGCC case showing the importance of continuing research and development activities in the energy field to deploy such advanced gasifiers and large scale fuel cells (Li et al., 2012). At ISO conditions, with the feedstock consisting of 66% Utah bituminous coal with the remainder consisting of equal amounts of corn stover and cereal straw (all on a dry weight basis), and exporting H_2 equivalent to 23% of the input fuel bound energy on an HHV basis, the IGFC case generates as much as 70% more electric power while the cost of producing the coproduct H_2 for the IGFC case is less than half of that for the IGCC case.

10.4.1 Current Technology Plant

The IGCC with currently available technology for the coproduction of electricity and H_2 is illustrated by the overall block flow diagram in Figure 10.7. The gasification plant is based on a dry feed (since the feed contains biomass that has a high moisture content) entrained bed gasifier. Received coal and biomass (corn stover and cereal straw in this case) is fed to commutation equipment and then dried in a fluid bed dryer. Biomass generally has higher moisture content than coal and requires more thermal energy. Fluidized bed drying with vapor recompression is employed. As explained in Chapter 8, the technology utilizes the latent heat of the vapor produced in the dryer by pressurizing the vapor and supplying it to the heat exchange tubes located within the fluidized bed dryer. This type of drying technology consisting of a "heat pump" has been successfully applied in the drying of high moisture content brown coals. The heat pump increases the quality of the thermal energy contained in the evaporated water vapor by using mechanical energy in the vapor compressor, this process being consistent with the second law of thermodynamics as the overall quality of energy leaving the system is less than that entering the system. The high quality electrical energy demand to drive the vapor compressor is quite small, however, as compared to the amount of energy recovered from the vapor that otherwise would be wasted. The dried feedstock is then further reduced in size to meet the specifications of the entrained bed gasifier. The energy required for this fine grinding operation is greatly reduced when the biomass feedstocks are predried. The air separation unit operating at elevated pressure supplies 95% by volume O_2 to the gasifier and for combustion of acid gas that is separated from the syngas during its cleanup downstream of the gasifier, and produces N_2 at an intermediate pressure for injection into the gas turbine as a thermal diluent for

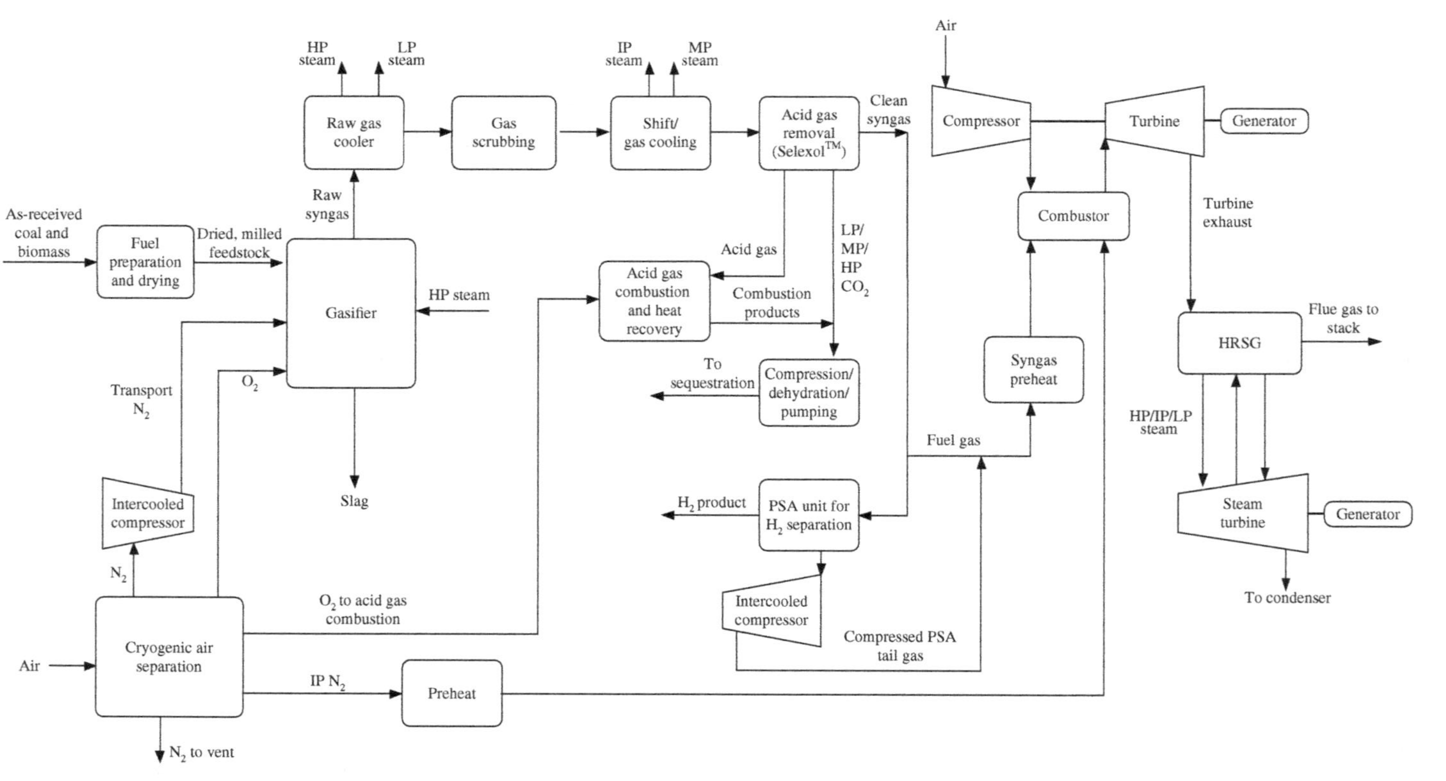

FIGURE 10.7 IGCC for coproduction of electricity and H_2

NO_x control. A portion of the intermediate pressure N_2 is further compressed for use as transport gas to pneumatically convey the feedstock to the gasifier.

The gasifier operates at a nominal temperature and pressure of 1370°C and 43 bar and partially oxidizes the coal and the biomass with the O_2 to generate hot raw syngas, slag, and char. The raw syngas is cooled in heat exchangers to provide heat for the steam system to generate high-pressure and low-pressure steam. The raw syngas is then water scrubbed to remove particulates, alkalis, chlorides, and NH_3. High-pressure steam is injected into the syngas to increase H_2O vapor content to a level sufficient for the downstream shift reactors. The syngas then enters the shift unit where most of the CO present in the syngas is reacted with the H_2O vapor to produce H_2 and CO_2. Heat generated by the exothermic shift reaction is recovered by generating intermediate pressure and medium pressure steam in heat exchangers located downstream of the reactors. The shifted gas leaving this unit further provides heat to the vacuum condensate from the steam cycle and is then cooled against cooling water. The cooled syngas is next superheated by about 11°C to avoid pore condensation and then fed to a sulfided activated carbon bed for removal of Hg. The effluent gas is then cooled in a trim cooler against cooling water and fed into the Selexol™ AGR unit. The high temperature condensate separated from the gas is recycled to the scrubber while the NH_3 laden colder condensate is fed to a sour water stripper. The sour gases stripped off from the water are routed to the acid gas combustor.

The Selexol™ process produces the clean decarbonized syngas (essentially H_2 with some CO, CO_2, CH_4, N_2, and Ar), acid gas, and CO_2 streams (at different pressures). The CO_2 streams are compressed to a pressure where the CO_2 liquefies near the cooling water temperatures, dehydrated, and then pumped to the pipeline pressure of 152 bar. For cases where the CO_2 stream can be sequestered close by, the acid gas from the Selexol™ process together with the sour gases from the sour water stripper after combusting with the air separation unit supplied O_2 may be co-sequestered with the CO_2 stream. The heat generated in the acid gas combustion unit can be recovered by producing intermediate-pressure and low-pressure steam while making this option slightly more efficient than making saleable byproducts such as elemental sulfur (described in Chapter 9) or H_2SO_4. The exhaust gas from this combustion unit would be compressed and then combined with the pressurized CO_2 stream for sequestration. If the acid gas is instead treated to make the saleable byproducts, then the CO_2 stream would maintain its higher purity making it suitable for use in enhanced oil recovery or long-distance transportation by pipeline without requiring any special metallurgy.

A portion of the clean decarbonized syngas leaving the Selexol™ unit with ultra-low sulfur content is treated in a PSA unit to produce a high-purity H_2 product that is compressed to 46 bar, which is the same pressure of the H_2 stream produced in the advanced (IGFC) case. The PSA also produces a tail gas stream that consists of the remaining fuel gas components (mostly CO, CO_2, H_2, N_2, and Ar). The PSA tail gas is compressed and then combined with the remainder of the clean decarbonized syngas. This combined fuel gas stream is preheated to 290°C using intermediate pressure steam and then fed to the gas turbine combustor. Intermediate pressure N_2 from the ASU (preheated by intermediate pressure steam) is also fed to the gas turbine combustor. The flow rate of this intermediate pressure N_2 added is such that the input syngas and N_2 has a combined LHV of 4730 kJ/Nm^3. The introduction of N_2 reduces the formation of NO_x within the combustor of the F class gas turbine (currently available for syngas fuels) by lowering the flame temperature. Note that because of the H_2 export, the relative amount of diluent N_2 generated by the air separation unit and available for gas turbine injection is high; thus, there is no need to humidify the fuel gas as in some of the power-only IGCC plant configurations (Black, 2010).

The exhaust gas exits the gas turbine at 564°C and enters the heat recovery steam generator (HRSG) to provide heat for the bottoming Rankine cycle consisting of a triple pressure reheat steam cycle. The flue gas is then discharged through the plant stack. Bulk of the steam generated within the plant is used in a steam turbine for power generation while the remainder to satisfy process steam demands. In addition to these on-site units, off-site units such as a cooling water system, instrument air, flare, waste water treatment, and so on, necessary for a stand-alone plant are provided.

10.4.2 Advanced Technology Plant

The IGFC with technology to be commercially available in the future for the coproduction of electricity and H_2 is illustrated by the overall block flow diagram in Figure 10.8. Although the O_2 blown entrained-bed gasifiers are by far the most proven and matured gasification technologies, the syngas produced by such gasifiers contains high CO and H_2 concentrations with only a small amount of CH_4 while in IGFC applications using either the solid oxide fuel cell (SOFC) or the molten carbonate fuel cell (MCFC), a syngas (the fuel cell anode inlet gas) rich in CH_4 is desirable. This is mainly because CH_4 can undergo endothermic reformation within the fuel cell. Thermodynamically, endothermic reactions have the potential to act as chemical heat sink and can reduce the air flow rate to the fuel cell stacks for the purpose of cooling, thus saving parasitic energy use for air compression. The thermal energy thus converted to chemical bond energy may then be recycled to the fuel cell anode to result in higher efficiency. Furthermore, higher CH_4 content in the syngas generally results from gasifiers operating at relatively lower temperatures, and with such gasifiers less fuel (coal or biomass) bound energy is degraded to heat to maintain the gasifier operating temperature, resulting in cold gas efficiencies higher than those for gasifiers operated at higher temperatures. Concerns, however, are that the reaction rates are relatively slow and the carbon conversion is low due to the lower operating temperature unless a suitable catalyst can be employed.

The gasification occurs in two sections, a "top stage" catalytic hydrogasifier and a "bottom stage" high-temperature slagging entrained-bed O_2 blown gasifier. The majority of the fuel (coal and biomass) are gasified in the catalytic hydrogasifier by a high-temperature gas stream containing H_2, H_2O, and CO. The development of this type of gasifier was being pursued by Exxon in the 1970s (Kalina and Moore, 1974) with the gasification agent consisting of steam, H_2 and CO being fed to a fluidized-bed gasifier to react with crushed coal in the presence of a potassium catalyst (KOH and K_2CO_3). In more recent years, GreatPoint Energy (Kang and Lee, 2013)

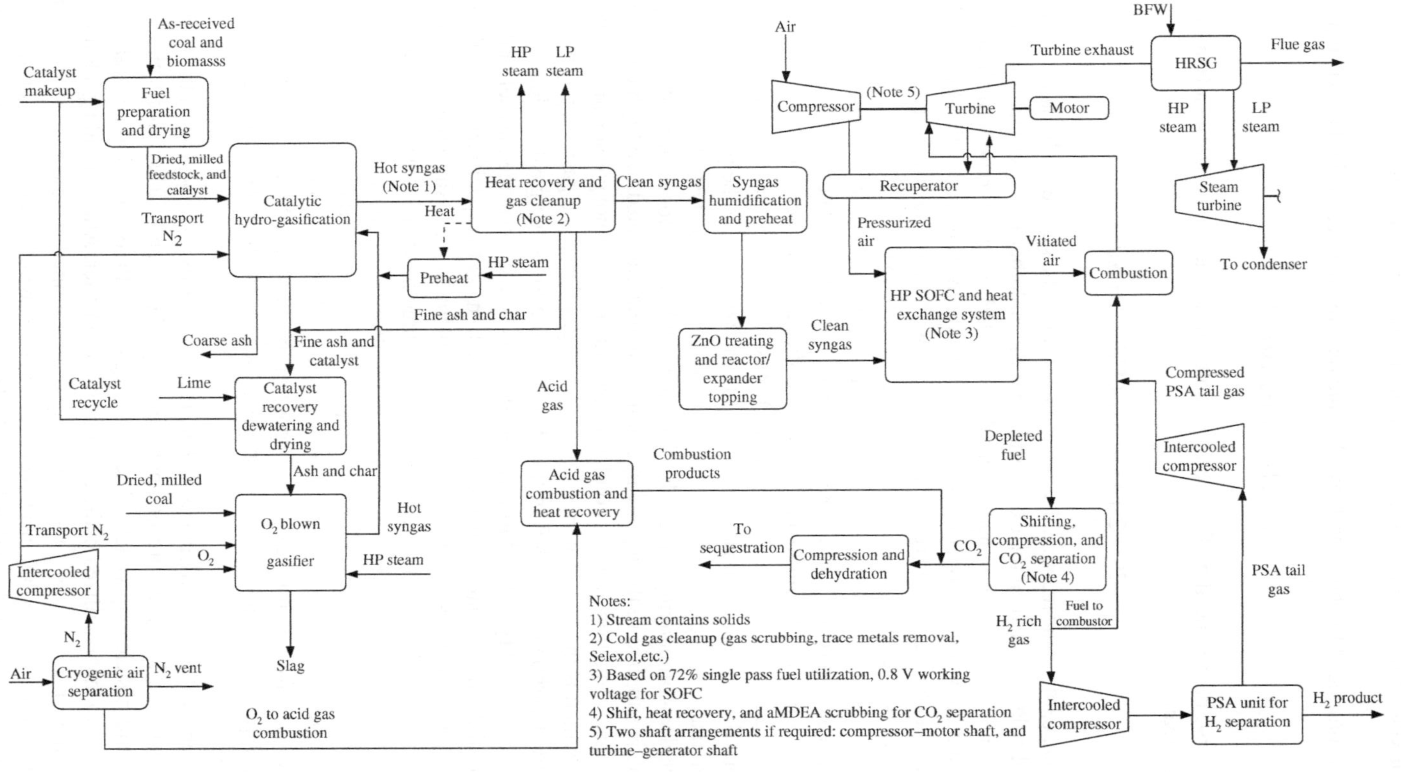

FIGURE 10.8 IGFC for coproduction of electricity and H_2

had resurrected the development of such a "hydrogasifier" but using their proprietary catalyst.

CH_4 is produced by the overall exothermic reaction between C in the coal and H_2:

$$C + 2H_2 = CH_4; \Delta H < 0$$

CH_4 can also be produced by the overall endothermic reaction between carbon in the coal and steam:

$$2C + 2H_2O = CH_4 + CO_2; \Delta H > 0$$

Another exothermic reaction that takes place within the gasifier is the shift reaction. A careful balance between the exothermic and the endothermic reactions is achieved by providing the gasification agent with the correct H_2 to steam ratio to maintain the gasifier at the appropriate operating temperature. Another characteristic of the SOFC that enables even bigger synergy when combined with such a gasifier is that the anode exhaust can be the source for the H_2 required by the gasifier.

The "as received" coal and biomass after size reduction are impregnated with potassium catalyst (in the form of KOH and K_2CO_3). Fluidized bed drying with vapor recompression is again employed here due to the high moisture content of the biomass. The catalyst requirement (on a K_2CO_3 basis) is approximately 15% by weight of the total dry coal and biomass input. Credit may be taken for the potassium present in the biomass. Some lime is required to regenerate the catalyst. The unconverted carbon and accompanying "fine ash" from the catalytic hydrogasifier, after catalyst recovery, is supplied to the high temperature slagging entrained-bed O_2 blown gasifier. The major function of this second gasifier is to produce a hot gas stream rich in H_2 and CO, which is then used in the catalytic hydrogasifier as the gasifying agent. The unconverted carbon content left in the char from the top stage gasifier alone cannot produce sufficient H_2 and CO to meet the demand of the top stage gasifier and thus a portion of the dried coal, further reduced in size to meet the specifications of the O_2 blown entrained bed gasifier, is provided as a supplement feed. Since the carbon content in the char and "fine ash" fed to the bottom stage gasifier is relatively low, only coal (which has higher carbon content than the biomass) is provided to the bottom stage gasifier. This O_2 blown gasifier helps increase the overall carbon conversion of the gasification process while converting the ash content into a vitrified non-leachable solid form that is easier to dispose of. N_2 produced by the ASU unit, after further compression, is used as transport gas to pneumatically convey the feedstock to the gasifiers.

After removal of the particulates that are carried over by the raw syngas, the syngas at a temperature and pressure of approximately 690°C and 70 bar enters the heat recovery and gas cleanup system. The raw syngas is initially cooled to generate high pressure steam that is utilized in the catalytic hydrogasifier. After providing additional heat for the steam system, the raw syngas is supplied to the syngas cleanup/low temperature gas cooling/heat recovery system, which includes water wash to remove particulates, alkalis, chlorides, and NH_3, a carbonyl sulfide hydrolysis reactor to catalytically convert the COS to H_2S[1] by the reaction $H_2O + COS = H_2S + CO_2$, and a sulfided activated carbon bed for capture of Hg, followed by a Selexol™ unit to remove the sulfur compounds.

Next, H_2O vapor is introduced into the clean CH_4-rich syngas leaving the Selexol™ unit. The added moisture prevents deposition of carbon in the downstream reactors and the SOFC anode compartment. Instead of using high pressure steam for humidification, here the water vapor is introduced by directly contacting the syngas with liquid water flowing down through a countercurrent column (columns for such operations were described in Chapter 5), which allows for recovery of low-temperature heat generated within the plant. Clean process condensate collected from within the plant is used as the makeup water for this humidifier.

The clean syngas is then supplied to a reactor/expander topping cycle (Rao, 1991) that includes a shift reactor where the syngas is heated up by the exothermic shift reaction; the high-temperature syngas is then expanded in a turbine to recover power.

Because an SOFC has more stringent requirements with respect to contaminants contained in the syngas than an F class gas turbine, a guard bed is included upstream of the expander as a final cleanup step to limit trace amounts of chlorides and sulfur compounds to less than 0.1 ppmv each. The guard bed consists of alternating layers of COS hydrolysis catalyst and ZnO for capture of the H_2S and the chlorides.

The syngas coming out of the expander is preheated to 650°C and fed to the anode side of SOFC stacks while ambient air is compressed to 11.3 bar, also preheated to 650°C and then sent to the cathode side of the SOFC stacks. The SOFC module design for this application consists of cascading four stages of identical SOFC stacks with air flowing in series and fuel flowing in parallel, and intra-stack introduction of fresh air to produce roughly identical operations for each stack[2] in the module. High single pass air utilization in each of the SOFC stacks (thus maintaining effective cooling of the SOFC stacks) and low overall air utilization at the same time (thus reducing the parasitic air compression power) may be achieved with this configuration (Li et al., 2011). About 70% of the fuel in the syngas is oxidized in the SOFC stacks. Anode exhaust leaving the SOFC stacks at 705°C is cooled by recovering the thermal energy to preheat (i) the syngas for the SOFC, (ii) the shift reactor feed gas (in the reactor/expander topping cycle), and (iii) the COS hydrolysis reactor feed gas. The gas stream is then fed to a catalytic reactor for the water gas shift reaction, in which most of the CO content in the anode exhaust is converted to CO_2 while producing H_2. The effluent from the shift reactor, after further heat recovery and cooling, is fed to an activated methyl diethanol amine (aMDEA) wash unit for CO_2 separation. As in the previous case, either the oxidized acid gas leaving the Selexol™ unit is co-sequestered with the CO_2 stripped from the aMDEA solvent or further processed to make the saleable byproducts such as elemental sulfur or H_2SO_4.

[1] H_2S being easier to remove in the downstream desulfurization unit. The catalyst consists of a Co-Mo, or a Ni-Mo formulation.

[2] A fuel cell stack consists of a number of individual cells stacked together to produce a more substantial voltage and/or current depending on how the individual cells are connected electrically (in series or in parallel).

A large fraction of the decarbonized anode exhaust leaving the aMDEA unit is fed to a PSA unit, while the remaining is sent to the combustor downstream of the SOFC stacks. The PSA unit produces high purity H_2 product at a pressure of 46 bar and a tail gas consisting of the remaining fuel gas components that is compressed (with intercooling) and also fed to the combustor. Oxidant for this combustor consists of the cathode exhaust from the SOFC stacks. The combustor exhaust is partially expanded in a turbine, fed to a recuperator to preheat the cathode inlet air supplied by the compressor, and then further expanded to near atmospheric pressure in the turbine to generate additional power before entering the HRSG. The exhaust gas from the HRSG is also used to provide heat for the fuel drying operation.

10.5 COMBINED HEAT AND POWER

Back pressure steam turbines, internal combustion engines such as diesel engines, gas turbines, and fuel cells in addition to generating electricity may also be used to provide thermal energy by recovery of the energy contained in the exhaust. Such a dual purpose application is called cogeneration or combined heat and power (CHP) and provides for very efficient utilization of the fuel-bound energy. Air conditioning may also be provided using the heat in an absorption refrigeration cycle such as the lithium bromide (LiBr) cycle for moderate temperature refrigeration as described later in this section or an NH_3 absorption cycle for deep refrigeration temperatures. The thermodynamic advantage of a CHP plant is illustrated in Figure 10.9, which compares the thermal performance of a boiler plant wherein a fuel is directly combusted to generate steam as the carrier of the thermal energy, and a CHP plant in which the fuel is instead utilized in a topping cycle (e.g., an internal combustion engine, a gas turbine, or a fuel cell) added ahead of the steam generator. Enough fuel is combusted (say additional x units of energy) in the topping cycle such that the same amount of thermal energy is recovered from the exhaust of the topping cycle for steam generation while cooling the gases to the same final temperature before they are exhausted to the atmosphere. It can be seen that the power generation by the topping cycle (say y units of energy) occurs at a very high efficiency, $y/x = 85\%$ in this case (obtained by equating $100 + x - y$ and $0.15(100 + x) + 85$) with the simplifying assumptions made.

Example 10.2. In a CHP plant, 100,000 kg/h of steam at 400°C and 42 bar are supplied to a back pressure steam turbine having an exhaust pressure of 8 bar. Assume the isentropic efficiency of the steam turbine is 80%, while the mechanical and generator losses amount to 5% of the power developed from the steam. Calculate the ratio of thermal energy to the electrical energy exported assuming that the entire power output of the turbine is exported, and that the exported steam is condensed at a pressure of 7 bar to provide the heating need.

Solution

The first step is to read off the specific enthalpy $\Delta h^{V}_{\text{turb, in}}$ and the entropy $\Delta s^{V}_{\text{turb, in}}$ of steam at the inlet conditions from steam tables (e.g., Keenan et al., 1969) at the given

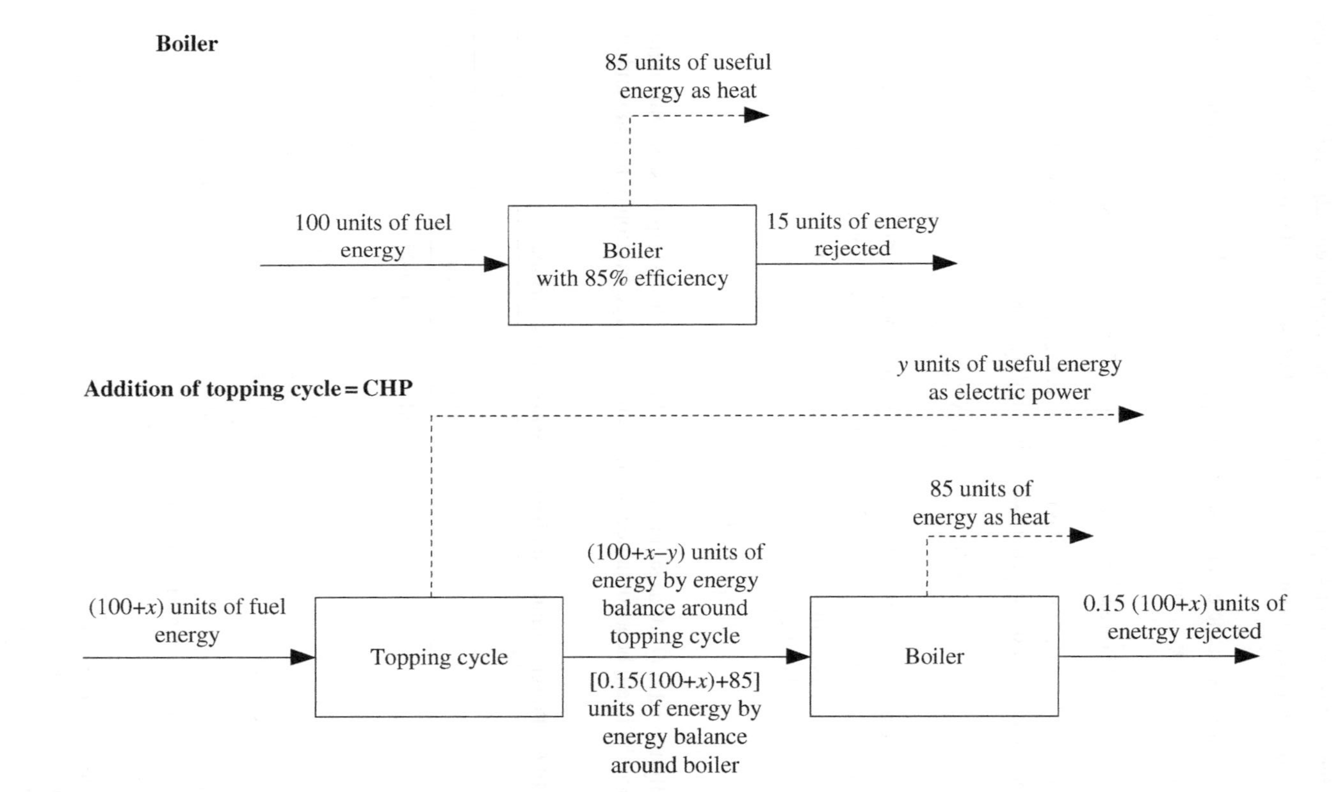

FIGURE 10.9 Thermodynamic advantage of a CHP plant

conditions of pressure and temperature. Since the entropy at the inlet is greater than the entropy at the outlet pressure for saturated steam, the exhaust has to be all vapor (single phase). So the next step is to read off the specific enthalpy $\Delta h^{V}_{turb,out}$ corresponding to the inlet entropy $\Delta s^{V}_{turb,in}$ of steam but at the outlet pressure.

The actual enthalpy of the exhaust steam $\Delta h^{act}_{turb,out}$ is then calculated using the turbine isentropic efficiency η_{turb} by:

$$\Delta h^{act}_{turb,out} = \Delta h^{V}_{turb,in} - \left(\Delta h^{V}_{turb,in} - \Delta h^{V}_{turb,out}\right)\eta_{turb}$$

The electric power output of the turbine $\dot{W}$ is then calculated using the mass flow rate of the steam $\dot{m}$ and the generator efficiency η_{gen} as

$$\dot{W} = \dot{m}\left(\Delta h^{V}_{turb,in} - \Delta h^{act}_{turb,out}\right)\eta_{gen}$$

The next step is to calculate the heat duty $\dot{Q}$ when the exported steam is condensed. The specific enthalpy $\Delta h^{L}_{cond,out}$ of the condensed steam may be read off from the steam tables corresponding to the saturated liquid at a pressure of 7 bar. Then the heat duty may be calculated by

$$\dot{Q} = \dot{m}\left(\Delta h^{act}_{turb,out} - \Delta h^{L}_{cond,out}\right)$$

and finally the ratio $\dot{Q}/\dot{W}$ as summarized in Table 10.6. As can be seen, this ratio is large since the latent heat contained in the steam is a large fraction of its total enthalpy

TABLE 10.6 Power output from back pressure steam turbine

			Outlet	
Turbine	Units	Inlet	Ideal	Actual
Flow rate	kg/h	100,000	100,000	100,000
Temperature	°C	400	186	221
Pressure	Bar	42	8	8
Enthalpy	kJ/kg	3,210.6	2,805.6	2,886.6
Entropy	kJ/kg-K	6.744	6.744	6.915
Saturated vapor entropy	kJ/kg-K		6.662	
Shaft power	kJ/h		32,399,000	
Electric power	kW		8,550	
Exhaust steam condenser (to supply heat)				
Temperature	°C			219
Pressure	Bar			7
Saturated vapor enthalpy	kJ/kg			2762.8
Saturated liquid enthalpy	kJ/kg			697.00
Duty (thermal energy exported)	MJ/h		206,575	
Thermal energy/electrical energy exported	Fraction		6.71	

and thus a CHP application is an effective strategy for recovery and use of this latent energy, whereas in plant generating only electricity as its product, most of the latent heat is rejected to the environment.

10.5.1 LiBr Absorption Refrigeration

As discussed in Chapter 2, in such a heat-driven refrigeration process where heat has to be transferred from a lower temperature to a higher temperature, additional heat at a higher temperature is required so that there is no spontaneous increase in the overall quality of energy. This higher temperature heat is provided by the exhaust of the topping cycle in a CHP plant.

Figure 10.10 illustrates a single stage LiBr absorption refrigeration cycle consisting of two vessels, an evaporator operating at sub-atmospheric pressure and the other a regenerator. The higher temperature heat supplied to the cycle is transferred to a dilute aqueous solution of LiBr contained in the regenerator to vaporize the water while concentrating the LiBr solution. The resulting water vapor flows into the condenser where a cooling medium is used to condense the water vapor to liquid water. The liquid water then flows over tubes (shell side) in the evaporator maintained at a much lower pressure and the liquid water vaporizes at a low temperature corresponding to the lower pressure. The fluid to be cooled (typically water) flows through the evaporator tubes while transferring heat to the vaporizing water on the shell side. The vaporized water then flows into the absorber to mix with the concentrated aqueous LiBr solution while the released heat is transferred to cooling water. Solution strength becomes weak as it absorbs the water vapor and this weak solution is pumped to the regenerator (after preheating in an interchanger) to be concentrated and repeat the cycle. Chilled water at approximately 7°C may be produced using saturated steam at a pressure lower than 2 bar while using cooling water supplied at approximately 30°C. The coefficient of performance or COP (ratio of refrigeration heat duty to the sum of the higher temperature heat duty supplied and electrical power input) can range from approximately 0.65 to 0.7. The COP may be increased by having interchangers within the cycle, and further improvements in the COP may be realized by using a two-stage cycle that utilizes two generators, with the higher temperature heat supplied to one regenerator while the heat to the second regenerator being recovered by condensing the water vapor formed in the first regenerator. The system cost on the other hand also increases.

Comparing this absorption refrigeration process to a compressor-driven refrigeration process (mechanical refrigeration), both processes require the evaporation of a liquid at low pressure to achieve the low refrigeration temperature and the condensation of this vapor at a higher pressure and a correspondingly higher temperature in order to reject the heat to the surroundings. The manner in which the low pressure vapor is pressurized before the condensing step differs for the two types of refrigeration, one involving absorption of the vapor into a liquid phase and the subsequent pumping of this liquid to the higher pressure followed by using thermal energy to release the vapor at a higher pressure, versus the single step mechanical compression of the low pressure vapor in the case of mechanical refrigeration.

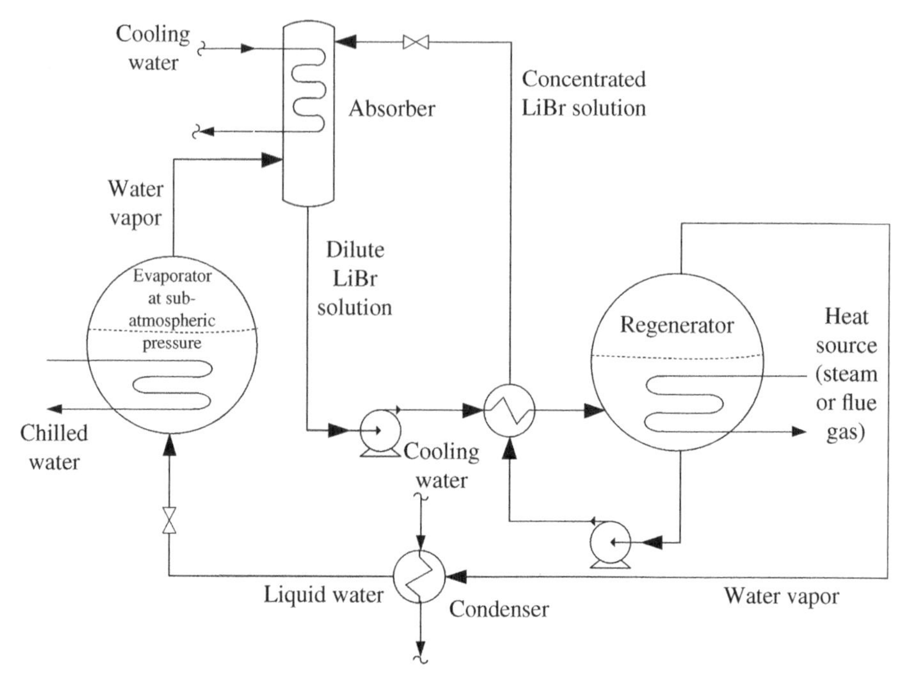

FIGURE 10.10 Single stage LiBr absorption refrigeration for chilled water

Example 10.3. In a CHP plant, 10,000 kg/h of steam at 400°C and 18 bar are supplied to a condensing steam turbine with an extraction port. Assume the isentropic efficiency of the steam turbine to the extraction point is 65% and that from the extraction point to the exhaust is 70% while the mechanical and generator losses amount to 5% of the power developed from the steam. The required amount of steam is extracted from the turbine at a pressure of 2.43 bar (based on supplying heat at a condensing temperature of 110°C while accounting for pressure drops on the steam side) to operate a single stage LiBr absorption refrigeration unit. The COP of this single stage system is 0.65 to export 50,000 kg/h of chilled water at a temperature of 7°C when the return water to the refrigeration unit is at 13°C. Assume cooling water to reject heat from the refrigeration unit and the steam cycle is supplied at 30°C and is returned to the cooling tower at 40°C (i.e., a 10°C rise), while a pinch temperature of 5°C is used in the vacuum steam condenser (difference in temperature of the cooling water leaving the condenser and temperature of the condensing steam). Calculate the percentage change in the net amount of electric power generated if a two-stage LiBr absorption refrigeration unit were used instead. This two-stage system has a higher COP of 1.0 but requires steam to be extracted from the turbine at a higher pressure of 5.74 bar (based on supplying heat at a higher condensing temperature of 150°C while accounting for pressure drops on the steam side). Neglect the electrical power required for operating the plant as well as in the calculation of the thermal energy required to operate the refrigeration units using the given COPs.

Solution
The first step is to calculate the refrigeration duty required which then may be used in calculating the amount of steam to be extracted from the turbine. These calculations are summarized in Tables 10.7 and 10.8.

Next, the calculation for the steam turbine is performed by modeling the turbine as two separate turbines, the first up to the extraction pressure (Turbine 1) and the second with flow rate deducted (to provide the steam for the refrigeration unit) up to the condenser inlet pressure (Turbine 2). The condenser pressure corresponds to the saturated pressure of steam at 45°C, which is calculated by adding the given pinch temperature of 5°C to the given cooling water exit temperature of 40°C. These calculations are summarized in Table 10.9. The total power generated for both cases is essentially the same, the savings in the thermal energy provided by the extracted steam for the two-stage refrigeration case is negated by the higher quality of this thermal energy required (i.e., steam at higher pressure has to be extracted). This result is not coincidental; if the steam cycle and the refrigeration unit were both ideal (from a second law standpoint), then the efficiency of the two alternate designs would be identical since quality and quantity of thermal energy entering and leaving each of the overall systems (by overall system is meant the steam cycle together with the refrigeration unit)

TABLE 10.7 Refrigeration duty

	Units	Water entering refrigeration	Chilled water leaving refrigeration
Flow rate	kg/h	$\dot{m} = 50{,}000$	
Temperature	°C	13	7
Enthalpy	kJ/kg	$h_{\text{in}}^{\text{chilled water}} = 54.0$	$h_{\text{out}}^{\text{chilled water}} = 28.8$
Refrigeration duty	kJ/h	$\dot{Q}_{\text{refrig}} = \dot{m} \times \left(h_{\text{in}}^{\text{chilled water}} - h_{\text{out}}^{\text{chilled water}}\right) =$ 1,259,000	

TABLE 10.8 Amount of steam to be extracted

	Units	Single-stage refrigeration	Two-stage refrigeration
COP	Fraction	0.65	1.00
Heat duty, $\dot{Q}_{\text{heat}} = \dot{Q}_{\text{refrig}}/\text{COP}$	kJ/h	$\dfrac{1{,}259{,}000}{0.65} = 1{,}937{,}000$	$\dfrac{1{,}259{,}000}{1.0} = 1{,}259{,}000$
Steam condensing temperature	°C	110	150
Enthalpy of saturated vapor/liquid	kJ/kg	$h_{\text{in}}^{\text{steam}} = 2690.8$ $h_{\text{out}}^{\text{condensate}} = 460.8$	$h_{\text{in}}^{\text{steam}} = 2745.7$ $h_{\text{out}}^{\text{condensate}} = 631.5$
Steam required $= \dfrac{\dot{Q}_{\text{heat}}}{h_{\text{in}}^{\text{steam}} - h_{\text{out}}^{\text{condensate}}}$	kg/h	868	595

TABLE 10.9 Power developed by turbines

		Single-stage refrigeration			Two-stage refrigeration		
			Outlet			Outlet	
Turbine 1 (up to extraction pressure)	Units	Inlet	Ideal	Actual	Inlet	Ideal	Actual
Flow rate	kg/h	10,000	10,000	10,000	10,000	10,000	10,000
Temperature	°C	400	149	233	400	244	295
Pressure	Bar	18	2.437	2.437	18	5.743	5.743
Enthalpy	kJ/kg	3251.3	2763.8	2934.4	3251.3	2946.0	3052.9
Entropy	kJ/kg-K	7.181	7.181	7.550	7.181	7.181	7.378
Saturated vapor entropy	kJ/kg-K		7.06			6.77	
Power	kJ/h		3,169,000			1,984,000	
			Outlet			Outlet	
Turbine 2 (up to condenser pressure)		Inlet	Ideal	Actual	Inlet	Ideal	Actual
Flow rate (remaining after extraction)	kg/h	10,000 – 868 = 9,132	9,132	9,132	10,000 – 595 = 9,405	9,405	9,405
Temperature	C	233	45		295	45	
Pressure	Bar	2.437	0.095		5.743	0.095	
Vapor enthalpy	kJ/kg		2582.2			2582.2	
Liquid enthalpy	kJ/kg		187.8			187.8	
Enthalpy	kJ/kg	2934.4	2386.2	2550.7	3052.9	2331.5	2547.9
Vapor entropy	kJ/kg-K		8.166			8.166	
Liquid entropy	kJ/kg-K		0.637			0.637	
Entropy	kJ/kg-K	7.550	7.550	8.067	7.378	7.378	8.058
Quality			0.918	0.987		0.895	0.986
Power	kJ/h		3,504,000			4,749,000	
Turbines 1 and 2							
Electric power	kW		1,761			1,777	

are fixed. Thus, the selection of a two-stage refrigeration unit with its higher capital cost versus a single stage refrigeration unit requires a careful system study that evaluates the various trade-offs from an overall plant standpoint.

REFERENCES

Bertoncini F, Marion MC, Brodusch N, Esnault S. Unravelling molecular composition of products from cobalt catalysed Fischer-Tropsch reaction by comprehensive gas chromatography: Methodology and application. Oil Gas Sci Technol Revue IFP 2009;64(1):79–90.

Black J. Cost and performance baseline for fossil energy plants, volume 1: Bituminous coal and natural gas to electricity. U.S. Department of Energy Report; 2010. Report nr. DOE/NETL-2010/1397. Morgantown, WV: U.S. Department of Energy.

Heydorn EC, Diamond BW, Lilly RD. Commercial-scale demonstration of the liquid phase methanol (LPMEOH^{TM}) process, Vol. 2. U.S. Department of Energy Final Report; June 2003. Morgantown, WV: U.S. Department of Energy.

Johnson BG, Bartholomew CH, Goodman DW. The role of surface structure and dispersion in CO hydrogenation on cobalt. J Catal 1991;128(1):231–247.

Kalina T, Moore RE, inventors; Exxon Research and Engineering Company, assignee. Catalytic coal hydrogasification process. US patent 3,847,567. November 12, 1974.

Kamali MR, Sundaresan S, Van den Akker HEA, Gillissen JJJ. A multi-component two-phase lattice Boltzmann method applied to a 1-D Fischer–Tropsch reactor. Chem Eng J 2012; 207:587–595.

Kang WR, Lee KB. Effect of operating parameters on methanation reaction for the production of synthetic natural gas. Korean J Chem Eng 2013;30(7):1386–1394.

Keenan JH, Keyes FG, Hill PG, Moore JG. *Steam Tables: Thermodynamic Properties of Water Including Vapor, Liquid, and Solid Phases*. New York: John Wiley & Sons, Inc.; 1969.

Kırtay E. Recent advances in production of hydrogen from biomass. Energy Convers Manage 2011;52:1778–1789.

Kurtz M, Wilmer H, Genger T, Hinrichsen O, Muhler M. Deactivation of supported copper catalysts for methanol synthesis. Catal Lett 2003;86(1–3):77–80.

Li M, Brouwer J, Rao AD, Samuelsen GS. Application of a detailed dimensional solid oxide fuel cell model in integrated gasification fuel cell system design and analysis. J Power Sources 2011;196:5903–5912.

Li M, Rao AD, Samuelsen GS. Performance and costs of advanced sustainable central power plants with CCS and H_2 co-production. Appl Energy 2012;91(1):43–50.

Rao AD, inventor, Fluor Corporation, assignee. Reactor expander topping cycle. US patent 4,999,993. March 19, 1991.

Soemardji A, Engineer P. January 2012. How green is the stamicarbon urea process? Available at http://www.stamicarbon.com/documents/ot/how-green-is-the-stamicarbon-urea-process-v2-w.pdf. Accessed February 8, 2014.

Tijm PJA, Waller FJ, Brown DM. Methanol technology developments for the new millennium. Appl Catal A General 2001;221(1):275–282.

11

ADVANCED SYSTEMS

Some of the promising advanced systems under development that will have a significant effect on the environmental and economic sustainability of energy plants such as membrane separators and reactors, and fuel cells along with hybrid cycles employing fuel cells, are discussed in this chapter. Also included is a brief discussion of the other promising technologies, chemical looping, and magnetohydrodynamics (MHD). Chemical looping which uses a metal carrier to chemically transfer oxygen required for combustion (in carbon capture and sequestration (CCS) applications) or gasification of a carbonaceous feedstock has the advantage of eliminating the expensive (from both cost and efficiency standpoint) air separation unit (ASU). MHD as a topping cycle has the potential for substantially increasing the efficiency of fossil fuel power plants and is gaining attention at the present time because its flue gas is more suitable for CCS as it uses O_2 instead of air for combusting the fuel to obtain the required very high temperature.

11.1 HIGH TEMPERATURE MEMBRANE SEPARATORS

Separation technologies utilizing high temperature membranes for applications in areas such as air separation to supply the O_2 to an integrated gasification combined cycle (IGCC) or for oxy-combustion, separation of H_2 from a syngas stream, say in an IGCC for either coproduction of H_2 or generating a decarbonized fuel gas for

Sustainable Energy Conversion for Electricity and Coproducts: Principles, Technologies, and Equipment, First Edition. Ashok Rao.

combusting in a gas turbine to limit CO_2 emissions are being developed aimed at increasing the system efficiency and reducing the plant cost. Ceramic materials as high temperature membranes are getting a major focus from the scientific and engineering community and in the following, a discussion of the basic materials science is provided before specific applications of ceramic membranes are discussed. Oxygen as an ion is transported in the case of ceramic membranes for air separation while protons are transported in the case of ceramic membranes for hydrogen separation. The oxygen ion transport materials also find use in a certain type of fuel cell also covered in this chapter, the solid oxide fuel cell (SOFC). Dense phase ceramic and metallic membranes for separating H_2 are also being developed and will be briefly discussed.

11.1.1 Ceramic Membranes

11.1.1.1 Conduction in Semiconductors Between perfect insulators and conductors lie a class of materials called semi-conductors which show conduction properties intermediate between the two. Semi-conductors consist of covalent crystals. Atoms in a covalent crystal are held together by covalent bonds, diamond being a typical example. Each of the four electrons in its outer most shell of a carbon atom forms a covalent bond with its nearest four neighbors. Since each of the valence electrons participates in the binding, none of the electrons is very mobile. However, at sufficiently high temperatures, increasing number of electrons are excited to empty levels in the conduction band.

Semiconductor materials can be doped with certain type of "impurity" elements in minute concentrations to dramatically increase the conducting properties of the semiconductor. There are two types of dopant atoms, those that can easily loose an electron from their outer shells which give rise to n-type of semiconductors and those that take up an electron from the original semiconductor atoms giving rise to p-type of semiconductors leaving a hole in the valence band. This hole is then "free to move" to facilitate the current flow by allowing an electron to move into this empty energy level. Electron holes are also formed when an electron moves from a valence band to the conduction band. It leaves behind an empty energy level in the valence band, the electron hole thus formed may be filled up by another electron with an energy level close to that of the unoccupied state.

11.1.1.2 Vacancies and Diffusion in Solids Crystal lattices may have internal defects such as vacant lattice sites where an atom is missing. These vacant sites make it possible for an atom such as oxygen in ionic form to diffuse through the solid, with the oxygen ion "hopping" from one vacancy to another. The vacancy may have positive charge(s) created by the dopant in case of a p-type of semiconductor. The nature of the hopping process is chemical in nature since the diffusing atom forms chemical bonds with the other atoms in the crystal lattice when it hops into the vacant site. Self-diffusion in solids occurs by random walk (in the case where the individual atoms of the diffusing specie do not interact with each other), with the ion or atom jumping out from one vacancy (leaving behind an empty acceptor site to receive another jumping ion or atom) into another site. In the case of interacting atoms, the chemical potential imposes a driving force and depending on the magnitude of the activity coefficient (or enhancement factor) this rate process of diffusion (chemical diffusion) may be increased by orders of magnitude over self-diffusion. The process of diffusion within a solid is similar to a three atom chemical reaction. The jumping ion or atom breaks the chemical bond with its neighboring atom(s) and jumps into the empty site where it forms new chemical bonds with its new neighboring atom(s). When the driving force is small, the rate of diffusion may be taken to be linearly proportional to the chemical activity gradient while the diffusion coefficient shows an Arrhenius dependence on temperature due to the chemical nature of the process.

11.1.1.3 Oxygen Transport Membranes Perovskites[1] have the chemical structure ABO_3 with the *A*-site and/or the *B*-site being possibly a mixture of two or more elements. $La_{0.4}Ca_{0.6}Fe_{0.75}Co_{0.25}O_{3\text{-}\delta}$ is an example of a perovskite suitable for oxygen ion transport with oxygen deficiency or vacancy represented by δ (Diethelm et al., 2003). Such perovskites can conduct both electrons and oxygen ions through the ceramic lattice and have high O_2 fluxes in the 700–1000°C range. There are a number of processes involved in the passage of oxygen through the ceramic membrane. After it reaches the membrane by diffusing through the gas film at the gas–membrane interface, it is adsorbed onto the surface and it is then transported by bulk transport through the ceramic lattice containing the oxygen ion vacancies at the required high temperatures. The oxygen flux across the membrane for temperatures >700°C may be obtained from the Nernst–Einstein equation (Foy and McGovern, 2005):

$$j_{O_2} = \frac{\sigma_i R_u T}{4 L n_e^2 \mathcal{F}^2} \ln\left(\frac{P_{O_2,in}}{P_{O_2,out}}\right)$$

where j_{O_2} is the oxygen flux through unit area, σ_i is the material's ionic conductivity, T is the absolute temperature, L is the membrane thickness, n_e is the charge on the charge carrier, which is always 2 for oxygen ions, $\mathcal{F}$ is Faraday's constant = 96,487 C/mol electron (Coulombs per g-mole of electrons, each g-mole of electrons containing Avogadro's number or 6.022×10^{23} electrons), $P_{O_2,in}$ is the oxygen partial pressure at the feed side, and $P_{O_2,out}$ is the oxygen partial pressure at the permeate side.

11.1.1.4 Hydrogen Transport Membranes Hydrogen transport through dense high temperature ceramic membranes is limited by mixed proton and electron conduction. Ceramic/ceramic composites incorporating a proton conducting perovskite phase with an electron conducting ceramic second phase are being developed (Roark et al., 2002). This combination of protonic and electronic conduction can be achieved by excessive doping of a parent perovskite with appropriate transition metal oxides. A transition metal dopant is introduced into the perovskite lattice during calcination which generates oxygen vacancies and promotes proton conductivity,

[1] Materials with the same type of crystal structure as the mineral perovskite, $CaTiO_3$.

while the remainder of the transition metal typically results in a metal oxide with high electron conductivity. Cermets formed by combining a proton-conducting ceramic phase with a metallic phase (metals and alloys with high hydrogen permeability) are also being investigated and have the advantage of very high electron conductivity with a small quantity of the metal and thus maximizing the amount of the proton conducting phase. The hydrogen flux may also be obtained from the Nernst–Einstein equation (Roark et al., 2002):

$$j_{H_2} = \frac{\sigma_{amb} R_u T}{L n_e^2 \mathcal{F}^2} \ln\left(\frac{P_{H_2,in}}{P_{H_2,out}}\right)$$

where j_{H_2} is the hydrogen flux through unit area, σ_{amb} is the mixed proton and electron conductivity referred to as ambipolar conductivity, T is the absolute temperature, L is the membrane thickness, n_e is the charge on the charge carrier, which is always 1 for protons, $P_{H_2,in}$ is the hydrogen partial pressure at the feed side, and $P_{H_2,out}$ is the hydrogen partial pressure at the permeate side.

Other types of high temperature membranes being developed include:

- Microporous inorganic membranes (Judkins and Bischoff, 2004) through which the hydrogen permeates in molecular form and its flux is directly proportional to the partial pressure gradient across the membrane.
- Pd-Cu alloy membranes (Enick et al., 2003) through which the hydrogen permeates in atomic form by the following steps: dissociative adsorption of H_2 onto the metal surface, then diffusion of the atomic hydrogen through bulk metal, and finally associative desorption of H_2 from the metal surface. The H_2 flux is proportional to the difference in the square root of the H_2 partial pressure on each side of the membrane.

11.1.2 Application of Membranes to Air Separation

As mentioned in Chapter 9, ASU power usage constitutes greater than half of the total power consumed or 10–20% of the total power produced by the IGCC. Air Products and Chemicals, Inc. (Armstrong et al., 2002) and Praxair, Inc. (Prasad et al., 2002) are developing oxygen ion transport membranes for large-scale O_2 production from air for use in gasification as well as oxy-combustion plants with the promise of reducing both power usage and capital cost by about 30% as compared to a conventional cryogenic ASU. Oxygen ions and electrons flow through a nonporous, mixed ion and electron conducting membrane typically at 800–900°C countercurrently. Air at 800–900°C temperature and 15–20 bar pressure flows on one side of the membrane while low to sub-atmospheric pressure is maintained on the oxygen permeate side to increase the O_2 partial pressure difference between the two sides of the membrane. In an IGCC application, the integrated system consists of hot pressurized air being extracted from the gas turbine compressor to supply the feed air to the membrane unit which separates a portion of the O_2 by transferring it in ionic form through the membrane wall while the depleted pressurized air is returned to the gas turbine. The extracted air has to be boosted in pressure in order to return the hot depleted air back to the gas turbine while the temperature of the extracted air after booster compression is increased to the required operating temperature of the membrane separator by directly combusting syngas into this air stream. The gas turbine must be capable of receiving the hot depleted air (800–900°C), and gas turbines with this required capability are not available at the current time. Both planar and tubular membrane designs are being developed and commercial offering of these systems may take place 10–15 years from the present time.

11.1.3 Application of Membranes to H_2 Separation

H_2 separating membranes are being investigated for use in "membrane reactors" which combine a water gas shift or a reforming reactor with product separation in a single unit. Not only does this lead to reduction in the number of equipment and their sizes but also higher thermal efficiency and conversion since a reaction product is continuously removed from the reaction mixture. With the tubular design, catalyst filled tubes are constructed out of the membrane material with appropriate support material to provide the necessary strength, and are contained within a shell into which the H_2 enters as it is formed. Commercial offering of these systems may take place 10–15 years from the present time.

11.2 FUEL CELLS

A disadvantage of heat engines is that they are dependent on the combustion process which produces pollutants and conversion efficiency of the fuel bound energy to electricity with the intermediate step involving combustion is limited by the temperature at which the heat may be utilized, a consequence of the second law of thermodynamics. In a heat engine, the materials of construction limit the temperature at which the heat is converted into power. A significant decrease in the quality of energy occurs when the chemically bound energy contained in the fuel is released and transferred to the working fluid of the heat engine at a significantly lower temperature. As discussed in Chapter 2, reduction in the quality of energy (due to utilizing the thermal energy at a lower temperature) is equivalent to generation of entropy, exergy loss, and, consequently, a decrease in efficiency.

Even if materials are developed to operate at the higher temperature, the efficiency is limited since the heat addition to the cycle from a combustion process does not occur at a constant temperature. Take for example, a fuel such as H_2. Its adiabatic flame temperature is 2483 K and the Carnot cycle efficiency >80% may be calculated using this temperature as the heat source temperature and typical ambient temperatures as the heat rejection temperature. This calculated efficiency, however, is misleading. In directly fired cycles such as the Brayton, Diesel, or Otto cycles, heat addition to the working fluid does not occur at constant temperature but starts at much lower temperature (temperature at the end of the compression step). In the case of indirectly fired cycles such as in the Rankine cycle where heat is transferred to a separate working fluid, not only does the heat addition start occurring at the inlet temperature of the economizer but the heat source temperature itself is not constant since combustion products (flue gases) cool off as heat is transferred to the working fluid. The true Carnot efficiency of these cycles has to take into account the variable temperature of heat addition and it may be found that it is considerably lower than the 80%.

A more efficient and environmentally responsible way of converting this fuel bound energy consists of utilizing fuel cells once the technological challenges discussed in this chapter have been overcome. Fuel cells convert a significant fraction of the fuel bound energy directly (electrochemically) into electricity without the intermediate step of converting it into thermal energy as in heat engines. Fuel cells however require clean fuels such as natural gas. Dirty fuels such as coal and biomass may be converted to clean fuels by gasification followed by cleaning the syngas thus derived, and then fueling a fuel cell with the clean syngas.

A fuel cell, as an electrochemical device is similar to a battery that converts chemically bound energy directly into electricity but unlike conventional batteries, the chemical energy to the cell is supplied on a continuous basis in the form of a fuel such as natural gas or syngas while the oxidant (air) is also supplied continuously.

The conversion process occurring in a fuel cell may be considered as electrolysis operating in reverse. As depicted in Figure 11.1, electrolysis uses electric current to

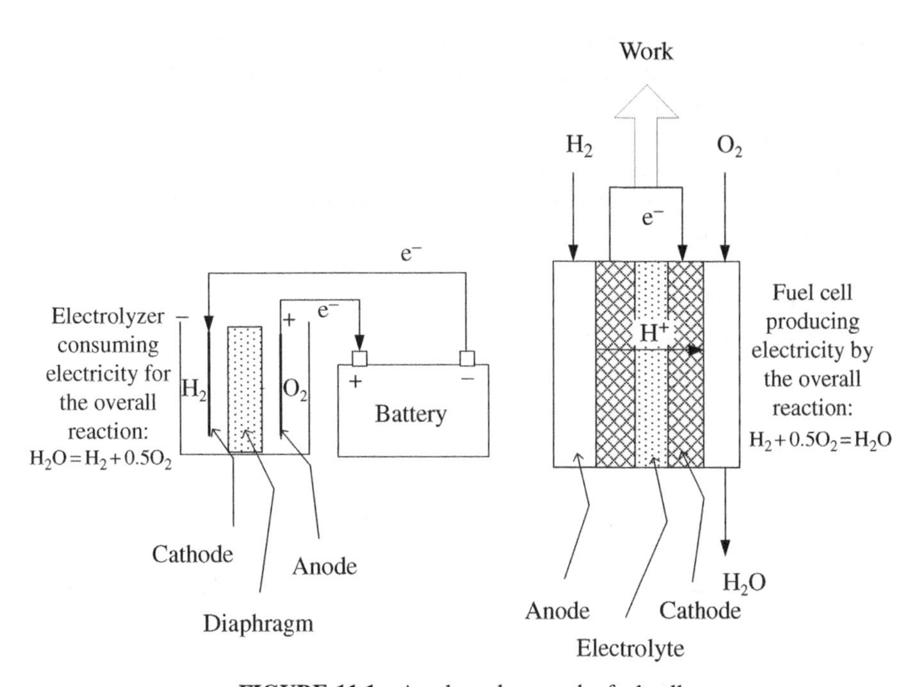

FIGURE 11.1 An electrolyzer and a fuel cell

disassociate, for example, water H_2O into H_2 and O_2 while the type of fuel cell shown here does the reverse of combining the fuel H_2 with the oxidant O_2 to produce H_2O and electric current. Fuel flows to the anode (negative electrode) of the fuel cell and oxidizes releasing electrons into the external circuit. Oxidant flows to the cathode surface and hydrogen ions flow across the fuel cell electrolyte to react with the O_2 while consuming electrons that are supplied from the external circuit after generating power.

Fuel cells were considered a curiosity in the 1800s, the first fuel cell being built as early as in 1839 by Sir William Grove, a lawyer and a scientist. Serious interest in fuel cells as practical electricity generators did not begin until the 1960s when the U.S. space program chose fuel cells over riskier nuclear power and more expensive solar energy. Fuel cells furnished power for the Gemini and the Apollo spacecraft. They also provided electricity and water for the space shuttle.

A fuel cell does have losses that limit the conversion of the fuel-bound energy to electricity. These losses are due to:

1. Irreversibilities caused by heating of the reactants to the reaction temperature and cooling of the products from the reaction temperature
2. The entropy change of the reaction
3. Irreversibilities caused by cell polarizations.

Fuel cell losses manifest as heat rejected or exhausted by the fuel cell system with a corresponding reduction in system efficiency. A large fraction of the rejected heat may be recovered for cogeneration or to produce additional power utilizing a heat engine to increase the overall system efficiency significantly. Combining a fuel cell with a heat engine is called a "hybrid cycle." A properly configured hybrid cycle can increase system efficiency, in excess of 60% with natural gas on LHV basis.

In addition to the efficiency advantage of being relatively independent of size, other major advantages of fuel cells when compared to other energy conversion devices are as follows:

- Good part load characteristics
- Modular design and flexibility of size
- Low environmental impact
- Siting ability due to favorable environmental signature
- Quick response to load changes.

Fuel cells however, do have certain disadvantages such as:

- Sensitivity to certain contaminants that may be present in the fuel such as sulfur compounds and chlorides
- Lack of field data on endurance/reliability
- Current installed plant costs on a $/kW basis are high.

11.2.1 Basic Electrochemistry and Transport Phenomena

There are various types of fuel cells but the major components in any type of fuel cell are the anode, the cathode, and the electrolyte which separates the anode and cathode. In the fuel cell type depicted in Figure 11.1, the fuel reacts or "oxidizes" in the anode releasing electrons, "reduction" of O_2 (usually from air) occurs in the cathode, and the electrolyte conducts ions from one electrode to other inside the cell. Thus, in contrast to combustion where the fuel (say H_2) and the oxidant (O_2 contained in air) are intimately mixed for the reaction to occur, in the case of a fuel cell the overall reaction is divided into the two "half reactions," reduction reaction, and oxidation reaction as shown below for the type of fuel cell under consideration, which are separated in space with electrons released at the anode and flowing in an ordered fashion through the external circuit to produce useful work before they enter the cathode.

Oxidation in the anode section: $H_2 \rightarrow 2H^+ + 2e^-$
Reduction in the cathode section: $\frac{1}{2}O_2 + 2e^- + 2H^+ \rightarrow H_2O$
Overall reaction: $H_2 + \frac{1}{2}O_2 \rightarrow H_2O$

The work produced by a reversible fuel cell may be shown to be the Gibb's free energy change for the overall reaction. When the reversible fuel cell is surrounded by a heat sink/source also at the temperature T, the heat transfer Q occurs reversibly but no heat engine work can be created because of zero temperature difference between the fuel cell and the heat sink/source. Because the fuel cell is operating reversibly, the total change in entropy of the fuel cell + the heat sink/source should be zero. The change in entropy of the heat sink/source $= Q/T$ and so the change in entropy of the fuel cell must be $= -Q/T$. From the first law, $W = \Delta H - Q = \Delta H - T\Delta S$, where ΔH and ΔS are the enthalpy and entropy changes between the streams entering and leaving the fuel cell (the reactants and products). This work has to be electrical since no other form of work is being realized. Thus, the reversible (electrical) work of fuel cell is given by the Gibb's free energy change for the overall reaction:

$$W_e = \Delta H - T\Delta S = \Delta G$$

ΔG may be calculated from the readily available standard Gibbs free energy change ΔG^o for a reaction at temperature T with reactants (e.g., H_2 and O_2) and products (e.g. H_2O) in their standard state of unit fugacity, by adjusting for the fugacities of each of the reactants and the products at the actual conditions of the reaction. For example, for the overall reaction $aA + bB \rightarrow cC + dD$,

$$\Delta G = \Delta G^o - RT\ln(\mathcal{Q}) \tag{11.1}$$

where

$$\mathcal{Q} = \frac{f_C^c f_D^d}{f_A^a f_B^b}$$

f_i is the fugacity of specie i.

Since electrical work is the product of the charge and the voltage across which the charge is moved, we have

$$\Delta G = -n_e \mathcal{F}\,(\text{Voltage}) \tag{11.2}$$

where n_e, number of electrons participating in a half reaction.

The ideal cell potential ($\mathcal{E}$ or the "open circuit" potential, i.e., without current flow) involving ideal gases may be calculated substituting Equation 11.2 into Equation 11.1 while replacing fugacities with partial pressures in Equation 11.1 resulting in the Nernst equation:

$$\mathcal{E} = \mathcal{E}_o - \frac{R_u T}{n_e \mathcal{F}} \ln\left(\frac{P_C^c P_D^d}{P_A^a P_B^b}\right) \tag{11.3}$$

where, $\mathcal{E}_o$ is the reference potential and P_i is the partial pressure of specie i.

Example 11.1. Calculate the H_2 flow rate required to generate 1.0 A (amp) of current in a fuel cell?

Solution

$$H_2 = 2H^+ + 2e^-$$

As seen from the half reaction above, 2 electrons are liberated at the anode per molecule of H_2 reacting within a fuel cell, or 2 mol of electrons per mol of H_2 while the charge carried by the electrons is given by Faraday's constant $\mathcal{F} = 96{,}487$ C/mol.

Then molar flow rate of H_2 to generate 1 A (or 1 C/s) of current is given by:

$$\dot{M}_{H_2} = \frac{\text{mol}\,H_2}{2\,\text{mol}\,e^-} \times \frac{\text{mol}\,e^-}{96{,}487\,\text{C}} \times \frac{\text{C/s}}{1\,\text{A}} \times \frac{3600\ \text{s}}{\text{h}} = 0.01865\frac{\text{mol/h}}{\text{A}}$$

Example 11.2. Calculate the standard potential ($\mathcal{E}_o$) of a H_2-O_2 fuel cell operating at 298 K (with the product H_2O as a liquid).

Solution

From Equation 11.2, $\mathcal{E}_o = -\Delta G^o/(n_e\mathcal{F})$.

For the overall reaction $H_2(g) + ½O_2(g) \rightarrow H_2O(l)$, $\Delta G^o = -237{,}141$ J/mol at 298 K.

This reaction may be written as the two half reactions showing $n_e = 2$.

$$H_2 \rightarrow 2H^+ + 2e^-$$
$$½O_2 + 2e^- + 2H^+ \rightarrow H_2O$$

Then, $\mathcal{E}_o = -\Delta G^o/(n_e\,\mathcal{F}) = 237{,}141\ \text{J/mol}/(2 \times 96{,}487\,\text{C/mol}) = 1.229\ \text{J/C}$ $= 1.229$ V.

If the product H_2O is considered as a vapor, then $\Delta G^o = -228{,}582$ J/mol at 298 K, and then a smaller value of 1.184 V is obtained.

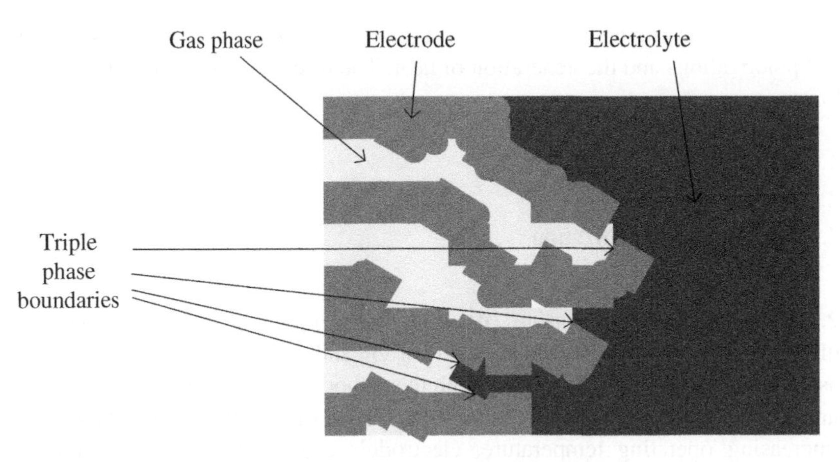

FIGURE 11.2 The triple phase boundary

11.2.2 Real Fuel Cell Behavior

As soon as the fuel cell does work (when electrons flow when the circuit is closed) there are observable drops in the potential or voltage as compared to the open circuit voltage due to irreversible processes and the actual work of a fuel cell is lower due to these losses, the three main losses mentioned previously. Losses caused by cell polarizations are due to the various transfer and chemical processes taking place within the fuel cell which require a sacrifice of the driving forces required to drive these processes described in the following at practical rates:

- Convective mass transfer of reactants and products to and from surface of catalyst on anode side (where applicable) for internal reforming and shift reactions:
  - Reforming: $CH_4 + H_2O = 3H_2 + CO$, $\Delta H > 0$
  - Shift: $CO + H_2O = CO_2 + H_2$, $\Delta H < 0$
- Convective mass transfer of reactants and products to and from surface of electrodes
- Mass transfer of reactants and products through porous electrodes
- Electrochemical reactions at or near interface between gas phase, electrode surface, and electrolyte surface also known as the triple phase boundary or TPB[2] (see Figure 11.2) for electronic electrodes (electronic conductors that pass electrons but not ions) involving a number of mass transfer mechanisms:
  - Adsorption
  - Desorption
  - Surface migration
- Conduction of electronic current through electrodes and current collectors
- Conduction of ions through electrolyte and electrodes (where applicable).

[2] Increasing available TPB increases power density and fuel cell performance requiring improvement in fabrication techniques.

These losses or irreversibilities manifest themselves as voltage losses also known as cell polarizations and the generation of heat. The three main irreversibilities are as follows:

1. Activation polarization
2. Ohmic polarization
3. Concentration polarization

11.2.2.1 Activation Polarization (η_{act}) It is the potential difference above the equilibrium value required to produce a net current and a portion of electrode potential is lost in driving the electron transfer. Both the cathode and the anode have activation polarizations which may be very different from each other but both may be reduced by increasing operating temperature, electrode's active surface area, activity of electrode through use of catalysts, and using a mixed ionic/electronic conducting electrode (materials that not only allow transfer of electrons but also ions). This loss may be estimated using the Equation 11.4 known as the Tafel equation when $\eta_{\text{act}} > 50 - 100\,\text{mV}$ with experimentally determined constants.

$$\eta_{\text{act}} = \frac{R_{\text{u}}\,T}{\alpha_{\text{e}}\,n_{\text{e}}\,\mathcal{F}} \ln\left(\frac{i}{i_{\text{o}}}\right) \tag{11.4}$$

where α_{e}, electron transfer coefficient, is a measure of the symmetry of the energy barrier for electron transfer; typically $\alpha_{\text{e}} = 0.5$ (i.e., the effect of changing the potential is identical on both sides of the barrier). i is current density (current per unit surface area of the electrode). i_{o} is exchange current density, the current density in the absence of net electric current (electron transfer processes go on in both directions at each electrode and when the current in one direction is balanced by the current in the opposite direction, that is, at equilibrium, then no net current is produced at that electrode while each of these current densities $= i_{\text{o}}$). Higher the i_{o}, faster the kinetics. It is experimentally determined by developing a Tafel plot (measured activation polarization plotted against the natural logarithm of the current density as depicted in Figure 11.3) and extrapolating the linear portion of the plot to zero over potential.

11.2.2.2 Ohmic Polarization (η_{ohm}) This polarization is due to the Ohmic loses caused by the resistance to electronic and ionic flows in the cell. The electrolyte is both an ionic and an electronic conductor although the loss due to electronic resistance is quite small; electrodes are typically electronic conductors (but materials are being developed that have also ionic conductivity), while the terminal connections in a fuel cell are electronic conductors. The Ohmic polarization may be reduced by using electrolytes with high ionic conductivity and/or thin electrolyte layer, and electrodes with high electronic conductivity and/or thickness (in cross section to electron flow). It may be estimated using Equation 11.5, the Ohms law.

$$\eta_{\text{ohm}} = i \sum \mathcal{R} \tag{11.5}$$

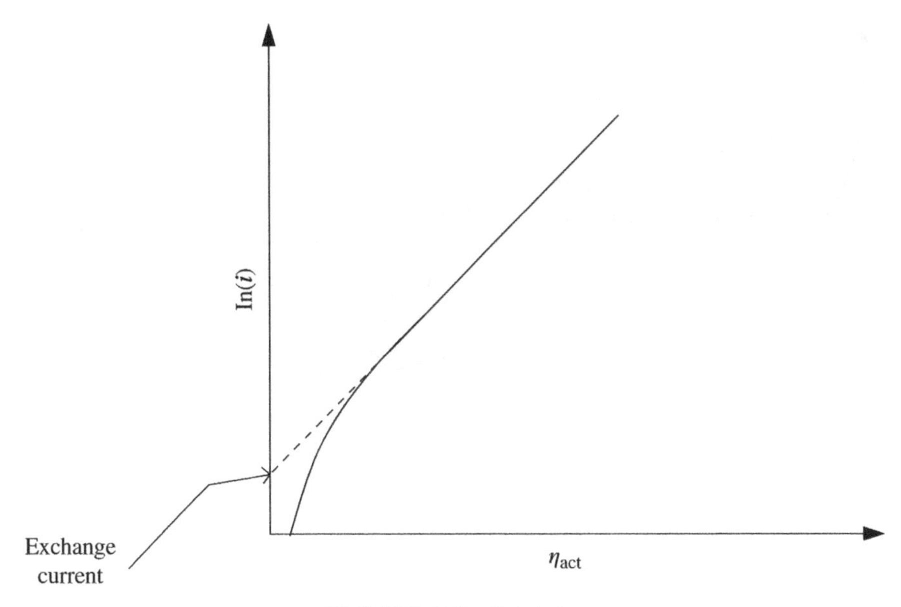

FIGURE 11.3 Tafel plot

where i is the current and $\Sigma\mathcal{R}$ is the sum of the electronic and ionic cell resistances determined experimentally as a function of temperature for electrodes, electrolyte, and terminal connections.

11.2.2.3 Concentration Polarization (η_{conc}) Mass transfer concentration gradients are established between bulk gas and at reaction sites for the reactant and the product species in both the cathode and the anode compartments. This polarization is a correction that is required for each electrode compartment for using bulk gas concentrations in the Nernst equation for calculating the cell voltage. Concentration polarization is significant at high power densities and can be reduced by operating at lower current densities to reduce the concentration buildup near the electrodes by supplying excess (sufficient) reactants to increase their concentrations as the reactants are depleted, thinner and/or more porous electrodes to reduce the resistance to mass transfer. This loss may be estimated using Equation 11.6.

$$\eta_{\text{conc}} = \frac{R_{\text{u}}T}{n_{\text{e}}\mathcal{F}}\ln\left(1-\frac{i}{i_{\text{L}}}\right) \tag{11.6}$$

where i_L is the limiting current density which is a measure of the maximum rate of reactant or product transfer to or from the electrode. It may be determined by modeling the mass transfer processes occurring in each of the cathode and anode compartments or by using experimental data.

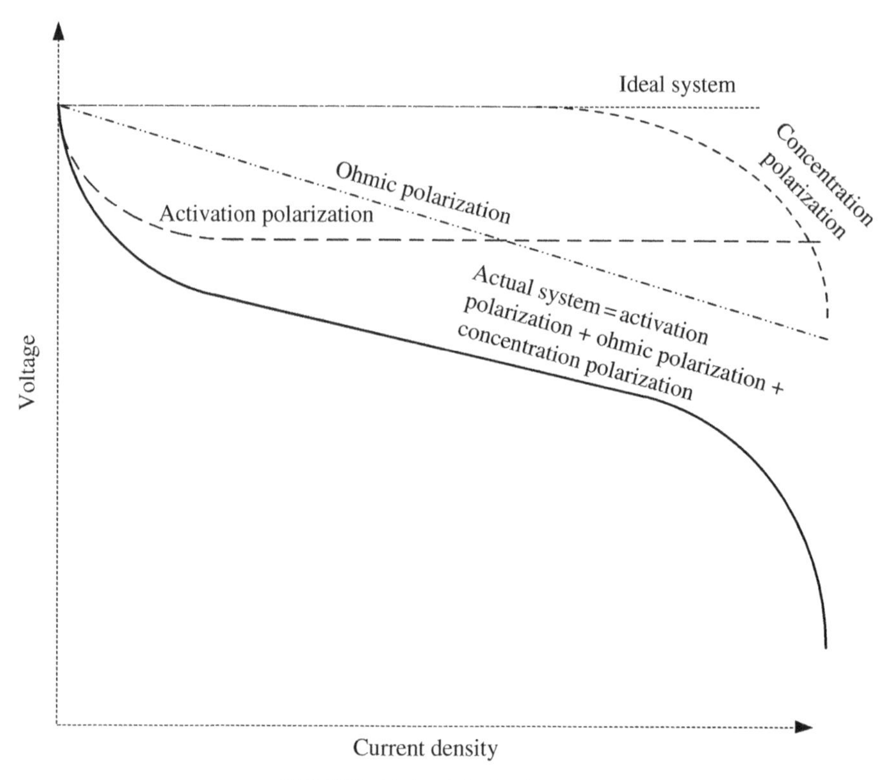

FIGURE 11.4 Polarization curve

11.2.3 Overall Cell Performance

The overall cell voltage which varies with the current is given by Equation 11.7 by subtracting the various polarizations from the voltage calculated using Equation 11.2, the Nernst equation (with bulk gas partial pressures for the reactants and products), and is graphically presented in Figure 11.4 as a "polarization curve."

$$V_{\text{cell}} = V_{\text{Nernst}} - \eta_{\text{act,cathode}} - \eta_{\text{conc,cathode}} - \eta_{\text{act, anode}} - \eta_{\text{conc, anode}} - \eta_{\text{ohm}} \tag{11.7}$$

The performance of a fuel cell depends on the operating conditions such as temperature, pressure, concentration of the fuel and oxidant species, and their utilizations within the cell.

11.2.3.1 Temperature Dependence Since $-n_e\mathcal{F}\,\mathcal{E} = \Delta G = \Delta H - T\Delta S$, while realizing that ΔH and ΔS are not strongly dependent on temperature T (i.e., considering them as constants with respect to T), we have at constant pressure P:

$$\left(\frac{\partial \mathcal{E}}{\partial T}\right)_{\text{P}} = \frac{\Delta S}{n_e\,\mathcal{F}}$$

Thus, the thermodynamic effect of operating temperature on the cell potential which is also a measure of its efficiency depends on whether $\Delta S > 0$ or < 0. However, there is typically a reduction in some overpotentials which can compensate for the thermodynamic reduction in $\mathcal{E}$ when T increases for reactions with $\Delta S < 0$.

11.2.3.2 Pressure Dependence Equation 11.2, the Nernst equation, may be written in terms of mole fractions y_i and total pressure P as:

$$\mathcal{E} = \mathcal{E}_o - \frac{RT}{n_e\mathcal{F}} \ln\left[\left(\frac{y_C^c y_D^d}{y_A^a y_B^b}\right)(P)^{(c+d-a-b)}\right]$$

which shows the following proportionality:

$$\left(\frac{\partial \mathcal{E}}{\partial P}\right)_T = -\frac{RT}{n_e\mathcal{F}P}(c+d-a-b)$$

Thus, the thermodynamic effect of the operating pressure on cell potential (also a measure of its efficiency) depends on whether there is an increase or decrease in moles during the overall reaction, while the effect becomes less significant as the pressure increases. However, gas solubility and mass transfer rates increase with pressure and some overpotentials associated with these processes are reduced. Also, electrolyte loss by vaporization is typically reduced when a molten electrolyte is used as in a molten carbonate fuel cell described later in this chapter. Materials and controls requirements may be more challenging with increased pressure, however. Transient pressure spikes on one side of the cell may have the potential to cause greater damage to the fuel cell at higher operating pressures.

11.2.3.3 Gas Concentration and Utilization Equation 11.2, the Nernst equation, shows that there is a strong effect of reactant concentrations in the fuel and oxidant on the fuel cell electric potential (and efficiency). Utilization is defined as the fractional amount of reactant consumed. Higher the utilization, lower is the mean reactant concentration in the fuel cell and thus lower is the Nernst potential. Fuel utilization U_{Fu} for a fuel stream containing both H_2 and CO, with the incoming CO being all "potential" H_2 with the shift reaction taking place in high temperature fuel cells, is then given by ($\dot{M}_{i,in}$ and $\dot{M}_{i,out}$ being moles of specie i entering and leaving the fuel cell):

$$U_{Fu} = \frac{\dot{M}_{H_2,in} - \dot{M}_{H_2,out} + \dot{M}_{CO,in} - \dot{M}_{CO,out}}{\dot{M}_{H_2,in} + \dot{M}_{CO,in}}$$

In the case of a fuel cell with internal reforming with a fuel stream containing in addition to H_2 and CO, also CH_4, with each mole of CH_4 converted being equivalent to four moles of H_2, the fuel utilization is given by:

$$U_{Fu} = \frac{\dot{M}_{H_2,in} - \dot{M}_{H_2,out} + \dot{M}_{CO,in} - \dot{M}_{CO,out} + 4\left(\dot{M}_{CH_4,in} - \dot{M}_{CH_4,out}\right)}{\dot{M}_{H_2,in} + \dot{M}_{CO,in} + 4\dot{M}_{CH_4,in}}$$

An oxidant utilization U_{Ox} may also be defined as:

$$U_{Ox} = \frac{\dot{M}_{O_2,in} - \dot{M}_{O_2,out}}{\dot{M}_{O_2,in}}$$

Typically, the U_{Fu} is held around 80–85% to maintain practical power densities since a finite concentration gradient of reactants between anode side and cathode side is required to drive the reaction. Neither U_{Fu} nor U_{Ox} can ever be 100% and would be equivalent to a heat exchanger with the temperature difference between the two streams exchanging heat equal to zero requiring an infinitely large heat transfer surface area. On the other hand, decreasing U_{Fu} or U_{Ox} increases the fuel or oxidant required and a systems analysis is required for evaluating the trade-off since unconsumed fuel components leaving the fuel cell may be utilized for other purposes such as generation of heat, or a fraction recycled to the fuel cell inlet, or used for coproduction of H_2 or syngas.

In high temperature fuel cells, from an electrochemical standpoint, it is primarily the H_2 that takes part in the reaction even if the fuel contains CO and CH_4. There is experimental evidence supporting this based on data collected with the SOFC described later and many experts believe that this is true with the other high temperature fuel cells such as the molten carbonate fuel cell (MCFC) also described later. Under practical operating conditions of a high temperature fuel cell fueled by such a gas stream, the steam to carbon ratio in the fuel is maintained high enough to avoid carbon deposition by reversal of the Boudouard reaction. Under such conditions Matsuzaki and Yasuda (2000) showed experimentally that the direct electrochemical oxidation of the CO is not significant. The CO is indirectly oxidized via the water gas shift reaction between the CO and H_2O. At the high operating temperatures of the fuel cell, the shift reaction readily occurs in the gas phase to form H_2 and CO_2, and the H_2 thus formed diffuses to the anode surface to be oxidized, its diffusion rate being significantly higher than that of CO. Ihara et al. (1999) showed experimentally that the direct electrochemical oxidation of the CH_4 under the practical operating conditions of high steam to carbon ratio is not significant either. A suitable catalyst is required for the utilization of CH_4 for reforming it within the anode section or for prereforming outside the fuel cell.

Example 11.3. Calculate the required flow rate of the fuel and the oxidant (air) for a 1.0 MW DC fuel cell stack (a stack consists of a number of individual cells stacked together to produce a more substantial voltage and/or current depending on how the individual cells are connected electrically, i.e., in series or in parallel) operated with a cell voltage of 0.75 V on pure H_2 with a fuel utilization, $U_{Fu} = 85\%$ and oxidant utilization, $U_{Ox} = 30\%$.

Solution

Assuming the stack contains a total of N cells in parallel, each with a voltage of 0.75 V, then with power developed per cell $\dot{W} = 1.0\ \text{MW}/N$ and $V\text{cell} = 0.75$ V, the current I through the fuel cell stack may be calculated as

$$I=\frac{\dot{W}}{V\text{cell}}=\frac{(1.0\ \text{MW})/N}{0.75\ \text{V}}\times\frac{1{,}000{,}000\ \text{W}}{\text{MW}}\times\frac{1\ (\text{V A})}{\text{W}}=\left(\frac{1{,}333{,}000}{N}\right)\text{A}.$$

Using the result obtained in Example 11.1, the required amount of fuel with $U_{Fu}=85\%$ for a total of N cells is then

$$\dot{M}_{H_2}=\frac{0.01865\dfrac{\text{mol}}{\text{h}}}{\text{A}}\times\frac{(1{,}333{,}000/N)\ \text{A}}{0.85}\times N=29{,}260\frac{\text{mol}}{\text{h}}=29.26\ \text{kmol/h}.$$

Then flow rate of air with $U_{Ox}=30\%$ assuming it consists of 21% O_2 and 79% N_2 (to oxidize 85% of the fuel calculated above) is

$$\dot{M}_{Ox}=\frac{\dot{n}_{O_2}}{0.21}=\frac{(0.5\,n_{H_2}\times 0.85)/0.3}{0.21}$$

$$=\frac{(0.5(29.26\ (\text{kmol/h}))\times 0.85)/0.3}{0.21}=197.4\,\text{kmol/h}.$$

Note that the same results are obtained if the cells are all assumed to be in series instead.

11.2.3.4 Cell Degradation Performance of a fuel cell can degrade over time and can vary significantly over period of time depending on the type of design. The degradation rate measured in terms of the cell voltage should be limited to about 0.1% per 1000 h for stationary applications (Fuel Cell Handbook, 2004). Depending on the type of fuel cell, contaminants in the fuel or oxidant streams can cause performance deterioration of the membrane or catalyst, or morphological changes of ceramic materials operating at high temperatures can occur over a period of time. This topic is discussed with specific reference to each type of fuel cell later in this chapter.

11.2.4 A Fuel Cell Power Generation System

A natural gas fired fuel cell system is composed of three primary subsystems:

1. Fuel processor consisting of
   a. A desulfurizer such as an activated carbon bed operating near ambient temperatures, or ZnO bed in cases where the gas is preheated (for the reformer) since it operates more effectively closer to 400°C. The sulfur content has to be limited to <0.1 ppmv to protect the downstream conventional Ni-based reforming catalyst. A deoxidizer may also be required consisting of a noble metal catalyst to react any free O_2 contained in the natural gas with the combustibles in the gas to avoid oxidizing the downstream Ni catalyst. As mentioned in Chapter 1, during peak demand months, the natural gas in

certain areas of the United States (the northeast) may contain O_2 as high as 4% by volume because during such seasons when there may be shortage in natural gas delivery, the gas company may blend in propane or butane to extend the fuel supply, and air would be added as a diluent to hold the Wobbe Index within limits. The volume of this deoxidizing catalyst reduces as the feed gas temperature increases.

b. An external reformer in case of fuel cells that are not capable of internally reforming the natural gas fuel (the fuel is reformed to produce H_2 which undergoes the electrochemical oxidation within the fuel cell), along with the heat exchange equipment and a blower or compressor for the fuel gas depending on the pressure at which the fuel gas is available. The external reformer within the fuel processor combines natural gas with steam (recovered from the heat recovered in the power section) to reform the fuel. Some of the fuel cells with internal reformers also require a prereformer to convert the higher hydrocarbons which otherwise cause coking of the internal reformer catalyst.

c. A pressure swing adsorption (PSA) when high purity H_2 is required by the fuel cell (e.g., when the residual CO cannot be tolerated by the fuel cell such as the proton exchange membrane fuel cell or PEMFC described later in this chapter), to separate out the high purity H_2 stream from the reformer effluent after it is cooled to near ambient temperature. The gas rejected by the PSA ("tail gas") contains significant amounts of combustibles which can be utilized for supplying at least a portion of the heat required by the endothermic reforming reactions.

2. Power section that includes the fuel cell stacks and the necessary blower or compressor to provide filtered oxidant (air) at the required pressure for the fuel cells, and the equipment for combusting the fuel remaining in the anode exhaust stream with the O_2 remaining in the cathode exhaust stream, and the equipment associated with providing heat for a reformer and/or for generating steam and/or hot water.
3. Power conditioning consisting of a direct current (DC) to alternating current (AC) power converter (called inverter) since the current produced by the fuel cell is DC. The DC voltage generated by small fuel cell stacks is low in magnitude being <50 V for a 5–10 kW system and <350 V for a 300 kW system and in such system a step up DC to DC converter is essential (before conversion to the more useful AC) to generate higher DC voltage, 400 V being typical for a 120/240 V AC output.

These subsystems are illustrated in Figure 11.5 (for those systems not requiring a pure H_2 stream as the fuel for the fuel cell). In the case of fuel cells that require cooling of the stack, a secondary subsystem for thermal management (a cooling module) is also required if recoverable thermal energy is not fully utilized in some form of cogeneration application. In the case of digester or landfill gas, in addition to the removal of the sulfur compounds, removal of the chlorine compounds and siloxanes is required and the

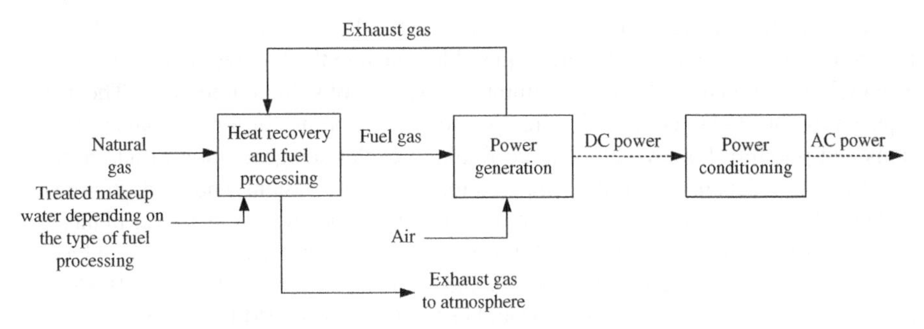

FIGURE 11.5 Schematic of a natural gas fuel cell system

treatment methods have been described in Chapter 6 for gas turbine applications although the specifications are more stringent for fuel cell applications. Chlorine compounds are poisonous for Ni catalysts while siloxanes can cause fouling of the catalysts and equipment in addition to PM emissions. In the case of landfill gas, as in the case of peak shave natural gas, the gas treatment will need to include a noble metal catalyst to react any free O_2 with the combustibles in the gas to avoid oxidizing the Ni catalyst. The volume of this deoxidizing catalyst reduces as the feed gas temperature increases.

11.2.5 Major Fuel Cell Type Characteristics

Over the past 50 years, various types of fuel cells have evolved and a description of the major types which are characterized by the type of electrolyte employed is presented in the following. At the present time, these fuel cell–based systems are for small-scale applications such as for providing electricity to a commercial building, but the SOFC and the MCFC are being investigated for larger-scale applications while operating in the hybrid mode (>40 MW).

11.2.5.1 Phosphoric Acid Fuel Cell or PAFC The electrolyte consists of concentrated phosphoric acid and a silicon carbide matrix is used to retain the acid while both the electrodes which also function as catalysts are made from Pt or its alloys. The operating temperature is maintained between 150 and 220°C; at lower temperatures, phosphoric acid tends to be a poor ionic conductor while CO poisoning of the Pt electrocatalyst in the anode becomes severe.

The electrochemical reactions occurring in a PAFC are:

At the anode: $H_2 \rightarrow 2H^+ + 2e^-$ (H^+ then transporting through the electrolyte)
At the cathode: $\frac{1}{2}O_2 + 2H^+ + 2e^- \rightarrow H_2O$
With the overall cell reaction: $\frac{1}{2}O_2 + H_2 \rightarrow H_2O$.

The PAFC operates on H_2 fuel while CO is a poison when present in a concentration greater than 0.5%. If a hydrocarbon such as natural gas is used as a fuel, reforming of the fuel by the reaction: $CH_4 + H_2O = 3H_2 + CO$ and shifting of the reformate by the reaction: $CO + H_2O = H_2 + CO_2$ are required to generate the fuel required by the cell. The fuel cell itself can tolerate a maximum of 50 ppmv of sulfur compounds. The heat rejected by the PAFC being at low temperatures may be useful for thermal applications and possibly for low temperature Rankine cycles for generation of additional electric power depending on the temperature, size of the system, and its economics.

United Technologies Corporation reports that the efficiency of its power plants at the beginning of life is 40% on natural gas LHV basis but the efficiency reduces rapidly to about 38% before it stabilizes and over its expected cell life (~40,000 h) averages to about 37% (Fuel Cell Handbook, 2004). The PAFC which comes in 100–400 kW modules is one of the most mature fuel cell types and the first to be used commercially. Its applicability is in small-scale stationary power applications such as for distributed power generation although some PAFCs have been used to power large vehicles such as city buses.

11.2.5.2 Molten Carbonate Fuel Cell or MCFC The electrolyte of a MCFC typically consists of a combination of alkali Na and K carbonates retained in a ceramic matrix of lithium aluminum oxide ($LiAlO_2$). The MCFC cell operates at temperatures of 600–700°C in order to keep the alkali carbonates in a highly conductive molten salt form with carbonate ions providing ionic conduction. In order to form the carbonate ions ($CO_3^=$) on the cathode surface, CO_2 must be supplied with the oxidant (typically air). The anode is made from Ni while the cathode is made from nickel oxide.

The electrochemical reactions occurring in the cell are as follows:

At the anode: $H_2 + CO_3^= \rightarrow H_2O + CO_2 + 2e^-$

At the cathode: $\frac{1}{2}O_2 + CO_2 + 2e^- \rightarrow CO_3^=$ (oxygen being transported to the anode by the carbonate ions $CO_3^=$ through the electrolyte)

With the overall cell reaction: $H_2 + \frac{1}{2}O_2 + CO_2 \rightarrow H_2O + CO_2$.

CO is not a poison, but rather a fuel for the MCFC. The CO is not electrochemically oxidized to any appreciable extent but produces additional H_2 by the water gas shift reaction ($CO + H_2O \rightarrow H_2 + CO_2$) in the gas phase of the anode compartment.

A fuel such as natural gas is either reformed externally or within the cell in the presence of a suitable catalyst to form H_2 and CO by the reaction: $CH_4 + H_2O \rightarrow 3H_2 + CO$. The high operating temperature of a MCFC allows internal reformation and the endothermicity of the reformation reaction provides a sink for the heat generated within the fuel cell. The MCFC itself can tolerate typically a maximum of 0.5 ppmv of sulfur compounds but again it is the reformer and the prereformer catalyst that set the limit.

As mentioned above, CO_2 must be supplied in the oxidant stream (cathode inlet). Typically the CO_2 generated at the anode is recycled to the cathode or CO_2 is produced by combustion of the anode exhaust gas and mixed with the cathode inlet gas. This requires additional equipment to either transfer the CO_2 from the anode exit gas or produce it. The management of the liquid electrolyte also poses operational challenges. FuelCell Energy Inc. reports that their power plants range in efficiency from 43 to 47% on natural gas LHV basis (FuelCell Energy Inc., 2013) and modules in the 300 kW to 3 MW range are available for small-scale electric utility or distributed power generation.

11.2.5.3 Proton Exchange Membrane Fuel Cell or PEMFC The PEMFC consists of a proton conducting membrane contained between two Pt impregnated porous electrodes. Perfluorosulfonic acid polymer is a typical membrane material which has good proton conducting properties. The back of the electrodes are coated with a hydrophobic compound such as Teflon® forming a wet proof coating which provides a gas diffusion path to the catalyst layer. Within the cell, H_2 at the anode provides protons and releases electrons which pass through the external circuit to reach the cathode. The protons solvate with water molecules and diffuse through the membrane to the cathode to react with the O_2 while picking up electrons and forming water.

Thus, the electrochemical reactions occurring in a PEMFC are as follows:

At the anode: $H_2 \rightarrow 2H^+ + 2e^-$ (H^+ then transporting through the electrolyte)
At the cathode: $\frac{1}{2}O_2 + 2H^+ + 2e^- \rightarrow H_2O$
With the overall cell reaction: $\frac{1}{2}O_2 + H_2 \rightarrow H_2O$.

The PEMFC operates on H_2 while only a few parts per million of CO may be tolerated by the Pt catalyst at its operating temperature of 80°C. A hydrocarbon fuel such as natural gas has to be reformed as in a PAFC, followed by shifting of the reformate, and removal of the unconverted CO to <10 ppmv. Ru added to the Pt catalyst provides tolerance against carbon oxides (Ralph and Hogarth, 2002). Research is being conducted to reduce the amount of Pt in order to reduce cost of the PEMFC.

Some of the advantages of the PEMFC are that it has a fast start capability, compact and light weight design, and lack of corrosive fluid spillage hazard because the only liquid present in the cell is water. Thus, a PEMFC is well suited for use in vehicles in addition to small-scale (<1 to100 kW) applications such as backup power, portable power, and distributed generation. A disadvantage associated with this type of fuel cell, however, is use of the expensive Pt catalyst. The ionic conductivity of the electrolyte increases with the water content and thus it is necessary to maintain a high enough water content in the electrolyte to avoid membrane dehydration and maintain proper ion conductivity but without flooding the electrodes. Thus, the rate of production of the water by the oxidation of the H_2 and its evaporation rate must be balanced. PEMFCs are capable of operation at pressures from 0.10 to 1.0 MPa. The conversion efficiency of fuel bound energy to electricity of a PEMFC ranges from 25 to 32% on a fuel (natural gas) LHV basis (Fuel Cell Handbook, 2004) while the heat rejected by the fuel cell is at a low temperature and is not useful for generation of additional electric power.

11.2.5.4 Solid Oxide Fuel Cell or SOFC The electrolyte in an SOFC consists of a solid, nonporous metal oxide, typically yttria (Y_2O_3) stabilized zirconia (ZrO_2) with the anode made from cobalt or nickel zirconia ($CoZrO_2$ or $NiZrO_2$) cermet (ceramic-metallic material), while the cathode is a perovskite with a dopant such as strontium-doped lanthanum manganite ($LaSrMnO_3$). The cell operates at temperatures >650°C such that diffusion of oxygen ions to the anode surface may occur. The SOFC with its solid state components may in principle be constructed in any configuration. Cells are being developed in the tubular, the flat plate, and the monolithic configurations. Current designs consist of the electrode, electrolyte, and interconnect material deposited in layers and sintered together to form a cell structure; the fabrication techniques differ however according to the type of cell configuration and developer.

The electrochemical reactions occurring within the cell for H_2 as the fuel are:

At the cathode/electrolyte: $\frac{1}{2}O_2 + 2e^- \rightarrow O^=$ ($O^=$ then transporting through the electrolyte)

At the anode/electrolyte: $H_2 + O^= \rightarrow H_2O + 2e^-$

With the overall cell reaction: $\frac{1}{2}O_2 + H_2 \rightarrow H_2O$.

CO and hydrocarbons such as CH_4 can also be used as fuels in an SOFC. At the high temperatures within the cell, it is feasible for the water gas shift reaction: $CO + H_2O = H_2 + CO_2$ to occur in the gas phase and in the case of natural gas, the steam reforming reaction: $CH_4 + H_2O = 3H_2 + CO$ to occur in the presence of a suitable catalyst, to produce H_2 that is easily oxidized at the anode. The fuel cell itself can tolerate H_2S at concentrations as high as 69 ppmv based on data presented by Marquez et al. (2006) which indicated that the H_2S at these concentrations had no inhibiting effect on the oxidation of H_2 but again it is the reformer and the prereformer catalyst that set the limit.

Advantages of the SOFC are that there is no liquid electrolyte with its associated corrosion and electrolyte management problems, and with an operating temperature greater than 650°C internal reforming can be achieved with the inclusion of a suitable catalyst. Overall efficiencies range from 40 to 50% for conversion of the fuel (natural gas) bound energy to electricity on a LHV basis while the exhaust heat from the SOFC is at high temperatures and may be recovered for useful purposes such as:

- Reforming the fuel (e.g., natural gas) where a major portion of the exhaust heat is converted into chemically bound energy of the reformed fuel due to the endothermicity of the reforming reaction ("chemical recuperation")
- Providing thermal energy to a bottoming cycle and/or
- Recovered for the generation of steam for cogeneration purposes.

The bottoming cycle may consist of a gas turbine in the case of an SOFC operating at high pressure as described later in this chapter. Another advantage of having the high operating temperature is that reaction rates are higher without requiring any precious metal catalysts. The high temperature of the SOFC however places special requirements on the materials of construction. Their current size in the range of 1 kW to 2 MW limits their use to small-scale electric power generation applications such as distributed power. Research aimed at reducing cost of the SOFC is being conducted in areas such as: (i) developing materials with high ionic/electronic conductivity (mixed conductivity for both ions and electrons) at lower operating temperatures, (ii) fabrication techniques, and (iii) constructing larger-scale stacks.

11.2.6 Hybrid Cycles

The high initial cost for fuel cell systems is a major barrier for widespread commercialization. By configuring the fuel cell in a hybrid cycle, a potential exists to not only further increase the overall system efficiency but also to reduce the plant specific cost ($/kW). A fuel cell–based hybrid cycle consists of combining a fuel cell with a heat engine to maximize the overall system efficiency. Overall system efficiencies greater than 60% on natural gas on an LHV basis may be achieved. High temperature fuel cells such as the SOFC and MCFC are most suitable for such applications. In the case of a high pressure fuel cell–based hybrid, the combustor of the gas turbine is replaced by the fuel cell system while in the case of a low (near atmospheric) pressure fuel cell–based hybrid, the heat rejected by the fuel cell may be transferred to the working fluid of the gas turbine through a heat exchanger (indirect cycle). Smaller-scale hybrids can provide distributed power that can outperform large-scale combined cycle plants with regard to efficiency and emissions of pollutants and greenhouse gases.

The integration of the fuel cell and the gas turbine can be achieved in various ways, but the most basic concept attempts to replace the combustor of a typical Brayton (gas turbine) cycle with a high pressure fuel cell equipped with an anode exhaust gas oxidizer as illustrated in Figure 11.6 in which the fuel cell consists of a SOFC. The anode exhaust gas containing the depleted fuel is oxidized with the remaining O_2 in the cathode exhaust gas. The system includes heat exchangers to preheat the fuel and air before entering the fuel cell using heat contained in the turbine exhaust. A representative performance of such a system is presented in Table 11.1 along with major design parameters, showing a net system efficiency of almost 60% on a natural gas LHV basis, which is remarkable for a small system. As can be seen, majority (>60%) of the electric power is produced by the fuel cell with the remainder coming from the gas turbine (after providing the power required by its compressor). Hybrid configurations with efficiencies >70% on a natural gas LHV basis have been identified (Rao and Samuelsen, 2003).

A hybrid configured with a MCFC operating at low (near atmospheric) pressure is illustrated in Figure 11.7 and requires much more elaborate heat exchange equipment. The exhaust heat from the anode gas oxidizer before it is recycled to the cathode side to provide the CO_2 is recuperated in a heat exchanger to heat up the working fluid (air) before it is expanded in the turbine. Heat from the cathode exhaust is also recuperated to preheat the working fluid (air) leaving the gas turbine compressor.

There are both advantages and disadvantages of a pressurized versus a low pressure fuel cell–based hybrid. A pressurized fuel cell–based system requires matching the

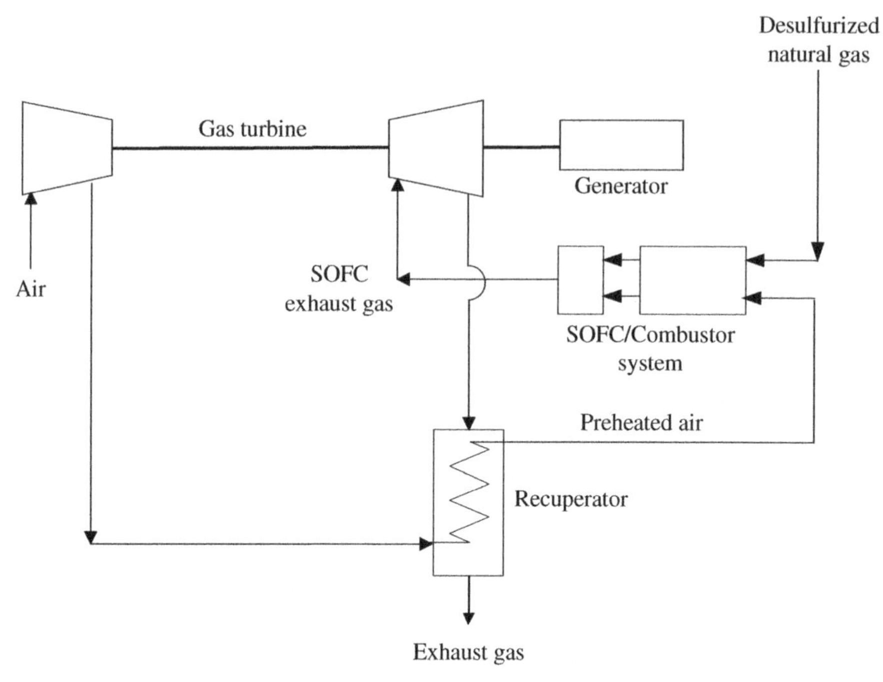

FIGURE 11.6 A pressurized fuel cell hybrid system

TABLE 11.1 Performance summary of a pressurized fuel cell hybrid system[a]

Compressor air intake flow rate	16.6 kg/s
Compressor pressure ratio	9.5:1
Gas turbine combustor exhaust temperature	1160°C
SOFC gross AC power	7.9 MW
Gas turbine gross AC power	4.8 MW
Power block net AC power	12.3 MW
Fuel flow rate to SOFC	0.31 kg/s
Fuel flow rate to gas turbine	0.14 kg/s
Power system efficiency (Net AC/LHV)	59.9%
CO_2 emission	340 kg/MWh
NO_x emission	0.04 kg/MWh
SO_x emission	Fuel is desulfurized to protect Ni-reforming catalyst and anode from forming NiS
Plant exhaust gas temperature	360°C
Plant exhaust gas flow rate	16.9 kg/s

[a] Data from Lundberg et al. (2003).

gas turbine to the fuel cell pressure and may require development of the customized gas turbine. Also, plant controllability can be more complex especially during an unplanned shutdown (trip) of the gas turbine leading to sudden depressurization of the fuel cell. On the other hand, high thermal efficiency and power density may be

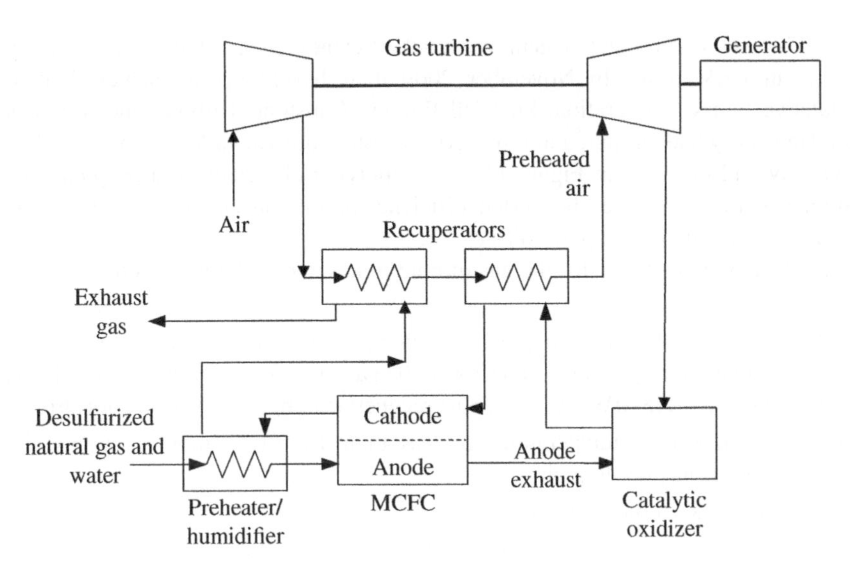

FIGURE 11.7 An atmospheric MCFC hybrid system

achieved with a pressurized fuel cell due to the higher Nernst potential and lower activation and concentration polarizations. Also the fuel cell and the heat exchange equipment would be more compact while also savings due to the smaller diameter piping may be realized.

An atmospheric pressure fuel cell decouples the gas turbine pressure ratio from the fuel cell. On the other hand, the operating temperature of the heat exchange equipment can be more severe.

11.2.6.1 Commercialization Status Siemens Power Generation, Southern California Edison, and the Advanced Power and Energy Program at the University of California, Irvine partnered to develop and test the world's first SOFC-based hybrid system from 1999 to 2003. This system operated for over 2900 h and produced up to 220 kW with net efficiencies up to 53% on natural gas on an LHV basis demonstrating the hybrid concept, revealing steps to improve future designs, and proving high efficiency and ultra-low emissions at this relatively small scale. This hybrid system was put together using an off-the-shelf gas turbine which limited fully realizing the efficiency potential. Rolls Royce Fuel Cell Systems has been working on SOFC gas turbine systems for more than 10 years. Their concept is based upon a dual-stage turbo-compressor with air flow rate that is reasonably matched to the air flow rates required by approximately 900 kW SOFC to produce a 1 MW output hybrid system. Rolls Royce has developed novel planar SOFC technology comprised of thin planar cells that are printed onto a flat tube support structure. Smaller-scale (250 kW) prototypes of the Rolls Royce hybrid design have been successfully demonstrated in the United Kingdom in 2008 and 2010. Mitsubishi Heavy Industries successfully tested a 229 kW hybrid system incorporating a SOFC and a micro-turbine generator in the 2007–2009 time frame that achieved a net efficiency greater than 52% on natural gas on an LHV basis. In November 2009 their hybrid system achieved 3000 cumulative hours of operation. FuelCell Energy, Capstone Turbine, and Montana State University had partnered to test a hybrid system utilizing a MCFC as described previously and depicted in Figure 11.7 for supplying electricity to a hospital. The system was able to achieve a net system efficiency of 56% on a natural gas LHV basis with net power output of up to 300 kW.

The development needs for the gas turbine in large hybrid applications are:

- Recuperation where the compressor discharge air is supplied to a heat exchanger rather than directly to the combustor of the gas turbine (currently only small gas turbines, i.e., <15 MW are offered as recuperated engines for generator drives)
- Low turbine inlet temperature[3] of less than 1000°C without requiring the parasitic turbine cooling air
- Pressure ratio in the range of 4–12
- Combustors accepting hot and depleted fuel and air when gas turbine combustors are utilized for oxidation of the fuel cell anode exhaust gas
- Oil free bearings to avoid carbon deposition in the anode section of the fuel cell.

11.2.7 A Coal-Fueled Hybrid System

An integrated gasification fuel cell (IGFC) system concept consisting of catalytic hydro-gasification which can be synergistically integrated with high temperature fuel cells such as the SOFC and the MCFC was described in Chapter 10 in conjunction with H_2 coproduction. For a plant producing electricity only (i.e., without the H_2 coproduction), an efficiency greater than 50% on a coal HHV basis is projected while capturing 90% of the CO_2 with a high rank coal such as a bituminous at ISO conditions (Li et al., 2010). Figure 11.8 depicts the overall block flow diagram of such an IGFC plant.

Implementing such plants requires the development of large-scale fuel cell modules in the 100 MW range versus the currently available units which have only few MW capacity or less.

11.3 CHEMICAL LOOPING

In chemical looping, a metal oxide (Me_xO_y) functions as an oxygen carrier for combustion or partial oxidation of a carbonaceous feedstock (e.g., natural gas, coal, and biomass). In the case of combustion, the flue gas produced is highly concentrated

[3] From an overall efficiency standpoint, it is advantageous to maximize the fuel utilization in the fuel cell (but without significantly lowering the overall Nernst potential). The corresponding temperature obtained by combusting the unconverted fuel leaving the fuel cell (and available for the gas turbine) is typically less than 1000°C.

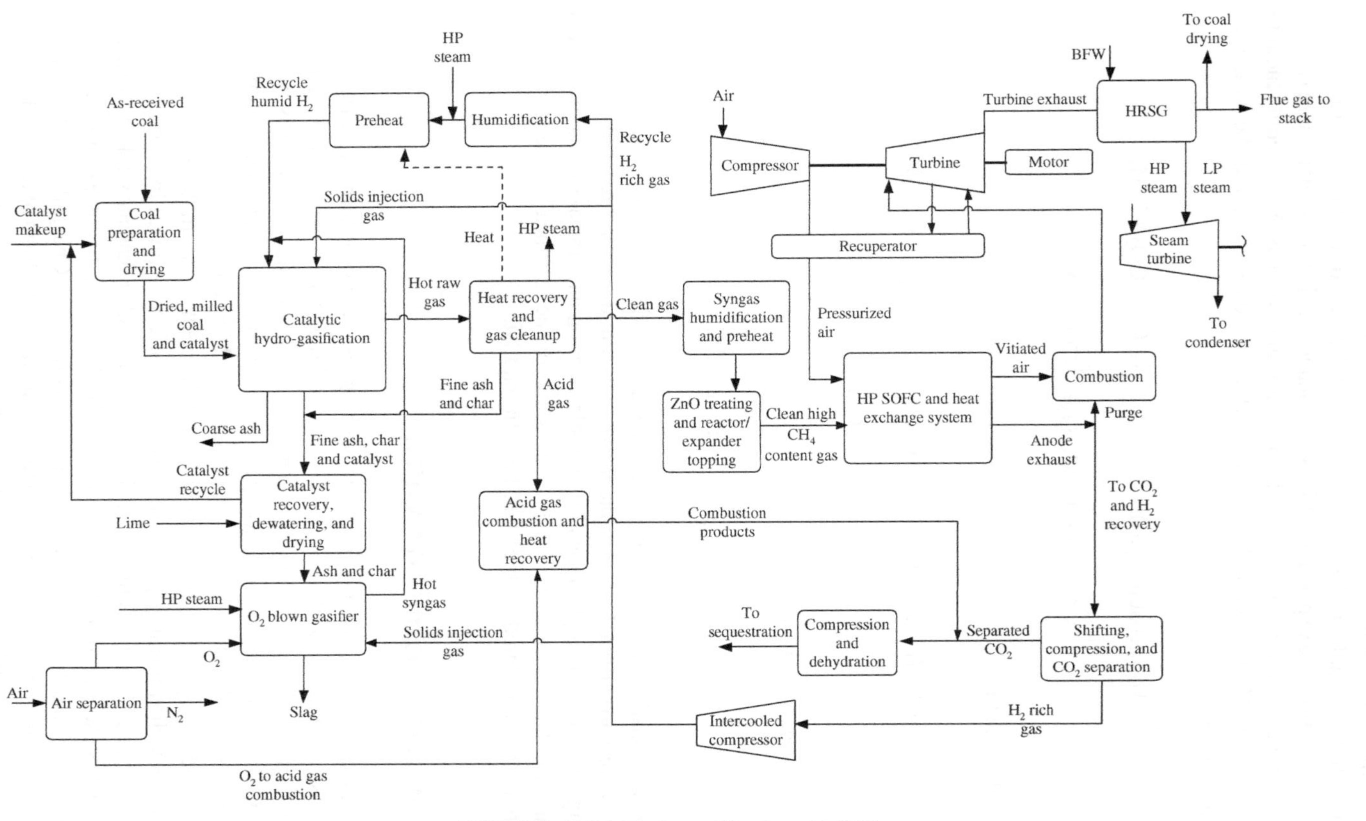

FIGURE 11.8 Hydrogasifier-based IGFC

in CO_2 without being diluted with N_2 as is the case when air is utilized to supply the O_2 making it more suitable for CCS applications. In the case of O_2-blown partial oxidation or gasification, the expensive ASU is eliminated making the syngas more suitable for CCS as well as for coproduction of chemicals and synfuels. The overall reactions of a metal oxide Me_xO_y with CH_4 and with carbon in the feedstock may be written as:

Combustion

$$\frac{y}{2}CH_4\ (\text{in natural gas}) + 2Me_xO_y = yH_2O + \frac{y}{2}CO_2 + 2xMe$$

$$\frac{y}{2}C\ (\text{in coal}) + Me_xO_y = \frac{y}{2}CO_2 + xMe$$

Partial Oxidation

$$\frac{y}{2}CH_4\ (\text{in natural gas}) + \frac{3}{2}Me_xO_y = yH_2O + \frac{y}{2}CO + \frac{3}{2}xMe$$

$$yC\ (\text{in coal}) + Me_xO_y = yCO + xMe$$

Metal carrier oxidation

$$xMe + \frac{y}{2}O_2\ (\text{from air}) = Me_xO_y$$

The combustion or partial oxidation reaction, and the metal carrier oxidation reaction are carried out in two separate vessels. Fluidized bed reactors are being investigated for these applications but particle attrition and transfer of the hot solids (Me and Me_xO_y) between the reactors are some of the technical challenges. Metal carriers being considered include Cu, Fe, Mn, and Ni.

11.4 MAGNETOHYDRODYNAMICS

The open cycle MHD generator under development for generating electric power from fossil fuels consists of moving a conductor through a magnetic field to generate electric current, like a conventional generator, except that the conductor is within a flowing gas stream, the combustion products containing charges. The combustion products have to be at very high temperatures (>2700°C) so that the gas is in the plasma state where ionization occurs by which the electrons from the atoms or molecules are detached forming positively charged ions. The required high temperatures may be realized by combusting the fuel with oxygen rather than air while making the process CCS friendly. If air is utilized, then preheating the air (using exhaust heat from the MHD generator) to very high temperatures is required. Charged particles may be introduced from an external source into the gas stream in the form of alkali salts which are more easily ionizable (in addition to the alkalis that may already be present in the coal ash) in order to reduce the temperature down to about 2500°C. The exhaust for the MHD generator may be used to generate steam for the Rankine cycle resulting in a combined or hybrid cycle. The closed cycle MHD generator either uses a noble gas (such as He or Ar) seeded with Cs or a liquid metal. This type of MHD generator is mainly being investigated for nuclear power plant applications and requires less severe operating temperatures.

The electric current produced by an MHD generator is DC as in fuel cells and requires an inverter to convert it into AC before supplying it to the grid. Overall plant efficiencies in excess of 60% on a coal HHV have been shown to be possible in studies with a bottoming steam cycle included. The MHD concept was proven in the 1980s with grid transferred power both in the United States and in the former U.S.S.R. but research and development activities in this area waned in later years. With current emphasis on CCS in fossil fired power generation, oxy-combustion–based MHD generation is getting renewed interest.

REFERENCES

Armstrong PA, Bennett DL, Foster EP, Stein VE. Ceramic membrane development for oxygen supply to gasification applications. Proceedings of the Gasification Technologies Conference; October 2002; San Francisco, CA.

Diethelm S, Van Herle J, Middleton PH, Favrat D. Oxygen permeability and stability of $La_{0.4}Ca_{0.6}Fe_{1-x}Co_xO_{3-\delta}$ (x = 0, 0.25, 0.5) membranes. J Power Sources 2003;118:270–275.

Enick RM, Cugini AV, Killmeyer RP, Howard BH, Morreale BM, Ciocco MV, Bustamante F. Towards the developments of robust water-gas-shift reactors. Am Chem Soc 2003;48(2).

Foy K, McGovern J. Comparison of ion transport membranes. Proceedings of the 4th Annual Conference on Carbon Capture and Sequestration; May 2–5, 2005. Morgantown, WV: U.S. Department of Energy.

FuelCell Energy, Inc. 2013. Types of fuel cells. Available at http://www.fuelcellenergy.com/why-fuelcell-energy/types-of-fuel-cells/. Accessed October 3, 2013.

Fuel Cell Handbook, seventh edition (EG&G Services, Inc). Morgantown, WV: U.S. Department of Energy, National Energy Technology Laboratory; November 2004. Contract nr DE-AM26-99FT40575.

Ihara M, Yokoyama C, Abudula A, Kato R, Komiyama H, Yamada K. Effect of the steam-methane ratio on reactions occurring on Ni/yttria-stabilized zirconia cermet anodes used in solid-oxide fuel cells. J Electrochem Soc 1999;146(7):2481–2487.

Judkins RR, Bischoff BL. Hydrogen separation using ORNL's inorganic membranes. Proceedings of the 29th International Technical Conference on Coal Utilization and Fuel Systems; April 18–23, 2004; Clearwater, FL.

Li M, Rao AD, Brouwer J, Samuelsen GS. Design of highly efficient coal-based integrated gasification fuel cell power plants. J Power Sources 2010;195:5707–5718.

Lundberg WL, Veyo SE, Moeckel MD. A high-efficiency solid oxide fuel cell hybrid power system using the Mercury 50 advanced turbine systems gas turbine. J Eng Gas Turbines Power 2003;125(1):51–58.

Marquez AI, Abreu YD, Botte GG. Theoretical Investigations of NiYSZ in the Presence of H_2S. Electrochem Solid-State Lett 2006;9(3):A163–A166.

Matsuzaki Y, Yasuda I. Electrochemical oxidation of H_2 and CO in a H_2- H_2O-CO-CO_2 system at the interface of a Ni-YSZ cermet electrode and YSZ electrolyte. J Electrochem Soc 2000;147(5):1630–1635.

Prasad R, Chen J, van Hassel B, Sirman J, White J, Apte P, Shreiber E. Advances in OTM Technology for IGCC. Nineteenth Annual International Pittsburgh Coal Conference; September 2002; Pittsburgh, PA.

Ralph TR, Hogarth MP. Catalysis for low temperature fuel cells. Platin Metals Rev 2002;46(3):117–135.

Rao AD, Samuelsen GS. A thermodynamic analysis of tubular solid oxide fuel cell based hybrid systems. J Eng Gas Turbines Power 2003;125(1):59–66.

Roark SE, Mackay R, Sammells AF. Hydrogen separation membranes for VISION 21 energy plants. Proc Intl Tech Conf Coal Utilization Fuel Syst 2002;27(1):101.

12

RENEWABLES AND NUCLEAR

Concern over energy usage is increasingly important as climate change caused by greenhouse gases becomes a global concern. Fossil fuels and coal in particular are a primary energy source; the world consumption of coal was as high as 162×10^{15} kJ in 2012 which accounted for about 30% of the world total energy usage and is predicted to increase to 232×10^{15} kJ in 2040 with an average annual growth rate of 1.3% while maintaining a similar share of the total energy usage. In the power generation sector, coal fired plants generated 8.4×10^{12} kWh of electricity in 2012 which is about 40% of the total world electricity (Conti and Holtberg, 2013). Coal fired power plants are the major emitters of CO_2 emissions, and various other pollutants, such as PM, SO_x, nitrogen oxides (NO_x and N_2O), and Hg. As an example, the United States emitted 5290×10^6 tonnes of CO_2 in 2012 of which electrical power generation accounted for almost 40% (U.S. Energy Information Administration, 2013). The options mentioned in this book for reducing the CO_2 emissions from fossil fueled plants have included increasing efficiency and carbon capture and sequestration. Use of renewable and nuclear energy is the other approach to reduce the CO_2 emissions and their contribution to the total energy mix should be increased for the sustainability of our planet. Table 12.1 (Edenhofer et al., 2011) shows the relative greenhouse gas emission intensities on an life cycle assessment (LCA) basis from various energy sources and wind, solar, geothermal, and nuclear are amongst the lowest and can make a significant impact toward the goal of reducing future

Sustainable Energy Conversion for Electricity and Coproducts: Principles, Technologies, and Equipment, First Edition. Ashok Rao.

TABLE 12.1 Relative LCA greenhouse gas emission intensities from power plants (data from Edenhofer et al., 2011)

Technology	g CO_2 equivalent/kWh
Hydroelectric	4
Wind	12
Nuclear	16
Biomass	18
Solar thermal	22
Geothermal	45
Natural gas (without CCS)	469
Coal (without CCS)	1001

greenhouse gas emissions. The National Renewable Energy Laboratory (NREL) conducted a comprehensive review of published LCA results and the data summarized in the table correspond to the emission intensities at the 50th percentile levels.

Biomass, also an important renewable energy, has been addressed along with coal and the intent of this chapter is to provide an introduction to energy recovery from these other important renewable and nuclear sources. Also included in this chapter is a discussion of the dependence of intermittent renewables such as solar and wind on fossil fueled plants for maintaining electrical power grid stability at least in the foreseeable future before large-scale energy storage devices are commercially available. The chapter ends with an introduction to the types of grid designs being considered to make them more compatible with environmentally preferred methods of generating power.

12.1 WIND

Turbines installed to convert the kinetic energy in wind to mechanical energy may be seen in many parts of the world, with total installations increasing from 14 GW in 1999 to 160 GW by the end of 2009 (Edenhofer et al., 2011). Much of this growth was due to reductions in cost of wind turbines and government incentives. Turbines with electrical power output as small as 1 kW to larger units of 2.5 MW output are available although much larger units with rated capacities greater than 6 MW are being tested. Several such units may be in one location connected to the electric power grid forming a "wind farm." They may be located onshore as well as offshore where the wind is steadier and stronger, and where there is a less visual impact but the construction and maintenance costs of such offshore units are much higher. Although wind power is very consistent from year to year, it can vary significantly over shorter timescales including throughout a day as can be seen from Figure 12.1 which illustrates the variability of power produced by wind farms in a day within the power grid managed by California Independent System Operators' (CAISO) that make up 80% of California's and a small part of Nevada's grid.

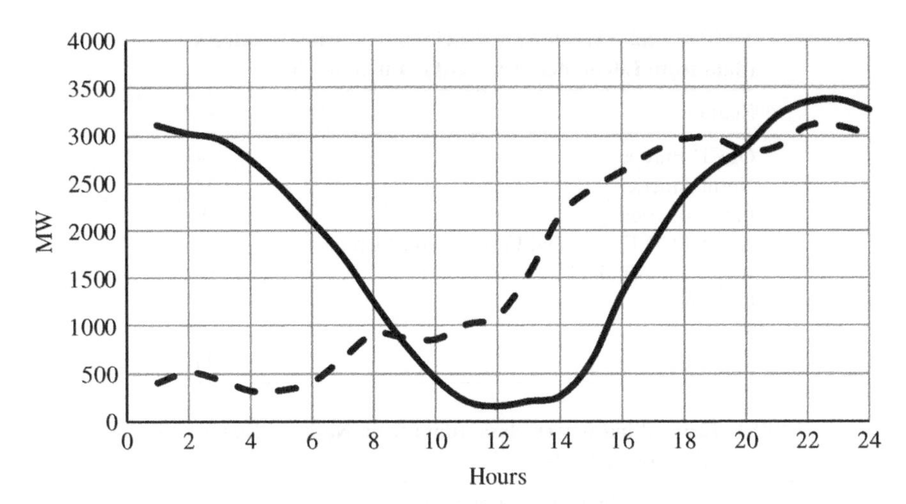

FIGURE 12.1 Variability of power produced by wind farms (CAISO data for April 27, 2013 as solid line and October 27, 2013 as dashed line)

Due to this variation, the maximum amount of power contributed by wind or wind energy penetration (energy produced by wind as a percentage of the total available generation capacity) may have to be limited before grid stability becomes an issue. Power management techniques may be utilized such as having energy storage, other type of power plants with excess capacity (e.g., fossil fueled units as discussed later in this chapter under "Electrical Grid Stability and Dependence on Fossil Plants") or dispatching other units such as gas turbine peakers to make up the difference. Typically, utility grids have some reserve generating and transmission capacity to account for equipment failures, and this capacity can then be utilized to compensate for some of the varying power generated by wind power plants depending on the wind and solar energy penetration. Thus, the penetration limit is grid-specific and studies have indicated that generally a wind energy penetration of 20–30% may be incorporated without much difficulty but economic constraints may limit going beyond these levels. Wind speed forecasting is important for a utility operator to be able to plan ahead with respect to brining online dispatchable units.

12.1.1 Wind Resources and Plant Siting

Wind is caused by areas of high and of low pressures in the earth's atmosphere due to the temperature gradients brought about by the uneven heating by the sun, the angle of incidence of sun's rays at the earth's surface being a function of the time of day and the latitude. The amount of thermal energy absorbed by the earth's surface also affects these temperature gradients and thus the type of surface, that is, whether it is open land or covered with vegetation, or it consists of a large body of water.

The worldwide technical potential for harnessing kinetic energy in wind is very large, with estimates ranging from 140 EJ/year to as high as 3050 EJ/year

TABLE 12.2 Worldwide distribution of wind resources (data from Edenhofer et al., 2011; Lu et al., 2009)

Region	%
OECD Europe	4
North America	22
Latin America	9
Non-OECD Europe and Former Soviet Union	26
Africa and Middle East	17
Oceania	13
Rest of Asia	9
Total	100

TABLE 12.3 Wind power class (data from National Renewable Energy Laboratory, 2014)

Power class	Wind speed at 50 m (m/s)	Quality of resource
1	0.0–5.6	Poor
2	5.6–6.4	Marginal
3	6.4–7.0	Fair
4	7.0–7.5	Good
5	7.5–8.0	Excellent
6	8.0–8.8	Outstanding
7	>8.8	Superb

(39,000–840,000 TWh/year) for both onshore and offshore installations together, depending on the assumptions made regarding the siting constraints (Edenhofer et al., 2011). The wind resource is not evenly distributed across the earth as seen by the data presented in Table 12.2 (Edenhofer et al., 2011; Lu et al., 2009), and is not necessarily close to population centers and is likely to limit how much of the technical potential is actually harnessed.

Table 12.3 classifies wind resources according to its speed (National Renewable Energy Laboratory, 2014) and for large utility scale wind farms, wind speeds greater than 7 m/s at the wind turbine's hub height are desirable for economic viability. Note that the power $\dot{W}$ recovered by a wind turbine is proportional to the cube of the wind velocity as seen by the relationship between the maximum power that may be recovered theoretically and density ρ of the air, wind speed v, and the swept area A of the turbine blades or the area normal to the wind direction through which the air passes (Betz, 1920):

$$\dot{W} = \frac{16}{27} \times \left(\frac{1}{2}\rho A v^3\right)$$

The actual power developed by the turbine which is lower due to various aerodynamic and mechanical losses is calculated by Equation 12.1 by introducing a coefficient of performance COP and introducing the mechanical (including gear) efficiency η_{mech} and the generator efficiency η_{gen}.

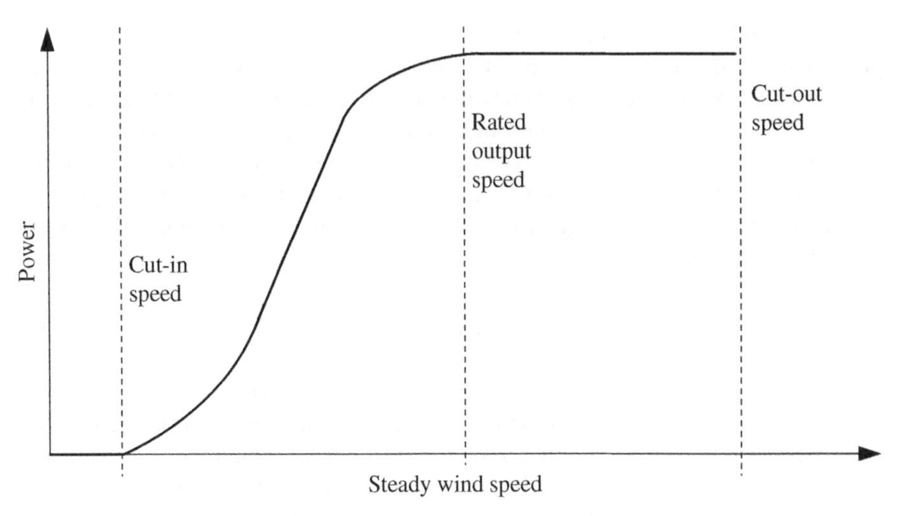

FIGURE 12.2 Wind turbine power curve

$$\dot{W} = \text{COP} \times \eta_{\text{mech}}\, \eta_{\text{gen}} \left(\frac{1}{2} \rho A v^3 \right) \tag{12.1}$$

The maximum value that COP can have is 16/27 or 0.596 and is a function of tip speed ratio (ratio of the tangential tip speed of a blade and the actual wind speed) for a given turbine design, initially increasing with the tip speed ratio to reach a maximum and then decreasing at the higher ratios.

Factors that should be taken into account in siting a wind farm are the distance the power has to be transmitted, land use compatibility, public acceptance of noise, the visual aspect, and impact on bird populations.

12.1.2 Key Equipment

Most large-scale wind turbines have a horizontal axis with three blades driving an induction generator and start producing electrical energy from the wind at speeds 3–4 m/s, the "cut-in speed." Figure 12.2 shows a typical relationship of the power developed by a turbine and the wind speed. At lower wind speeds, the torque exerted by the wind on the turbine blades is too low to rotate the turbine. The power developed by the turbine increases rapidly above the cut-in speed due to its dependence on the cube of the wind speed. However, there is an upper limit, typically in the 12–17 m/s wind speed range where the power output reaches the rated power output limited by the electrical generator. The turbine system is designed to limit the power to its rated output, for example, by adjusting the blade angles (pitch) in the case of large turbines. Finally, a braking system is employed for very high wind speeds known as the "cut-out speed," usually greater than 25 m/s to avoid damaging the rotor due to excessive forces exerted by the wind.

Smaller wind turbines can achieve higher capacity factors if designed to operate at lower wind speeds but would operate less efficiently in high winds and also have higher installed specific costs ($/kW), while large turbines due to their larger inertia would stall at the lower wind speeds but have the advantage of economies of scale.

Large onshore wind turbines are typically installed on towers 50–100 m high with rotors that are typically 50–100 m in diameter but units with rotor diameters greater than 125 m are operating, and even larger machines are under development. Note that the wind velocity v increases with height as approximately:

$$v \propto (\text{height})^{1/7} \tag{12.2}$$

Crests of hills however do not follow this 1/7th power rule. The density of air on the other hand decreases with elevation and should be accounted for when calculating the power developed. Another factor that should be taken into account while developing the layout of a wind farm is the interference caused by adjacent turbines to reduce the recoverable energy flow field. The turbines within the wind farm should be optimally positioned so that the wake effects are minimized. In addition, site-specific conditions such as the local topology as well as construction and other logistical issues should be taken into account in developing the layout.

The electrical conversion system is an important part of the system reliability. The most commonly used generator for larger wind turbines is an AC generator while DC generators are utilized for smaller wind turbines and work well when charging batteries. The shaft on which the blades are mounted is connected to the generator via a gearbox in the case of larger turbines while smaller turbines may have a simple direct drive. The gearbox allows the better matching of the generator speed with the turbine speed but is subjected to wear while in the case of direct drive, the generator shaft and bearings are subjected to the full torque produced by the wind acting on the blades. Synchronous generators tied to the utility grid need to operate at a constant speed to synchronize with the grid frequency and a major disadvantage of such single speed operation is that the turbine does not always operate at its peak efficiency as the wind speed varies. Power electronics however have made it possible to build wind turbines with variable speed.

12.1.3 Economics

Wind energy is capital intensive at the present time and incentives offered by local, state, and federal governments are important for the economic viability in many countries. The levelized cost of electricity depends on the capacity factor of a power plant and in the case of wind farms, typical capacity factors may range from 15 to 50% whereas in the case of a fossil fueled power plant such as a coal combustion-based plant, capacity factors are as high as 85%. Continued advances in technology are expected to reduce the cost of electricity from onshore wind energy plants by 10–30% while for offshore wind energy plants, the expected reduction in the cost of electricity is 10–40% by 2020 (Edenhofer et al., 2011). Some of the advances include improvements in tower design that reduce the need for large cranes and minimize materials; more efficient rotor blade, drive train, and generator designs; advanced manufacturing methods and materials of construction; and improved capacity factors by implementing advanced controls and power electronics.

In order to take advantage of economies of scale, longer blade lengths with reduced fatigue loading are being developed while limiting the amount of material used. Variable pitch rotors are also being investigated to optimize the angle of attack of the wind but are quite a challenge since wind velocity (direction and speed) can vary significantly.

12.1.4 Environmental Issues

Wind energy has a relatively small environmental footprint but impacts do exist. As illustrated in Table 12.1 (Edenhofer et al., 2011), the relative greenhouse gas emissions on an LCA basis for wind energy is among the lowest, but wind energy has some detrimental impacts such as bird and bat collisions. For offshore wind energy plants, impact on the fisheries and marine life should also be taken into account. Impact on the landscape, noise, and possible radar interference are also considerations to be accounted for.

Example 12.1. A wind turbine with diameter of 20 m and COP of 40% produces 2.5 MW in a wind speed of 25 m/s. What is the power output of a wind turbine with diameter of 44 m and COP of 50% in a wind speed of 20 m/s? Assume that the wind density and the mechanical and generator losses are the same in both cases.

Solution

From Equation 12.1, $(\dot{W}_1/\dot{W}_2) = (\text{COP}_1/\text{COP}_2) \times (D_1/D_2)^2 \times (v_1/v_2)^3$

Substituting the given values, $(\dot{W}_1/2.5) = (50/40) \times (44/20)^2 \times (20/25)^3$

Then the power output of the larger turbine, $\dot{W}_1 = 7.74\,\text{MW}$

12.2 SOLAR

Solar energy may be utilized directly for providing low temperature heat such as hot water or for building heating, converting to electricity using solar cells, or providing high temperature heat to operate a heat engine or a thermochemical process. Note that wind as well as ocean thermal energy also use solar energy but after it has been absorbed on the earth. This section is limited to the direct use of solar energy to operate a heat engine for large-scale electric power generation where the solar energy is concentrated to heat up the working fluid either directly or indirectly using a heat transfer fluid. Such concentrated solar power plants are expected to have a cumulative installed capacity of 10.8 GW in 2014, up from just 0.29 GW in 2009. The capacity factors for a number of the plants to be added are expected to range from 25% to as high as 75% with thermal storage included.

TABLE 12.4 Worldwide technically potential solar energy resources (data from Edenhofer et al., 2011; Rogner et al., 2000)

	Estimates	
Region	Minimum (EJ/year)	Maximum (EJ/year)
North America	181	7,410
Latin America and Caribbean	113	3,385
Western Europe	25	914
Central and Eastern Europe	4	154
Former Soviet Union	199	8,655
Middle East and North Africa	412	11,060
Sub-Saharan	372	9,528
Pacific Asia	41	994
South Asia	39	1,339
Centrally Planned Asia	116	4,135
Pacific OECD	73	2,263
Total	1575	49,837

12.2.1 Solar Resources and Plant Siting

Solar radiation includes the direct component coming straight from the sun ("beam radiation") and a diffuse component reflected off the clouds, atmospheric moisture vapor and particulates ("sky radiation"), as well as the earth's surface ("ground reflected radiation"). About 40% of the total energy reaching the earth's surface is the direct or beam radiation. The potential for producing electric power from solar energy is huge with estimates ranging from 1,575 to 49,837 EJ/year, roughly equivalent to as much as 3–100 times the world's primary energy usage in 2008 (Edenhofer et al., 2011). In the direct use of solar energy to operate a heat engine, only the direct beam component of solar irradiation is utilized, however, and thus limited geographical regions can realize this technology's full benefit, the angle of incidence of the sun's rays at the earth's surface being a function of the latitude. The worldwide solar energy resources are summarized in Table 12.4 (Edenhofer et al., 2011; Rogner et al., 2000).

12.2.2 Key Equipment

The key equipment includes the concentrator that reflects the sun's rays on to the receiver or absorber which includes heat exchange equipment, and a heat engine. There are two types of concentrators that reflect the solar rays, one to a point focus using dish systems and the other to a line focus using trough or linear Fresnel systems.

12.2.2.1 Point Focus Reflectors The central receiver or the "power tower" has an array of mirrors or heliostats which track the sun focusing the sun's rays to a receiver installed on top of a tower consisting of a fixed inverted cavity and/or tubes in which the heat transfer fluid circulates and is heated to temperatures as high as 1000°C, much higher than those with line focus systems. The heat transfer fluid is then used for heating the working fluid such as steam for the Rankine cycle. A major advantage of this system is due to the higher temperature which allows superheating the steam to higher temperatures (or even utilizing a supercritical steam cycle) resulting in higher efficiencies. There is also flexibility in plant construction because the heliostats do not all have to be installed on an even surface.

For smaller-scale applications (in the size range of 10–25 kW of electric power), a single paraboloidal tracking reflector focuses the rays onto a receiver that is about one dish diameter away and moves with the dish, and temperatures as high as 900°C may be reached. The heat may be transferred directly into the working fluid of a Stirling engine mounted at the focus. This indirectly fired recuperative engine, which lends itself to small-scale applications, uses a compressible working fluid and the ideal cycle consists of constant volume heat addition and rejection (Walker, 1980). The compression and the expansion steps are carried out either in two separate cylinders which also serve as heat exchangers and each is equipped with a piston, or in a single cylinder with a piston that pushes the working fluid back and forth between the heated and the cooled sections of the cylinder.

A 400 MW plant went into commercial operation in California in 2013. For large-scale applications where steam turbines are utilized, a central receiver or power tower demonstration unit for northerly latitudes with the capability of generating 1.5 MW of electric power has been built in northwest Germany and the system went online in 2009. The heat transfer fluid consists of air despite its low heat transfer coefficient and generates steam for the steam turbine with inlet conditions of 27 bar and 480°C.

12.2.2.2 Line Focus Reflectors Long rows of tracking parabolic reflectors or trough concentrators concentrate the sun's rays by 70–100 times onto a blackened[1] inner steel pipe incased in a glass outer tube with the annular space between them evacuated to minimize convective and conductive heat losses. The troughs track the sun along one axis, typically north–south. A heat transfer fluid is circulated through the steel pipe where it is heated to about 400°C which in turn is used for generating the working fluid such as steam for the Rankine cycle. Use of a heat transfer fluids such as an oil allows limiting the design pressure of tubes but safety is a concern due to its flammability. Other heat transfer fluids such as molten salts which can be stored to provide the heat to generate the steam during unfavorable weather conditions or at night time, as well as generating steam directly in the pipe to eliminate the indirect heat transfer loop are being developed. This trough collector design is the most common type with commercial operating experience of more than 20 years but the linear Fresnel system which uses long parallel tracking flat mirror strips to concentrate the sun's rays onto the fixed linear receiver is gaining popularity due to simplified plant design and lower capital, operating, and maintenance costs. The reflector design is based on the Fresnel lens effect which allows construction of a concentrating mirror (or lens) with a large aperture and short focal length. The flat mirrors are less expensive than the parabolic reflectors while the structural costs,

[1] For example, electrochemically deposited black chrome to optimize the surface selectivity (ratio of energy absorbed to energy emitted).

wind loads, and energy losses are all lower and outweigh the lower temperature limitation of the linear Fresnel system.

The earliest commercial plants utilizing the parabolic troughs were installed between 1985 and 1991 in California capable of producing 354 MW of power. An example of a more recent parabolic trough-based plant is the 64 MW nominal electric power output system in operation since 2007 in Nevada, USA. It has 760 troughs consisting of more than 180,000 mirrors to heat up a thermal oil, the heat transfer fluid, to a temperature of 390°C which in turn is used to provide steam at 90 bar and 371°C to a steam turbine. Two units operating in Spain also utilizing this technology have electric power generation capacities of 50 MW each with similar steam turbine inlet conditions of 100 bar and 377°C. Each of these plants in Spain covers a field of 1.95 km^2 with the mirror field size of about 510,000 m^2. Another example of the parabolic trough-based plant is the integrated solar combined cycle operating in Egypt since 2010. It consist of a solar field capable of generating about 110 MW thermal energy at a temperature of 400°C to provide additional steam for the steam turbine of a natural gas fired gas turbine–based combined cycle. The gas turbine produces 74 MW while the steam turbine can produce 77 MW. The additional steam from the solar facility is combined with the saturated steam generated in the HRSG of the gas turbine and superheated in the HRSG to enter the steam turbine at 92 bar and 560°C. An advantage of such an integrated plant is the reduction in the overall capital cost by sharing components between the two systems. Duct firing in the HRSG may also be included to compensate for times when the solar energy intensity is lower or not available but then the HRSG will have to be oversized to provide the additional steam. Integrating the steam generated by a solar plant with that of a coal or biomass fired boiler plant may also be accomplished in a similar manner.

A linear Fresnel demonstration plant with the capability of generating 1.4 MW electric power has been operating in Spain since 2009. Each receiver uses 16 parallel lines of mirrors with a total surface of 18,662 m^2. The steam turbine operates on 55 bar saturated steam (270°C). A heat storage system consisting of hot water and saturated steam is utilized for steam buffering. Since 2012, a utility scale plant using this technology has been operating also in Spain with the capability of generating 30 MW.

12.2.3 Economics

Concentrating solar power plants are capital intensive at the present time and incentives offered by local, state, and federal governments are important for the economic viability in many countries, although currently the cost of electricity is about half of that when produced by solar photovoltaic cells. The levelized cost of electricity depends on the capacity factor of a power plant and in the case of solar plants, typical capacity factors as noted previously without thermal storage may range from 20 to 30% whereas in the case of a fossil fueled power plant (coal combustion based), capacity factors are as high as 85% as noted previously. The capacity factor of a solar thermal plant may be increased depending on the extent of thermal storage added but will also increase the capital cost and a careful trade-off analysis is required.

Continued advances in technology are expected to reduce the specific plant cost ($/kW basis) and increase the capacity factor resulting in a reduction in the cost of electricity by more than two-thirds by 2020.

Research and development areas include identifying and developing reflector materials that are low cost while maintaining greater specular reflectance under the severe outdoor environments, receivers for the supercritical CO_2 cycle without using an intermediate heat transfer fluid, high temperature fluidized bed particle receivers to transfer and store thermal energy at higher temperature (650°C) than molten salt storage, and advanced coatings for the receiver to increase solar absorption and decrease emissivity.

Partially or completely replacing fossil fuel use in an "indirectly fired" Brayton cycle with solar energy is a promising technology being pursued by a number of developers. Novel heat exchangers as external gas turbine "combustors" are under development to heat the pressurized air to as high a temperature as 1000°C (Heller et al., 2006) before expansion. By including the capability of combusting a fuel such as natural gas, the plant output may be maintained during periods when the solar energy is insufficient. When both solar energy and fossil fuel are used with the solar energy providing the thermal energy to preheat the air before it enters the combustor, NO_x formation may be significantly increased due to the higher flame temperature however, and either post combustion control such as the use of a SCR or novel low NO_x combustor design will be required.

12.2.4 Environmental Issues

Solar energy has a relatively small environmental footprint but impacts do exist. As illustrated in Table 12.1 (Edenhofer et al., 2011), the relative greenhouse gas emissions on an LCA basis for solar energy is amongst the lowest but solar energy technologies may create other types of air, water, land, and ecosystem impacts depending on how they are managed. For example, environmentally sensitive lands should be avoided. Since this technology utilizes a heat engine to convert the thermal energy, a fraction of the heat has to be rejected to the surroundings as a consequence of the second law. There are different methods for heat rejection which are described in Chapter 8 and it can have environmental impacts depending on the manner in which heat is rejected. For example, if the heat is rejected to a large body of water such as a lake, ocean, or river, impact on the aquatic habitat should be taken into account. Ingesting and destroying the tiny aquatic fauna (in their larvae state) in the intake cooling water that may pass through the screens, and shocking the surrounding aquatic habitat when the water is returned are concerns as discussed in Chapter 7. In the case of using wet cooling towers, significant quantities of water, a valuable resource, are lost to the ambient air.

Example 12.2. The useful rate of energy $\dot{Q}_u$ absorbed by a receiver, that is, after its heat losses are deducted from the energy reflected to it by concentrating solar collectors is given by (Kalogirou, 2004)

$$\dot{Q}_u = F'[\eta_{collector} A_{collector}\, \dot{q}_{in} - \mathcal{U}_{loss} A_{reciever}(T_{reciever} - T_{ambient})] \qquad (12.3)$$

F' is the collector efficiency which is essentially a constant for a given collector design and fluid flow rate. $\eta_{collector}$ is the optical efficiency which defines the ratio of the energy absorbed by the receiver to the energy incident on the collector's aperture area (collector's useful area) and depends on the optical properties of the materials, the geometry of the collector including the incident angle of the solar irradiation, and the imperfections arising during fabrication of the collector. $\dot{q}_{in}$ is the direct solar irradiation per unit collector aperture area, $A_{collector}$ is the total collector aperture area, $\mathcal{U}_{loss}$ is the overall heat transfer coefficient for the energy loss from the receiver to the ambient, $A_{reciever}$ is the receiver area, $T_{reciever}$ is the receiver temperature, and T_{amb} is the ambient temperature. Derive an expression for the first law efficiency for generating electric power from direct solar irradiation ($\dot{q}_{in}$) using the closed cycle gas turbine of Example 9.3. Then calculate the first law efficiency using a diatomic gas as the working fluid operating under the same cycle conditions as in Example 9.3 and the following data for the solar plant:

$$F' = 0.9,\ \eta_{collector} = 0.8, \text{concentration ratio} = A_{collector}/A_{reciever} = 10,\ \dot{q}_{in} = 300\ (\text{W/M}^2),\ \mathcal{U}_{loss} = 2.7(\text{W/M}^2\,\text{K}), T_{reciever} = 1150\ \text{K},\ T_{amb} = 25\ \text{K}.$$

Solution

Equation 12.3 may be written as

$$\frac{\dot{Q}_u}{A_{collector}} = F'\left[\eta_{collector}\, \dot{q}_{in} - \mathcal{U}_{loss}\left(\frac{A_{reciever}}{A_{collector}}\right)(T_{reciever} - T_{amb})\right]$$

Then power developed by substituting Equation 9.10 into the above

$$\frac{\dot{W}}{A_{collector}} = F'\left[\eta_{collector}\, \dot{q}_{in} - \mathcal{U}_{loss}(T_{reciever} - T_{amb})\left(\frac{A_{reciever}}{A_{collector}}\right)\right]$$
$$\times \eta_{gen} \times \frac{T_{turb,in}\,\eta_{turb}\left[1 - (1/\pi_{turb})^{(\kappa-1)/\kappa}\right] - T_{comp,in}/\eta_{comp}\left[(\pi_{comp})^{(\kappa-1)/\kappa} - 1\right]}{T_{turb,in} - T_{comp,in}\left\{1 + (1/\eta_{comp})\left[(\pi_{comp})^{(\kappa-1)/\kappa} - 1\right]\right\}}$$

The first law efficiency is then given by dividing the above expression by $\dot{q}_{in}$

$$\eta_I = F'\left[\eta_{collector} - \mathcal{U}_{loss}(T_{reciever} - T_{amb})\left(\frac{A_{reciever}}{A_{collector}}\right)/\dot{q}_{in}\right]$$
$$\times \eta_{gen} \times \frac{T_{turb,in}\,\eta_{turb}\left[1 - (1/\pi_{turb})^{(\kappa-1)/\kappa}\right] - (T_{comp,in}/\eta_{comp})\left[(\pi_{comp})^{(\kappa-1)/\kappa} - 1\right]}{T_{turb,in} - T_{comp,in}\left\{1 + 1/\eta_{comp}\left[(\pi_{comp})^{(\kappa-1)/\kappa} - 1\right]\right\}}$$

Substituting the given numerical values and the cycle efficiency of 24.27% calculated in Example 9.3, $\eta_I = 9.2\%$.

12.3 GEOTHERMAL

Geothermal energy is stored in both rocks and as trapped steam or liquid water beneath the earth's surface and may be utilized directly for providing hot water, for building heating, or converting to electricity using a heat engine, depending on the resource temperature. Geothermal energy in 2008 supplied only about 0.1% of the global energy demand but is projected to meet roughly 3% of the electricity demand and 5% of the heating and cooling demand globally by 2050 (Edenhofer et al., 2011). This section is limited to the use of geothermal energy to operate a heat engine where the geothermal energy contained in liquid water or steam is utilized directly as the working fluid or indirectly to heat up the working fluid. Such power plants can be operated to supply base load power unlike wind and solar, and can also be dispatched to meet peak demand complementing wind and solar.

12.3.1 Geothermal Resources and Plant Siting

Geothermal resources are available in various forms, as natural hydrothermal systems which spontaneously provide hot water (brine) and/or steam (at <3 km depths), geopressured systems which are sediment filled reservoirs containing hot pore fluid under lithostatic pressure which is greater than the local hydrostatic head, hot dry rock either of low permeability or without naturally contained fluids (at 3–10 km depth), ultra low-grade systems available at shallow depths, and magma which is partially or completely molten rock. The earth's core which is in excess of 5000°C maintained by the heat liberated by the gradual radioactive decay of unstable isotopes of the elements Th, U, and K (and is not heat left over from the planet's formation as calculations have shown that the earth would have cooled down long time ago) supplies this energy. In this context, geothermal energy may be considered renewable, the conversion of a small quantity of mass m being required to give a very large quantity of energy E, governed by Einstein's famous equation, $E = mc^2$ where c is the speed of light having a very large magnitude. Heat transfer occurs continuously from the hot core to the surface by conduction and by convection currents in the molten mantle beneath the earth's crust, and tends to be strongest along tectonic plate boundaries. Volcanic activity transports hot material to near the earth's surface but only a small fraction of molten rock actually reaches the surface, most being left at depths of 5–20 km beneath the surface. Hydrological convection by groundwater can bring this energy closer to the surface. Hot springs (liquid water), geysers (mixtures of liquid water and steam), and fumaroles (steam) are all due to hydrological convection. The groundwater is replenished by rainfall but at a finite rate and thus the geothermal resource when recovered from natural hydrothermal systems should be exploited at such a rate that a steady state is maintained between replenishment and recovery.

Thermal energy accessible from hot dry rocks is estimated to range from 34×10^6 EJ for depths down to 3 km depth, 56 to 140×10^6 EJ for depths down to 5 km depth, and 110 to 403×10^6 EJ for depths down to 10 km; estimates for hydrothermal resources for electric power generation range from 118 to 146 EJ/year (at 3 km depth)

TABLE 12.5 Worldwide geothermal energy potential at various depths for electric power generation (data from Edenhofer et al., 2011)

	3 km		5 km		10 km	
Region	EJ/year (lower)	EJ/year (upper)	EJ/year (lower)	EJ/year (upper)	EJ/year (lower)	EJ/year (upper)
OECD North America	25.6	31.8	38	91.9	69.3	241.9
Latin America	15.5	19.3	23	55.7	42	146.5
OECD Europe	6	7.5	8.9	21.6	16.3	56.8
Africa	16.8	20.8	24.8	60	45.3	158
Transition Economies	19.5	24.3	29	70	52.8	184.4
Middle East	3.7	4.6	5.5	13.4	10.1	35.2
Developing Asia	22.9	28.5	34.2	82.4	62.1	216.9
OECD Pacific	7.3	9.1	10.8	26.2	19.7	68.9
Total	117.5	145.9	174.3	421	317.5	1108.6

to 318 to 1109 EJ/year (at 10 km depth); and for direct uses (as heat) range from 10 to 312 EJ/year (Edenhofer et al., 2011). Table 12.5 summarizes the lower and upper estimates for global geothermal energy on a worldwide basis which is quite large. Most of the active geothermal resources are found along major plate boundaries and are concentrated along the Pacific Ocean bounded by Japan; the Philippines; the Aleutian Islands; and North, Central, and South America, known as the "Ring of Fire."

12.3.2 Key Equipment

Geothermal energy is currently extracted using wells that are drilled to depths of up to 5 km using drilling methods similar to those used in the oil and gas industry. Electricity generation from hydrothermal reservoirs with high permeability has been practiced for a century now and entails recovering the thermal energy from the naturally occurring hot fluids brought to the surface. Enhanced or engineered geothermal systems are under development consisting of pumping an artificial fluid (water while CO_2 is also being considered) in a closed loop to recover the geothermal energy through the high temperature subsurface regions fractured to create a network of fluid pathways connecting the production well (providing the hot fluid) and the injection well (returning the fluid after heat recovery). When the hot fluid extracted from a natural reservoir is in the form of steam ("dry steam plant" typically at 4–8 MPa and 180–225°C), it is supplied to a condensing steam turbine to produce the power while the discharge after condensation in a condenser is pumped back into ground. These plants installed in the intermediate and high temperature resources have capacities often between 20 and 110 MW. When the hot fluid is in the form of hot water or is a mixture of hot water and steam, either a single flash or double flashes are utilized to produce the optimum combination of steam pressure and flow rate. Flashing consists of introducing the fluid mixture into a vessel located at the surface where the fluid pressure is reduced adiabatically such that a fraction of the liquid water changes phase to steam quite rapidly (water "flashes"). The steam is separated from water in the same vessel (KO drum) and the steam utilized in a steam turbine. The specific plant cost ($/kW) of such plants is higher than the dry steam plants since additional equipment associated with flashing the brine is required as well as due to the lower temperature of the resource. The condensed steam along with the water leaving the flash vessel is pumped back into ground. In a double flash system, the second flash occurs at a lower pressure than the first flash with the liquid leaving the higher pressure flash vessel entering the second flash vessel to generate additional steam but at a lower pressure. This lower pressure steam is also introduced into the steam turbine but at the appropriate lower pressure stage. Depending on the resource temperature, a double flash system can increase the power output by as much as 20–25% for an approximate 5% increase in plant cost. Another option is to use a combination of two Rankine cycles in series, the first using steam as the working fluid and the second using an organic fluid as the working fluid. The steam turbine used in such a cycle has an exhaust pressure that is much higher than atmospheric (instead of the typical vacuum) such that its discharge condenses at a temperature high enough to be economically suitable in providing the thermal energy to the bottoming organic Rankine cycle (ORC) fluid. The ultimate heat rejection from this combined cycle occurs from the organic vapor condenser downstream of the organic vapor turbine. Figure 12.3 illustrates such a combined cycle.

Steam as a working fluid in Rankine cycles is a preferred working fluid due to safety considerations (nonflammable, nontoxic) and thermal stability at the higher temperatures. However, from a turbine design (e.g., number of stages required) and isentropic efficiency standpoint, it has the disadvantage of having a relatively low molecular weight (unlike the ORC fluids) which becomes an issue for small turbine sizes. Also due to its relatively low vapor pressure, vacuum conditions need to be maintained in the condenser at normal heat rejection temperatures (set by the cooling water or the ambient air depending upon the mode of heat rejection used).

Careful selection of materials of construction for the equipment and piping are required since geothermal fluids contain oxygen (generally from aeration), hydrogen ions (which effects the pH), chloride ions, sulfide species, CO_2, carbonate and bicarbonate ions, NH_3 and ammonium species, and sulfate ions. Scale formation is another factor that should be taken into account since geothermal fluids frequently contain $CaCO_3$ and Ca silicate (Ca_2SiO_4). Metal (Zn, Fe, Pb, Mg, Sb, and Cd) silicate and sulfide scales are often observed in higher temperature resources, as well as amorphous silica scales that are not associated with other cations. Since temperature affects the solubility, the brine return temperature to the reservoir and hence the amount of thermal energy recovered can be limited. Scale control additives are often used.

In addition to these contaminants, steam turbines need to withstand the geothermal steam conditions consisting of saturated or slightly superheated low pressure steam. Low pressure turbine stages may require a special (stellite, a Co-Cr alloy) coating to limit the erosion by water droplets since the steam entering the turbine is saturated or has little superheat. Special materials of construction are required for the turbine blades as well as the shaft due to the presence of the corrosive contaminants.

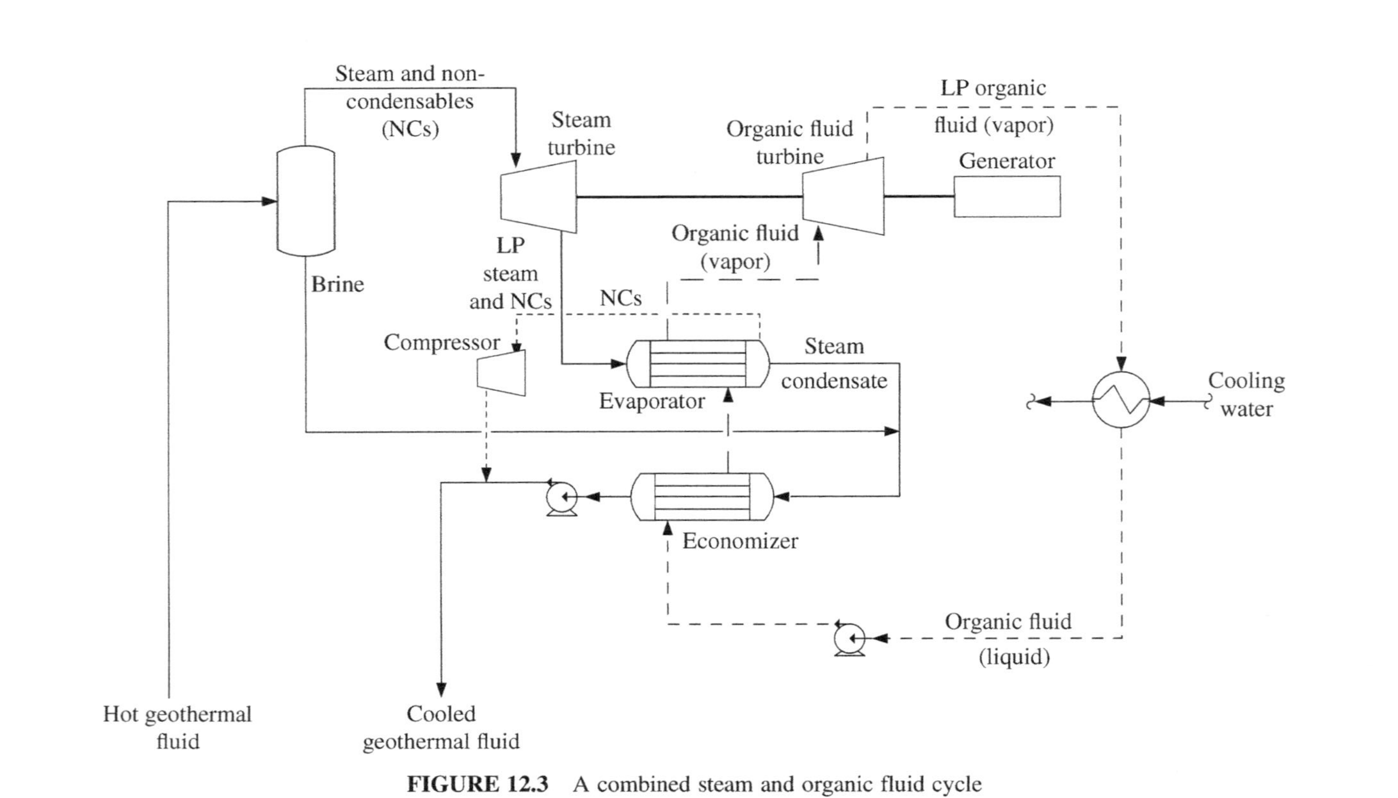

FIGURE 12.3 A combined steam and organic fluid cycle

Steam provided to the turbine by the geothermal well may be varying with time and this variable flow should be accounted for in the design of the controls such that the efficiency loss is minimized, for example, at lower steam flow rates, partial arc admission to the turbine might be used.

When the hot geothermal fluid is in the form of hot brine (with no accompanying steam) in the 100–150°C range, a "binary cycle" may be employed. Heat from geothermal fluid is transferred to a working fluid with a vapor pressure much higher than that of water through a heat exchanger to operate an ORC consisting of a single component or a mixture of two components; for example, the Heber geothermal plant built in 1985 in California used a mixture of i-C_4H_{10} and i-C_5H_{12} (Sones and Krieger, 2000). The Kalina cycle (Kalina and Leibowitz, 1989) which uses a mixture of NH_3 and H_2O as the working fluid has also been employed for such low temperature applications. Since mixtures vaporize and condense at variable temperatures, it is possible to more closely match the variable heat source temperature (as the brine is cooled) and the variable heat sink temperature (as the cooling water is heated) to reduce the irreversibilities. Efficiencies of binary plants have ranged from 7 to 12% and systems as small as 100 kW to as large as 40 MW are common.

Kalina Cycle. Figure 12.4 illustrates the Kalina cycle at the geothermal power plant in Húsavík, Iceland (Mlcak et al., 2002). The NH_3–H_2O stream leaving the water

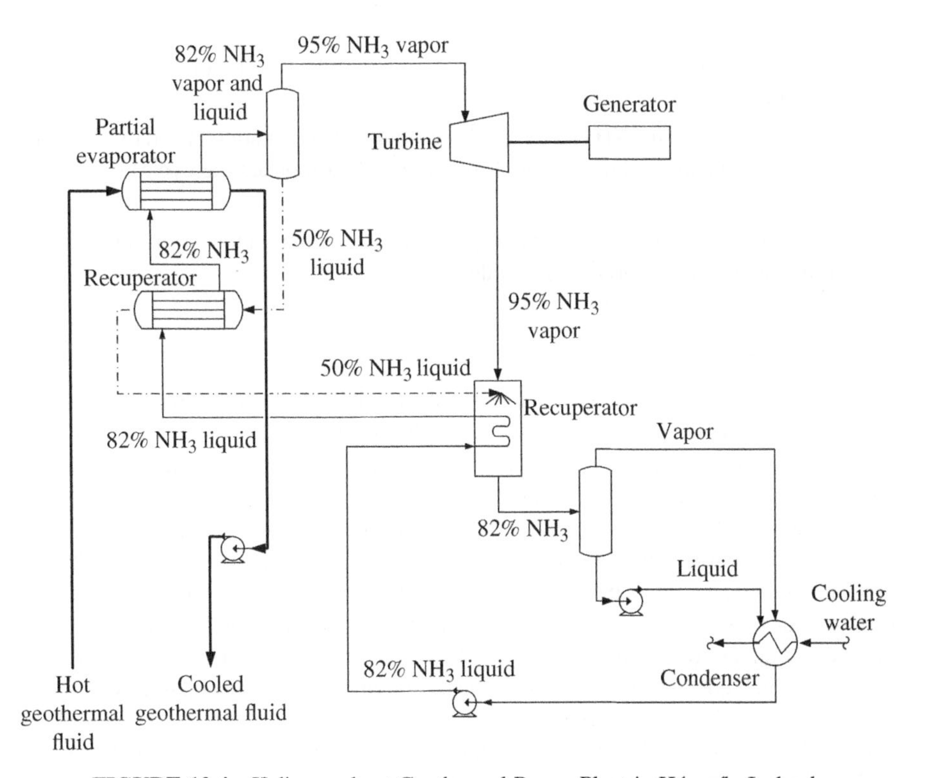

FIGURE 12.4 Kalina cycle at Geothermal Power Plant in Húsavík, Iceland

cooled condenser consisting of 82% by weight NH_3 and at an initial pressure of 5.4 bar and temperature of 12.4°C is pumped through a low temperature recuperator, then into a high temperature recuperator to be preheated to a final temperature of 68°C. It then enters the evaporator where it is heated by the brine to 121°C and 75% of the NH_3–H_2O stream is vaporized. This two-phase fluid then enters a vapor–liquid separator and the NH_3-rich H_2O vapor fraction enters the turbine while the liquid fraction which is 50% by weight NH_3 after providing heat for preheating the 82% NH_3 stream in the high temperature recuperator enters the low temperature recuperator where it combines with the turbine exhaust. Liquid exiting the low temperature recuperator is collected and pumped into the water cooled condenser as a spray contacting the vapor stream. Absorption of the NH_3-rich vapor into the NH_3 lean liquid occurs in the condenser while the heat of solution is rejected to the cooling water.

Thus, the basic idea behind this type of absorption cycle is:

1. Using a high vapor pressure (low boiling point) working fluid such as NH_3 which can be generated using a low temperature heat source.
2. Instead of condensing the vapor after expansion in a turbine to pump the condensate to the higher pressure before generating the vapor at the higher pressure (as is done in a Rankine cycle) which would require a heat sink also at low temperature (lower than the ambient), an absorbent is used as follows.
3. The absorbent which in this case is liquid H_2O is contacted with NH_3 to absorb it, followed by pumping of the liquid (aqueous NH_3 solution) to the higher pressure.
4. The higher pressure NH_3 vapor working fluid is then generated from this higher pressure aqueous solution (step 1 above which completes the cycle).

NH_3 is a toxic substance and leakage, especially of the vapor from the turbine (around the shaft seals), should be avoided.

12.3.3 Economics

Like the other renewable energy plants, geothermal power plants typically require high initial investment costs while the operating costs are low. In addition to the costs of purchasing and installing the power plant equipment, geothermal plants have costs associated with exploration and resource confirmation which could be as much as 10–15% of the total cost, and drilling of the wells which could be as much as 20–35% of the total cost (Edenhofer et al., 2011). The resulting cost of electricity is very site-specific being dependent on the quality and quantity of the geothermal resource but power plants using hydrothermal resources are often competitive with the more conventional power plants. Plants employing engineered geothermal resources which are currently in the demonstration phase are estimated to cost higher than hydrothermal plants, however.

Capacity factor which also affects the cost of electricity had averaged on a worldwide basis in 2008 at 74.5% for existing geothermal power plants but newer installations have had capacity factors much higher, greater than 90% (Edenhofer et al., 2011). Stimulation of geothermal reservoirs, use of improved materials of construction, and scaling and corrosion inhibition can all increase the capacity factor and reduce the cost of electricity. Unlike wind and solar, geothermal plants with their higher capacity factors can provide base-load electricity and thus the integration of new geothermal capacity into existing power systems can be done without major challenges.

Other areas of research and development to reduce the cost include advanced geophysical surveys to make exploration efforts more effective, improved drilling and well construction technology to increase the rate of penetration when drilling through hard rock, and use of CO_2 as the heat transfer fluid (which could also provide a means for CO_2 sequestration by mineralization of a fraction of CO_2 pumped into the reservoir) and possibly as the power cycle working fluid itself in engineered geothermal systems.

12.3.4 Environmental Issues

Environmental issues concerning the use of geothermal energy are very site- and technology-specific. For example, there can be CO_2 emissions directly associated with the operation of a geothermal facility since the geothermal fluid may contain CO_2. Another issue is land use which may lead to vegetation loss and soil erosion which in turn may cause landslides. Emissions of H_2S, NH_3, and possibly Hg could occur and the plant should be engineered to capture these pollutants. Seismicity has been an issue in some sites leading to tremors. Like solar thermal power plants, water usage can be an issue when wet cooling towers are utilized. Plant wastes generated by any facility should be properly disposed but in the case of geothermal plants, the proper disposal of the collected wastes may require transportation over wide distances.

Example 12.3. What is the maximum amount of electric power that could be generated from a 250°C geothermal fluid flowing at 40 kg/s and returned to the ground at 150°C? Cooling water is supplied at 15°C and returned to the cooling tower at a 10°C higher temperature. Assume the specific heat of the geothermal fluid to be that of pure water (4.184 kJ/kg K). List the inefficiencies in the plant that limit from realizing the maximum power calculated above.

Solution

The maximum power output occurs at the best efficiency which is that of an ideal Carnot cycle as defined by Equation 8.3 for the case where the heat source and sink temperatures do not remain constant.

Absolute temperature of geothermal fluid supplied $= 250 + 273 = 523$ K

Absolute temperature of geothermal fluid returned $= 150 + 273 = 423$ K

Log-mean temperature of heat source $(\Delta T_H)_{ln} = (523 - 423)/[\ln(523/423)] = 471$ K

Absolute temperature of cooling water supplied $= 15 + 273 = 288$ K

Absolute temperature of cooling water returned $= 15 + 10 + 273 = 298$ K

Log-mean temperature of heat sink $(\Delta T_C)_{ln} = (298 - 288)/[\ln(298/288)] = 293$ K
Carnot cycle efficiency $= 1 - (\Delta T_C)_{ln}/(\Delta T_H)_{ln} = 1 - 293/471 = 0.378$
Maximum power output $= 0.378 \times 40\ \text{kg/s} \times 4.184\ \text{kJ/kg K} \times (523-423)\ \text{K} \times 10^{-3}\ \text{MW/(kJ/s)} = 6.33\ \text{MW}$

Inefficiencies in plant that limit from realizing the maximum power calculated above are due to:

- ΔT between the heat source or sink fluids (geothermal fluid or cooling water) and the working fluid in heat exchangers
- Isentropic and mechanical inefficiencies of heat engine/generator
- Isentropic and mechanical inefficiencies of pumps
- Other auxiliary power consumptions and losses such as the fan when a mechanical draft cooling tower is used, transformer losses.

12.4 NUCLEAR

Electric power from nuclear energy involving nuclear fission reactions has been commercially practiced for several years. Fission reactions occur when the nucleus of certain isotopes of elements such as uranium (U), plutonium (Pu), and thorium (Th) captures a neutron. U-233 (where 233 represents the sum of neutrons and protons in the nucleus), U-235, Pu-239, and Pu-241 all undergo fission while U-238, Th-232 require fast (much higher kinetic energy) neutrons. The captured neutron upsets the internal force balance between neutrons and protons resulting in the nucleus splitting into two lighter nuclei. Two or three neutrons are also emitted, while a large quantity of energy released per Einstein's equation, $E = mc^2$ which was referred to in the previous geothermal section. The excess neutrons generated in the nuclear reaction sustain a chain reaction where additional nuclei are split making it possible to design self-sustaining nuclear reactors. A nuclear reactor is said to achieve criticality when each fission causes an average of one additional fission leading to a constant level of fission. Control rods that can absorb neutrons are inserted into a nuclear reactor to regulate such nuclear fission chain reactions. In nuclear weapons on the other hand, the number of fission reactions increases (exponentially), and this condition is called supercriticality. The heat generated in a nuclear fission reactor is transferred to a working fluid to operate a steam Rankine cycle. There are more than 400 nuclear reactors operating in more than 30 countries, and produce close to 15% of world's electricity requirement. Many new nuclear power plants are on order or being planned, mostly for Asia.

Another type of nuclear reaction is fusion which also releases large quantity of energy according to the above referred Einstein's equation when two nuclei with combined masses lower than Fe fuse. The two isotopes of hydrogen, deuterium (H-2), and tritium (H-3) undergo fusion to form He while releasing a neutron. H-2 may be extracted from water which makes it almost an inexhaustible fuel supply when considering the large quantity of energy released by conversion of a very small amount of mass. H-3 has to be produced from Li inside a nuclear reactor but vast known reserves of Li exist in the earth's crust and could potentially supply the fuel required for more than a thousand years even if all of world's electricity demand were met by fusion reactors. Nuclear fusion however is still in the research phase, a major challenge being the containment of the plasma at extremely high temperatures required for nuclear fusion to occur. The potential for generating large quantities of energy with a number of advantages over fission exist such as no long-lived radioactive materials being produced, and no runaway reactions can occur as will be discussed for fission reactions.

12.4.1 Nuclear Fuel Resources and Plant Siting

World's proven nuclear fuel resources that can be economically recovered at the current market price (reasonable assured resources or RAR) are shown in Figure 12.5 (Uranium 2011: Resources, Production, and Demand, 2012) and represent enough reserves to supply the needs for another 80 years of demand. Further exploration and market higher prices should open up further resources.

The first step in the process of producing fuel rods from uranium is mining the ore which contains both U-235 and U-238 as oxides. The ore is milled and then leached

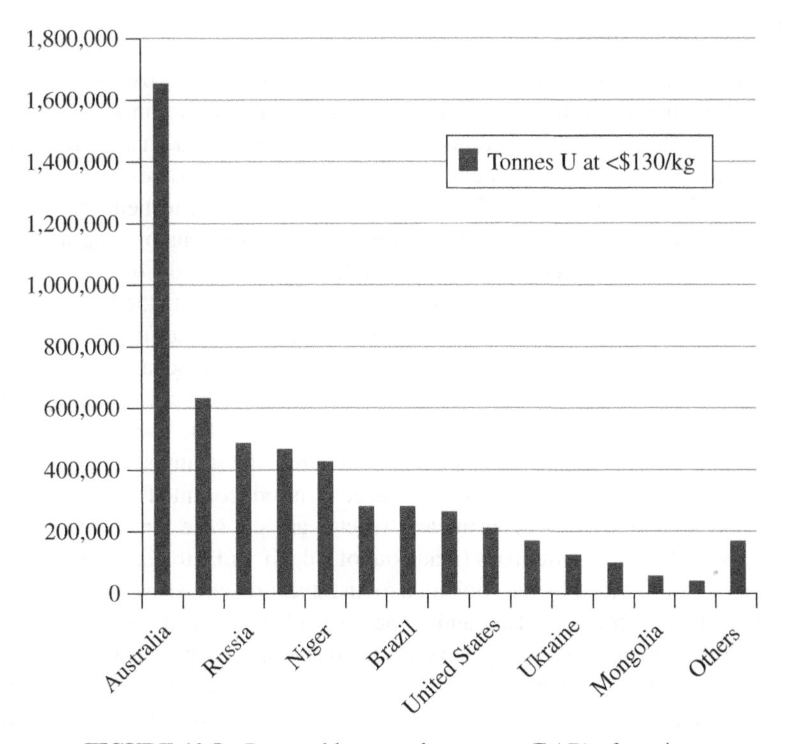

FIGURE 12.5 Reasonable assured resources (RAR) of uranium

with H_2SO_4 or sometimes a strong alkaline solution to concentrate these oxides from about 0.1% by weight concentration to about 80% concentration. The resulting concentrate is shipped for further processing as dried "yellow cake." About 200 tonnes are required for a1000 MW nuclear power plant operating for 1 year. Tailings, the remainder of the ore containing low concentrations of long life radioactive materials and also heavy metals need to be isolated from the environment and are placed in facilities near the mine.

It is the U-235 that is the fissile isotope of uranium in conventional nuclear reactors and an enrichment process is utilized to increase the U-235 concentration from about 0.7% to about 3.5–5% by first converting the uranium oxide to a fluoride by reaction with HF and centrifuging the gas using density difference due to the difference in molecular weights of the two isotopes. A small number of reactors such as the Canadian and Indian heavy water (deuterium oxide) moderated and cooled reactors do not require enriched uranium.

Reactor fuel in the form of ceramic pellets are formed by sintering pressed uranium dioxide at a temperature greater than 1400°C and then encasing in zirconium alloy tubes to form fuel rods. This zirconium alloy can withstand high temperatures, nuclear radiation, and corrosion. About a third of the rods are replaced every year to year and a half in order to maintain efficient reactor performance. Some of the U-238 is turned into Pu which in turn (approximately half of it) also undergoes fission.

12.4.2 Key Equipment

The two main types of commercial nuclear reactors are the pressurized water reactor (PWR) and the boiling water reactor (BWR) based on the manner in which the heat liberated within the reactor core is transferred to the working fluid. In a PWR, the heat is first transferred into heavy water circulating at high pressure to avoid boiling (~140 bar) through the primary circuit. The heavy water in addition to being the reactor coolant functions also as a moderator to slow down the neutrons making them into "thermal neutron," which can sustain the nuclear chain reaction of U-235. Next, the heat from the heavy water is transferred into a heat exchanger to the Rankine cycle working fluid to generate steam (~70 bar) for the Rankine cycle. In the BWR, the steam is generated in the reactor itself without having the two separate loops. High purity (light) water moves upward through core absorbing heat and changing phase and the steam–water mixture leaves the top of the core and enters two stages of moisture separation where water droplets are removed before the steam is allowed to enter the steam line. The saturated steam thus produced is introduced into the steam turbine.

The main components of such nuclear reactor are the core that holds the fuel assembly as well as the control rods (made out of Cd, Hf, or B) to control the nuclear chain reaction, surrounded by the coolant, a robust steel pressure vessel containing the reactor core and moderator/coolant, and in the case of PWR, a steam generator as part of the cooling system. PWRs may have up to four loops, each with its own steam generator. The reactor and the steam generators are housed within a meter thick reinforced concrete containment structure built to contain radioactive materials from escaping to the outside in case of any serious malfunction inside the reactor, as well as protect the reactor system from any outside intrusion. The reactor vessel may be surrounded by a bio shield to protect the workers from radiation consisting of more than a meter thick leaded concrete wall with steel lining on both the inside and outside.

An advantage of the PWR is that it has a strong negative void coefficient, that is, the reactor cools down if the heavy water starts boiling (the steam bubbles causing the voidage), water being the moderator. Another advantage is that it has the inner loop (consisting of the heavy water) separated from the Rankine cycle working fluid which keeps any radioactive materials from reaching the steam turbine making maintenance easier. A disadvantage is that the coolant which is at high pressure can leak out rapidly in case of a pipe rupture requiring backup cooling systems. The advantages of a BWR include simpler piping which reduces cost, and power levels that can be changed more easily. The change in power level is accomplished by changing the supply of the water which changes the voidage caused by the steam bubbles and thus the moderation.

Nuclear power plants with saturated steam entering the steam turbine can have as much as 15% moisture by weight in the exhaust from the high pressure section of the turbine. Since excessive moisture content in the steam can lead to corrosion and erosion in the low pressure section of the turbine, the low pressure wet steam leaving the high pressure section of the turbine is passed through a moisture separator (e.g., employing vanes that the moisture droplets impinge upon and separate out from the steam) and preheated against a fraction of the high pressure steam in a heat exchanger(s) before feeding it to the low pressure section of the turbine.

12.4.3 Economics

The nuclear fuel cost is relatively low with large fuel reserves on a worldwide basis. Capital costs are high, however, while plant construction periods are long largely due to regulatory delays, and plant decommissioning are also high. Despite these factors, electricity generated from nuclear plants tends to be one of the cheapest options but without accounting for the cost of permanent long-term storage of the high level radioactive waste generated (discussed later in this section under "Environmental issues") and any catastrophic accidents especially at the power plant although rare. Current designs are for large-scale base loaded power generation and do not complement the intermittent renewables.

Advanced reactor designs are being developed such as the supercritical water reactor which has the potential for as much as a one-third increase in efficiency over today's BWRs. Increase in power plant efficiency typically reduces its capital cost on a $/kW basis as long as equipment fabrication and materials of construction are not too exotic. An international effort lead by Japan aims to resolve the most pressing materials and system design issues. The high temperature gas cooled reactor using He as the coolant as well as the working fluid in an intercooled, recuperative Brayton cycle is another advanced design being pursued. It has the major advantage of being inherently safer having a negative reactivity temperature coefficient (i.e., fission reaction slows as temperature increases). The fuel is fabricated into spherical shapes about the size of tennis balls made out of pyrolytic graphite (as the moderator) and containing thousands of micro-fuel particles of a fissile material such as U-235, and coated with a layer of SiC to provide structural integrity and containment of the fission product. Advantages include high plant efficiency, and from a safety standpoint, the reactor has a large thermal inertia due to the graphite moderator, and there is no phase change with its accompanying large increase in pressure possible since the coolant remains a gas. He has no neutron absorption cross section (a measure of the likelihood of interaction between incident neutron and target nucleus) and remains inert with no reactivity effects. Another advantage of using He is that the gas turbine–specific power (kW/kg/s) at low pressure ratios is very high since He has a high C_p to C_v ratio (see Example 9.3) compared to diatomic (N_2, O_2) or triatomic (CO_2) gases.

The breeder reactor generates more fissile material than it consumes. By breeding fissile material (as fuel for the conventional fission reactors) from fertile materials such as U-238 and Th-232, it holds the promise of increasing the nuclear fuel supply options. The reactor operates at very high temperatures and a suitable coolant is liquid Na metal which has a low vapor pressure to limit the operating pressure of the reactor.

12.4.4 Environmental Issues

Today's reactors are designed to withstand large aircraft impacts by locating the reactor core in particular below grade and protecting by more than 3 m reinforced concrete containment. The reactor building consists of reinforced concrete walls that average almost 2 m. The control rooms are also located below grade while the reactor emergency cooling system includes gravity driven cooling rather than depending on motor or engine driven pumps which could lose their supply of electricity or fuel during an emergency.

Nuclear energy, as illustrated in Table 12.1 (Edenhofer et al., 2011), has low greenhouse gas emissions compared to fossil fuel power plants without CO_2 capture and sequestration on an LCA basis. No criteria air pollutants are emitted during operation and the next generation reactors with their passive safety features hold the promise of higher safety, the release of radioactive materials to the environment having been a major issue. Disposal and management of the radioactive spent fuel is a major challenge however. The waste generated is categorized as low level, intermediate level, and high level. The low level waste includes tools, protective clothing, and some plant hardware and disposal procedure consists of placing them in large metal containers (similar to shipping containers), filling with cement, and placing in concrete-lined vaults. The intermediate level waste, which includes reactor fuel cladding and contaminated materials from reactor decommissioning, is immobilized in cement-based materials within stainless steel drums or boxes. The high level waste which consists of the spent reactor fuel and waste materials remaining after the spent fuel is reprocessed can remain harmful for thousands of years. The spent fuel when removed from the reactor emits nuclear radiation and heat, and is transferred into a cooling water circulating storage pond adjacent to the reactor and stored to allow cooling and decrease the radiation levels which may be followed by onsite dry storage. Ultimately, the spent fuel must either be reprocessed or prepared for permanent disposal. At the present time, there are no permanent disposal facilities in operation but deep geological repositories may be possible sites after immobilizing the radioactive material so that it is not leached out by groundwater. Vitrification to form a glassy material is one approach for immobilizing the radioactive material while another approach consists of forming a ceramic. The French government is testing underground storage (about 500 m below surface) in a sparsely populated town in France. Reprocessing of the spent fuel can reduce its amount for ultimate long-term storage while recovering valuable U and Pu. U and Pu may be recovered from the fuel rods after breaking them apart and dissolving in an acid. The U can be recycled to enrich the U-235 content while the Pu along with U may be used in mixed oxide fuel reactors where the Pu substitutes for U-235.

Example 12.4. Calculate the energy released when 1 g of matter is converted using the relationship $E = mc^2$ and then compare it against the energy released by combusting 1 g of a bituminous coal using the LHV provided in Table 1.4.

Solution

Speed of light $c = 299{,}792{,}458\ \mathrm{m/s}$

Mass converted $m = 0.001\ \mathrm{kg}$

Energy released $E = 0.001\ \mathrm{kg} \times (299{,}792{,}458\ \mathrm{m/s})^2 = 89.88 \times 10^{12}\ \mathrm{J}$ or $89.88\ \mathrm{TJ}$

LHV of the bituminous coal =26,172 kJ/kg =26,172 J/g

Ratio of energy released by converting matter into energy to that by combustion of the coal $= 89.88 \times 10^{12}\ \mathrm{J}/26{,}172\ \mathrm{J}$ which is greater than three billion times!

12.5 ELECTRIC GRID STABILITY AND DEPENDENCE ON FOSSIL FUELS

Electric power generated by solar and wind energy is variable and tends to be unpredictable especially in the case of wind, and maintaining frequency and voltage tolerances of the AC through the grid system becomes an issue when the supply and demand do not match up. One option is to maintain additional "spinning reserve" to make up the shortcomings of the intermittent renewables. Spinning reserve is the unused but available capacity of power plant units that are already supplying power to the grid, and their output can be ramped up rapidly. Such power plants with the excess capacity tend to be fossil fueled. Peakers consisting of simple cycle gas turbines operating on distillate oil or natural gas as well as specially designed gas turbine–based combined cycles operating on natural gas with fast start capability will also be helpful although they may take longer time to make up the difference than the spinning reserves.

In the case of such combined cycles, special design features are incorporated to keep them ready for the required quick start-up. Pre-purging the system and then isolation of the HRSG is utilized while the stack is equipped with an automatic damper that closes upon plant shutdown to reduce heat loss which reduces thermal stress of the start-up operation. HRSG with specially designed steam drums and overhung superheater and reheater coils with one upper header spring supported are also used to reduce the thermal stresses.

In the case of coproduction plants as described in Chapter 10, the split of syngas going to the power block as fuel versus that going to the coproduct synthesis unit can be changed quickly to ramp up or down the power supplied to the grid but the synthesis unit should be designed such that its stability and coproduct purity are not compromised as its capacity is quickly ramped down or up.

Solar and wind power can be complementary to each other in some geographic areas by compensating for each other. On a daily to weekly basis, high pressure areas tend to bring low surface winds but also clearer skies which are conducive to solar power. On the other hand, low pressure areas tend to be windier which are conducive to wind power but are cloudier. Next, on a seasonal basis, solar power peaks during the summer months while in many locations wind tends to be lower, and the reverse tends to occur during the winter months. Utility companies should consider this synergy between the two renewables while planning such power plants.

Spinning reserve may not always be fast enough to provide the power shortfall and energy storage would also be required where the renewable energy penetration is high or for isolated and remote grids. Energy storage plants can take excess power from the grid, say during nonpeak hours, and provide it back to the grid when there is the power shortfall. A number of options are currently available for large-scale energy storage, such as pumped hydro and compressed air energy storage.

A pumped hydro system contains two water reservoirs at different elevations. Energy is stored as potential energy in water by pumping the water from the lower reservoir into the higher reservoir while drawing power from the grid. During power demand, the water from the higher reservoir flows by gravity through a hydraulic turbine into the lower reservoir to supply power for the grid. This is a well-proven technology for large-scale energy storage.

Compressed air energy storage uses a motor driven compressor to compress air to a high pressure and store it in a large underground space such as a salt cavern or aquifer. During power demand, the compressed air flows through a turbine to supply power to the grid. The turbine may be fired with a fuel such as distillate oil or natural gas and a recuperator may be installed downstream of the turbine to preheat the air before it enters the turbine or the combustor to increase the efficiency, depending on whether the turbine is fired or not. Such plants have been built and operated successfully, while using reheat combustors.

H_2 storage, batteries, and electrochemical capacitors ("supercapacitors") are other options being investigated for the large-scale storage applications economically. The H_2 may be produced by electrolysis of water while electricity may be generated using the stored H_2 in a fuel cell, while SOFCs that can also function as electrolyzers are under development to reduce the cost of having separate units. Two of the methods for storing and supplying H_2 consist of using H_2 storage alloys and liquid H_2 but have their challenges; current H_2 storage alloys have low storage density while liquid H_2 requires refrigeration. Other methods of H_2 storage being investigated include synthesizing NH_3 and organic hydrides but the drawback with NH_3 is its toxicity while the drawback with the current organic hydrides is the requirement for large amount of heat for the dehydrogenation reaction to release the H_2.

12.5.1 Super and Micro Grids

Another option is to construct wide area high voltage “super grids” with DC lines, DC transmission having the advantage of lower cost and losses than conventional AC transmission. This would allow connecting renewable energy resources to loads that may be located far distances away. Superconducting transmission lines are being investigated for such applications which can transmit over several hundred kilometers at nearly no line loss.

On the other hand, microgrids are gaining significant interest to complement small-scale distributed electric power producers. Examples of such distributed power producers are fuel cells, gas turbines (or microturbines), and reciprocating engines operated on natural gas or gas derived from biomass, steam turbines in biomass fired boilers, as well as wind turbines and solar photovoltaic cells as well as heat engines driven by solar thermal energy. Since natural gas–based fuel cells and renewable energy–based producers have a better carbon footprint than conventional fossil fuel fired power plants (i.e., without CCS), monetary incentives are being provided by regulators to encourage their deployment and negate the economic penalties of smaller-scale plants. Many of the CHP plants also tend to be small scale while their contribution to the electric power grid is increasing. A prudent approach to realizing this emerging potential of distributed generation consisting of small-scale generators is to consider the generators and loads together as subsystems to form a microgrid allowing for local control of the generators which will be useful during disturbances. Batteries which are amenable to small-scale storage at the current time connected to the microgrid can smooth out any mismatch between the generators and the loads.

REFERENCES

Betz A. Das Maximum der theoretisch möglichen Ausnützung des Windes durch Windmotoren. Z Gesamte Turbinenwesen 1920;26:307–309.

Conti J, Holtberg P. International Energy Outlook 2013. Washington, DC: Energy Information Administration (EIA); 2013.

Edenhofer O, Madruga RP, Sokona Y, Seyboth K, Matschoss P, Kadner S, Zwickel T, Eickemeier P, Hansen G, Schlömer S, von Stechow C, (eds.) *IPCC Special Report on Renewable Energy Sources and Climate Change Mitigation*. Cambridge: Cambridge University Press; 2011.

Heller P, Pfänder M, Denk T, Tellez F, Valverde A, Fernandez J, Ring A. Test and evaluation of a solar powered gas turbine system. Solar Energy 2006;80(10):1225–1230.

Kalina AI, Leibowitz HM. Application of the Kalina cycle technology to geothermal power generation. Geothermal Resources Council Transactions 1989;13:605–611.

Kalogirou SA. Solar thermal collectors and applications. Progress in Energy and Combustion Science 2004;30(3):231–295.

Lu X, McElroy MB, Kiviluoma J. Global potential for wind-generated electricity. Proceedings of the National Academy of Sciences 2009;106(27):10933–10938.

Mlcak H, Mirolli M, Hjartarson H, Húsavíkur O, Ralph M. Notes from the North: a report on the debut year of the 2 MW Kalina Cycle® geothermal power plant in Húsavík, Iceland. Transactions of the Geothermal Resources Council 2002;26:715–718.

National Renewable Energy Laboratory, U.S. Department of Energy. 2014. Available at http://www.nrel.gov/wind/facilities_site_tour.html. Accessed May 28, 2014.

Rogner HH, Barthel F, Cabrera M, Faaij A, Giroux M, Hall D, Kagramanian V, Kononov S, Lefevre T, Moreira R, Nötstaller R, Odell P, Taylor M. Energy resources, world energy assessment, energy and the challenge of sustainability. World Energy Council Report. New York: United Nations Development Programme, United Nations Department of Economic and Social Affairs; 2000.

Sones R, Krieger Z. Case history of the binary power plant development at the Heber, California geothermal resource. Proceedings of the World Geothermal Congress; May 28–June 10, 2000; Kyushu-Tohoku, Japan. Bochum, Germany: International Geothermal Centre, Bochum University of Applied Sciences. p 2217–2219.

U.S. Energy Information Administration, Office of Energy Analysis, U.S. Department of Energy. 2013. September 2013 monthly energy review. Available at http://www.eia.gov/totalenergy/data/monthly/#environment. Accessed October 10, 2013.

Uranium 2011: Resources, Production and Demand. A Joint Report by the OECD Nuclear Energy Agency and the International Atomic Energy Agency; 2012.

Walker G. *Stirling Engines*. Oxford: Clarendon Press; 1980.

Appendix

ACRONYMS AND ABBREVIATIONS, SYMBOLS AND UNITS

ACRONYMS AND ABBREVIATIONS

AACE	Association for the Advancement of Cost Engineering
AC	Alternating current
AGR	Acid gas removal
aMDEA	Activated methyl diethanol amine
API	American Petroleum Institute
ASME	American Society of Mechanical Engineers
ASU	Air separation unit
ATR	Autothermal reformer
BACT	Best available control technology
BFD	Block flow diagram
BGL	British Gas Lurgi
CCS	Carbon capture and sequestration
CF	Capacity factor
CFLs	Compact fluorescent lamps
CHP	Combined heat and power
COP	Coefficient of performance
DC	Direct current
EAF	Equivalent availability factor
EOR	End of run

Sustainable Energy Conversion for Electricity and Coproducts: Principles, Technologies, and Equipment, First Edition. Ashok Rao.

U.S. EPA	U.S. Environmental Protection Agency
ESP	Electrostatic precipitator
FGD	Flue gas desulfurization
(g)	Gas phase
HHLL	High–high liquid level
HHV	Higher heating value (kJ/kg)
HLL	High liquid level
HRSG	Heat recovery steam generator
IGCC	Integrated gasification combined cycle
IGFC	Integrated gasification fuel cell
ISO	International Organization for Standardization
KO	Knock out (vapor/liquid separation)
LAER	Lowest achievable emission rate
LCA	Life cycle assessment
LEDs	Light emitting diodes
LHV	Lower heating value (kJ/kg)
LLL	Low liquid level
LLLL	Low–low liquid level
LMTD	Log-mean temperature difference (°K or °C)
LPMEOH™	Low pressure methanol
MCFC	Molten carbonate fuel cell
MHD	Magnetohydrodynamics
MSW	Municipal solid waste
MWI	Modified Wobbe Index
M-100	Automobile fuel grade methanol
NAAQS	National Ambient Air Quality Standard
NASA	National Aeronautics and Space Administration
NERC	North American Electric Reliability Council
NLL	Normal liquid level
NO_x	NO and NO_2
NPSH	Net positive suction head for pump
$NPSH_R$	Minimum NPSH requirement
NSPS	New Source Performance Standards
NSR	New Source Review
OECD	Organisation for Economic Co-operation and Development
ORC	Organic Rankine cycle
P&ID	Piping and instrumentation diagram
PAFC	Phosphoric acid fuel cell
PEMFC	Proton exchange membrane fuel cell
PFD	Process flow diagram
PID	Proportional-integral-derivative
PM	Particulate matter
ppb	Parts per billion
ppm	Parts per million

ppmv	Parts per million by volume
PSA	Pressure swing adsorption
PSD	Prevention of significant deterioration
PV	Process variable
RDF	Refuse-derived fuel
RO	Reverse osmosis
(s)	Solid phase
SCR	Selective catalytic reduction
SMR	Steam methane reformer
SOR	Start of run
SO_x	SO_2 and SO_3
SOFC	Solid oxide fuel cell
SRU	Sulfur recovery unit
TEMA	Tubular Exchanger Manufacturers Association
TOC	Total organic compounds
TRIG™	Transport Integrated Gasification
T/T	Tangent to tangent length
TVC	Trapped vortex combustor
VOCs	Volatile organic compounds
Wg	Gauge pressure in height of water column

SYMBOLS AND UNITS[1]

a	Interfacial area per unit volume of packing (m^2/m^3)
A	Area (m^2)
$\dot{B}$	Molar flow rates of side-draws (kmol/s)
c	Speed of light (299,792,458 m/s)
c_1	Constant (exponent)
c_2	Constant (exponent)
C	Constant
C_{-1}	Constant
C_0	Constant
C_1	Constant
C_2	Constant
C_D	Drag coefficient (dimensionless)
$\mathcal{C}_F$	Tray or packing factor
C_F	Mass or molar concentration in fluid F ($kmol/m^3$ or kg/m^3)
C_p	Constant pressure-specific heat (kJ/kg/°K)
C_{pi}	Constant pressure-specific heat of specie i (kJ/kg/°K)
C_v	Constant volume-specific heat (kJ/kg/°K)

[1] SI units are used except where noted.

C_{vi}	Constant volume-specific heat of specie i (kJ/kg/°K)
D	Diameter (m)
D_i	Inside diameter (m)
D_o	Outside diameter (m)
D_P	Droplet or particle diameter (m)
$\dot{D}$	Molar flow rates of distillate (kmol/s)
$\mathcal{D}_{ij}$	Diffusivity of specie i in j (m^2/s)
E	Energy (kJ)
$\mathcal{E}$	Ideal cell or open circuit potential (V)
$\mathcal{E}_o$	Reference cell potential (V)
f	Fanning friction factor (dimensionless)
f_T	Fanning friction factor in zone of complete turbulence (dimensionless)
f'	Darcy's (or Moody's) friction factor (dimensionless)
f'_T	Darcy's (or Moody's) friction factor in zone of complete turbulence (dimensionless)
f_i	Fugacity of specie i (dimensionless)
$\vec{F}$	Force (N)
F	LMTD correction factor (°K or °C)
$\dot{F}$	Molar flow rate of feed (kmol/s)
F_{ij}	Configuration or view factor between areas A_i and A_j (dimensionless)
F'	Solar collector efficiency (dimensionless)
g	Gibbes free energy (kJ/kmol) or local gravitational acceleration (m/s^2)
g_i^o	Standard-state Gibbs free energy for component i (kJ/kmol)
g_c	Conversion factor in Newton's law of motion (kg.m/s^2/N)
G	Gibbs free energy (kJ)
G^o	Gibbs free energy (kJ)
Gr	Grashof number (dimensionless)
h	Specific enthalpy (kJ/kmol or kJ/kg)
h^o	Standard state enthalpy (kJ/kmol or kJ/kg)
h_i^o	Standard state enthalpy (kJ/kmol or kJ/kg) of specie i
$\hbar$	Heat transfer coefficient (kJ/m^2/s/°C)
$\hbar_r$	Radiant heat transfer coefficienct (kJ/m^2/s/°C)
H	Enthalpy (kJ)
H_i	Enthalpy of specie i (kJ)
H_n	Enthalpy at statepoint n (kJ)
ΔH_{expan}	Enthalpy change by expansion (kJ)
ΔH_{chem}	Enthalpy change due to chemical reaction (kJ)
H^o	Standard state enthalpy (kJ)
H_i^o	Standard state enthalpy (kJ) of specie i
H_{pump}	Pump or compressor head (m)
$\mathcal{H}_i$	Henry's constant (kPa)
i	Current density (A/m^2)
i_o	Exchange current density (A/m^2)
I	Current (A)

j_H	Chilton–Colburn factor for heat transfer (dimensionless)
j_M	Chilton–Colburn factor for mass transfer (dimensionless)
J_A	Molar flow rate per unit area ($kmol/m^2/s$)
κ	Ratio of the specific heats, C_p/C_v
k	Thermal conductivity (kJ/m/s/°K) or rate constant (kmol/s)
k_L	Mass transfer coefficient in liquid phase ($kmol/m^2/s$)
k_V	Mass transfer coefficient in vapor phase ($kmol/m^2/s$)
K	Resistance coefficient or vapor–liquid distribution (dimensionless)
K_i	Reaction quotient or equilibrium constant for reaction or system i
K_{pump}	Constant specific to a pump
K_{sonic}	Constant
K_y	Equilibrium constant based on mole fractions
K_p	Equilibrium constant based on partial pressures
KE	Macroscopic kinetic energy of system as a whole (kJ)
$\mathcal{K}_p$	Proportional gain
$\mathcal{K}_i$	Integral gain
$\mathcal{K}_d$	Derivative gain
L	Length or height (m)
L_e	Equivalent length of straight pipe (m)
$\dot{L}$	Molar flow rate of liquid (kmol/s)
P	Pressure (kPa)
P^o	Reference pressure (kPa)
P_i	Partial pressure of specie i (kPa)
Pr	Prandtl number (dimensionless)
m	Mass (kg)
$\dot{m}$	Mass flow rate (kg/s or kg/h)
$\dot{m}_i$	Mass flow rate of specie i (kg/s)
$\bar{m}$	Mass velocity ($kg/s/m^2$)
$\dot{M}_i$	Molar flow rate of specie i (kmol/s)
M_w	Molecular weight
n	Number of mol (kmol)
n_1	Constant
n_2	Constant
n_{years}	Payback period (years)
n	Number of intercoolers (dimensionless)
N	Number of solid particles or cells (dimensionless)
N_i	Molar fluxes of specie i with respect to stationary coordinates ($kmol/m^2/s$)
N_s	Pump-specific speed (dimensionless)
N_{suct}	Pump suction-specific speed (dimensionless)
$\mathcal{N}$	Number of tubes (dimensionless)
$\dot{N}$	Rotational speed (s^{-1})
Nu	Nusselt number (dimensionless)
p	Tube pitch (m)
P_o	Dead state pressure (kPa)

P_C Critical pressure (kPa)
P_i' Vapor pressure of component i (kPa)
PE Macroscopic potential energy of system as a whole (kJ)
Q Heat (kJ)
$\dot{Q}$ Heat transfer rate (kJ/s or kJ/h)
$\mathcal{Q}$ Reaction quotient
$\dot{q}_{in}$ Heat transfer rate per unit area (kJ/m^2/s)
r Radial distance (m)
r_i Rate of conversion of specie i (kmol/s)
R Individual ideal gas constant (kJ/kg/°K)
Re Reynolds number (dimensionless)
R_H Hydraulic radius (m)
R_u Universal ideal gas constant (8.314462 kJ/kmol/°K)
$\mathcal{R}$ Sum of electronic and ionic cell resistances (ohm)
s Specific entropy (kJ/kmol/°K or kJ/kg/°K)
S Entropy (kJ/°K)
Sc Schmidt number (dimensionless)
Sh Sherwood number (dimensionless)
S_P Mean individual particle surface area (m^2)
SG Specific gravity (dimensionless)
t Time (s) or thickness (m)
T Temperature (°K)
T_o Dead state at temperature (°K)
u Specific internal energy (kJ/kmol or kJ/kg)
U Internal energy of system (kJ)
U_{Fu} Fuel utilization (dimensionless)
U_{Ox} Oxidant utilization (dimensionless)
$\mathcal{U}$ Overall heat transfer coefficienct (kJ/m^2/s/°K)
v Magnitude of Velocity (m/s)
v_e Magnitude of the Erosional velocity (m/s)
v_{sonic} Magnitude of Sonic velocity, that is, when Mach number = 1 (m/s)
v_V Magnitude of the Allowable velocity for vapor (m/s)
v_T Magnitude of the Terminal velocity of the droplets (m/s)
$\bar{v}_S$ Magnitude of the Superficial fluid velocity (m/s)
V Specific volume (m^3/kmol)
V_P Mean individual particle volume (m^3)
V_{cell} Cell voltage (V)
V_{Nernst} Ideal cell voltage (Nernst) (V)
$\dot{V}$ Volumetric flow rate (m^3/s) or molar vapor flow rate (kmol/s)
W Work (kJ)
$\dot{W}$ Power (kW)
W_s Shaft work (kJ)
W_e Electrochemical work (kJ)
W_f Mechanical energy loss due to friction (kJ)
x Distance or length (m)

x_i	Mole fraction of specie (kmol/kmol), i
$\mathcal{X}$	Mass fraction (kg/kg)
X	Fraction molar conversion (dimensionless)
y_i	Mole fraction of specie (kmol/kmol), i
z_i	Mole fraction of specie (kmol/kmol), i
Z	Elevation or distance (m)
$\mathcal{Z}$	Compressibility factor (dimensionless)

GREEK SYMBOLS AND UNITS

α	Fraction of radiant energy absorbed (dimensionless)
α_c	Correction factor in Bernoulli equation (dimensionless)
α_e	Electron transfer coefficient (dimensionless)
α_p	Probability of chain growth (dimensionless)
α_{ij}	Relative volatility of component i with respect to component j
β	Coefficient of thermal expansion (1/°K)
β_{ij}	Solvent selectivity (dimensionless)
ε_S	Emissivity of the surface (dimensionless)
ε	Pipe roughness (m) or bed porosity (m^3/m^3)
η	Efficiency (dimensionless)
η_I	First law efficiency (dimensionless)
η_{II}	Second law efficiency (dimensionless)
η_{act}	Activation polarization (V)
η_{conc}	Concentration polarization (V)
η_{Carnot}	Carnot efficiency (dimensionless)
η_{gen}	Generator efficiency (dimensionless)
η_{ohm}	Ohmic polarization (V)
δ	Thickness (m)
Δ	Difference
γ_i	Activity coefficient for component i (dimensionless)
λ	Fractional excess air (dimensionless)
λ_P	Particle shape factor
λ_{con}	Latent heat of condensation (kJ/kg)
λ_{vap}	Latent heat of vaporization (kJ/kg)
λ'	Effective difference in heat content between vapor and liquid (kJ/kg)
μ	Viscosity ($N{\cdot}s/m^2$)
π	Pressure ratio (dimensionless)
ρ	Fraction of radiant energy reflected (dimensionless) or density (kg/m^3)
ρ_P	Density of solid particles (kg/m^3)
σ	Surface tension between liquid and vapor phases (N/m)
σ_i	Ionic conductivity (S/m)
σ_{amb}	Mixed proton and electron conductivity (S/m)
σ_{SB}	Stefan–Boltzmann constant (5.67×10^{-11} $kJ/s/m^2/°K^4$)

τ Fraction of radiant energy transmitted (dimensionless)
$\mathcal{T}$ Time constant (s)
θ Quality of thermal energy ($-1/T$)
Θ Fraction of sites covered with adsorbed molecules at equilibrium (dimensionless)
φ_i Impeller size (m)
$\emptyset_i$ Fugacity coefficient for component i
χ Exergy (kJ)
$\dot{\chi}$ Exergy rate (kJ/s)
ν Kinematic viscosity (m^2/s)
Ψ_A Fractional loading in adsorbent of its equilibrium value (dimensionless)
ξ Index (or measure) of heterogeneity (dimensionless)

SUBSCRIPTS

A Adiabatic or adsorbent
acc Acceleration
act Actual (non-isentropic)
Ad Adsorbent
Adorp Adsorption
C Cold
chem Chemical reaction
comp Compressor
cond Condenser
desorp Desorption
disch Discharge
DG Digester gas
DP Dew point
e Error
expan Expansion
f Fouling
F Fluid
gen Generator
G Gas
Hy Hydraulic
H Hot
i Specie i or inside
in Inlet
I First law or insulation
II Second law
j Specie j
k Stage number
l Specie l

L	Liquid
M	Mixture
mech	Mechanical
min	Minimum
NG	Natural gas
o	Outside
out	Outlet
Ox	Oxidant
P	Polytropic or pipe
Pr	Products
R	Required or desired
reax	Reaction
reb	Reboiler
refrig	Refrigeration
S	Steam or solid surface
sat	Saturated
sink	Heat sink
sonic	At sonic conditions (Mach number = 1)
source	Heat source
suct	Suction
TOT	Total
turb	Turbine
u	Useful
V	Vapor
W	Water

SUPERSCRIPTS

act	Actual
int	Interface
L	Liquid
mix	Mixture
V	Vapor
*	Equilibrium value

INDEX

Sustainable Energy Conversion for Electricity and Coproducts: Principles, Technologies, and Equipment, First Edition. Ashok Rao.

Printed and bound by CPI Group (UK) Ltd, Croydon, CR0 4YY

16/04/2025

14658426-0002